Synthesizable VHDL Design for FPGAs

Eduardo Augusto Bezerra
Djones Vinicius Lettnin

Synthesizable VHDL Design for FPGAs

Springer

Eduardo Augusto Bezerra
Djones Vinicius Lettnin
Department of Electrical and Electronic Engineering
Universidade Federal de Santa Catarina
Florianópolis, Santa Catarina
Brazil

ISBN 978-3-319-37733-9 ISBN 978-3-319-02547-6 (eBook)
DOI 10.1007/978-3-319-02547-6
Springer Cham Heidelberg New York Dordrecht London

Softcover reprint of the hardcover 1st edition 2014

Printed on acid-free paper

Springer is part of Springer Science+Business Media (www.springer.com)

Contents

Chapter 1
Digital Systems, FPGAs and the Design Flow

This chapter introduces the target technology used in the laboratory sessions, and provides a step-by-step guide for the design and simulation of a digital circuit. At the end of this chapter, the reader should be able:

- to have a basic understanding of the FPGA technology (internal blocks, applications);
- to understand the basic logic gates operation;
- to design a logic circuit using a schematic editor tool;
- to simulate and test a digital circuit;
- to understand the basic flow of an Electronic Design Automation (EDA) tool.

1.1 Digital Systems

Digital systems are composed of two basic components: Datapath and Control Unit. As shown in Fig. 1.1, the control unit has as its main function the generation of control signals to the datapath unit (also known as "operational block"). The control signals "command" the desired operations in the datapath. Furthermore, the control unit may receive control inputs from the external environment; for instance, it can be a simple "start" or even an operation code ("opcode" in microprocessors). Finally, this unit can also generate one or more output control signals to communicate with other digital systems (e.g., "done, bus request, nack").

The datapath unit performs operations on data received, usually, from the external environment. The operations are performed in one or more steps, where each step takes a clock cycle. The datapath generates "status" signals (sometimes also called as "flags") that are used by the control block to define the sequence of operations to be performed. Examples of datapath blocks include:

- Interconnection network—wires, multiplexers, buses, and tri-state buffers;
- Functional units—adders, subtractors, shifters, multipliers, and Arithmetic and Logic Unit (ALU);

E. A. Bezerra and D. V. Lettnin, *Synthesizable VHDL Design for FPGAs*,
DOI: 10.1007/978-3-319-02547-6_1,

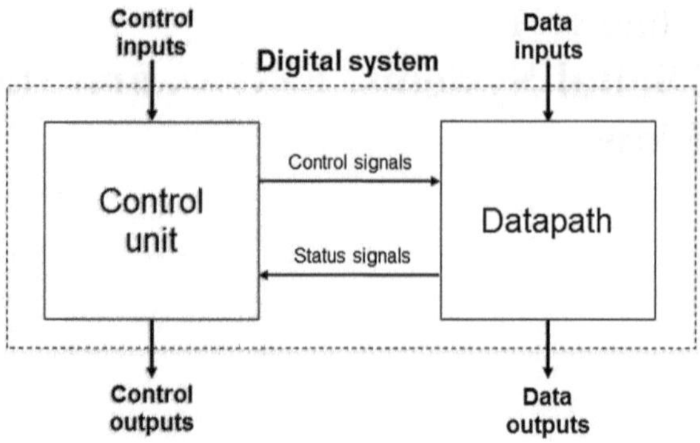

Fig. 1.1 Digital system (adapted from [ManoKime99])

- Memory elements—registers, and Random Access Memory (RAM).

The datapath and control units are based on two types of circuits: combinational and sequential circuits.

Combinational circuits have I_m inputs and O_n outputs, as shown in Fig. 1.2. In these circuits the output pins depend only on the values that are presented to the input pins and each output is defined by a different Boolean logic equation.

A combinational circuit can also be classified according to its application as:

- Interconnection circuit—e.g. multiplexers, demultiplexers, encoders, and decoders;
- Logical and arithmetic circuit—e.g. adders, subtractors, multipliers, shifters, comparators, and ALUs (circuits that combine more than one arithmetic or logical modules).

Sequential circuits have I_m inputs and O_n outputs. However, in these circuits the output pins depend not only on the values that are presented to the input signals (i.e. m signals), but they depend also on the current state stored in the memory module, as shown in Fig. 1.3. The number of both next state (i.e. N_j) and current state signals (i.e. S_j) depend on the encoding style of a Finite State Machine (FSM) states, such as, sequential binary, gray code or one hot.

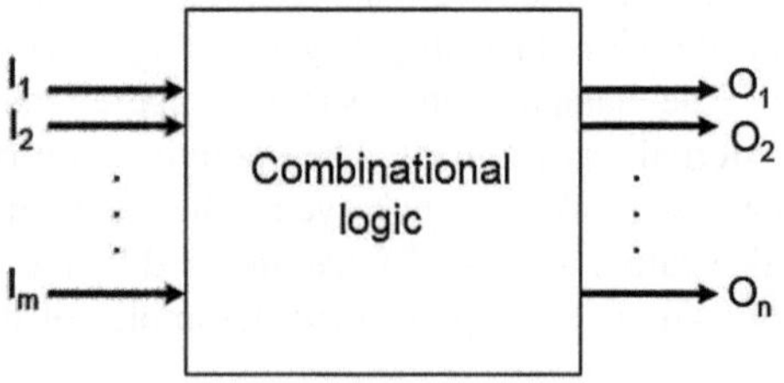

Fig. 1.2 Combinational circuit

Fig. 1.3 Sequential circuit

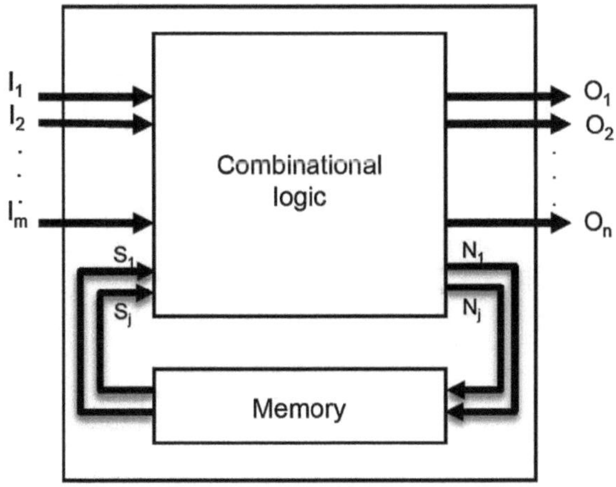

1.2 Field Programmable Gate Array

The concept behind the Field Programmable Gate Array (FPGA) technology is better understood through a digital circuit design example.

Problem:
You are asked to design a circuit that implements the following logic function:

$$f(A,B) = A \text{ AND } B$$

In this circuit, inputs *A*, *B*, and output *f* are all 4 bits wide.

Solution 1: Using a 7408 TTL device (Application-specific Integrated Circuit—ASIC solution)
The operations to be performed are:

$$f(1) = A(1) \text{ AND } B(1)$$
$$f(2) = A(2) \text{ AND } B(2)$$
$$f(3) = A(3) \text{ AND } B(3)$$
$$f(4) = A(4) \text{ AND } B(4)$$

The 7408 TTL device, shown in Fig. 1.4, has four 1-bit AND gates. Therefore, a single 7408 chip is sufficient to implement the 4 bits function. In Fig. 1.4, the *A* inputs are connected to pins 1, 4, 9 and 12. The *B* inputs are connected to pins 2, 5, 10 and 13. The *f* output can be obtained from pins 3, 6, 8 and 11. The chip must be placed (soldered) on a printed circuit board (PCB) or protoboard, and all pin connections should be made, including the power lines (Vcc and GND).

This is a nice solution for the proposed problem, but it has some drawbacks. TTL is an obsolete technology, presenting high-energy consumption, and the chip itself takes considerable space on the PCB. Another drawback is in case of changes

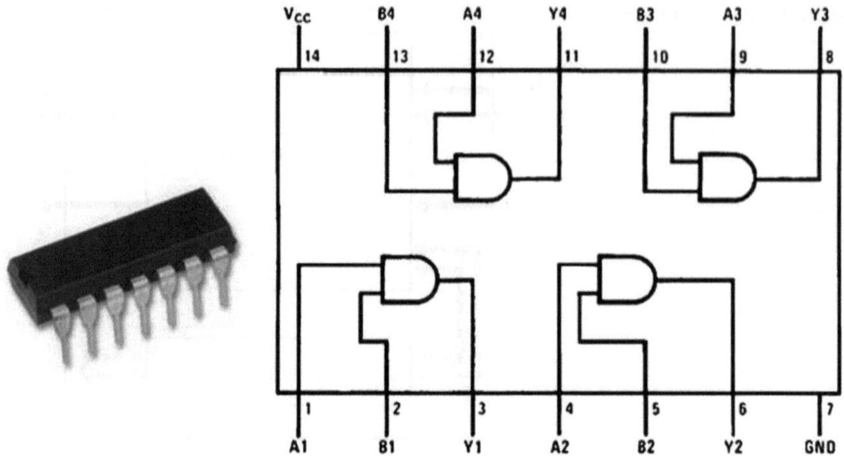

Fig. 1.4 7408 TTL chip

or upgrades in the design requirements. In these situations, a recall may be needed in order to replace the components (7408 chip) by the new ones. Depending on the required changes, a full PCB redesign may be needed.

Solution 2: Microprocessor (software solution)
A microprocessor (or microcontroller) based system is not the best solution for this sort of "simple problem", but it has been included here in order to show the differences between the hardware and software approaches.

In a similar way to the first solution, a microprocessor also needs to be placed on a PCB, but it has more flexibility for changes in the system functionality. As the solution is implemented at the "software level", any changes in the design requirements may be performed in the system with no need for hardware or PCB modifications. A possible solution for the problem is shown in the C language program as follows:

```
int main(){
    int A, B, F;
    F = A & B; //F = A AND B
}
```

However, this solution also has some drawbacks. A special C compiler may be required in order to have a 4-bits wide *int* type. Another difficulty would be to assign the inputs and outputs (A, B, F) to the microprocessor physical pins. Considering a standard C compiler, the developer would use *scanf* or *getc* input functions to obtain the *A* and *B* values, and *printf* to output the *F* result. Depending on the target systems, several adaptations would be necessary in order to associate the functions to the device's pins. In a microcontroller based C compiler, usually, special functions are provided to the programmer in order to access the device's pins.

Solution 3: FPGA
An FPGA device could be seen as an intermediary solution, between the hardware and the software approaches. In a similar way to the hardware (TTL chip) and the software (microprocessor chip) approaches, the FPGA chip also has to be placed on a PCB, and all its input and output pins should be connected. An important difference from the microprocessor approach is that the solution for the problem is not implemented in software. It is a hardware solution, similar to the 7408 chip, but with the same flexibility for changes as the microprocessor based approach. In case of bugs or design changes, a new circuit can be "placed" inside the FPGA, with no need for PCB modifications. A solution for the problem could be written in the VHDL language as follows:

```
library IEEE;
use ieee.std_logic_1164.all;
entity AND_VHDL is
   port (A, B : in std_logic_vector (3 downto 0);
   F : out std_logic_vector (3 downto 0)
   );
end entity;
architecture rtl of AND_VHDL is
begin
   F <= A AND B;
end rtl;
```

The microprocessor solution drawbacks are easily outwitted when using an FPGA. Different bit lengths for the inputs and outputs are defined by standard language constructions, as well as the assignment of physical pins.

However, an interesting advantage of Solution 3 (FPGA technology) is the flexibility for hardware changes, after the product has been completed. Assuming that a requirement mistake has been identified, and the circuit were supposed to perform an OR operation and not the implemented AND operation. In Solution 1, the 7408 chip would have to be replaced by the 7432 TTL device (four 1-bit OR gates), which means to remove the chip from the board, and to place the new one. In Solutions 2 and 3, there will be no need for chips replacement. In Solution 2 the software would have to be changed and reloaded in the microprocessor's memory. In Solution 3, the VHDL code would have to be changed, using OR instead of AND, and a new FPGA configuration should be generated. It is important to notice that in Solution 3 a "new hardware" is generated, for the FPGA, while in Solution 2 a new piece of "software" is loaded in the microprocessor.

1.3 FPGA Internal Organization

An FPGA is a device (a chip) whose internal logic can be changed by the developer (user) after the chip has been manufactured—it is "field programmable". From the developer point of view, its functionality is similar to a microprocessor, as it is possible to change its operation according to the application requirements. In the ASIC approach, the chip has its functionality defined by the manufacturer (permanently), and it cannot be changed. For instance, the 7408 device introduced in "Solution 1" has four 1-bit AND gates, and this hardware arrangement is fixed and it is not possible to be modified. The microprocessor ("Solution 2"), also has a fixed hardware that cannot be changed (registers, ALU, …), but its functionality can be changed by the user at the "software level". The internal hardware of a microprocessor has been designed in order to execute instructions of a program, and different programs can be written according to the application requirements. An FPGA, on the other hand, has an internal logic that can be changed by the developer. The main similarity to a microprocessor is that the FPGA developer also uses a language to describe the application's functionality, and a tool to generate the new hardware to be configured inside the FPGA. However, as an FPGA is not a microprocessor, it has not been designed to execute a program written by the developer. The source code written by the developer is in fact a "hardware description", and it becomes the FPGA internal logic configuration.

In order to have its internal logic changed, the architecture of an FPGA chip employs "writable" technologies, such as SRAM or FLASH memories. FPGA devices have three main configurable components: the logic blocks; the routing elements; and the input/output (I/O) blocks. The three types of components are shown in Fig. 1.5, where the I/O blocks are located in the periphery of the chip, and the logic blocks are the larger squares, interconnected by the routing elements.

The "logic blocks" are configured in order to perform the logic operations required by the application, and so they are the "processing elements" of an FPGA. The logic blocks are interconnected through the configurable routing elements, which are used also for the connection to/from I/O blocks. The I/O blocks may be configured in order to function as input pins, output pins, or both. Figure 1.6 shows a possible hardware configuration for the Solution 3 described before. A single logic block is used to perform the AND operation. Eight I/O blocks are configured as

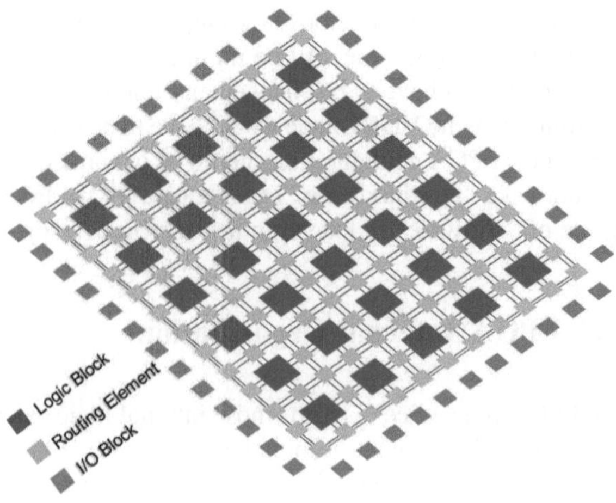

Fig. 1.5 A typical FPGA internal components

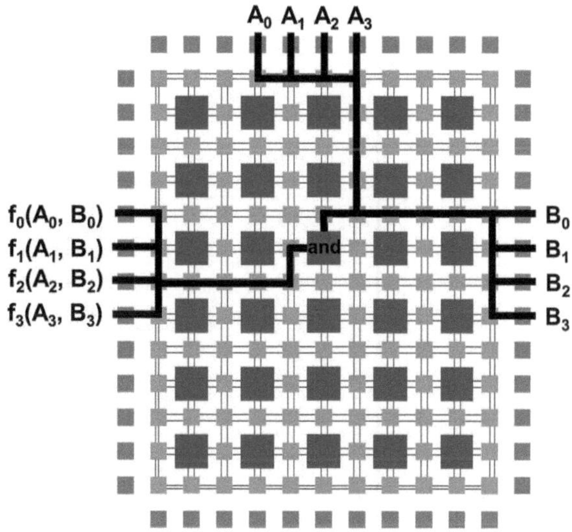

Fig. 1.6 Possible FPGA configuration for Solution 3

input, and a path (routing) is configured between them and the AND logic block. Four I/O blocks are configured as output, and they are connected to the output of the logic block.

1.4 Configurable Logic Block

As stated before, FPGAs are made of three major configurable resources: Logic Blocks; Routing Elements; and I/O Blocks. The internal organization of all three elements is very interesting and deserves a more detailed explanation, but this is not the main goal of this book. However, in order to better understand the developing tools functionality, it is important to have a basic idea of how logic functions are implemented in FPGAs. Most FPGAs are based on look-up tables (LUTs), which can be used as "universal function generators". An LUT is a configurable resource capable of implementing any n-input truth table. Figure 1.7 shows an abstract view of an LUT contents generation from a logic function.

In Fig. 1.7, the logic function in the upper left hand corner is written in an input format accepted by the selected Electronic Design Automation (EDA) tool (e.g. schematic diagram, VHDL, Verilog, ...). The EDA tool generates the truth table for the logic function (I). During the programming process, the results column of all truth tables is written to the FPGA configuration memory (II). More precisely, in a 4-input LUT based FPGA, a truth table for a 4-input logic function (A, B, C, D) is written to a LUT of a logic block. The FPGA does not need to have any physical logic gates, as only the results of a truth table are stored in the LUTs (FPGA configuration memory). In order to perform the equation, the inputs (A, B, C, D) are used as the MUX selection (III), and the provided output is the final result (IV). This functionality

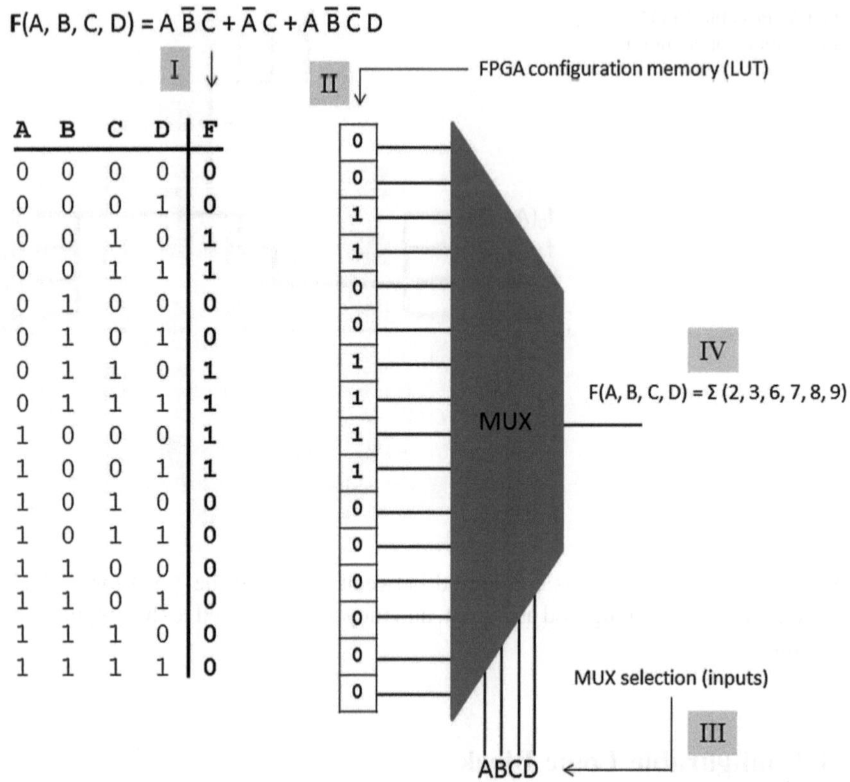

Fig. 1.7 LUT working model

explains how Solution 3 described before can be implemented in an FPGA. As no logic gates are used to implement a logic function, each time a new functionality is required in a design, the circuit developer just needs to change the hardware description, and the EDA tool generates a new truth table to be used in the LUT configuration. An LUT is the building block of the Logic Block shown in Fig. 1.5.

1.5 Electronic Design Automation and the FPGA Design Flow

In order to develop a digital circuit targeting an FPGA device, it is compulsory to use Electronic Design Automation tools. Figure 1.8 shows a typical FPGA design flow, where several EDA tools are employed.

The first step is the "Design Entry", where the developer uses a design-editing tool to provide their circuits description. Usually, EDA tools take a number of description formats as, for instance, schematic diagrams, hardware description languages (HDLs), state machine diagrams, among others.

The next step in Fig. 1.8 is the circuit simulation. There are several simulation tools available, and the most used method in all of them is the waveform analysis. The circuit's logic and states can be fully simulated using tools that provide a detailed waveform generation and analysis. It is important to notice that this is a "functional simulation", as at this stage of the design flow there is no timing information yet. The functional simulation process tests only the design's logic operation, as it has no further information regarding the internal circuit in the target device (FPGA) or routing path delays. This step should be performed in order to verify the circuit functionality disregarding timing constraints and other performance parameters. In this step, in case of functional errors, the designer may change the circuit using the design entry tool, and perform new simulations until the design shows the expected behavior.

After have finished the functional simulation, the designer can perform the synthesis activity and a *netlist* representing the provided circuit is generated. This

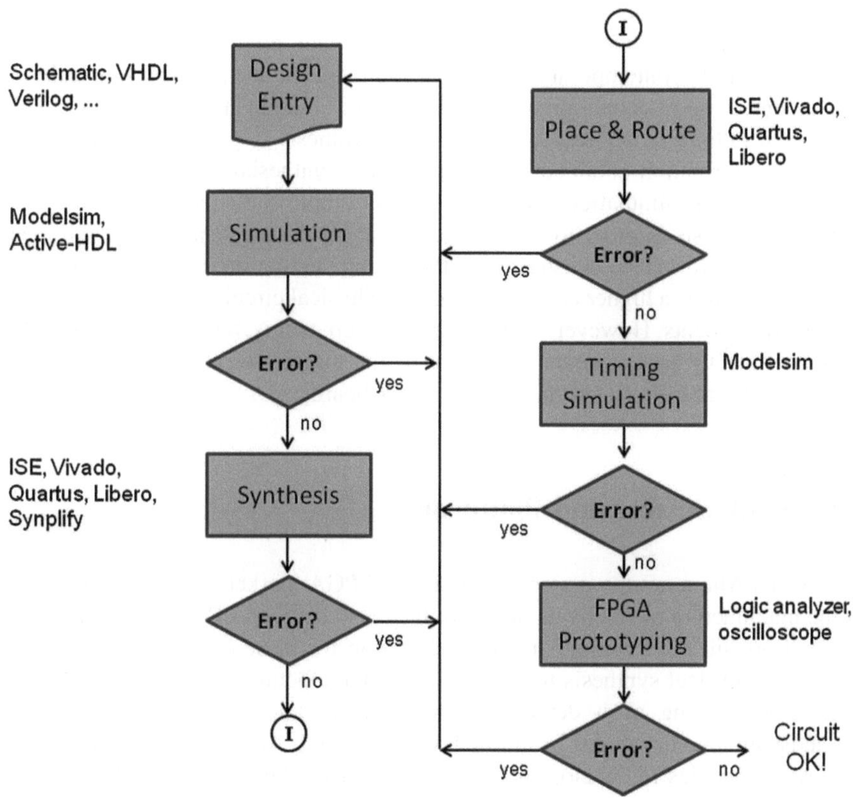

Fig. 1.8 Typical FPGA design flow

activity is split in two operations: the logic synthesis; and the physical synthesis. Traditionally, the logic synthesis role is to map the provided circuit to a set of gates (netlist) performing the same functionality as the original circuit description. This netlist has not yet any physical information such as power consumption or routing delays for the final physical implementation. The "design entry", the "functional simulation", and the "logic synthesis" are activities performed by a professional called the "Frontend Engineer".

The remaining activities shown in Fig. 1.8 are performed by the "Backend Engineer". In FPGA designs, it is not unusual to have the same engineer performing both, frontend and backend, activities. The first backend activity in an FPGA design flow is the "placement and routing". At the end of the process, in case of errors the designer may go back to the design entry step in order to fix the problems pointed out during the synthesis. The physical synthesis output is a netlist and the respective physical design data. The design data, used in the timing simulation step, include performance figures such as maximum clock rate, area usage, power estimation, and throughput. The netlist has all the physical design information needed for the FPGA configuration, including: I/O pins, routing paths, and operational blocks for LUTs configuration, among others. In the physical synthesis step activities as placement, routing and circuits optimizations are accomplished. In contemporary EDA tools, the logic and the physical synthesis are implemented in a cooperative way, employing incremental algorithms. This strategy results in better results, as the logic synthesis is performed considering feedback information obtained from the physical synthesis.

In the "timing simulation" step, the designer employs the data provided by the physical synthesis in order to verify whether or not the physical design will meet the defined performance constraints. If the timing simulation results in acceptable figures, so there is a higher confidence that the physical circuit will work according to the requirements. However, only in the final step, that is, the FPGA prototyping, it is possible to have a better assurance of the circuit correctness. In this step, usually, the circuit is tested in a development board.

1.6 FPGA Devices and Platforms

Xilinx and Altera are the major players in the FPGA market. Their main business is the manufacture of FPGA devices, but they also develop EDA tools. The FPGA internal organization complexity is a motivation for these companies to produce their own physical synthesis tool. The FPGA manufacturer has, undoubtedly, the best understanding of a device's internal organization and, consequently, is the most suitable developer for a low level EDA tool. Another motivation for the FPGAs companies to develop their own physical synthesis tools is related to the confidential information regarding the device's internal organization. Holding back this information is important for a company as it may result in technology advantages over competitors. Other important players in the FPGA market include:

Microsemi Corporation (formerly Actel); Lattice Semiconductor; and SiliconBlue Technologies.

These companies target very specific markets, where the most common applications are: networking (switches and routers); video and image processing; cryptography; space systems (satellites and deep space missions); avionics; defense; ASIC prototyping; audio equipment; medical solutions; motor control; high performance computing; and automotive systems. In order to attend the requirements of these applications, current FPGA devices have extra hardware resources as, for instance, DSP operators, RAM blocks, and even embedded microprocessors.

FPGA devices are used not only as basic components in the design of boards for a variety of applications, but also in the design of development and educational boards. The laboratory assignments described throughout this book have been elaborated targeting, mainly, the following educational boards:

- The Altera DE2 Development and Education board shown in Fig. 1.9 has an Altera Cyclone II 2C35 FPGA, and several peripherals. The board has been designed for teaching digital systems classes, in laboratory-based courses.
- The Mercurio IV board has been developed by Macnica-DHW Engineering and it has an Altera Cyclone IV FPGA. It also has a set of peripherals, and was developed aiming university courses.

Fig. 1.9 DE2 Development and Education board [Altera]

1.7 Writing Software for Microprocessors and VHDL Code for FPGAs

Applications for FPGAs and microprocessor based systems can be modeled and conceived at different levels of abstraction, and using different design entry formats. A compiler used to generate object code (machine code) for a microprocessor can have as input, for instance, C/C++ source code (computer program) and also other pieces of software and components available in libraries. A synthesis tool (hardware compiler) used to generate a netlist (bitstream, circuit) to an FPGA has as input a code written in an HDL (VHDL, Verilog, …), as well as components available in libraries. An important distinction here is that a piece of software written in the C language is at a level of abstraction many times higher than a hardware description written in VHDL. Even the assembly language can be considered as in a higher abstraction level than VHDL. The VHDL developer is in fact describing the behavior of a piece of hardware to be used to implement some desired functionality (application), while with the assembly language (or any other high level languages) the developer is describing the application's functionality that will be performed by an existing piece of hardware (microprocessor). The VHDL developer uses the language to design the hardware and the application to be performed, while a microprocessor programmer has to worry, basically, with the application functionality to be implemented.

Considering the low abstraction level of VHDL descriptions, an attractive approach used by software developers when targeting an FPGA implementation, is the use of software programming languages followed by the automatic conversion to VHDL. Problems appear when circuits with a certain level of optimization are expected. In this case programmer needs to use special keywords and special instructions to be used by the high level software compiler to produce a more optimized code. Depending on the design complexity, it may be harder to realize than developing the system straight in VHDL. Nested loops, the variety of data types, and some arithmetic operators are examples of restrictions for the translation. The main point is that a compiler for a high level language can translate the program to a low-level language (assembly language), which has similar instructions, making the translation process not too difficult. On the other hand, synthesis tools have to translate the high level programs to a much reduced set of low level "instructions". For instance, while a compiler can use a *je* (conditional jump in assembly) instruction as the target for the translation of a high level *if* constructor, a synthesis tool may have to use D flip-flops and multiplexers.

Over the years some "programming language" to "HDL" translators have been developed, but with no major impact in the market. Developers with a software background starting in the field of HDL design, usually try the way that appears to be the easiest one, which is the use of a language translator. As soon the design challenges start, and the developer must learn very specialized commands and constructions in order to produce a proper code suitable for the chosen translator, so there is a realization that learning a hardware description language may be a better option. Learning a new programming paradigm is not easy, but the syntaxes of languages like Verilog or VHDL are not so different, for instance, from the C language. In addition, a C developer has to use only a sub-set of the language, because of the synthesis tools limitations. For example, the translation of dynamic structures (pointers) from C to VHDL is not useful, as this class of data type is not suitable for static hardware generation.

1.8 Laboratory Assignment

The laboratory objectives are:

- practicing the use of basic logic gates in the design of a digital circuit;
- to use a design entry tool based on schematic editing;
- to have a first contact with an EDA tool;
- to have a basic understanding of a complete FPGA design flow, starting from the schematic design entry step, performing the circuit simulation, and prototyping it in an FPGA development board.

1.8.1 Logic Gates

Figure 1.10 shows the basic logic gates: OR; AND; XOR; and NOT. The gate functionality is also described through different textual representations, and by its truth table.

A digital circuit can be designed as a collection of logic gates, connected in order to perform the required function. For instance, in Fig. 1.11 it is shown the design of a circuit that implements an add operation. This circuit is called a

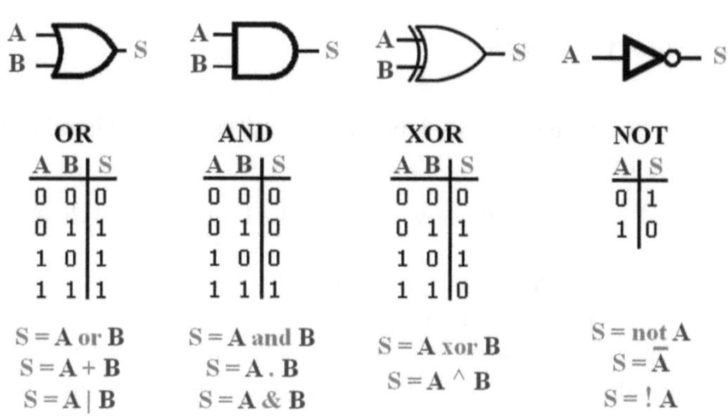

Fig. 1.10 Basic logic gates

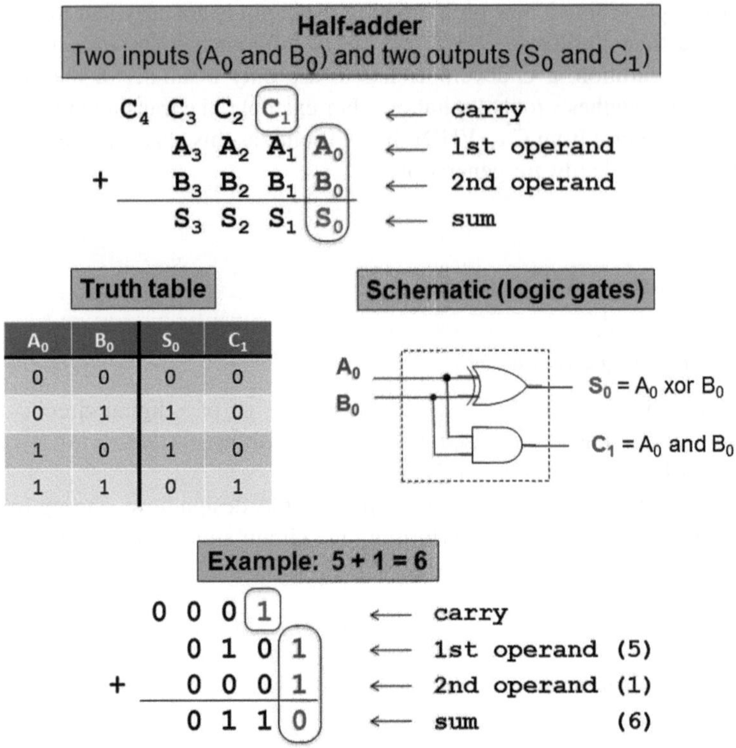

Fig. 1.11 Half-adder circuit implemented with an AND gate and an XOR gate

"half-adder" and it is used for the summation of the least significant bits (LSB) of two n-bits operands, providing as a result the sum and a carry. Figure 1.11 shows the sum of two 4-bit values, and the half-adder circuit used to add their LSBs.

1.8.2 Laboratory Session

The tasks to be completed in this laboratory session, using Altera's Quartus II EDA tool, are as follows:

- Create the half-adder circuit shown in Fig. 1.11, using the schematic design entry tool.
- Perform the synthesis (logic and physical).
- Perform the functional simulation.
- Fix simulation and synthesis errors.
- Prototype the half-adder circuit in the development board, in order to check its functionality.

For those who do not have the Quartus II software installed, there is a free version available on Altera's website at http://www.altera.com.

Quartus II 12.1 was the latest version available when this book was written, and this version has been used in all laboratory sessions.

In this first laboratory session, the idea is to follow the step-by-step instructions in order to get used to the EDA tool facilities and resources. This tutorial can be used as a guide to the next chapters as well.

Step 1: Creating a New Project

Start Quartus II, close all initial windows, and open the menu *File→New Project Wizard*, as shown in Fig. 1.12.

Next, choose the project's folder and name. In a VHDL design, it is important to provide the top-level entity's name. Do not use names or paths with spaces or special characters (e.g., à, ç, …) in any of the text boxes (Fig. 1.13).

In "page 2 of 5", just click on the "Next" button. In "page 3 of 5", choose the DE2's FPGA, which is the Cyclone II ("Family") EP2C35F672C6 ("Device"), as shown in Fig. 1.14. In Fig. 1.15, choose ModelSim-Altera as the simulation tool.

In "page 5 of 5" it is shown a summary of the new project. Just click "Finish".

Step 2: Design Entry (Schematic)

Chose *File→New* and *Block Diagram/Schematic File*, as shown in Fig. 1.16. This will open an empty editor, used to create the new circuit schematic.

The schematic to be created should implement the half-adder circuit shown in Fig. 1.11. The schematic editor tool has several pre-defined components stored in libraries. The logic gates are available in the Primitives/Logic library. To add a

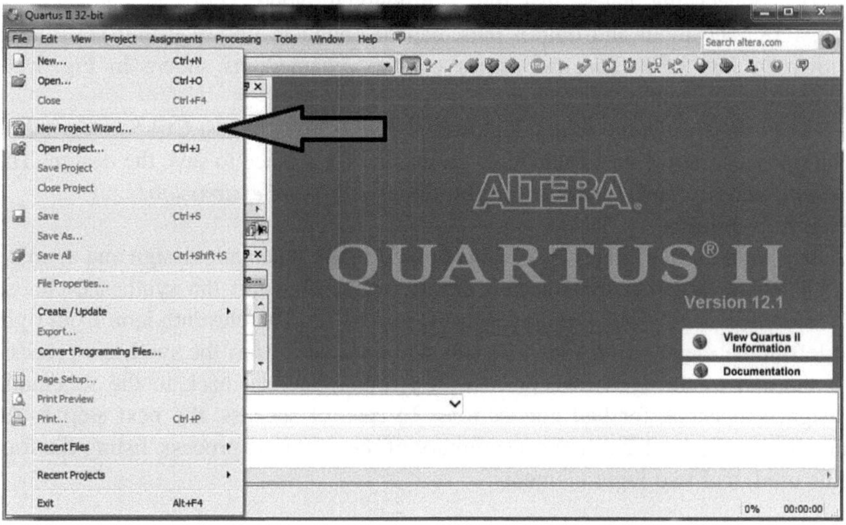

Fig. 1.12 Quartus II—New project wizard

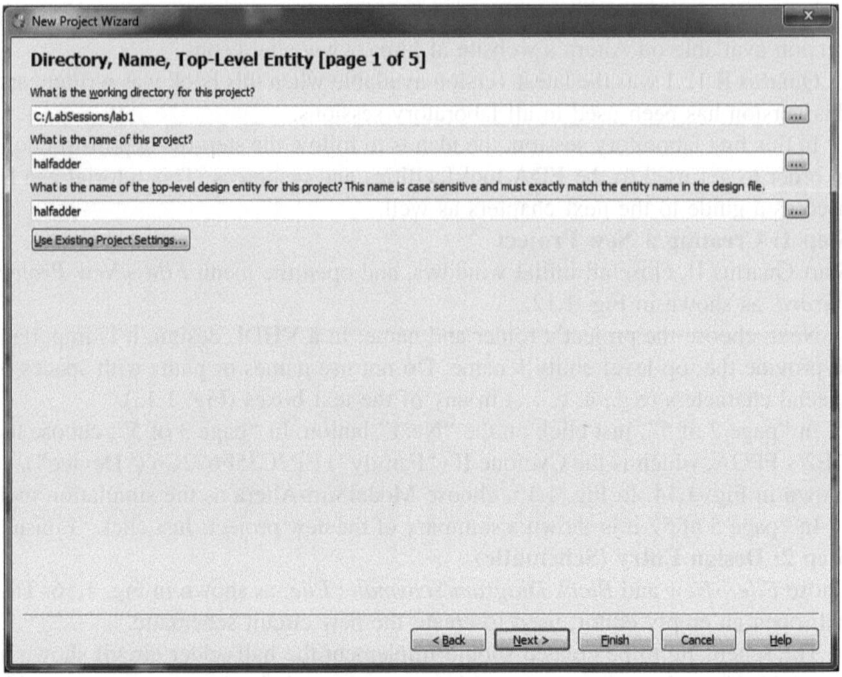

Fig. 1.13 Directory, name and top-level entity

logic gates to the schematic, click on the "Symbol Tool" icon highlighted in Fig. 1.17.

To connect the gates, use the "Orthogonal Node Tool" as indicated in Fig. 1.18. The input and output pins, can be found in library Primitives/Pin, as shown in Fig. 1.19. The pin names can be renamed, as shown in Fig. 1.20, completing the half-adder design.

When done, save the schematic design: *File→Save* (or just Ctrl-S). A window will open asking for the schematic's name, and for a place to save the design. The default name is *halfadder.bdf*, and the folder is *c:\LabSessions\lab1*.

Step 3: Synthesis

In the synthesis step, an analysis is performed in the schematic design and a circuit is generated targeting the selected FPGA device. To start the synthesis, choose *Processing→Start Compilation* (or just type Ctrl-L). The eleventh icon from right to left in the menu bar in Fig. 1.21 can also be used to start the synthesis process. At the end of the synthesis, in case of errors found, go back to the schematic design, and check for bad connections. In case of success, the next step is the simulation. Figure 1.22 shows a summary of the synthesis process, listing the four pins used, and two logic elements.

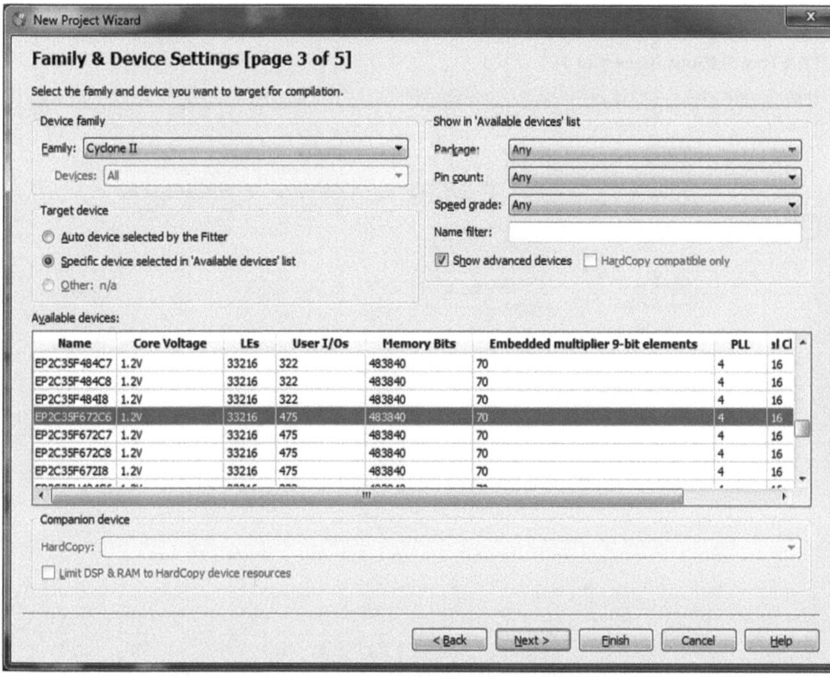

Fig. 1.14 Selecting Altera FPGA—Cyclone II EP2C35F672C6

Step 4: Simulation

Quartus II (v.10 and following) does not have its own simulator, and ModelSim, a third party software tool is used. The ModelSim-Altera tool is installed together with Quartus II, but before the first simulation it is recommended to verify that the tool is properly configured. In Quartus II, as shown in Fig. 1.22, open *Tools→Options*.

Click on EDA Tool Options, and type ModelSim path in the text box shown in Fig. 1.23.

To start the simulation tool, select *Tools→Run Simulation Tool → Gate Level Simulation* (or press the sixth icon from right to left shown in Fig. 1.21). This will run ModelSim software.

When ModelSim starts, just close the presentation window, and initiate the simulation by putting ModelSim in "simulation mode": select *Simulate→Start Simulation*. The "Start Simulation" window in Fig. 1.24 will show. This window has several tabs, but for this functional simulation just the first tab (Design) will be used. The Design tab lists all designs ready for simulation, including the half-adder circuit that has been synthesized.

The target design is located in the "work" library. Open this library clicking on the "+" sign. As a next step, click on "halfadder Entity" to select this module for simulation, as shown in Fig. 1.24. Click OK to start the simulation, and ModelSim

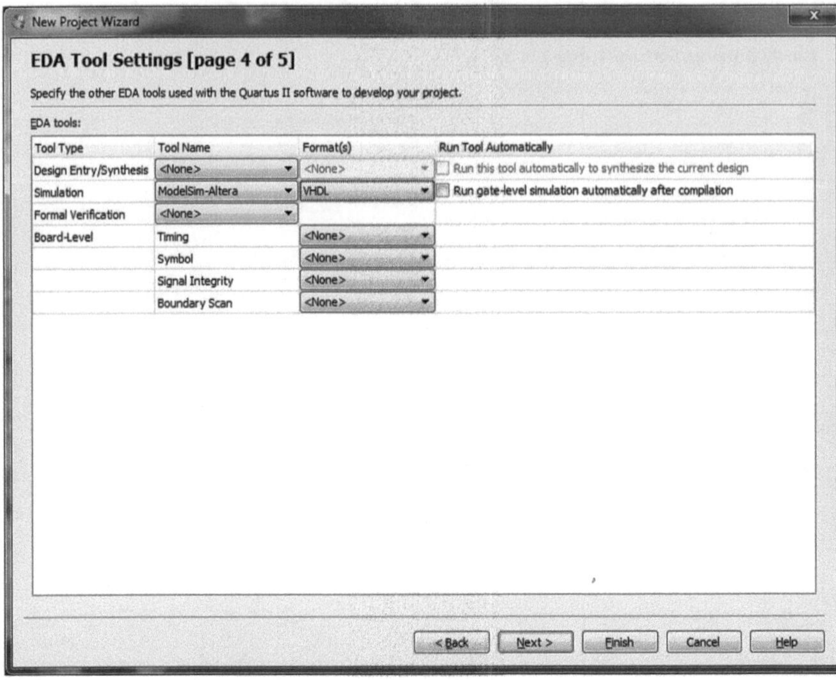

Fig. 1.15 Selecting the simulation tool—ModelSim-Altera

will open several new windows on the left side of Fig. 1.25. The "sim" window lists the hierarchical view of the circuit, and it is used in order to select the signals to be shown during the simulation. When the half-adder module is highlighted, the Objects window lists the signals that can be observed and changed in the simulation process. The signals of interest for the half-adder simulation include A0, B0, S0 and C1.

As shown in Fig. 1.26, to add signals to the simulation window, select the signals in the Objects window using the mouse pointer, and with a right click select *Add→To Wave→Selected Signals* from the pop-up menu. After have added signals A0, B0, S0 and C1, they will show in a waveform window.

In ModelSim the simulation process can be done manually or through scripts. The script language is relatively intuitive, but for the basic circuits introduced in most laboratory sessions it would be overwhelming. The manual option is described next.

The waveform window can be undocked using the button in the upper right hand corner, as shown in Fig. 1.27. To organize the signals (it is optional), place the mouse on the A0 signal, hold down the left button, and drag the signal to the desired position.

To set a signal to 0 or 1, right click on the signal and select the Force option from the pop-up menu. In Fig. 1.27, the B0 signal is "forced" to 1 (one). Next,

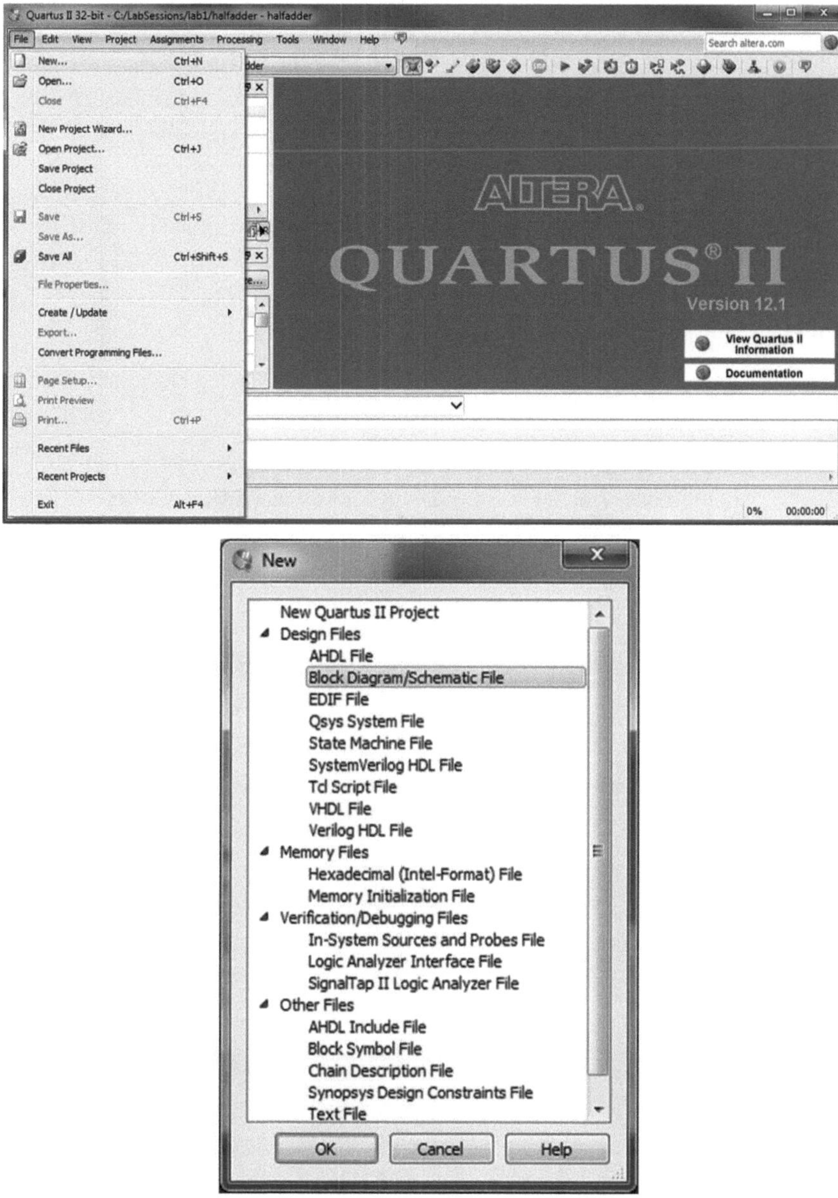

Fig. 1.16 New schematic file

repeat the operation for signal A0, but forcing it to 0 (zero). Set the simulation run length to 100 ps, and press the Run button. This will make the simulation run for 100 ps. In Fig. 1.27 there is an arrow pointing at the Run button. The remaining simulation buttons are:

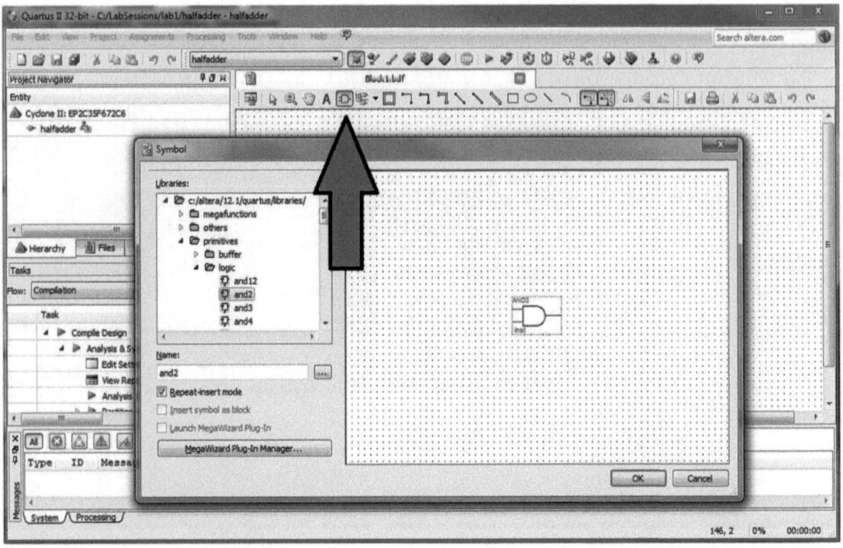

Fig. 1.17 Symbol tool icon

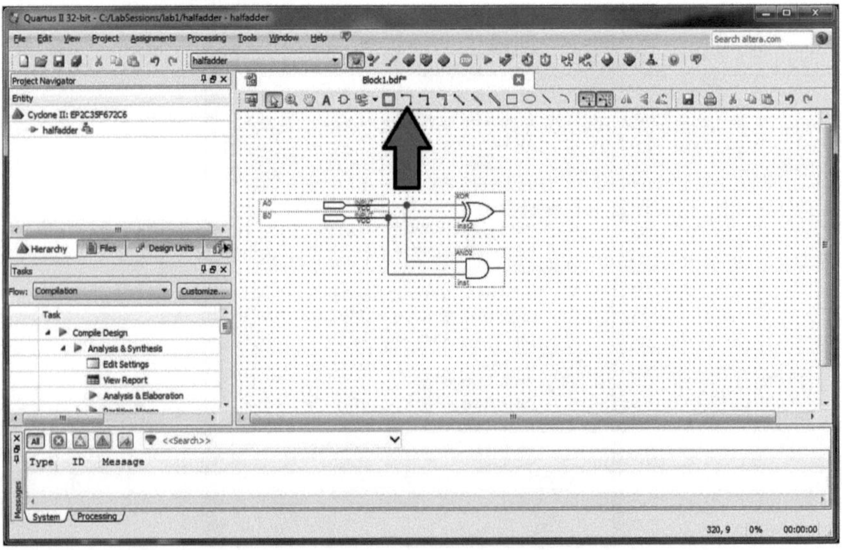

Fig. 1.18 Schematic entry

- Restart simulation button—it is located on the left of "Run length" (100 ps);
- Run length text box—defines for how long the simulation will run;
- Run button—runs the simulation for the duration specified in the Run length text box;

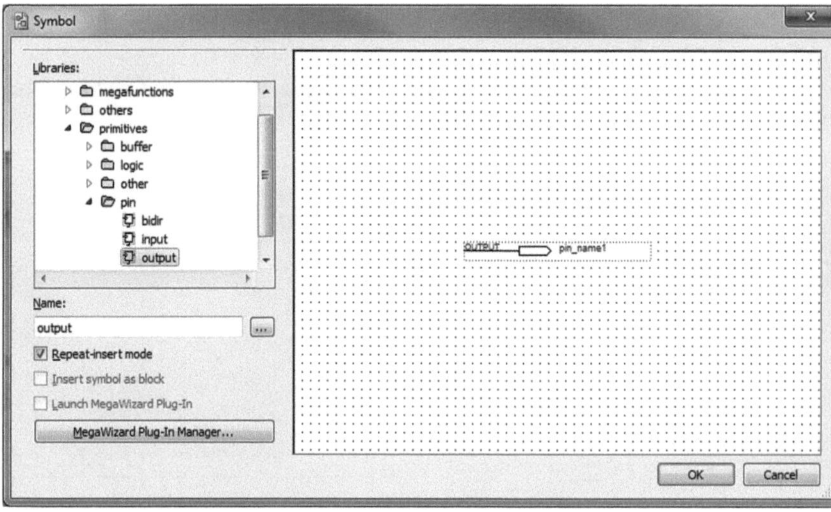

Fig. 1.19 Input and output pins

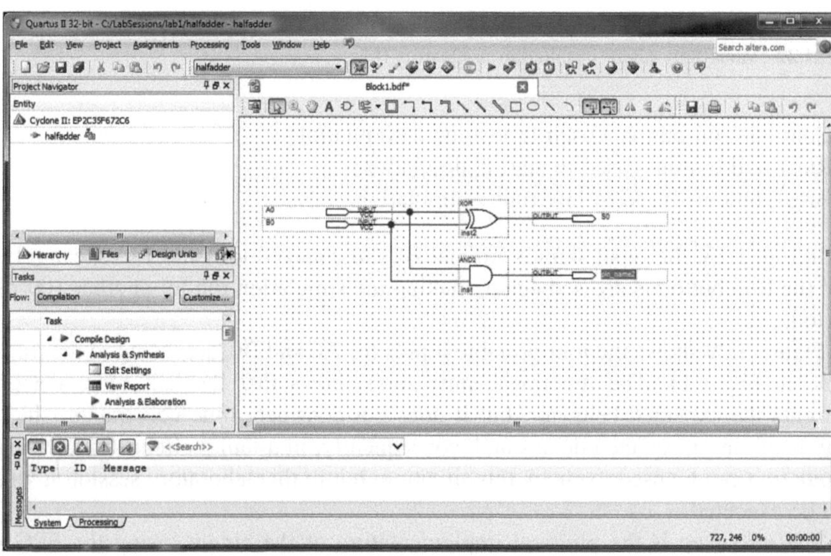

Fig. 1.20 Renaming the input and output pins

Fig. 1.21 Quartus II menu bar

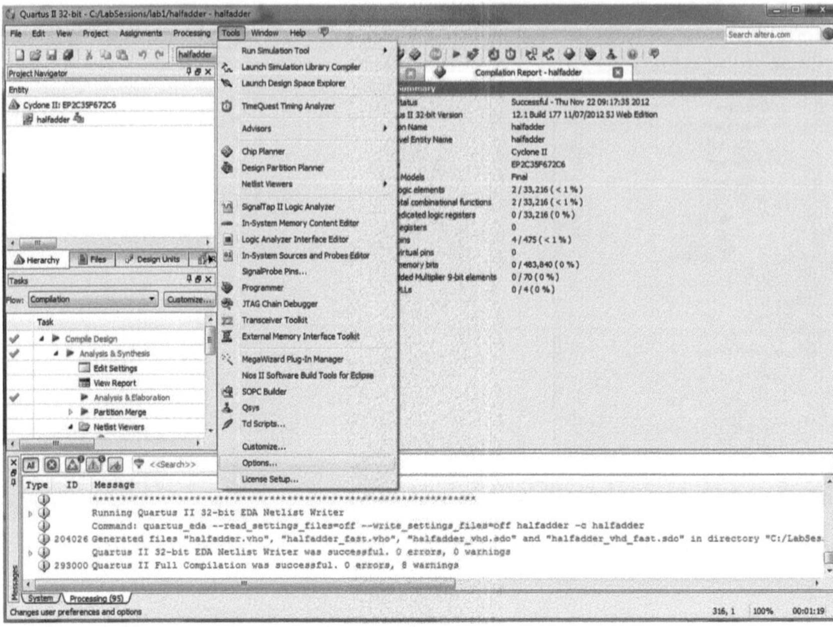

Fig. 1.22 ModelSim-Altera installation in Quartus II

- Continue run button—used to resume a simulation;
- Run—All button—used to start a simulation;
- Break button—used to interrupt a simulation.

In Fig. 1.28, the input signals A0 and B0 are changed to 0 and 1, respectively, and the Run button is pressed again, making the simulation run for another 100 ps period.

Figure 1.29 shows the four possible inputs for A0 and B0, according to the truth table in Fig. 1.11. The functional simulation shows that the half-adder design presents the expected behavior, but it does not considers time constraints which may result in unexpected delays and, consequently, in a faulty circuit. The timing simulation can be used to check this situation, but in this laboratory session it will not be executed.

The next step is the physical implementation of the circuit, using the FPGA board. Close (quit) all ModelSim windows, and go back to the Quartus II environment.

Step 5: FPGA Prototyping

Using Quartus II, choose menu *Assignments→Pin Planner*. As shown in Figs. 1.30 and 1.31, this tool is used to associate the input and output signals defined in the half-adder schematic design, to the actual FPGA pins. The list of FPGA pins and their connections to the DE2 resources (peripherals) can be found in the

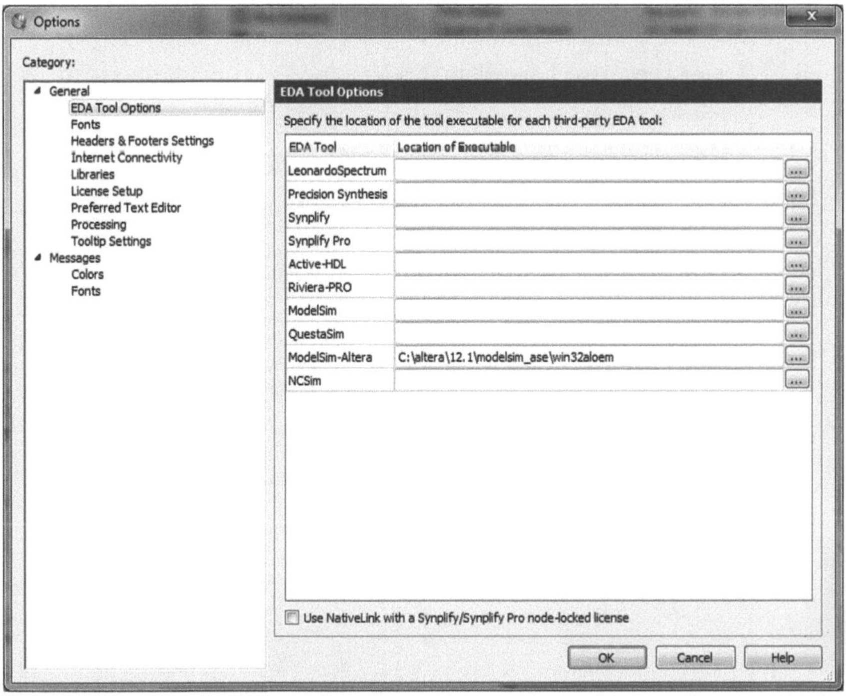

Fig. 1.23 ModelSim-Altera path configuration

board's user manual available on the DE2 folder: DE2\DE2_user_manual\ DE2_UserManual.pdf.

In DE2 user manual there is a list of all available pins, with the respective connection between the FPGA pins and the board's peripherals. For instance, SW[0] and SW[1] switches are connected to the N25 and N26 FPGA pins (see DE2_UserManual.pdf Table 4.1 in page 28). The half-adder output signals can be shown on LEDR[0] (sum) and LEDR[1] (carry), which are connected to FPGA pins AE23 and AF23 (see DE2_UserManual.pdf Table 4.3 in page 29).

Using the Pin Planer tool, perform these pin assignments in the Location column, as shown in Fig. 1.31, and close the Pin Planner.

Perform a new synthesis (button Compile in Quartus II menu), in order to assign the physical pins to the design. It will generate the final netlist that can be downloaded to the FPGA.

After the user has powered on and connected the DE2 board to the USB (use the "blaster" USB connector), press the red button to switch on the board, and run the programming tool: Menu *Tools→Programmer* (or fourth icon from right to left in Fig. 1.21). The window in Fig. 1.32 shows the Programmer tool interface. In order to perform a download of a configuration file to the FPGA, be sure to have the

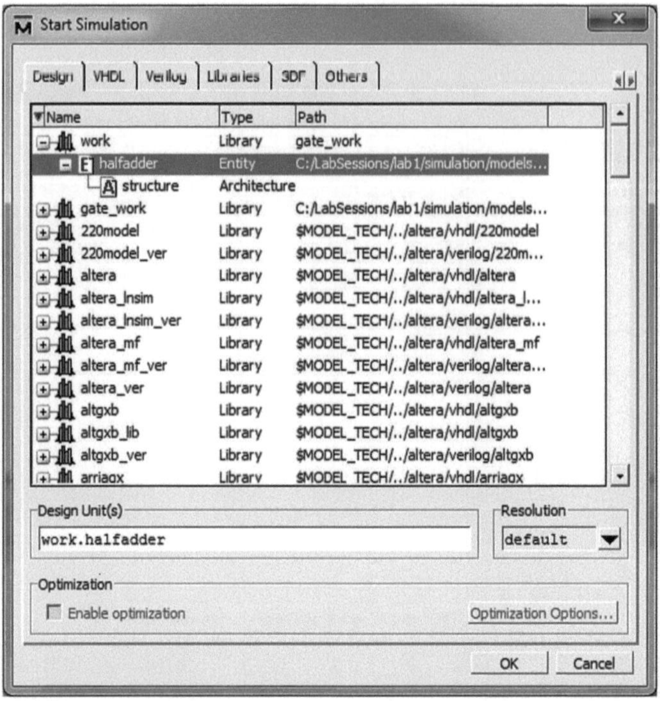

Fig. 1.24 Start simulation window

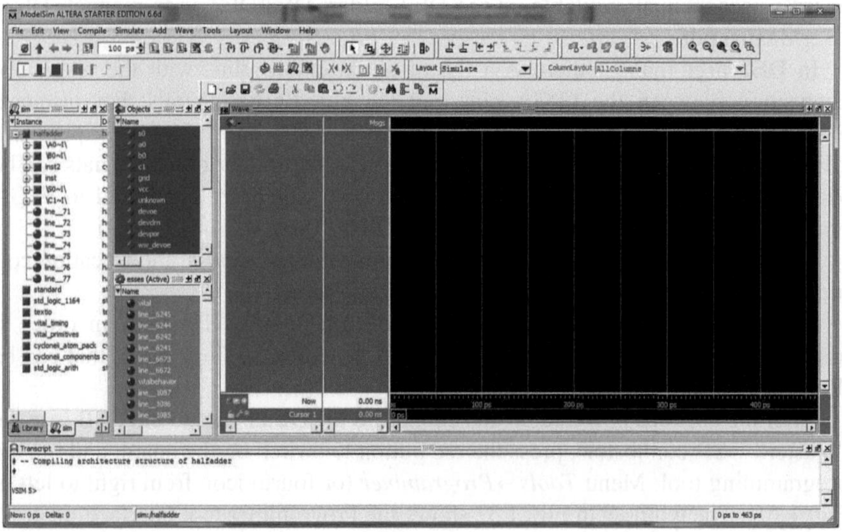

Fig. 1.25 Simulation environment

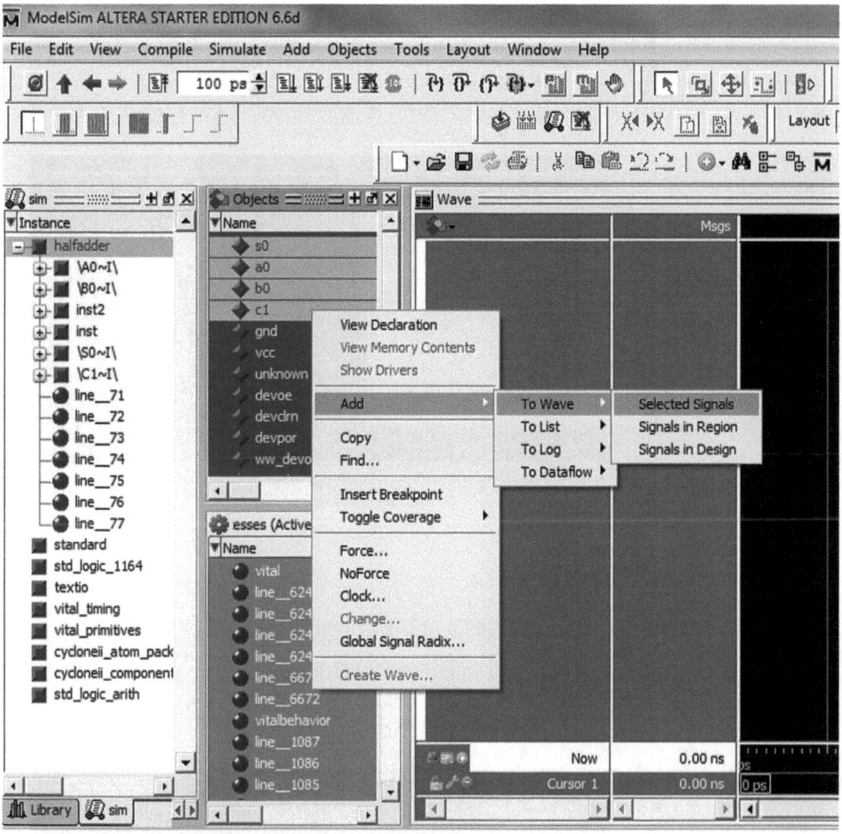

Fig. 1.26 Selecting signals for simulation

"USB-Blaster [USB-0]" message showing next to the "Hardware Setup" button, and that the configuration file "halfadder.sof" is listed in the File column.

The "Start" button is used to perform the download of the halfadder.sof to the FPGA configuration memory (Fig. 1.33).

Some problems that may happen regarding the FPGA Programmer tool are listed next:

- The USB port has not been found (DE2 board not detected). Solution: start "Devices and Printers" in Windows start menu and proceed with the new driver installation. The DE2 driver is located in the Quartus II folder: C:\altera\12.1\quartus\drivers. For Linux systems, follow the standard procedure for driver installation.
- The USB-Blaster [USB-0] is not shown. Solution: certify that the USB cable is connected to the "Blaster" input. Press the "Hardware Setup" button, double-click on "USB-Blaster", and close the Hardware Setup window.

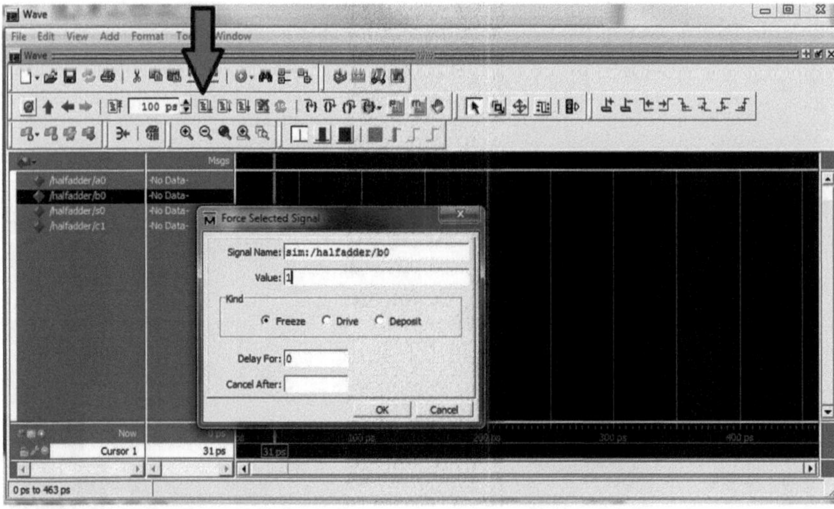

Fig. 1.27 Waveform window

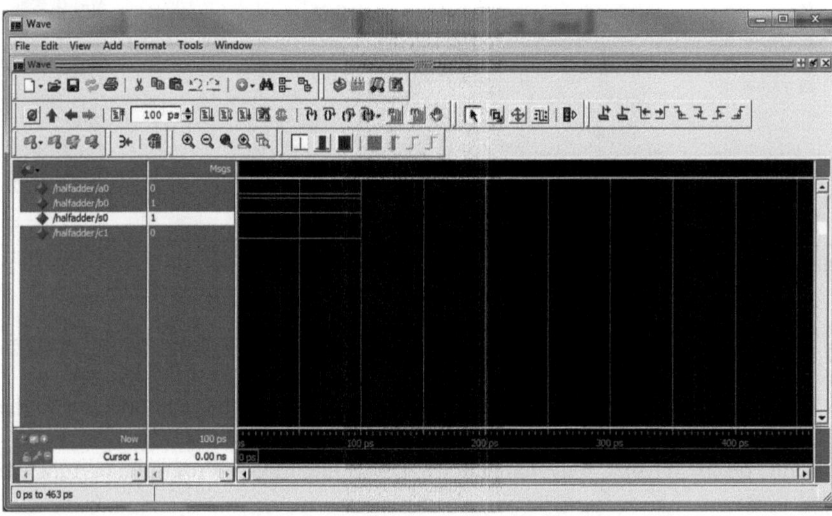

Fig. 1.28 A0 = 0 and B0 = 1, resulting in S0 = 1 and C1 = 0

- The halfadder.sof file is not shown. Solution: press the "Add File" button, go to c:\LabSession\lab1\output_files, and double-click on halfadder.sof

After have finished the download to the FPGA, in order to check the circuit functionality in the board, just switch inputs SW0 (A0) and SW1 (B0) "on" and

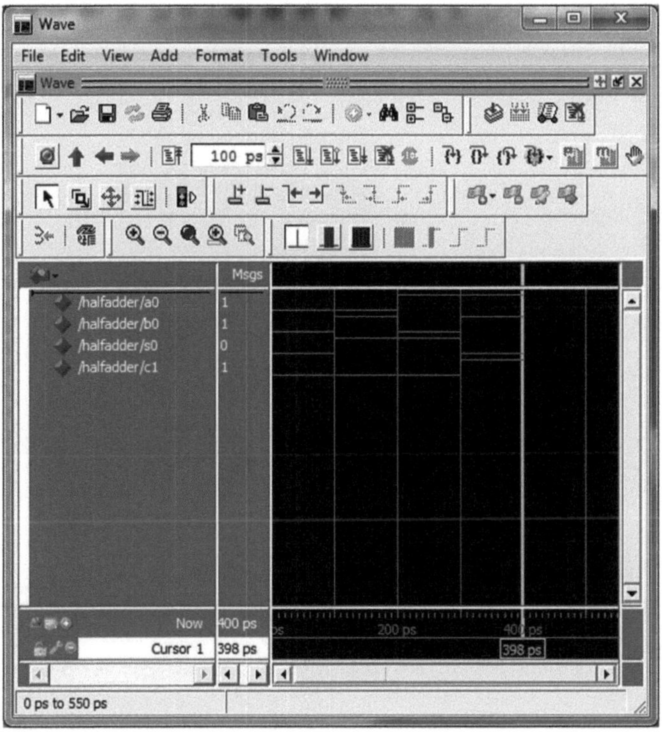

Fig. 1.29 All four combinations for A0 and B0

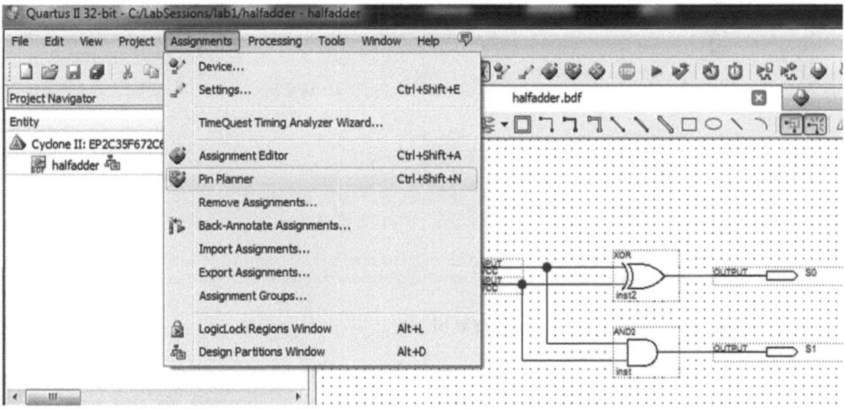

Fig. 1.30 FPGA pin assignments

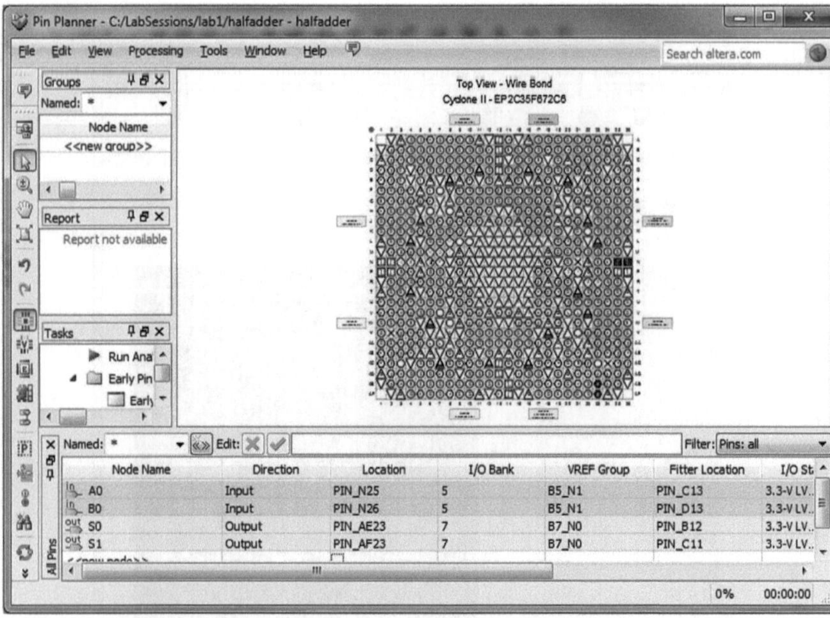

Fig. 1.31 Half-adder pin assignment

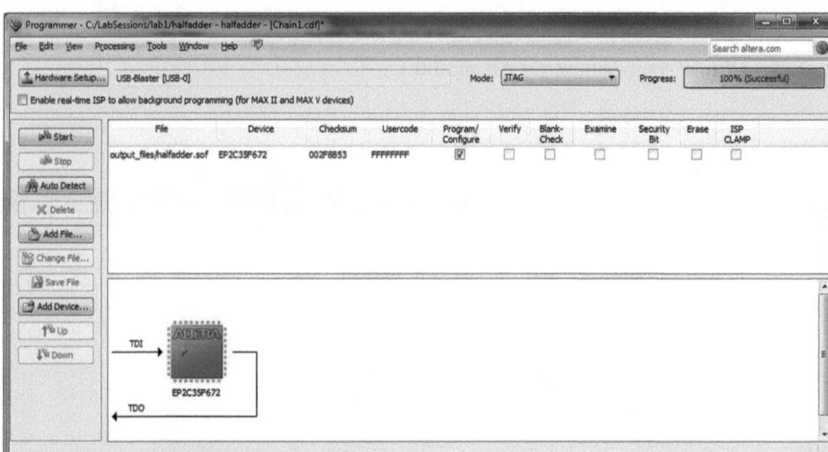

Fig. 1.32 Quartus II FPGA programming tool

"off", and make a comparison between the results showed in the red LEDs (Light-Emitting Diode) with the truth table shown in Fig. 1.11.

A summary of the activities developed in this laboratory session is as follows:

1. Project creation *File→New Project Wizard*

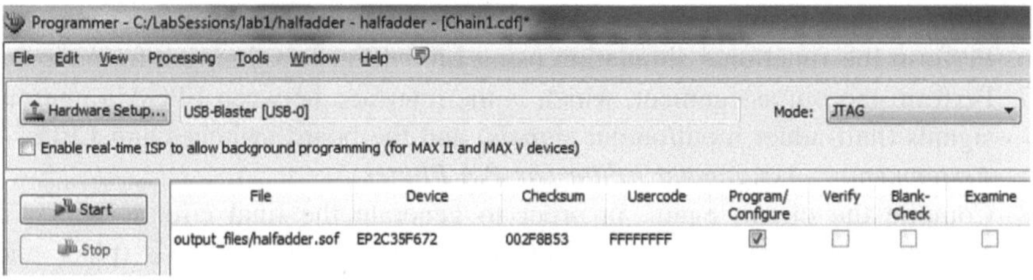

Fig. 1.33 FPGA programmer

2. In "project wizard", follow *exactly* the steps listed in this tutorial. Any oversight may result in errors in the hardware generation for the FPGA.
3. Design Entry (schematic) *File→New→Block Diagram*
4. Draw the half-adder schematic (Fig. 1.11).

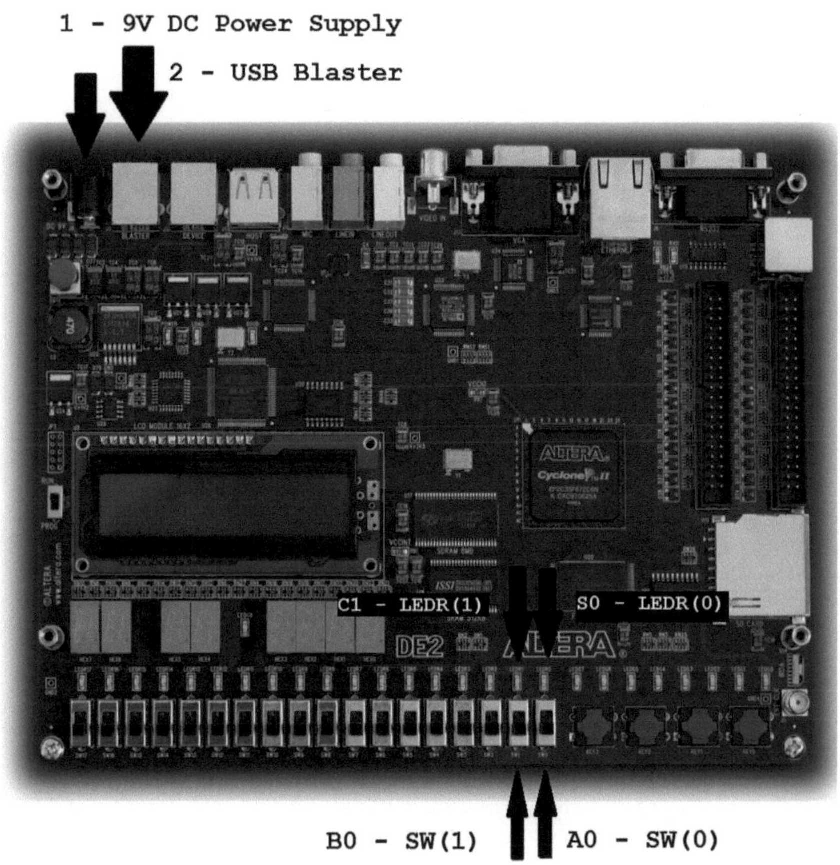

Fig. 1.34 DE2 board

5. Compile the circuit (synthesis).
6. Perform the functional simulation using ModelSim (no timing information).
7. Perform the pin assignment, which is the interface between FPGA's internal signals (half-adder input/output signals) and the board switches and LEDs—*Assignments—Assignment Editor (or Pin Planer).*
8. Compile the circuit again, in order to generate the final circuit (with pin assignments) to be used in the FPGA. In the physical synthesis (place and route), the logic elements defined in the logic synthesis are placed using the physical resources found in the FPGA. In this process, all connections (routing) between logic elements are performed.
9. Temporal Analysis—Analysis of propagation delays of signals after physical synthesis resulting in a report consisting of the expected performance.
10. Programming—The synthesized circuit is downloaded to the FPGA—*Tools—Programmer. Hardware Setup—USB-Blaster. Start!*
11. Testing the prototyped circuit—Fig. 1.34 shows the power connector (1), the USB blaster port (2), the inputs SW(0) and SW(1), and the outputs LEDR(0) and LEDR(1). Switch the inputs on and off, and observe the outputs (sum and carry) in the LEDs, according to the truth table in Fig. 1.11.

Chapter 2
HDL Based Designs

In this chapter the FPGA design flow is revisited, but using a hardware description language for the design entry instead of the schematic diagram discussed in Chap. 1. At the end of the chapter, the reader should be able:

- to understand the basic structure of a VHDL description;
- to model a simple digital circuit in VHDL;
- to follow the VHDL design flow;
- to simulate and test a digital circuit designed in VHDL;
- to prototype a circuit described in VHDL, using an FPGA board.

2.1 Theoretical Background

A Hardware Description Language (HDL) is a modeling language used to describe the structure and behavior of a digital circuit. In addition, an HDL allows the test of the designed circuits through simulations. Hardware description languages express a temporal behavior and circuit structure in a text format. Additionally, the syntax and semantics of HDLs include notations to express temporal sequences and concurrences, as required by a hardware module.

A good example of hardware description language is VHDL, which is an acronym for VHSIC Hardware Description Language. VHSIC stands for Very High Speed Integrated Circuits and it was a government program of USA in the early 1980s. Later, the VHDL language became an IEEE (Institute of Electrical and Electronic Engineers) standard and nowadays there are several tools to simulate and synthesize (i.e., generate hardware) circuits described in VHDL. Other hardware description languages are SystemC, Handel-C, Verilog, and System Verilog.

In VHDL, the design of a digital circuit can be described in two abstraction levels, that is, structural or behavioral. Descriptions at the register transfer level (RTL) are widely used to develop digital systems. VHDL is not equivalent to a

E. A. Bezerra and D. V. Lettnin, *Synthesizable VHDL Design for FPGAs*,
DOI: 10.1007/978-3-319-02547-6_2,

software programming language and the synthesis tools are not equal to the compiler tools, since they do not generate executable code from a VHDL description. Additionally, these descriptions can be used to generate a piece of hardware, for example, a file for configuring an FPGA.

VHDL descriptions can be simulated, that is, run in a simulator. The testing process can be performed via *testbenches* where stimuli are generated for simulating VHDL descriptions. A testbench defines the external stimulus to be used as test cases to a circuit input. The testbenches can be written in VHDL or in many other languages (e.g. C, C++, and SystemC).

The VHDL models consist of two main parts, the *entity* and the *architecture*. As shown in Fig. 2.1, the entity part of a VHDL description has only the input and output pins (interface) to be used by the circuit. It has no information at all regarding the circuit's internal logic (circuit operation). In the architecture part, as shown in Fig. 2.2, there is the operational description of the circuit. Figure 2.3 shows the complete VHDL description for the half-adder circuit. It is important to notice that in this example, the *std_logic* type has been used and, for this reason, its library has to be included in the description. As a result, a VHDL description usually has the following sessions:

- **Library** definition—includes the required libraries and packages, for instance, the IEEE package (see lines 1–2 in Fig. 2.3);
- **Entity**—defines the "pins" of the digital circuit (signals), i.e. the interface between the implemented logic and the external world (see lines 4–11 in Fig. 2.3);
- **Architecture**—defines the functionality of a digital circuit, using the input and output "pins" that are listed at the entity part (see lines 13–17 in Fig. 2.3). Thus, an entity may have different implementations (i.e., architectures) for the same function.

ENTITY

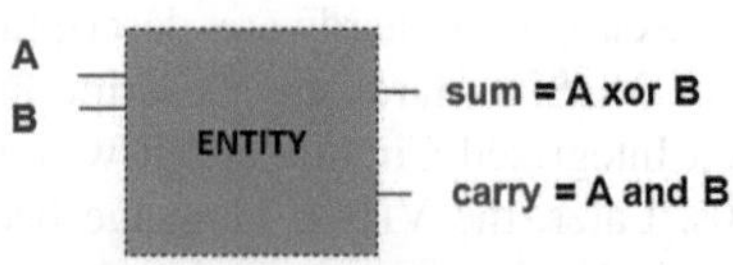

A	B	F1	F2
0	0	0	0
0	1	1	0
1	0	1	0
1	1	0	1

```
entity halfadd is
port (A: in std_logic;
      B: in std_logic;
      sum: out std_logic;
      carry: out std_logic
      );
end halfadd;
```

Fig. 2.1 "Entity" declaration for the half-adder circuit

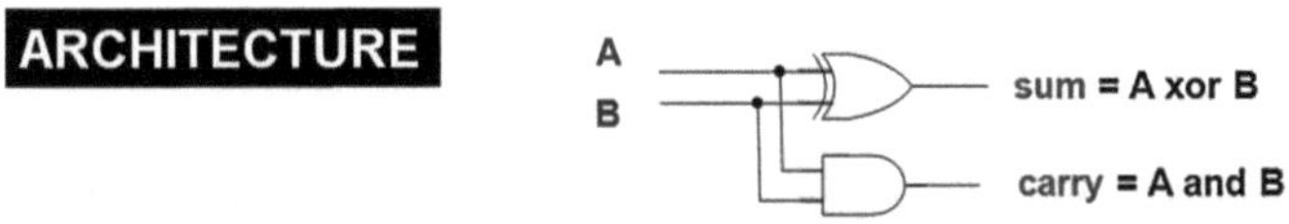


```
architecture ha_stru of halfadd is
begin
  sum <= A xor B;
  carry <= A and B;
end ha_stru;
```

Fig. 2.2 "Architecture" declaration for the half-adder circuit

Fig. 2.3 Example of entity and architecture in VHDL. Sum and carry are assigned in parallel

```
library IEEE;
use IEEE.Std_Logic_1164.all;

entity halfadder is
port (
      A    : in std_logic;
      B    : in std_logic;
      sum  : out std_logic;
      carry    : out std_logic
      );
end halfadder;

architecture ha_stru of halfadder is
begin
      sum <= A xor B;
      carry <= A and B;
end ha_stru;
```

2.2 Laboratory Assignment

The laboratory objectives are:

- to allow the reader to have a first contact with the VHDL language;
- to provide a first contact with an EDA tool;
- to continue the training and understanding of a complete FPGA design flow, but this time starting from a VHDL description design entry.

2.2.1 Laboratory Session

The tasks to be completed in this laboratory session, using Altera's Quartus II EDA tool, are as follows:

- Create the half-adder circuit shown in Fig. 2.3 using the VHDL design entry editor;
- Perform the synthesis;
- Fix simulation and synthesis errors;
- Prototype and test the half-adder circuit in the FPGA board.

Step 1: Creating a New Project
This step is exactly the same as "Step 1" described in Chap. 1. It is important to remember to type "halfadder" as the project's name (page 1 of 5 in the New Project Wizard—see Fig. 1.13).
Step 2: Design Entry (VHDL)
Chose *File* → *New* and *VHDL File*, as shown in Fig. 2.4. This will open an empty text editor.

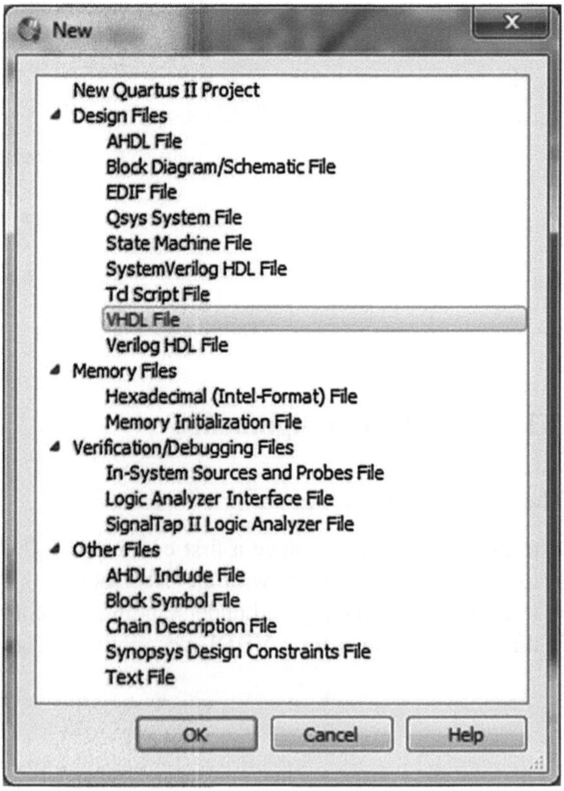

Fig. 2.4 New VHDL file

Type in the text editor the half-adder VHDL code shown in Fig. 2.3, and save it: *File* → *Save* (or just ctrl-S). The default name to save the file is halfadder.vhd, as it is the name provided as the project name in Step 1. The file will be saved, also in the default location, in the project folder defined in Step 1. Be sure that the check box "Add file to current project" is checked. Again, at this point it is important to be sure that the entity's name is also "halfadder", otherwise, the synthesis tool will not be able to find the entity to be synthesized.

Step 3: Synthesis

This step is the same as "Step 3" described in Chap. 1, that is, choose *Processing* → *Start Compilation* (or Ctrl-L). In case of errors, read carefully the "Processing" messages (bottom window), and fix them in the VHDL source code (in the text editor). There will be some warning messages, and at this point look just for the "critical" ones.

Step 4: Simulation

Before starting the simulation, see "Step 4" in Chap. 1 in order to configure simulator options. After that, to run ModelSim, in Quartus II menu choose *Tools* → *Run Simulation Tool* → *RTL Simulation* (or press the 7th icon from right to left shown in Fig. 1.21). Next, follow the instructions provided in Chap. 1, "Step 4":

- Simulate → Start Simulation;
- In the "Start Simulation" window, look for the "Design" tab (see Fig. 1.24);
- In the "work" Library, click on the "+" sign, and select the "halfadder" entity as shown in Fig. 1.24;
- Click OK to start the simulation;
- Select the *A*, *B*, *sum* and *carry* signals in the Objects window, and add them to the Wave window, as shown in Fig. 2.5;
- Set the *A* signal to 0, and the *B* signal to 0, using the "Force" option as explained in Chap. 1, "Step 4". Set the simulation run length to 100 ps, and press the *Run* button. Next, change *A* to 0 and *B* to 1, and run for another 100 ps. Change *A* to 1 and *B* to 0, and run for 100 ps. Finally, change *A* to 1 and *B* to 1, and run for 100 ps.
- The expected simulation results are shown in Fig. 2.6.

Step 5: FPGA Prototyping

The first task in order to build a physical design to prototype in an FPGA board is to specify which entity signal goes to which FPGA pin. For instance, in the half-adder design, if nothing is done regarding the *A*, *B*, *sum* and *carry* signals, the synthesis tool will randomly allocate pins in the FPGA device, and the design may not work in the target board. In Chap. 1 the half-adder entity signals were assigned to the physical (FPGA) pins using the Pin Planner tool (see Fig. 1.31). This is an adequate strategy for designs with few external interface pins, which is the case of the half-adder circuit. However, this is not the case in real world designs were dozens or even hundreds of signals are used in their top-level entities. For this reason, for the VHDL version of the half-adder design an automatic pin assignment strategy will be adopted.

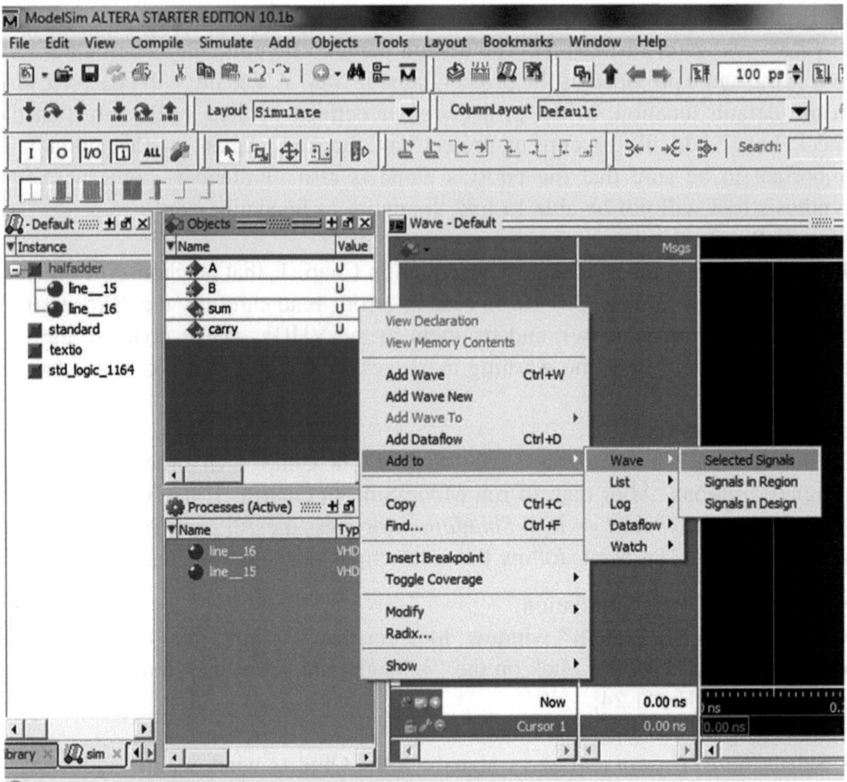

Fig. 2.5 Selecting signals from the half-adder design

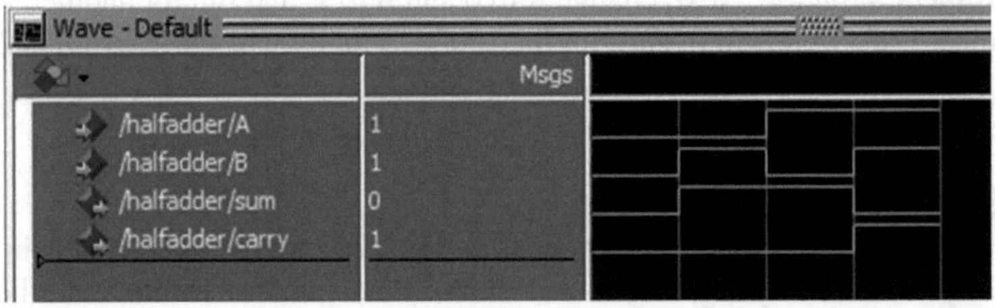

Fig. 2.6 Simulation results for the half-adder in VHDL

The listing of all pins available in the DE2 board is included in a text file. In this listing, each pin has a specific name, and the signals of the half-adder entity should be renamed according to the FPGA board nomenclature. Alternatively, the pin names in the file can be renamed to match the signal names in the design entity, but this is not a common practice.

Each FPGA vendor has its own format, but usually only the relation between the signal name and the target pin is specified. There is no indication regarding the pin direction (input, output).

Xilinx pin assignment files have the ".ucf" extension, and lines such as:

NET "sum" LOC = "P28";
NET "carry" LOC = "P110";

In this case, the sum and carry signals of the half-adder entity will be assigned to FPGA pins 28 and 110, respectively.

Altera has a different format for their pin assignments files, as shown next:

set_location_assignment PIN_N25 -to A
set_location_assignment PIN_N26 -to B
set_location_assignment PIN_AE23 -to sum
set_location_assignment PIN_AF23 -to carry

In this example, *A* and *B* are assigned in the DE2 board to switch 0 and switch 1, respectively. The signals *sum* and *carry* are assigned to LED red 0 and LED red 1, respectively.

The full list of available pins on the board is included in the *DE2_pin_assignments.qsf* file, which can be found in the official website for this book. The Quartus II Setting File (".qsf") can be downloaded from the book's website and copied to the project's folder (e.g., C:\LabSessions\halfadder), the FPGA pins listed in this file must be imported into the design: menu *Assignments → Import Assignments*. Browse to find the folder, and press OK as shown in Fig. 2.7.

This file has 425 pins listed, and each pin is assigned to a pre-defined label. The N25, N26, AE23 and AF23 FPGA pins are assigned to the labels SW[0], SW[1], LEDR[0] and LEDR[1], as shown next:

set_location_assignment PIN_N25 -to SW[0]
set_location_assignment PIN_N26 -to SW[1]
…
set_location_assignment PIN_AE23 -to LEDR[0]
set_location_assignment PIN_AF23 -to LEDR[1]
…

In the DE2 board Printed Circuit Board (PCB) design, the FPGA pin N25 is connected to the switch SW_0 shown at the bottom of Fig. 1.34. The FPGA pin N26 is connected to switch SW_1, and so on.

In the *DE2_pin_assignments.qsf* file, the label SW represents a vector with 18 positions (0–17). LEDR is also a vector with 18 positions.

To perform the connection between the physical world (DE2 switches and LEDs) and internal FPGA signals, a VHDL entity should include the input and output pins available on the board. As shown in Fig. 2.8, on line 6 the label SW is used. SW is an 18 bits vector, and it has the same name (and size) as the SW

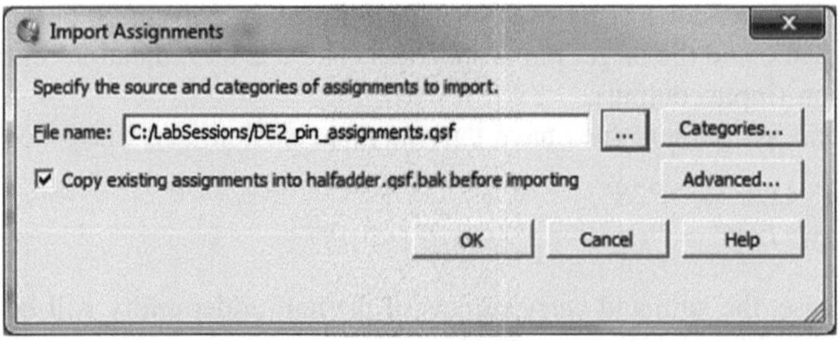

Fig. 2.7 Import assignments for the half-adder design

symbol connected to the FPGA switch pins listed in the *DE2_pin_assignments.qsf* file. On line 7, label LEDR is used in order to match the name used in the pin assignments file. The VHDL commands in lines 13–15 write to the red LEDs: switch values (line 13); a constant value (line 14), and the results of a logical operation (line 15).

In order to adapt the half-adder circuit to the DE2 board, the design's entity should be changed according to the *DE2_pin_assignments.qsf* file contents. The changes consist in renaming the *A*, *B*, *sum* and *carry* signals to *SW(0)*, *SW(1)*, *LEDR(0)*, and *LEDR(1)*, respectively. The changes are shown in Fig. 2.9.

After have concluded the modifications in the VHDL source code, perform a new synthesis (Compile button in Quartus II menu), and download the generated configuration file to the FPGA. The steps to perform the download are the same as "Step 5" in Chap. 1 (see Figs. 1.32 and 1.33).

To test the circuit, follow the instructions provided at the end of Chap. 1, using the switches and LEDs as shown in Fig. 1.34, where *A* is *SW(0)*, *B* is *SW(1)*, *sum* is *LEDR(0)*, and *carry* is *LEDR(1)*.

```
1   library IEEE;
2   use IEEE.Std_Logic_1164.all;
3
4   entity DE2_pins is
5   port (
6         SW   : in  std_logic_vector(17 downto 0);
7         LEDR     : out std_logic_vector(17 downto 0)
8         );
9   end DE2_pins;
10
11  architecture logic_circuit of DE2_pins is
12  begin
13        LEDR(7 downto 0) <= SW(15 downto 8);
14        LEDR(15 downto 8) <= "01010101";
15        LEDR(17) <= (SW(17) AND SW(0)) OR SW(16);
16  end logic_circuit;
```

Fig. 2.8 DE2 pins declared in an entity

```
library IEEE;
use IEEE.Std_Logic_1164.all;

entity halfadder is
port (                                                  -- A -> SW(0)
      SW   : in  std_logic_vector(17 downto 0);         -- B -> SW(1)
      LEDR: out std_logic_vector(17 downto 0)           -- sum   -> LEDR(0)
      );                                                -- carry -> LEDR(1)
end halfadder;

architecture ha_stru of halfadder is
begin
    LEDR(0) <= SW(0) xor SW(1);         -- sum   <= A xor B
    LEDR(1) <= SW(0) and SW(1);         -- carry <= A and B
end ha_stru;
```

Fig. 2.9 Half-adder design adapted to DE2 board pin names

2.2.2 Going Beyond

The DE2 board has an LCD which can be programmed and written manually, using the available switches. The VHDL code to access the LCD pins is shown in Fig. 2.10.

After have synthesized and downloaded the generated configuration file to the FPGA board, follow the next steps to write control and data to the LCD.

```
library ieee;
use ieee.std_logic_1164.all;
entity LCD is
port (
      LCD_DATA    : out std_logic_vector(7 downto 0);
      LCD_RW      : out std_logic;
      LCD_EN      : out std_logic;
      LCD_RS      : out std_logic;
      LCD_ON      : out std_logic;
      LCD_BLON    : out std_logic;
      SW          : in std_logic_vector(17 downto 0)
);
end LCD;
architecture LCD_WR of LCD is
begin
      LCD_ON   <= SW(17);
      LCD_BLON <= SW(16);
      LCD_DATA <= SW(7 downto 0);
      LCD_RS   <= SW(8);
      LCD_EN   <= SW(9);
      LCD_RW   <= SW(10);
end LCD_WR;
```

Fig. 2.10 VHDL code for LCD writing in DE2 board

Step 1: Write Command 1 (38H)—Switches the LCD ON, Cursor ON, and Blink ON

Switch (SW)	Value (switch position)	Effect
17	1	LCD_ON
16	1	LCD_BLON
7..0	0011 1000	Command
8	0	LCD_RS (0 = control)
9	0 → 1 → 0	LCD_EN
10	0	LCD_RW

Step 2: Write Command 2 (0FH)—Switches the LCD ON, Cursor ON, and Blink ON

Switch (SW)	Value (switch position)	Effect
17	1	LCD_ON
16	1	LCD_BLON
7..0	0000 1111	Command
8	0	LCD_RS (0 = control)
9	0 → 1 → 0	LCD_EN
10	0	LCD_RW

Step 3: Write Command 3 (06H)—Switches the LCD ON, Cursor ON, and Blink ON

Switch (SW)	Value (switch position)	Effect
17	1	LCD_ON
16	1	LCD_BLON
7..0	0000 0110	Command
8	0	LCD_RS (0 = control)
9	0 → 1 → 0	LCD_EN
10	0	LCD_RW

Step 4: Write Data to LCD

Switch (SW)	Value (switch position)	Effect
17	1	LCD_ON
16	1	LCD_BLON
7..0	0100 0001	'A' or 41H (ASCII)
8	1	LCD_RS (1 = data)
9	0 → 1 → 0	LCD_EN
10	0	LCD_RW

Step 5: Clear the LCD

Switch (SW)	Value (switch position)	Effect
17	1	LCD_ON
16	1	LCD_BLON
7..0	0000 0001	Clear command
8	0	LCD_RS (0 = control)
9	0 → 1 → 0	LCD_EN
10	0	LCD_RW

Chapter 3
Hierarchical Design

This chapter introduces the concept of hierarchical design in digital systems. A full VHDL project made of several components is developed as a case study, and its components are glued together using the VHDL reserved words *component* and *portmap*. These VHDL statements are used in all remaining projects throughout this book. At the end of the chapter, the reader should be able:

- to identify and comprehend the use of components in VHDL;
- to understand the concept of hierarchical design, and its implementation in VHDL;
- to add a component to the proposed case study;
- to simulate and test both, the original and the modified version of the case study;
- to prototype the case study using an FPGA board.

3.1 Hierarchical Design in VHDL

A digital system can be seen as a set of components interconnected in order to perform some required functionality. As applications grow in complexity, so do the number of components in a design. To deal with system complexity, VHDL offers mechanisms for the design organization. The language has appropriate structures allowing complex components to be built from less complex components. The hierarchical design methodology is not a novelty in traditional hardware and software systems. In a similar way, in VHDL a component oriented solution results not only in a more organized design, but also in more opportunities for components reusability. Components already employed in other systems may have been tested and verified, shortening the design cycle.

In order to introduce the VHDL structure and statements used in the hierarchical design methodology, consider the development of a digital system to implement the following equation:

$$F = (F1 \; and \; F2) \; or \; F3 \tag{3.1}$$

E. A. Bezerra and D. V. Lettnin, *Synthesizable VHDL Design for FPGAs*,
DOI: 10.1007/978-3-319-02547-6_3,

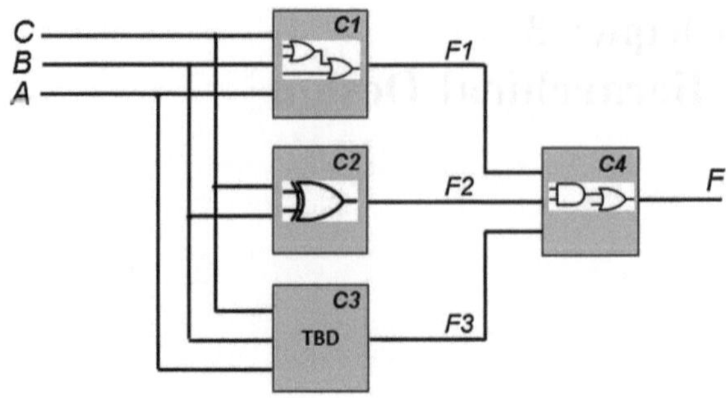

Fig. 3.1 Digital system design based on components

where $F1$, $F2$ and $F3$ are outputs in the following equations:

$$F1 = A\,or\,B\,or\,C \tag{3.2}$$

$$F2 = B\,xor\,C \tag{3.3}$$

$$F3 = \text{``}\textit{to be defined}\text{''} \tag{3.4}$$

As shown in Fig. 3.1, the circuit to be designed in VHDL has four components: $C1$ implements Eq. (3.2); $C2$ implements Eq. (3.3); $C3$ implements Eq. (3.4) (TBD—*to be defined*); and $C4$ implements Eq.3.1. Component $C1$ has three inputs (A, B, C) and one output ($F1$). Component $C2$ has two inputs (B, C) and one output ($F2$). Component $C3$ has three inputs (A, B, C) and one output ($F3$). Component $C4$ has three inputs ($F1$, $F2$, $F3$) and one output (F).

In VHDL, each of these components is implemented through an *Entity/Architecture* construction. Component $C1$ is shown in Fig. 3.2, component $C2$ in Fig. 3.3, and component $C4$ in Fig. 3.4. Component $C3$ functionality will be defined next in Sect. 3.2.

```
library IEEE;
use IEEE.Std_Logic_1164.all;

entity C1 is
port (A: in std_logic;
      B: in std_logic;
      C: in std_logic;
      F: out std_logic
    );
end C1;

architecture c1_estr of C1 is
begin
  F <= A or B or C;
end c1_estr;
```

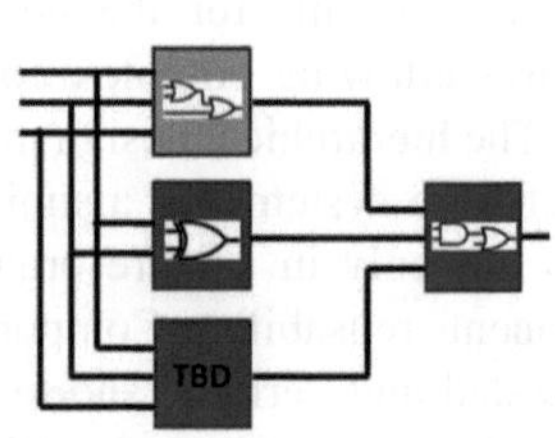

Fig. 3.2 VHDL implementation for component $C1$

```
library IEEE;
use IEEE.Std_Logic_1164.all;

entity C2 is
port (A: in std_logic;
      B: in std_logic;
      F: out std_logic
     );
end C2;

architecture c2_estr of C2 is
begin
  F <= A xor B;
end c2_estr;
```

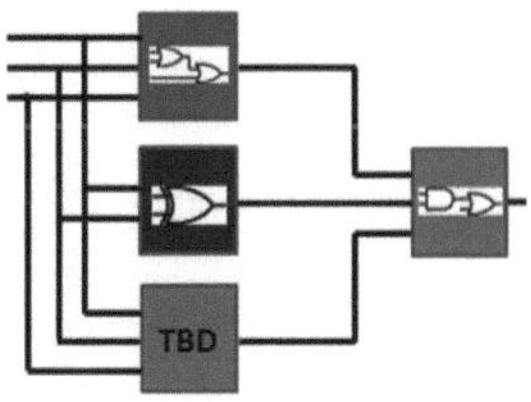

Fig. 3.3 VHDL implementation for component *C*2

```
library IEEE;
use IEEE.Std_Logic_1164.all;

entity C4 is
port (A: in std_logic;
      B: in std_logic;
      C: in std_logic;
      F: out std_logic
     );
end C4;

architecture c4_estr of C4 is
begin
  F <= (A and B) or C;
end c4_estr;
```

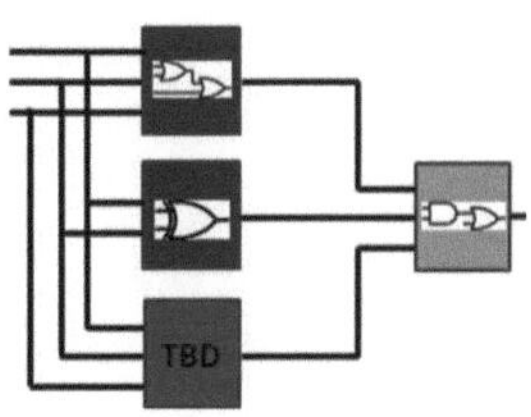

Fig. 3.4 VHDL implementation for component *C*4

Having all the required components, the next step in a hierarchical design is to glue the components together. As can be observed in Fig. 3.1, the four components are connected through wires. The three inputs to *C*4 are the wires named *F*1, *F*2 and *F*3, which are also the outputs for *C*1, *C*2 and *C*3. The inputs for *C*1, *C*2 and *C*3 are the signals *A*, *B* and *C*, and the circuit output is signal *F*. In VHDL, all these connections are defined using the *port map* statement. In order to use this statement, first it is necessary to list the components to be connected, using the *component* statement. Figure 3.5 has the VHDL code used to generate the circuit shown in Fig. 3.1. This VHDL code is usually known as the "top" component in a design hierarchy. The "top" component is the higher level one, and it has the purpose of instantiating and connecting lower hierarchy level components.

Fig. 3.5 VHDL code for inferring and connecting Fig. 3.1 components (top-level component)

```
1  library ieee;
2  use ieee.std_logic_1164.all;
3
4  entity top is
5    port ( A, B, C : in  std_logic;
6           F       : out std_logic
7    );
8  end top;
9
10 architecture top_map of top is
11    signal F1, F2, F3: std_logic;
12    component C1
13       port (  A : in  std_logic;
14               B : in  std_logic;
15               C : in  std_logic;
16               F : out std_logic
17       );
18    end component;
19    component C2
20       port (  A : in  std_logic;
21               B : in  std_logic;
22               F : out std_logic
23       );
24    end component;
25    --
26    -- ADD C3 HERE
27    --
28    component C4
29       port (  A : in  std_logic;
30               B : in  std_logic;
31               C : in  std_logic;
32               F : out std_logic
33       );
34    end component;
35 begin              -- begin architecture
36    L0: C1 port map (A, B, C, F1);
37    L1: C2 port map (B, C, F2);
38    --
39    -- ADD C3 CONNECTIONS HERE
40    --
41    L3: C4 port map (F1, F2, F3, F);
42 end top_map;     -- end architecture
```

In Fig. 3.5, components *C*1 to *C*4 are listed in the *architecture* section, just before the *begin* statement (lines 12–34). The VHDL syntax for defining the components is similar to the entity declaration. In fact, a good practice in order to avoid coding mistakes is to copy the desired component's entity section to the top-level file (Fig. 3.5), replacing the word “entity” by “component”, removing the “is” word, and adding an “end component” at the end of the declaration. Changes in a component header should be performed just in its entity, in order to avoid double changes mistakes. After have made the needed changes in an entity, just copy the whole entity section to the architecture section of the top-level file, and perform the modifications as stated before.

The component declarations in the architecture section of a top-level file inform the synthesis tool which components are to be used in the design. However, so far no component instantiation of connection has been made. In order to “place” the components in the design, and to connect them, the *port map* VHDL statement must be used. In Fig. 3.5, line 36, an instance (one copy) of component *C*1 is placed in the circuit. On line 37, an instance of *C*2 is place in the design, and on line 41 an instance of *C*4 is placed in the design. The connections are established according to the arguments in the *port maps* parameters list. On line 36, the first parameter (*A*) tells the synthesis tool to connect the first input signal of component *C*1 (*A* on line 13) to the *A* signal listed in the top-level entity (line 5). The parameter *B* on line 36 results in the connection between the second input signal of *C*1 (*B* on line 14) and the *B* signal listed in the top-level entity (line 5). Parameter *C* on line 36 is the connection from the *C* signal listed in the top-level entity (line 5) to the third input signal of *C*1 (*C* on line 15). Finally, the *F*1 parameter connects *C*1 output to one of *C*4 inputs (see line 41).

An important remark is that in the parameters list of *port map* statements, only entity and architecture signals are allowed. This means that all signals listed in the components declarations cannot be used. For instance, the *A*, *B*, and *C* parameters used on line 36 are the signals listed on line 5 of the entity, and not the *A*, *B* and *C* signals that are listed on lines 13, 14 and 15, respectively. The *F*1 parameter is defined as an internal signal, in the architecture section.

The synthesis tool makes the relation between the informed parameter in the *port map* and the respective component’s signal, according to their order in the argument’s list and in the component’s declaration. For instance, on line 37, *C*2 inputs *A* and *B* are connected to the top-level inputs *B* and *C* (listed on line 5). *F*2 on line 37 is connected to *F*2 on line 41. Therefore, the signals are mapped in the same order as defined in the component declaration.

VHDL has another syntax for signals mapping, which is the “explicit mapping”. For example, line 37 can be rewritten as follows:

```
L1: C2 port map (F=>F2, A=>B, B=>C);
```

In this alternative syntax, there is no need to keep the parameters order, as the => operator forces the signals mapping. Thus, *A*, *B* and *F* signals of *C*2 are explicitly connected to *B*, *C* and *F*2, where *B* and *C* are top-level entity signals, and *F*2 is an internal signal.

Another important remark is that each entity/architecture construction needs its own library declaration. Also, it is a good coding practice to keep each component in a separate file. The top-level entity/architecture (see Fig. 3.5) should also be kept in another file, and the synthesis tool can find all components of a design just looking for the "component" statements in this file.

Figure 3.6 shows the circuit generated from the VHDL code in Fig. 3.5. As stated on line 36, *C*1 has *A*, *B* and *C* as inputs, and *F*1 as its output. On line 37, *B* and *C* are defined as the inputs for *C*2, and *F*2 is its output. Component *C*4 is instantiated on line 41, having *F*1, *F*2 and *F*3 as inputs, and *F* as output. All 4 components are in the box named "top" in Fig. 3.6, and this box represents the VHDL code in Fig. 3.5 (see the entity's name).

3.2 Laboratory Assignment

The laboratory objectives are:

- to practice the basic concepts of hierarchical design in VHDL;
- to build a basic design to be used as a skeleton in the remaining lab sessions in this book;
- to carry on the training and understanding of a complete FPGA design flow.

3.2.1 Laboratory Session

This lab has as its main task the design of the circuit shown in Fig. 3.6, including all the aforementioned components. The whole circuit is built from five VHDL files, each of them comprising an entity and an architecture section. The four components *C*1, *C*2, *C*3 and *C*4, are saved in files c1.vhd, c2.vhd, c3.vhd, and c4.vhd, respectively. Component *C*3 has to be implemented according to the specifications provided next. The top-level component (see Fig. 3.5), performs all the instantiations and connections and it is saved in file *top.vhd*.

The tasks to be completed in this laboratory session, using Altera's Quartus II EDA tool, are as follows:

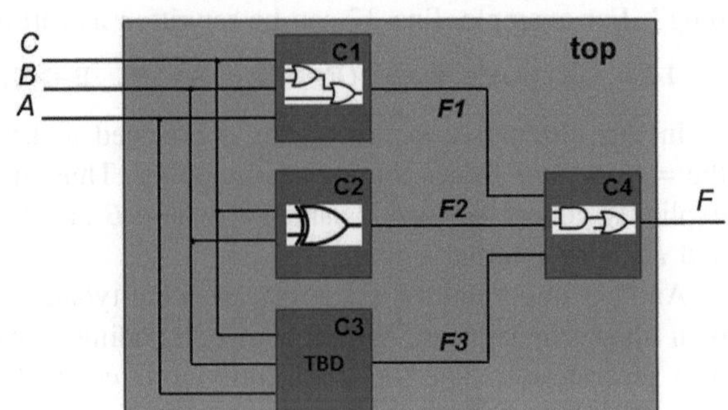

Fig. 3.6 Final circuit generated from VHDL code shown in Fig. 3.5

- Create a new project with all the components shown in Fig. 3.6—Steps 1, 2;
- Create the missing component (*C*3) using the VHDL design entry editor—Step 3;
- Modify the top-level component provided in Fig. 3.5, in order to add *C*3 to the circuit—Step 4;
- Modify the top-level component provided in Fig. 3.5, in order to adapt its inputs and outputs to the signal labels used in the DE2 (prototyping board) interface—Step 5;
- Perform the synthesis—Step 6;
- Perform the simulation and fix errors, if any—Step 7;
- Prototype and test the circuit in the FPGA board—Step 8.

Step 1—Creating a New Project

This step is essentially the same as "Step 1" described in Chap. 1. The major difference is in the project's name. In this new design, instead of "halfadder" use "top" as the project's name (page 1 of 5 in the New Project Wizard—see Fig. 1.13), as this is the name of the entity in Fig. 3.5 (top-level entity).

Step 2—Adding VHDL Files to the Project

Chose *File* → *New* and *VHDL File*, as shown in Fig. 3.7. This will open an empty text editor.

Type in the text editor the VHDL code for *C*1 as shown in Fig. 3.2, and save it: *File* → *Save* (or just Ctrl-S). The default name to save the file is top.vhd, as it is the name provided as the "project name" in Step 1. Before saving it, change the name to c1.vhd. The file will be saved, also in the default location, in the project folder defined in Step 1. Be sure that the check box "Add file to current project" is checked. Again, at this point it is important to be sure that the entity's name is also "*C*1" otherwise, the synthesis tool will not be able to find the entity to be synthesized.

Repeat the same procedure to create c2.vhd, c4.vhd, and top.vhd:

1. *File* → *New* → *VHDL File*;
2. Type de VHDL code for each component and the top-level—use Fig. 3.3 for c2.vhd, Fig. 3.4 for c4.vhd, and Fig. 3.5 for top.vhd;
3. Use *File* → *Save* for saving each file, always changing the name according to what is being saved;
4. Be sure to select the "Add file to current project" checking box;

Alternatively, if you already have c1.vhd, c2.vhd, c3.vhd, c4.vhd and top.vhd, just select *Project* → *Add/Remove Files in Project*, as shown in Fig. 3.8. This option will open a window allowing the import of VHDL files into the new project.

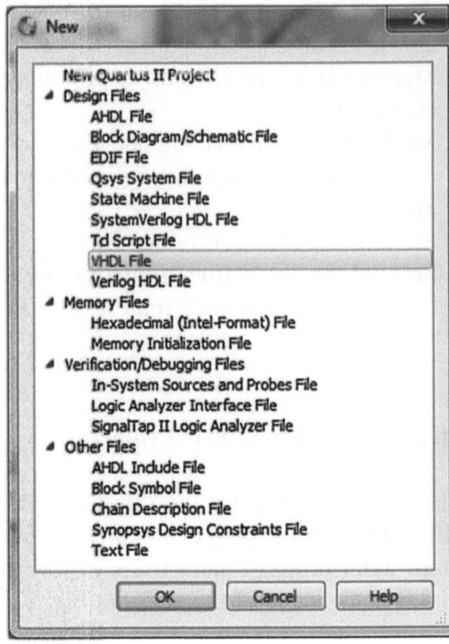

Fig. 3.7 New VHDL file

Step 3—Creating and Adding *C*3 to the Project

Using *C*1, *C*2 or *C*4 as a model, write in VHDL the circuit presented in Fig. 3.9. Follow "Step 2" instructions to create and add this component as c3.vhd to the project.

Step 4—Editing top.vhd to Add *C*3 to the Project

As shown in Fig. 3.10, select *Project Navigator* (1), *Files* (2), and double-click on top.vhd (3). This will open the top-level component in the VHDL text editor. Add *C*3 component interface (entity) to the designated location in Fig. 3.5 (see line 26), and make the necessary modifications in order to have the wright syntax for components definition in VHDL. Use components *C*1, *C*2 and *C*4 of Fig. 3.5 as examples.

Create an instance of *C*3 using the port map statement on line 39 of Fig. 3.5. Use the port maps of components *C*1, *C*2 and *C*4 as examples, but be careful to make the correct connections according to Figs. 3.6 and 3.9. Notice that in Fig. 3.9, *B* and *C* are the inputs for the OR gate, and *A* is the input for the NOT gate. *F*3 is the component output. All these connections should be performed by the port map statement.

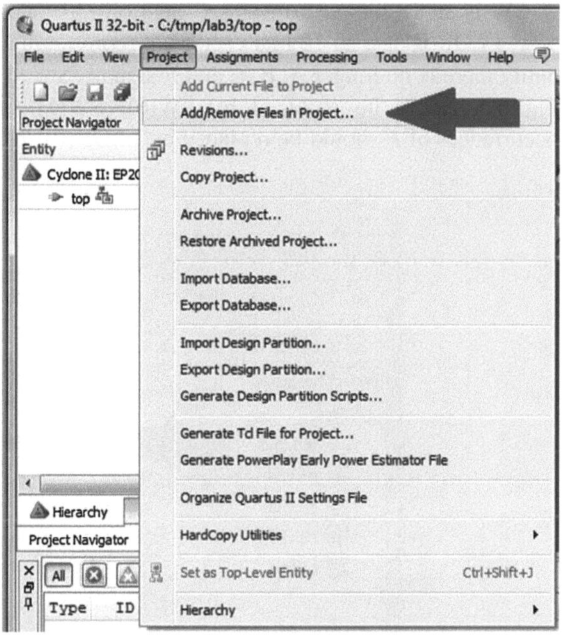

Fig. 3.8 Adding existing files to the project

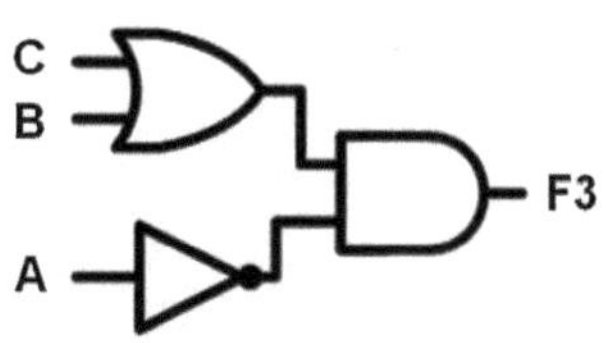

Fig. 3.9 *C*3 schematic diagram

Step 5—Adapting the Design to Work in the DE2 FPGA Board

At this point, the circuit is ready for synthesis (see Chap. 2, Step 3) and simulation (see Chap. 2, Step 4), but it will not work on DE2 FPGA board. As explained in Chap. 2, and introduced in Fig. 2.8, a circuit to be prototyped in DE2 board should employ specific signal names in its entity.

Figure 3.11 shows the available switches and LEDs on DE2 board. In the design shown in Fig. 3.12, *A*, *B* and *C* inputs will be provided by DE2 switches 0, 1 and 2, respectively. The *F* output will be presented in red LED 0.

In order to adapt inputs and outputs of top.vhd component to the DE2 board, replace its entity signals in Fig. 3.5, lines 5 and 6 by:

```
4:    entity top is
5:      port ( SW   : in  std_logic_vector(17 downto 0);
6:             LEDR: out std_logic_vector(17 downto 0)
7:      );
8:    end top;
```

SW is an 18 bits vector, where each bit represents a DE2 switch. In a similar way, LEDR is also an 18 bits vector but representing the 18 red LEDs. In this new entity there is no longer *A*, *B*, *C* and *F* signals. So, all occurrences of *A*, *B* and *C* in Fig. 3.5 should be replaced by SW(0), SW(1) and SW(2), respectively. All occurrences of *F* should be replaced by LEDR(0). The port map statements for *C*1

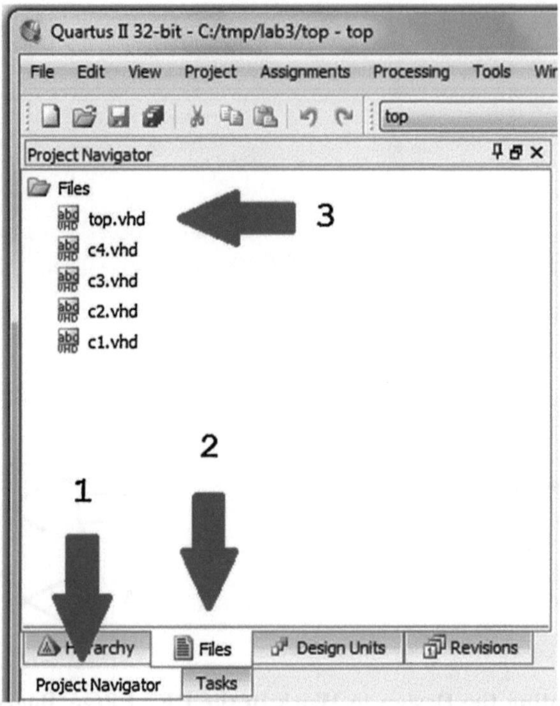

Fig. 3.10 Making changes to top.vhd

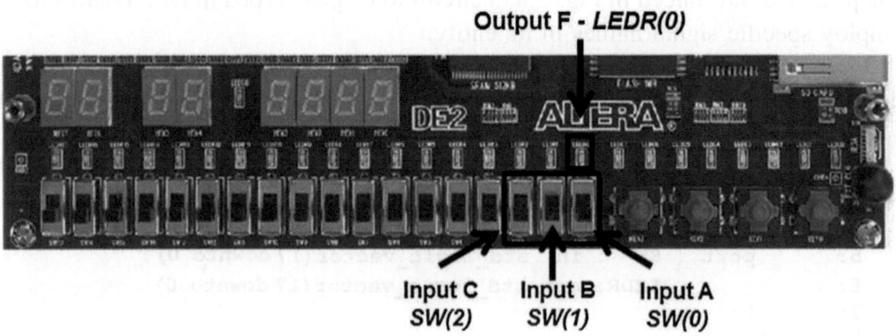

Fig. 3.11 DE2 input and output resources

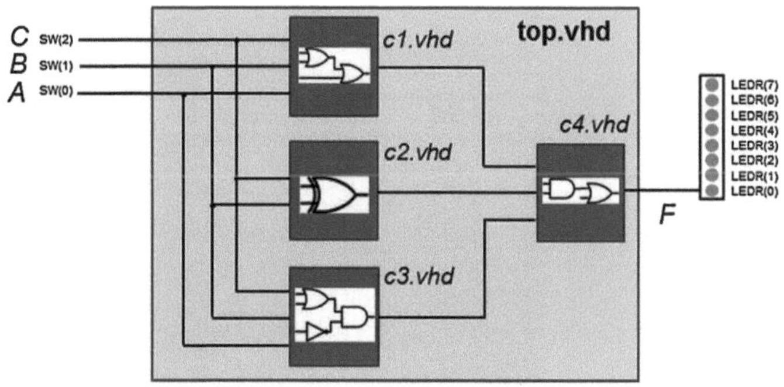

Fig. 3.12 Final circuit with DE2 board connections

and *C*4 for instance, should look like:

```
L0: C1 port map (SW(0), SW(1), SW(2), F1);
L3: C4 port map (F1, F2, F3, LEDR(0));
```

Only the top-level entity signals *A*, *B*, and *C*, and the internal signal *F* should be replaced by SW and LEDR, respectively. The remaining *A*, *B*, *C* and *F* signals belonging to the components should not be changed.

Step 6—Synthesis

To generate the configuration file (bitstream) to be downloaded to the FPGA board, the first step is to import the *DE2_pin_assignments.qsf* file, as discussed in Chap. 2, Step 5, and shown in Fig. 2.7. To perform the synthesis, use the Compile button in Quartus II menu. Fix any errors pointed out by the synthesis tool.

Step 7—Simulation

If the simulation tool has not been configured already, follow the instructions in Chap. 1, "Step 4". To start the simulator in Quartus II, select *Tools → Run Simulation Tool → RTL Simulation*. Wait for ModelSim to open and perform the following steps:

- Simulate → Start Simulation;
- In the "Start Simulation" window, look for the "Design" tab;
- In the "work" Library, click on the "+" sign, and select the "top" Entity;
- Click OK to start the simulation;
- Look for the Objects window and click on the "+" sign next to the SW label. This will open all 18 bits of vector SW, as shown in Fig. 3.13. Select SW(0), SW(1), and SW(2) in the Objects window, drag and drop them in the Wave

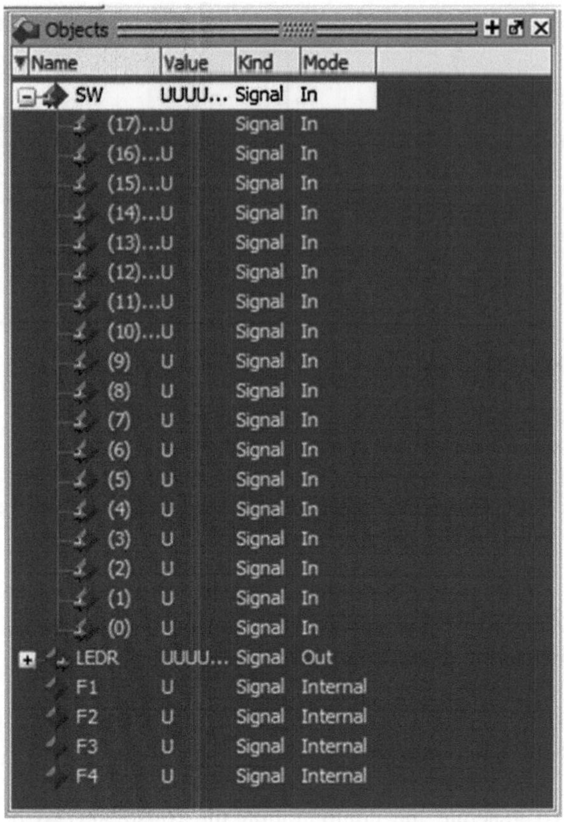

Fig. 3.13 Simulation signals selection

window. Do the same for LEDR(0), *F*1, *F*2, and *F*3. At the end, the Wave window should look like Fig. 3.14.

- Set SW(0), SW(1) and SW(2) to 0 using the "Force" option, as explained in Chap. 1, "Step 4". Set the simulation run length to 100 ps, and press the *Run* button. Next, change SW(0), SW(1) and SW(2) to the remaining seven possible combinations, always running for 100 ps after each input combination.
- The expected simulation results are shown in Fig. 3.15.

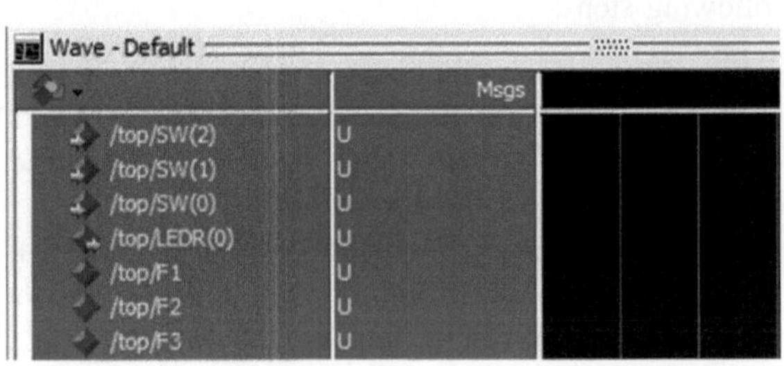

Fig. 3.14 Wave window

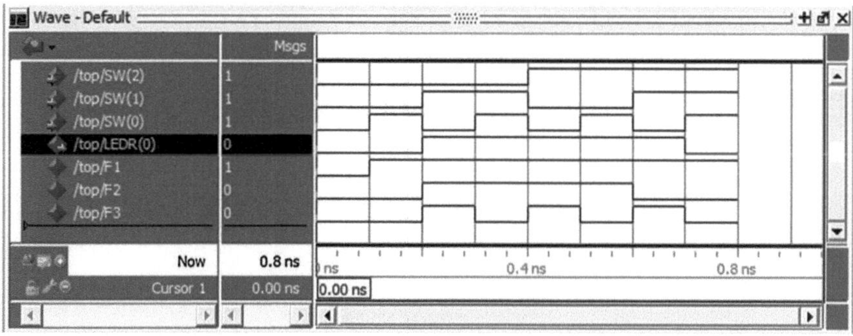

Fig. 3.15 Expected simulation results

Step 8—Prototyping the Circuit in the FPGA Board

Power on and connect the FPGA board to the host computer through the USB port (always using the "blaster" USB connector). Press the red button to switch on the board, and run the programming tool: Menu *Tools* → *Programmer*.

Follow the instructions provided in Chap. 1, Step 5, observing the "hardware setup" instructions in Figs. 1.32 and 1.33.

Test the circuit in the FPGA board, using switches SW(0), SW(1) and SW(2) as inputs, and observing the results in LEDR(0).

Fill in the truth table provided next in Table 3.1. The second column should be filled in with the results obtained from the evaluation of Eq. (3.1). The third column can be obtained straight from the simulation results shown in Fig. 3.15. The fourth column should be filled in with the FPGA execution results.

In case of mismatches in any line of the truth table, the circuit should be fully revised in order to find the error.

Table 3.1 Truth table for the case study

Input SW(2..0) C B A	Output *F* = (*F*1 *and F*2) or *F*3 *where:* *F*1 = *A or B or C* *F*2 = *B xor C* *F*3 = *(B or C) and (not A)*	Output Obtained in Step 7 simulation LEDR(0)	Output Obtained in Step 8 FPGA LEDR(0)
0 0 0			
0 0 1			
0 1 0			
0 1 1			
1 0 0			
1 0 1			
1 1 0			
1 1 1			

Fig. 3.16 Quartus II RTL viewer tool

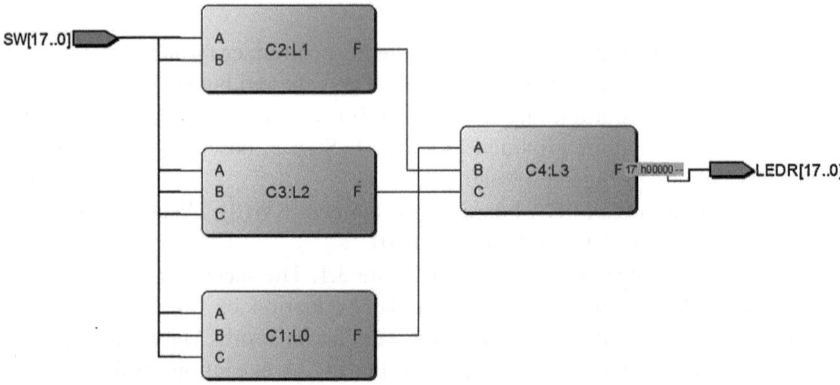

Fig. 3.17 Final circuit in RTL viewer

The Quartus II tool "RTL viewer" can be used to check if a circuit has been generated according to its requirements. Figure 3.16 shows the menu used to start the tool, and Fig. 3.17 shows the circuit's block diagram generated by Quartus II. In this block diagram, the developer can travel through the design hierarchy by double-clicking on the components.

Chapter 4
Multiplexer and Demultiplexer

In this chapter components for building decision maker circuits are introduced. The reader will be guided through the design of multiplexer components in VHDL using only logical gates (i.e., structural level), and also in a higher level of abstraction using *when/else* statements (i.e., behavioral level). As a case study the hierarchical design described in Chap. 3 will be re-used and modified in order to include a multiplexer. At the end of the chapter, the reader should be able:

- to understand the concept of multiplexer and demultiplexer;
- to implement a multiplexer in VHDL using only Boolean functions;
- to implement a multiplexer in VHDL using *when/else* statements;
- to design, simulate and prototype a proposed case study using an FPGA board.

4.1 Theoretical Background

A multiplexer (Mux) component has several inputs, a control signal, and one output. The output receives the value present in one of its inputs, according to the control signal. The block diagram of a general mux in Fig. 4.1a has n input signals (I) and m control signals (S). A 2^n inputs mux, needs n control signals.

The mux component in Fig. 4.1b has four 1 bit inputs A, B, C and D, set to the values 0, 1, 1, and 0, respectively. The selection is set to allow the third input to be connected to the output and, consequently, the F signal presents the value 1. Every time C has its input value changed, F will also be changed. To connect another input to the F output, it is necessary to change the selection signal.

A mux circuit, also known as "data selector", only allows one input to be available at its output at a time, according to the control signals ("Selection" in Fig. 4.1). As examples of applications, this component can be used for data selection, for data routing, in parallel to serial conversions, and also to implement truth tables.

E. A. Bezerra and D. V. Lettnin, *Synthesizable VHDL Design for FPGAs*,
DOI: 10.1007/978-3-319-02547-6_4,

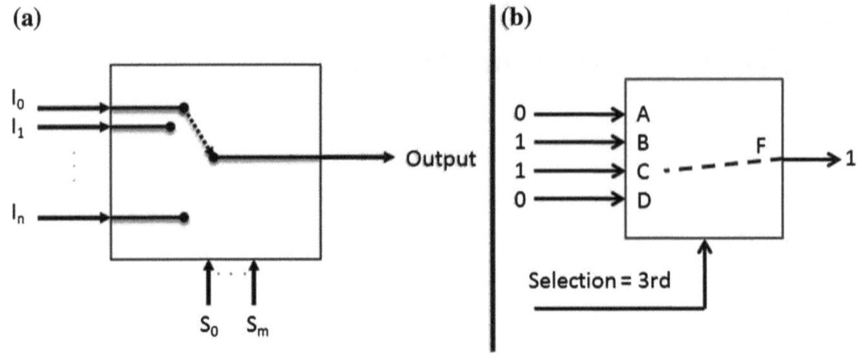

Fig. 4.1 Mux component block diagram. **a** n:1 Mux. **b** 4:1 Mux

A demultiplexer (Demux) circuit, also known as "data distributor", has the opposite function of the multiplexer, that is, it allows an input to be available at only one output at a time. This decision is also defined by control signals. The block diagram of a general demux shown in Fig. 4.2a has *n* output signals (*O*) and *m* control signals (*S*). A 2^n outputs demux needs *n* control signals. In Fig. 4.2b the value 0 is provided to input A, and the control signal "Selection" is set to the second output. As a result, the output F2 is connected to the input A, providing also the 0 value. If the selection is changed, for instance, to the first output, so F1 will also provide the 0 value. When the A input receives a new value, the corresponding output will also be updated.

The 2:1 multiplexer (or 2:1 selector) has the function to select one of its two inputs, making the selected entry appear on its output. In Fig. 4.3, if selector *s* is equal to 0, then *m* output receives the *x* input. If $s = 1$, then the *m* output receives the *y* input.

This multiplexer can be implemented in VHDL using only Boolean functions (i.e., structural VHDL) as well as using *when/else* statements (i.e., behavioral VHDL). In

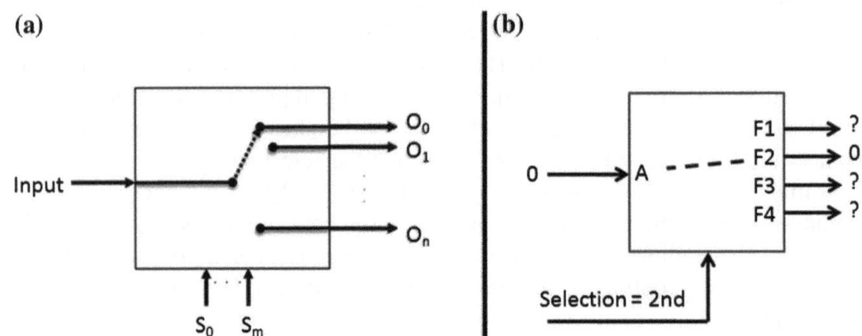

Fig. 4.2 Demux component block diagram. **a** 1:n Demux. **b** 1:4 Demux

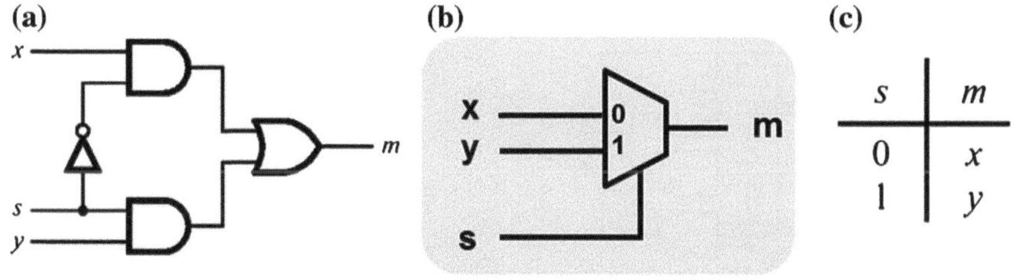

Fig. 4.3 2:1 Multiplexer. **a** Schematic diagram. **b** Block diagram. **c** Truth table

Fig. 4.4, in the structural VHDL version only logic gates are used to implement the functionality described in the truth table shown in Fig. 4.3c. In the algorithmic version, the behavior of the truth table is modeled using the *when/else* VHDL statements.

The 4:1 multiplexer selects one of its four inputs, making the selected entry appear on its output. In a four inputs mux, its selector must be 2 bits long. In Fig. 4.5, if selector *s* is equal to "00", then the *m* output receives the *w* input. If $s =$ "01", then the *m* output receives the *x* input, and so on. In Fig. 4.5a, the 2 bits of the *s* selector are represented explicitly by S0 and S1. In Fig. 4.5b, c, the *s* symbol is used to represent the 2 bits needed to select one of the four inputs.

4.2 Laboratory Assignment

The laboratory objectives are:

- to implement a multiplexer in VHDL;
- to learn and to practice the use of *when/else* VHDL statements;
- to add a multiplexer to the circuit of the case study design.

4.2.1 Laboratory Session

In this laboratory session the reader will have the opportunity to design a multiplexer in VHDL in two levels of abstraction: structural; and behavioral. At the structural level, the multiplexer is written in VHDL using only logic gates

Structural VHDL (logic gates):
```
m <= (NOT (s) AND x) OR (s AND y);
```

Behavioral VHDL (algorithm):
```
m <= x when s = '0' else
     y;
```

Fig. 4.4 2:1 Multiplexer in VHDL

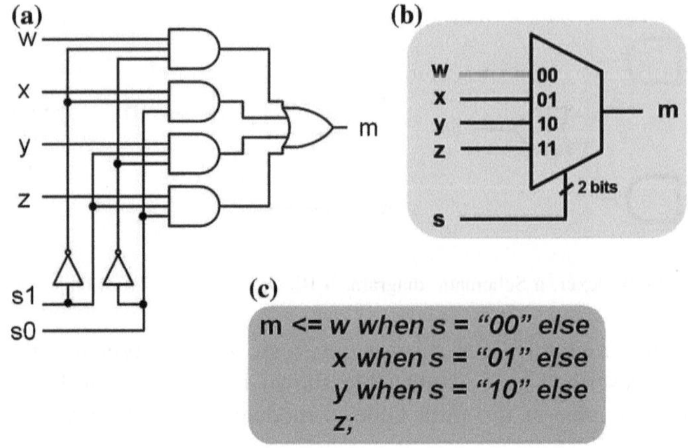

Fig. 4.5 4:1 Multiplexer. **a** Schematic diagram. **b** Block diagram. **c** Truth table

(see Chap. 1). At the behavioral level, the *when/else* statements in VHDL are used in a more algorithmic approach to the design of digital systems. Figure 4.6 shows the two versions to be implemented. In version I, the two bits of the selector input are explicated presented, as the figure shows the internal gate connections of the multiplexer. In version II, the multiplexer is described using the *when/else* VHDL statements, in a higher level of abstraction, with no information regarding the circuit's internal gate connections.

The case study discussed in previous chapters is reused, but with the C4 component shown in Fig. 3.6 replaced by a 4:1 multiplexer component. In the circuit shown in Figs. 4.7 and 4.8, the multiplexer is used to direct one of its inputs F1, F2 or F3 (w, x, y, z), to the F output (m). Thus, LEDR(0) will light on when the

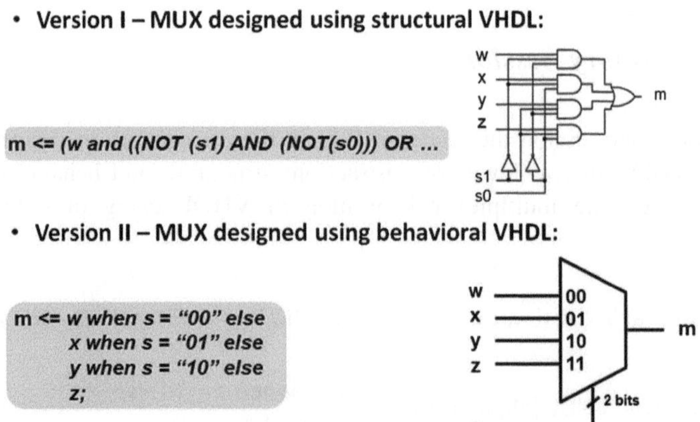

Fig. 4.6 Multiplexer described in two abstraction levels

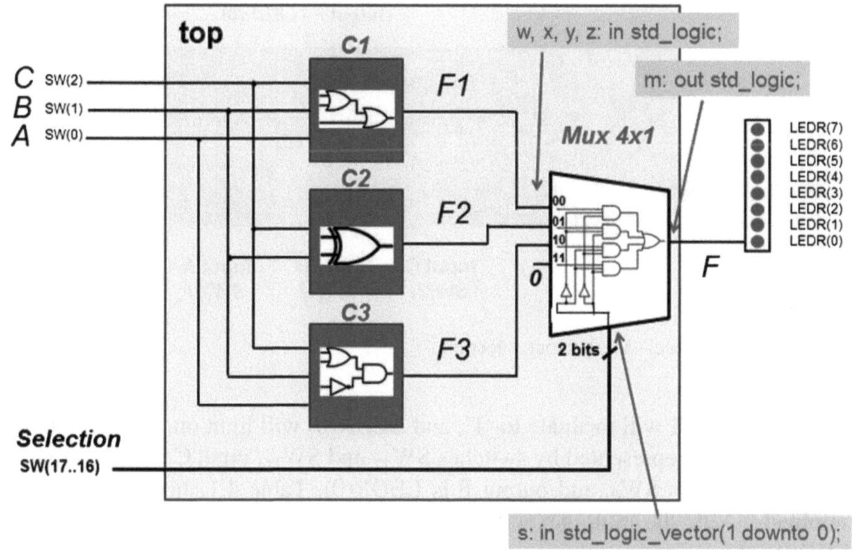

Fig. 4.7 Using a multiplexer instead of C4, version I—structural VHDL

2 bits selection input is set to a position whose component C1, C2 or C3 provides an '1' logic on its output.

For instance, when the selection is set to "00" (SW_{17} and SW_{16} are switched off), the F output receives F1. In this case, if one of the three inputs C1, C2 or C3 is

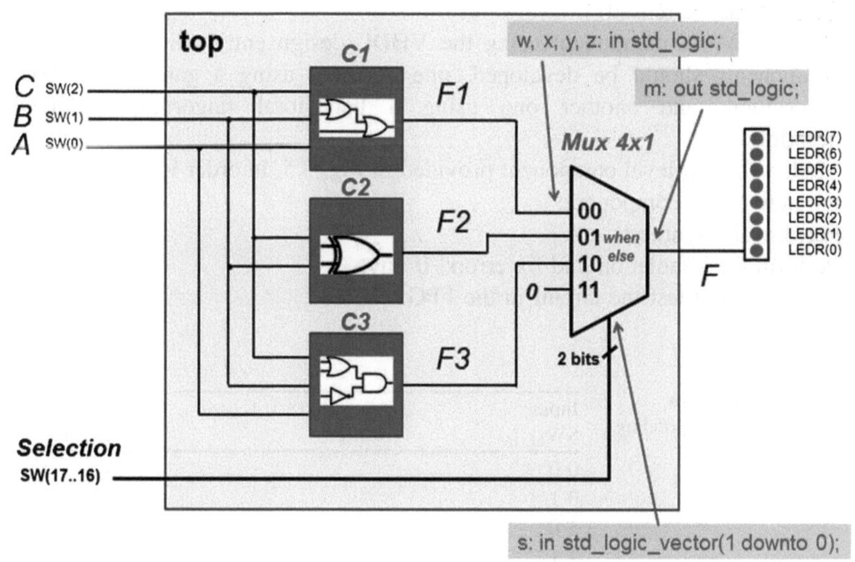

Fig. 4.8 Using a multiplexer instead of C4, version II—behavioral VHDL

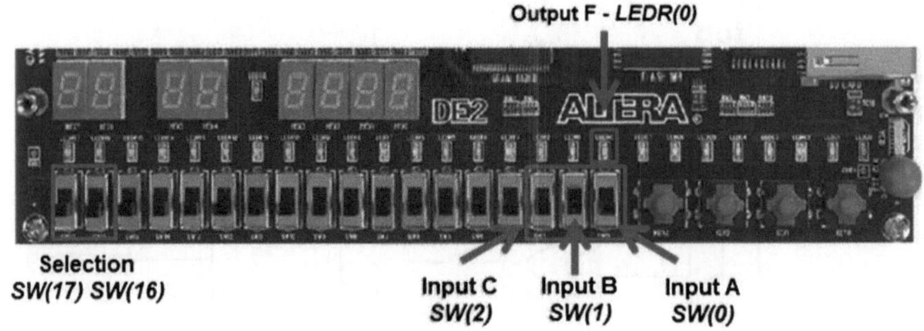

Fig. 4.9 User interface—MUX input selection

switched on, so C1 will evaluate to '1', and LEDR(0) will light on. In Fig. 4.9, the selection bits are represented by switches SW_{17} and SW_{16}, input C is SW_2, input B is SW_1, input A is SW_0, and output F is LEDR(0). Table 4.1 shows all possible combinations for the multiplexer selection bits, and the corresponding output in LEDR(0).

In this laboratory assignment, two versions for the multiplexer based design are to be implemented. Version I, i.e., multiplexer implemented in structural VHDL is shown in Fig. 4.7. In version II, shown in Fig. 4.8, the multiplexer is to be implemented in behavioral VHDL. Therefore, in this laboratory assignment, two different VHDL designs should be created. The tasks to be completed in this laboratory session, using Altera's Quartus II EDA tool are as follows:

- Create a project with all the components shown in Fig. 4.8 (last chapter components can be reused);
- Create a MUX component using the VHDL design entry editor—two MUX components should be developed, one of them using a gate level VHDL description, and another one using a behavioral (algorithmic) VHDL description;
- Modify the top-level component provided in Fig. 3.5, in order to replace C4 by the new MUX component;
- Perform the synthesis;
- Perform the simulation and fix errors, if any;
- Prototype and test the circuit in the FPGA board.

Table 4.1 Mux input selection and corresponding outputs

Input $SW_{17..16}$	Output $LEDR_0$
0 0	F1
0 1	F2
1 0	F3
1 1	0 (LED off)

4.2.2 Version I: Multiplexer in Structural VHDL

Step 1—Creating a New Project

This step is essentially the same as "Step 1" described in previous chapters. Just remember to use "top" as the project's name (see Fig. 1.13), as this is the name of the top-level entity. Remember also to create the folder for this new project in a place such as c:\LabSessions\mux4x1, where "mux4x1" is the name of the new folder.

Step 2—Adding VHDL Files to the Project and Creating the Mux Component

Using File Explorer (formerly known as Windows Explorer), or any other file manager program, locate the folder where the five VHDL files developed in last chapter are located (c1.vhd, c2.vhd, c3.vhd, c4.vhd, and top.vhd), and copy all files to the new folder c:\LabSessions\mux4x1. This procedure is shown in Fig. 4.10.

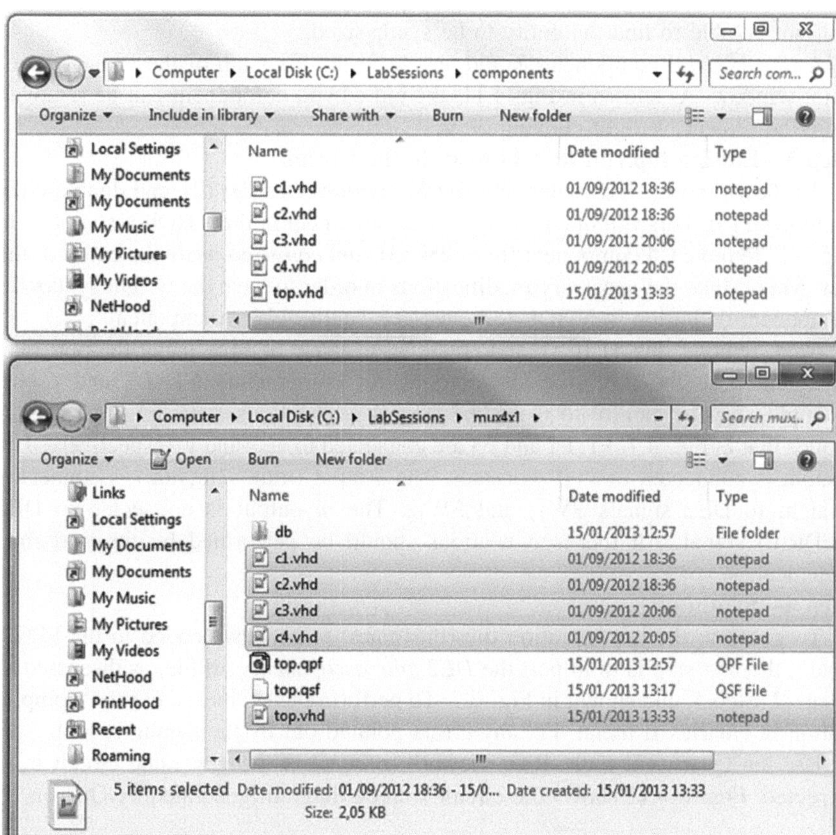

Fig. 4.10 Copying files using File Explorer

In Quartus II, choose *Project → Add/Remove Files in Project*, as shown in Fig. 3.8. A window will open allowing c1.vhd, c2.vhd, c3.vhd, c4.vhd and top.vhd, to be imported into the project. In this window, after adding the files press the *Apply* button.

In the Project Navigator window, select the Files tab, and double-click on c4.vhd in order to open it in the text editor. Make the following changes in c4.vhd, in order to turn it into a multiplexer component:

- Change the entity's name from "C4" to "Mux" (three changes are needed);
- Change the architecture's name to "mux_stru";
- Add the new entity's signals w, x, y, z, s, and m, as shown in Fig. 4.7;
- Replace the expression F <= (A and B) or C), by the multiplexer described in Fig. 4.6, version I;

Chose *File → Save as*, and rename c4.vhd to mux4x1.vhd. The file will be saved, also in the default location, in the project folder defined in Step 1. Be sure that the check box "Add file to current project" is checked. At this point it is important to be sure that the entity's name is "mux" otherwise the synthesis tool will not be able to find the entity to be synthesized.

A mux4x1.vhd component should appear in the Files tab of the Project Navigator window. As shown in Fig. 4.11, the old c4.vhd component can be removed from the project by right-clicking on it.

Step 3—Editing top.vhd to Add Mux to the Circuit

As shown in Fig. 3.10, select *Project Navigator* (1), *Files* (2), and double-click on top.vhd (3). This will open the top-level component in the VHDL text editor. In Fig. 3.5, remove C4 component (lines 28–33), and *copy and paste* the entity of the new Mux. Make the necessary modifications in order to have the wright syntax for component definition in VHDL (e.g., replace "entity" by "component", …).

Next, remove line 41 (C4), and create an instance of Mux using the port map statement. Use the port map declarations for components *C1*, *C2* and *C3* as examples, but be careful to make the correct connections according to Fig. 4.7. Notice that in Fig. 4.7, F1, F2 and F3 are connected to *w*, *x* and *y*, respectively. The *z* input is connected to a '0' constant. The *s* input (mux selection) is connected straight to DE2 signals SW_{17} and SW_{16}. The *m* output is connected to DE2 LEDR(0) signal. All these connections should be performed by the port map statement.

Step 4—Synthesis

To generate the configuration file (bitstream) to be downloaded to the FPGA board, the first step is to import the *DE2_pin_assignments.qsf* file, as discussed in Chap. 2, Step 5, and shown in Fig. 2.7. To perform the synthesis, use the Compile button in Quartus II menu. Fix any errors pointed out by the synthesis tool.

Use the Quartus II tool "RTL viewer" to check if the generated circuit is as expected. Figure 4.12 shows the circuit's block diagram generated by Quartus II.

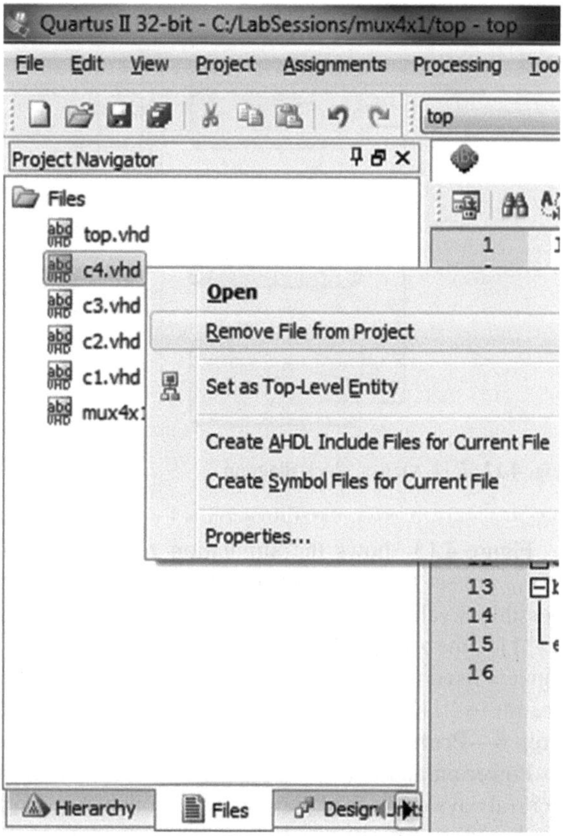

Fig. 4.11 Removing c4.vhd from the project

Step 5—Simulation

To start the simulator in Quartus II, select *Tools* → *Run Simulation Tool* → *RTL Simulation*. Wait for ModelSim to open and perform the following steps:

- Simulate → Start Simulation;
- In the "Start Simulation" window, look for the "Design" tab;
- In the "work" Library, click on the "+" sign, and select the "top" Entity;
- Click OK to start the simulation;
- Look for the Objects window and click on the "+" sign next to the SW label. This will open all 18 bits of vector SW, as shown in Fig. 3.13. Select SW(0), SW(1), SW(2), SW(16) and SW(17) in the Objects window, drag and drop them in the Wave window. Do the same for LEDR(0), F1, F2, F3 and F4.
- Set SW(0), SW(1), SW(2), SW(16) and SW(17) to 0 using the "Force" option, as explained in Chap. 1, "Step 4". Set the simulation run length to 100 ps, and press the *Run* button. Next, change SW(0), SW(1), SW(2), SW(16) and SW(17) to the remaining 31 possible combinations, always running for 100 ps after each input combination.

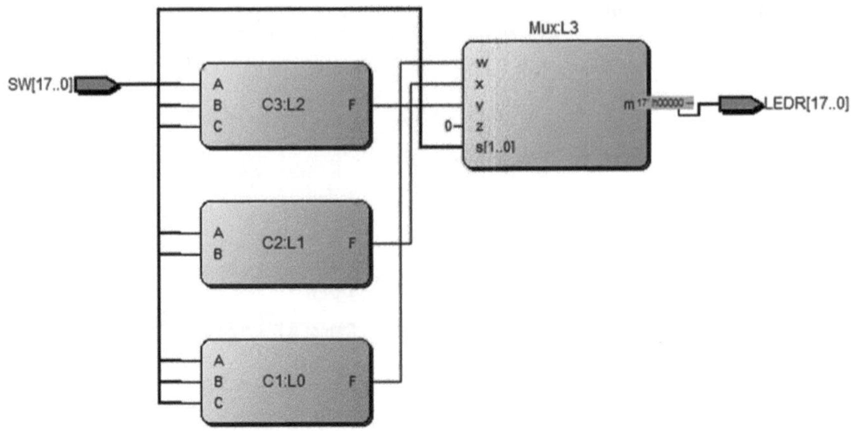

Fig. 4.12 RTL viewer block diagram

Figure 4.13 shows the simulation results when the selection bits SW_{17} and SW_{16} are set to "00", and the inputs A, B and C (SW_0, SW_1, SW_2) receive all possible 8 values ("000", "001", ..., "111"). Every time SW_{17} and SW_{16} are set to "11", the z input is selected and, as shown in Fig. 4.7, for this circuit this mux input is fixed in zero. As shown in the simulation results, at 0.8 ns SW_{17} and SW_{16} are set to "11", and the $LEDR_0$ output results in "0".

Step 6—Prototyping the Circuit in the FPGA Board

Power on and connect the FPGA board to the host computer through the USB port (always using the "blaster" USB connector). Press the red button to switch on the board, and run the programming tool: Menu *Tools* → *Programmer.*

Follow the instructions provided in Chap. 1, Step 5, observing the "hardware setup" instructions in Figs. 1.32 and 1.33.

Test the circuit in the FPGA board, using switches SW_0, SW_1, SW_2 as inputs, switches SW_{16} and SW_{17} as selection bits, and observing the results in $LEDR_0$.

Fill in the truth table provided next in Table 4.2. The second column is filled in with the selection bits ("00", "01", "10", and "11"). The third column should be filled in with the results obtained from the manual evaluation of equations F1, F2

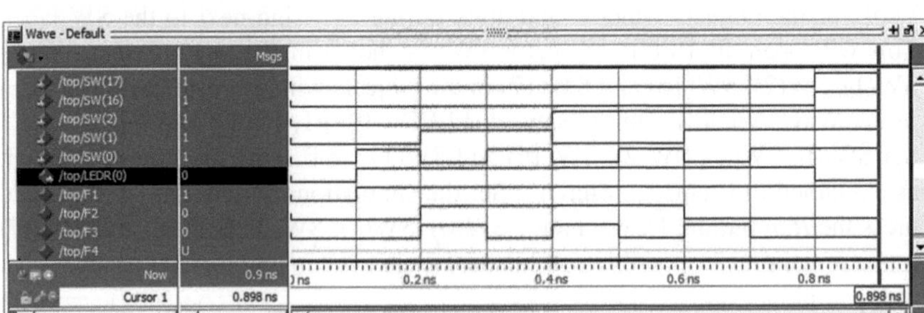

Fig. 4.13 Expected simulation results

and F3. In the third column, when the selection bits are "11", the output is always '0' disregarding the input values. The fourth column can be obtained straight from the simulation results, and it is partially filled in according to the waveforms in Fig. 4.13. The fifth column should be filled in with the FPGA execution results. In case of mismatches in any line of the truth table, the circuit should be fully revised in order to find the error.

4.2.3 Version II: Multiplexer in Behavioral VHDL

Follow, basically, the same steps as before in order to implement the behavioral VHDL version of the multiplexer based circuit. There are just a couple of remarks to be considered:

Table 4.2 Truth table for the MUX based circuit

Inputs		Outputs		
$SW_{2..0}$ C B A	$SW_{17..16}$ Selection	$F1 = A$ or B or C $F2 = B$ xor C $F3 = (B$ or $C)$ and (not A)	Simulation Step 5 $LEDR_0$	FPGA Step 6 $LEDR_0$
0 0 0	00	F1 =	0	
0 0 1	00	F1 =	1	
0 1 0	00	F1 =	1	
0 1 1	00	F1 =	1	
1 0 0	00	F1 =	1	
1 0 1	00	F1 =	1	
1 1 0	00	F1 =	1	
1 1 1	00	F1 =	1	
0 0 0	01	F2 =		
0 0 1	01	F2 =		
0 1 0	01	F2 =		
0 1 1	01	F2 =		
1 0 0	01	F2 =		
1 0 1	01	F2 =		
1 1 0	01	F2 =		
1 1 1	01	F2 =		
0 0 0	10	F3 =		
0 0 1	10	F3 =		
0 1 0	10	F3 =		
0 1 1	10	F3 =		
1 0 0	10	F3 =		
1 0 1	10	F3 =		
1 1 0	10	F3 =		
1 1 1	10	F3 =		
X X X	11	0	0	

Step 1

A new folder may be created in order to hold the behavioral version of the project. A suggestion of a folder to be created is c:\LabSessions\mux4x1_beh.

Step 2

Copy the version I VHDL files (.vhd) to the new folder, and add them to the project as explained before.

Edit the mux4x1.vhd file, and make the necessary changes in order to have a behavioral implementation of the multiplexer. The VHDL code based on *when/ else* statements is listed in Figs. 4.5c and 4.6.

Steps 3, 4, 5 and 6

There are no need for changes in the top.vhd file, as the new component has exactly the same interface (entity) and functionality as the structural version (version I).

Follow the same instructions for synthesis, simulation and FPGA implementation, as defined for version I.

Chapter 5
Code Converters

This chapter introduces a category of circuits known as "code converters". This type of circuit includes the encoders and the decoders. Encoders are circuits used in the conversion of information in one format to another. Decoders are complementary circuits, used to undo a conversion performed by an encoder. The concepts of code converter circuits are investigated in this chapter through their implementation in VHDL.

In previous chapters, a single LED has been employed in the circuits as their outputs. An LED was a feasible output interface, as so far the case study circuits in this book have been conceived in order to perform single bit functions. In this chapter a more real world case study is developed. It is a basic calculator that performs 8 bits operations, and presents the results in 7-segment displays. A decoder circuit is used to convert the binary result provided by the calculator, into a hexadecimal value to be shown in the 7-segment displays. In order to perform 8 bits operations, the single bit signals are arranged in VHDL as an array of signals, also known as a "bus". For a better understanding of the proposed case study, in this chapter there is also a discussion on "array of signals in VHDL" and "7-segment displays decoders". At the end of the chapter, the reader should be able:

- to understand the concepts of encoder and decoder;
- to implement a decoder in VHDL;
- to design, to simulate and to prototype the proposed case study using an FPGA board.

5.1 Arrays of Signals

As shown in Fig. 4.8, A, B, C and F are all 1 bit signals. In Fig. 5.1 the same signals are represented as arrays of 4 bits signals. This means that A, B, C and F can now hold values in the range from 0 to 15 in decimal or, in binary, from 0000 to 1111.

E. A. Bezerra and D. V. Lettnin, *Synthesizable VHDL Design for FPGAs*,
DOI: 10.1007/978-3-319-02547-6_5,

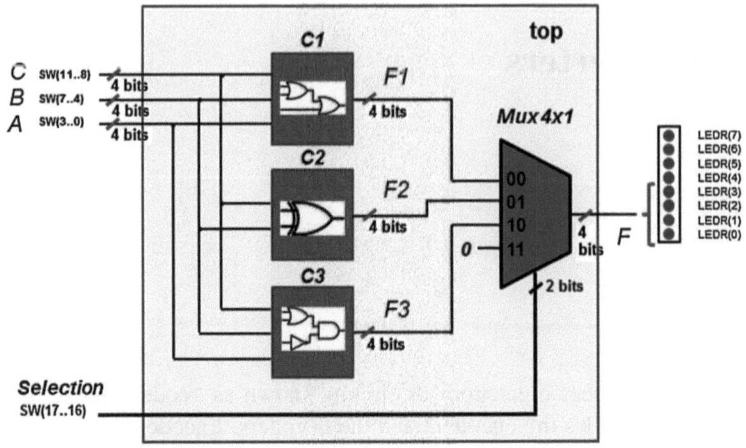

Fig. 5.1 Block diagram of a 4 bits circuit

Considering in Fig. 4.8 and in Fig. 5.1 that:

- C1 performs the function F1 = A or B or C
- C2 performs the function F2 = B xor C
- In Fig. 4.8 (1 bit operands), assume A = 0, B = 1, C = 0
- In Fig. 5.1 (4 bits operands), assume A = 0101, B = 1000, C = 0001.

Figure 5.2 shows the results obtained considering 1 bit and 4 bits wide operands. For 1 bit operands, F1 and F2 result in 1. For the 4 bits version, F1 results in 1101 (13 in decimal) and F2 results in 1001 (9 in decimal).

In VHDL the array of signals used in Fig. 5.1 can be defined through the *std_logic_vector* declaration:

- *std_logic* is used to declare a single signal, a single bit, a wire.
- *std_logic_vector* is used to declare an array of signals, a set of bits, a bus.

Considering a multiple bit implementation (array of signals), the components shown in Fig. 5.1 have been rewritten, and they are listed next. It is important to notice in Figs. 5.3, 5.4, 5.5 and 5.6, that only the entity has changed. The architecture, where the functions are performed, has not been changed.

In Fig. 5.7, the A, B and C inputs are now 4 bits signals represented, respectively, by $SW_{3..0}$, $SW_{7..4}$, and $SW_{11..8}$. The output is now also a 4 bits signals represented by $LEDR_{3..0}$.

Fig. 5.2 C1 and C2 components using 1 bit and 4 bits operators

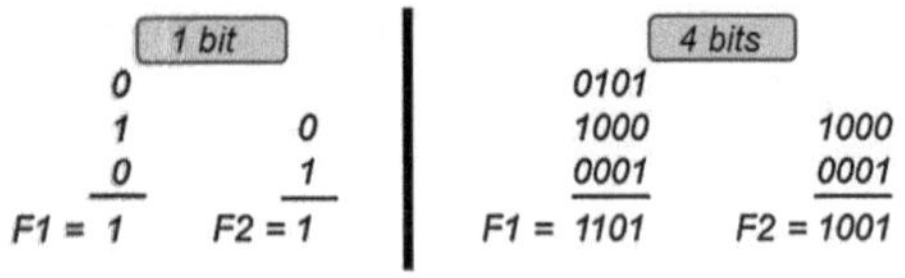

```
library IEEE;
use IEEE.Std_Logic_1164.all;

entity C1 is
port (A: in std_logic_vector(3 downto 0);
      B: in std_logic_vector(3 downto 0);
      C: in std_logic_vector(3 downto 0);
      F: out std_logic_vector(3 downto 0)
     );
end C1;

architecture c1_str of C1 is
begin
  F <= A or B or C;
end c1_str;
```

4 bits

4 bits

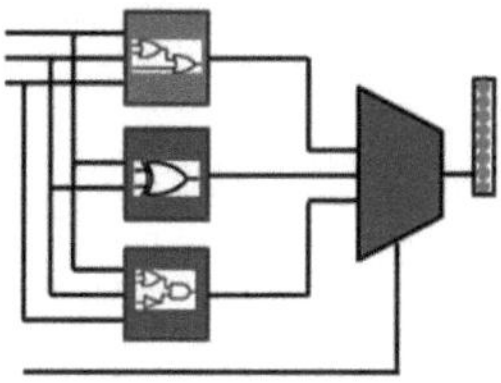

Fig. 5.3 VHDL implementation for component C1 (4 bits)

```
library IEEE;
use IEEE.Std_Logic_1164.all;

entity C2 is
port (A: in std_logic_vector(3 downto 0);
      B: in std_logic_vector(3 downto 0);
      F: out std_logic_vector(3 downto 0)
     );
end C2;

architecture c2_str of C2 is
begin
  F <= A xor B;
end c2_str;
```

4 bits

4 bits

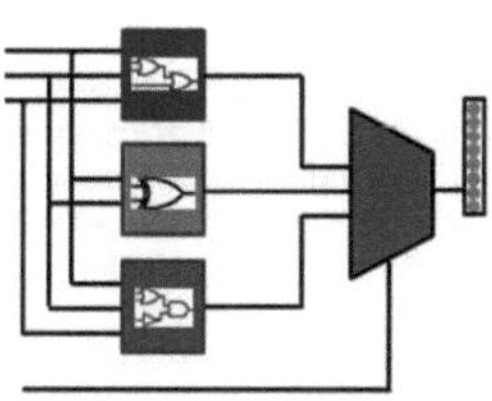

Fig. 5.4 VHDL implementation for component C2 (4 bits)

```
library IEEE;
use IEEE.Std_Logic_1164.all;

entity C3 is
port (A: in std_logic_vector(3 downto 0);
      B: in std_logic_vector(3 downto 0);
      C: in std_logic_vector(3 downto 0);
      F: out std_logic_vector(3 downto 0)
     );
end C3;

architecture c3_str of C3 is
begin
  -- proposed lab exercise
end c3_str;
```

4 bits

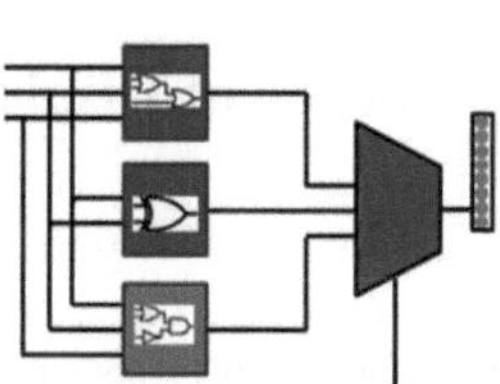

Fig. 5.5 VHDL implementation for component C3 (4 bits)

4 bits

```
library IEEE;
use IEEE.Std_Logic_1164.all;

entity Mux is
port (w: in std_logic_vector(3 downto 0);
      x: in std_logic_vector(3 downto 0);
      y: in std_logic_vector(3 downto 0);
      z: in std_logic_vector(3 downto 0);
      s: in std_logic_vector(1 downto 0);   ← 2 bits
      m: out std_logic_vector(3 downto 0)
     );
end Mux;

architecture mux_bhv of Mux is
begin
  -- see previous lab.
end mux_bhv;
```

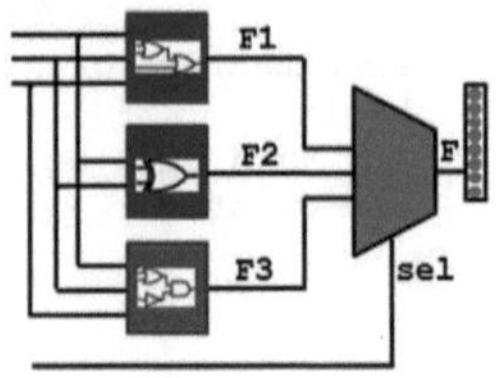

Fig. 5.6 VHDL implementation for component Mux (4 bits inputs/outputs)

```
library ieee;
use ieee.std_logic_1164.all;
entity top is
 port ( SW : IN STD_LOGIC_VECTOR(17 downto 0);
        LEDR : OUT STD_LOGIC_VECTOR(17 downto 0)
 );
end top;
architecture top_stru of top is
  signal F1, F2, F3: std_logic_vector(3 downto 0);
  component C1
    port (A : in std_logic_vector(3 downto 0);
          B : in std_logic_vector(3 downto 0);
          C : in std_logic_vector(3 downto 0);
          F : out std_logic_vector(3 downto 0));
  end component;
  component C2
    port (A : in std_logic_vector(3 downto 0);
          B : in std_logic_vector(3 downto 0);
          F : out std_logic_vector(3 downto 0));
  end component;
  component C3
    port (A : in std_logic_vector(3 downto 0);
          B : in std_logic_vector(3 downto 0);
          C : in std_logic_vector(3 downto 0);
          F : out std_logic_vector(3 downto 0)
    );
  end component;
  -- Mux component declaration goes here
```

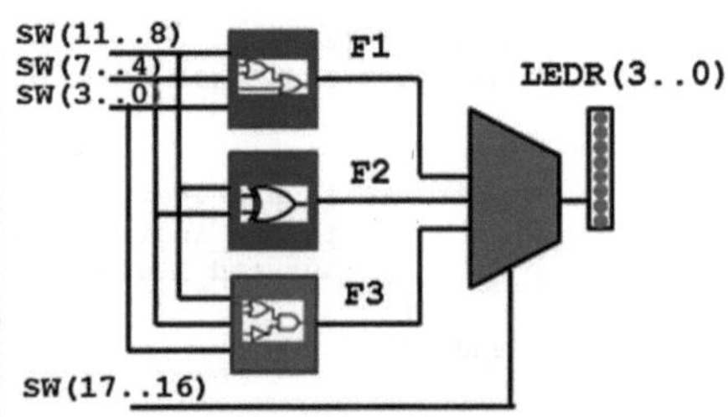


```
begin

 L0: C1 port map (SW(3 downto 0),
          SW(7 downto 4), SW(11 downto 8), F1);

 L1: C2 port map (SW(3 downto 0),
          SW(7 downto 4), SW(11 downto 8), F2);

 L2: C3 port map (SW(3 downto 0),
          SW(7 downto 4), SW(11 downto 8), F3);

 L3: Mux port map (F1, F2, F3,
          SW(17 downto 16), LEDR(3 downto 0));

end top_stru;   -- END architecture
```

Fig. 5.7 VHDL implementation for the top circuit of Fig. 5.1

5.2 Seven Segment Displays

In digital circuits, the 7-segment display is a widely used component. This component has seven LEDs that can be turned on and off independently. As shown in Fig. 5.8, there are two types of 7-segment displays:

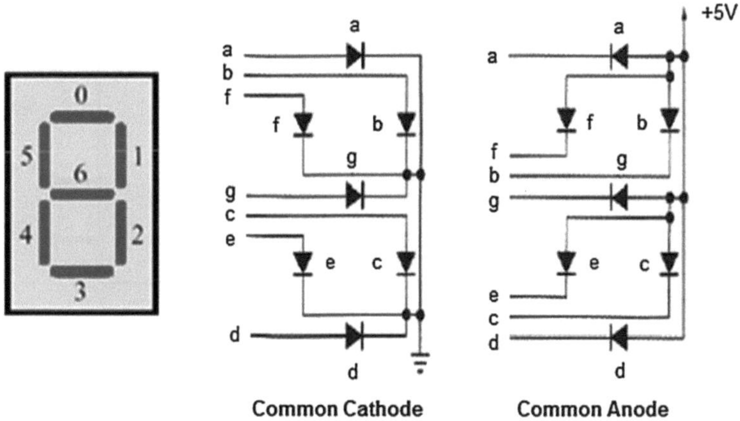

Fig. 5.8 Seven segment display

- **Common cathode**—the LEDs cathode terminals are connected together to the ground and each anode terminal is connected individually to a Vcc source.
- **Common anode**—the LEDs anode terminals are connected together to Vcc, and each cathode terminal is connected individually to ground.

In Fig. 5.1, the 4 bits result is shown in four LEDs (4 bits). A 7-segment display could have been used to show the result in another format as, for instance, in decimal or hexadecimal. In this case, a decoder circuit would be necessary in order to convert the binary output into the appropriate format and coding expected by the 7-segment display.

5.3 Encoders and Decoders

As stated before, encoders are used to convert a piece of information from one format to another, and decoders are used to undo the conversion. Figure 5.9 shows an example of encoder, used to convert the 4-bits input value I to a 2-bits coded

(a)

I_3	I_2	I_1	I_0	O_1	O_0
0	0	0	1	0	0
0	0	1	0	0	1
0	1	0	0	1	0
1	0	0	0	1	1

(b)

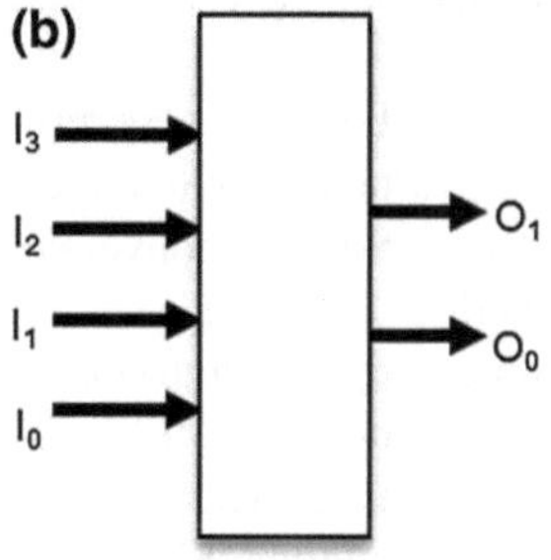

Fig. 5.9 Encoder. **a** Truth table. **b** Symbol

(a)

I_1	I_0	O_3	O_2	O_1	O_0
0	0	0	0	0	1
0	1	0	0	1	0
1	0	0	1	0	0
1	1	1	0	0	0

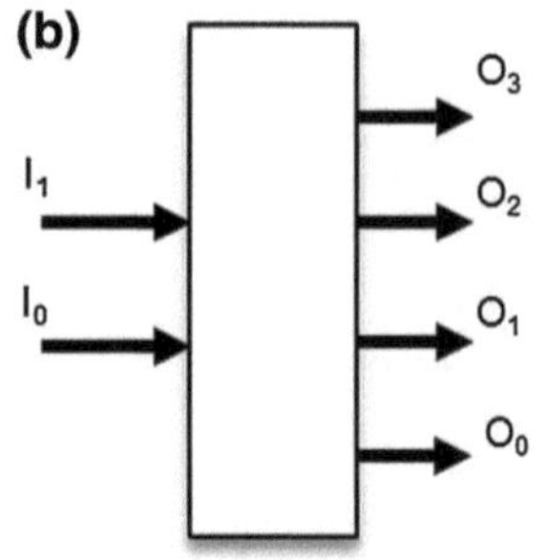

Fig. 5.10 Decoder. **a** Truth table. **b** Symbol

output O. In this example, the input shown in Fig. 5.9a presents just 1 bit set at a time, which is also known as the "one-hot" code. When comparing to the decoder, an encoder component can be used to reduce the amount of bits required to represent any piece of information.

The decoder shown in Fig. 5.10 was designed to convert back the code provided by the encoder shown in Fig. 5.9, that is, it receives a 2-bits code on its input, and provides a one-hot output. This decoder activates only one O_i output at a time, according to the combination of the I input values. In this sense, there is a direct relationship between the number of outputs (N_o) and the number of inputs (N_i), i.e., $N_o = 2 \times N_i$.

5.4 Designing a Seven Segment Decoder

As shown in Fig. 5.8, algorisms can be exhibited in a 7-segment display by turned on and off the different segments. For instance, in Fig. 5.8 when providing the value "0000000" in binary to the seven inputs (a, b,…, g), all LEDs are lighted on, and the '8' value will show on the display. In this case, in order to display the number '8', the code "1000" (eight in binary) should be decoded into "0000000". A 7-segment decoder is a very useful component, as it can be used in digital instruments that have a numerical output, for instance, multimeters, frequency-meters, etc.

The DE2 board has seven 7-segment displays, all of them common anode (LEDs are turned on with a logic zero in the inputs). To show a binary value in one of DE2's displays, a conversion from binary code to 7-segment code should be performed.

As an example of a decoder design, consider a circuit whose functionality is to display the letters 'U', 'F', 'S', and 'C' in a 7-segment display. The truth table in Fig. 5.11a is conceived considering 5 codes: "000" for letter 'U'; "001" for letter 'F'; "010" for letter 'S'; "011" for letter 'C'; and "111" to clear the display (all LEDs are off). The remaining 3 possible codes to represent in 3 bits ("100",

(a)

C2	C1	C0	7 bits *6543210*	Display
0	0	0	1000001	U
0	0	1	0001110	F
0	1	0	0010010	S
0	1	1	1000110	C
1	1	1	1111111	

(b)

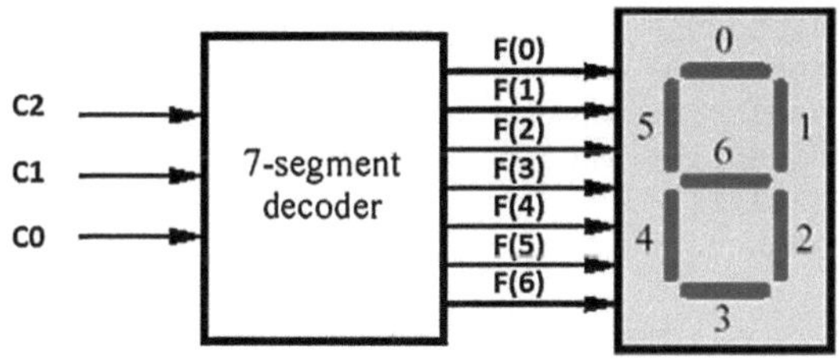

Fig. 5.11 A 7-segment decoder. **a** Truth table, **b** block diagram

"101", and "110") are not used in this circuit. The 7-segment decoder has to be designed in order to convert the 3-bit input codes into 7-bit output codes shown in the 4th column of Fig. 5.11a, which are used to turn on and off the LEDs of the display. For instance, to show the letter 'U', LEDs 0 and 6 (a and g in Fig. 5.8) should be turned off. So, in the 4th column of Fig. 5.11a, there is a '1' on this positions (common anode).

Considering the truth table in Fig. 5.11a, a component to implement this 7-segment decoder could be designed using different methods as, for instance:

- Using the Sum of Products synthesis method

```
F(0)   <= C2′ C1′ C0′ + C2 C1 C0
F(1)   <= C2′ C1′ C0 + C2′ C1 C0′ + C2′ C1 C0 + C2 C1 C0
F(2)   <= ...
```

- Using behavioral analysis

```
F <=   ″1000001″ when C2C1C0 = ″000″ else
       ″0001110″ when C2C1C0 = ″001″ else
       ...
       else ″1111111″
```

The full VHDL implementation for the proposed decoder component, based on the behavioral analysis method is shown in Fig. 5.12.

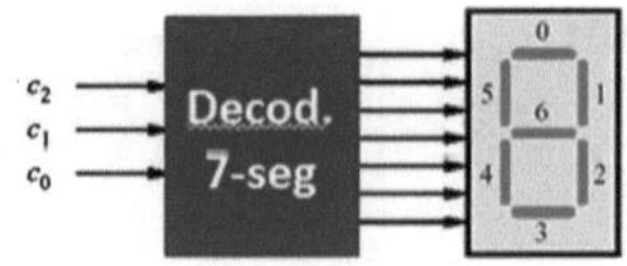


```
library IEEE;
use IEEE.Std_Logic_1164.all;

entity decodUFSC is
port (C:   in std_logic_vector(2 downto 0);      ← 3 bits
      F:   out std_logic_vector(6 downto 0)
      );                                          ↖ 7 bits
end decodUFSC;

architecture decod_bhv of decodUFSC is
begin
  F <= "1000001" when C = "000" else      -- U
       "0001110" when C = "001" else      -- F
       "0010010" when C = "010" else      -- S
       "1000110" when C = "011" else      -- C
       "1111111";
end decod_bhv;
```

Fig. 5.12 VHDL code for the UFSC 7-segment decoder

5.5 Case Study: A Simple but Fully Functional Calculator

A case study, adapted from previous chapters is presented next. The case study follows the same structure as shown in Fig. 4.8, but considering a multi-bit implementation as shown in Fig. 5.1. Furthermore, instead of components implementing arbitrary logic functions, in the proposed case study the components are conceived targeting the design of a simple but actual calculator.

It is an 8 bits calculator (operands are 8 bits long) that performs 4 operations, one arithmetic (F1, summation) and three logic functions (F2, F3 and F4), as follows:

- F1 = A + B
- F2 = A or B
- F3 = A xor B
- F4 = not A

Figure 5.13 shows the switches, displays and LEDs used by the calculator in the DE2 board.

The calculator's usage is straightforward:

1. Operand A is provided in switches $SW_{7..0}$
2. Operand B is provided in switches $SW_{15..0}$
3. One of the four operations is provided in switches $SW_{17..16}$
4. Result is shown in hexadecimal on displays HEX_1 and HEX_0 and in binary on $LEDR_{7..0}$ (red LEDs).

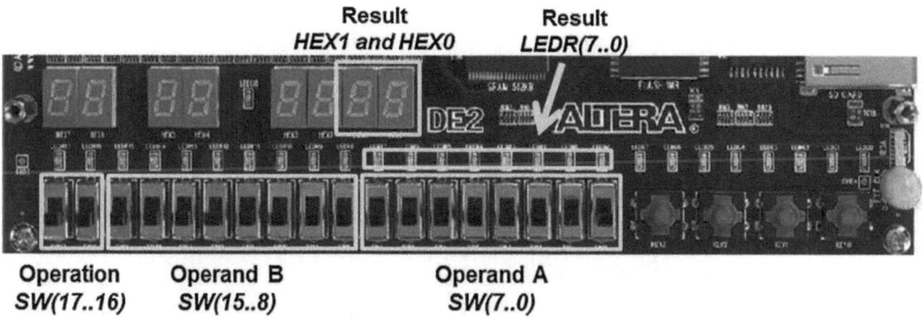

Fig. 5.13 Inputs and outputs for the calculator using DE2 board

In Table 5.1, all possible combinations for the selection bits SW_{17} and SW_{16} are listed. For instance, if the user wants to perform an exclusive or (xor) operation between inputs A (provided in $SW_{7..0}$) and B (provided in $SW_{15..8}$), SW_{17} should be switched on ('1') and SW_{16} should be switched off ('0'). The operation result is shown on the 7-segment displays and LEDs straightaway.

The block diagram in Fig. 5.14 shows the calculator input/output (I/O) signals, and the top-level component "top_calc", which is used to glue together the internal components. It shows also the two 7-segment displays and the eight LEDs used to output the calculator results. The internal components and all the connections are shown in Fig. 5.15.

The listing in Fig. 5.16 is a VHDL implementation for the top-level component shown in Fig. 5.14. In this implementation there are some important remarks:

- The "entity" section lists the signals used to identify the switches, 7-segments displays and LEDs, according to the *DE2_pin_assignments.qsf* file, as discussed in Chap. 2, Step 5;
- The internal signals, used to connect the internal components, are declared right below the "architecture" statement;
- Components C1 and C4 are listed next, but the VHDL code for components C2 and C3 are not included, as they are exactly the same as component C1;
- The Decod7seg component declaration is also not included, as this is the task to be achieved in this laboratory session;
- All the connections shown in Fig. 5.15 are performed by the VHDL code listed in the body of the architecture (right after the "begin" statement);

Table 5.1 Operations performed by the calculator

Input $SW_{17..16}$	Output $LEDR_{7..0}$ and 7-seg displays
0 0	A + B
0 1	A or B
1 0	A xor B
1 1	Not A

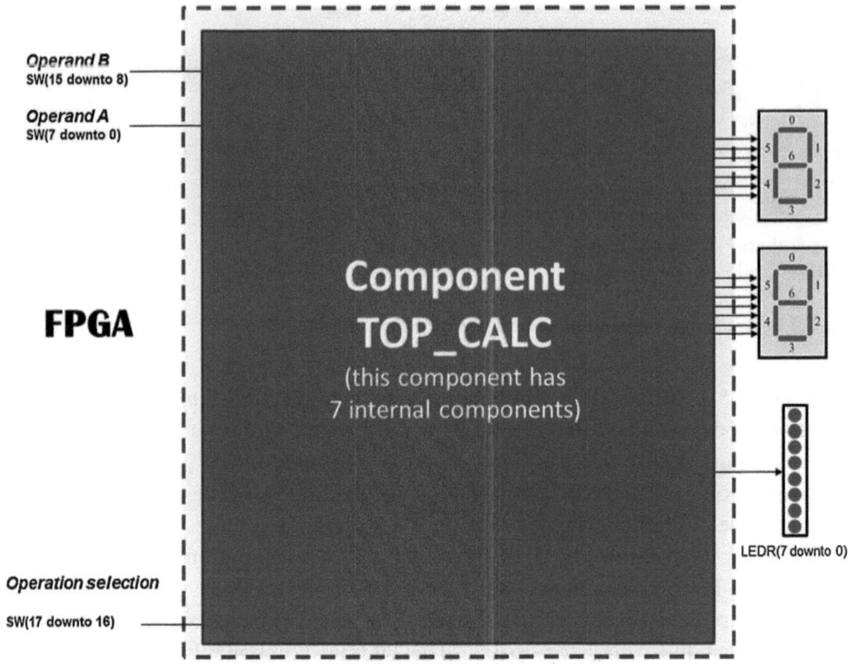

Fig. 5.14 Top-level view of the calculator's circuit

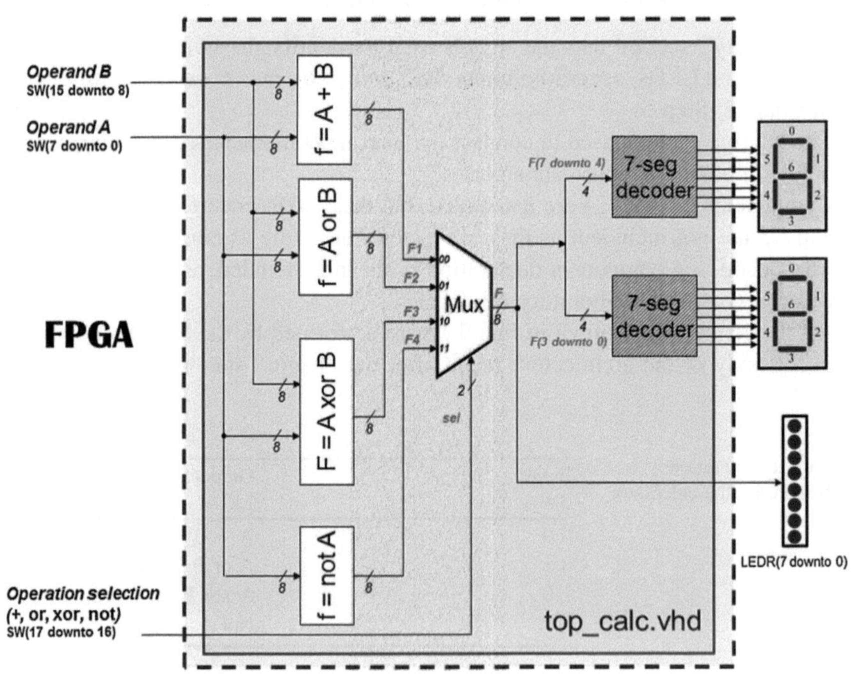

Fig. 5.15 Calculator internal components

```
library ieee;
use ieee.std_logic_1164.all;
entity top_calc is
  port ( SW : in std_logic_vector (17 downto 0);
         HEX0, HEX1: out std_logic_vector (6 downto 0);
         LEDR : out std_logic_vector (17 downto 0)
  );
end top_calc;
architecture top_stru of top_calc is
  signal F, F1, F2, F3, F4: std_logic_vector (7 downto 0);
  component C1
    port (A : in std_logic_vector(7 downto 0);
          B : in std_logic_vector(7 downto 0);
          F : out std_logic_vector(7 downto 0));
  end component;
  -- components C2 and C3, same as C1
  component C4
    port (A : in std_logic_vector(7 downto 0);
          F : out std_logic_vector(7 downto 0)
    );
  end component;
 component mux4x1
   port (w, x, y, z: in std_logic_vector(7 downto 0);
                  s: in std_logic_vector(1 downto 0);
                  m: out std_logic_vector(7 downto 0)
   );
 end component;
 -- Add component Decod7seg here
 -- Attention!! Only one declaration for Decod7seg
```

```
begin

  L1: C1 port map (SW(7 downto 0),
          SW(15 downto 8), F1);

  L2: C2 port map (SW(7 downto 0),
          SW(15 downto 8), F2);

  L3: C3 port map (SW(7 downto 0),
          SW(15 downto 8), F3);

  L4: C4 port map (SW(7 downto 0), F4);

  L5: mux4x1 port map (F1, F2, F3, F4,
          SW(17 downto 16), F);

→ L6: Decod7seg port map (F(3 downto 0),
          HEX0);

→ L7: Decod7seg port map (F(7 downto 4),
          HEX1);

  LEDR(7 downto 0) <= F;

end top_stru;  -- END architecture
```

Fig. 5.16 VHDL implementation for the top circuit of Fig. 5.14

- Components C1, C2, C3 and C4 provide signals F1, F2, F3 and F4 as their outputs;
- The mux4 × 1 component has F1, F2, F3 and F4 as its inputs. $SW_{17..16}$ is the operation selection, and the internal signal F is its output;
- The Decod7seg component was declared just once in the architecture's declarations section, and in the architecture's body the two copies of the component are instantiated (see the two pointers to L6 and L7).

The VHDL implementations for the remaining components, but for the 7-segments decoder, are provided next.

5.6 Laboratory Assignment

The laboratory objectives are:

- to implement a 7-segments decoder in VHDL;
- to add the implemented decoder to the calculator.

Using the components provided in Sect. 5.5, write the VHDL implementation for the 7-segments decoder, and make the necessary modifications in the top_calc component as indicated in Fig. 5.16.

The whole design has in total 7 files (8 components), as follows:

- *c1.vhd*, provided in Fig. 5.17;
- *c2.vhd*, provided in Fig. 5.18;
- *c3.vhd*, provided in Fig. 5.19;
- *c4.vhd*, provided in Fig. 5.20;
- *mux4x1.vhd*, provided in Fig. 5.21;
- *decod7seg.vhd*, to be designed (see Sect. 5.4 and Fig. 5.12);
- *top_calc.vhd*, partially provided in Fig. 5.16.

Before starting the modifications in the top-level file (*top_calc.vhd*), make sure to have a good understanding of component declaration and instantiation procedures in VHDL. As pointed out in Fig. 5.16, just one *decod7seg* should be

```
library IEEE;
use IEEE.Std_Logic_1164.all;
use IEEE.std_logic_unsigned.all; -- to use the + operator

entity C1 is
port (A: in std_logic_vector(7 downto 0);
      B: in std_logic_vector(7 downto 0);
      F: out std_logic_vector(7 downto 0)
     );
end C1;

architecture circuit of C1 is
begin
  F <= A + B;
end circuit;
```

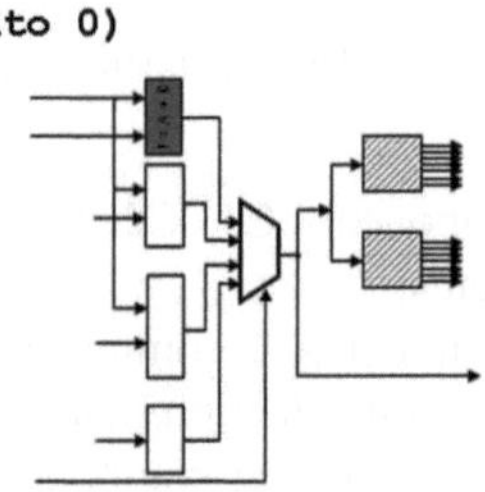

Fig. 5.17 VHDL implementation for the C1 component

```
library IEEE;
use IEEE.Std_Logic_1164.all;

entity C2 is
port (A: in std_logic_vector(7 downto 0);
      B: in std_logic_vector(7 downto 0);
      F: out std_logic_vector(7 downto 0)
     );
end C2;

architecture circuit of C2 is
begin
  F <= A or B;
end circuit;
```

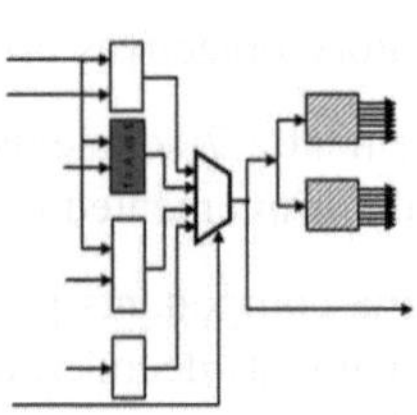

Fig. 5.18 VHDL implementation for the C2 component

```
library IEEE;
use IEEE.Std_Logic_1164.all;

entity C3 is
port (A: in std_logic_vector(7 downto 0);
      B: in std_logic_vector(7 downto 0);
      F: out std_logic_vector(7 downto 0)
     );
end C3;

architecture circuit of C3 is
begin
  F <= A xor B;
end circuit;
```

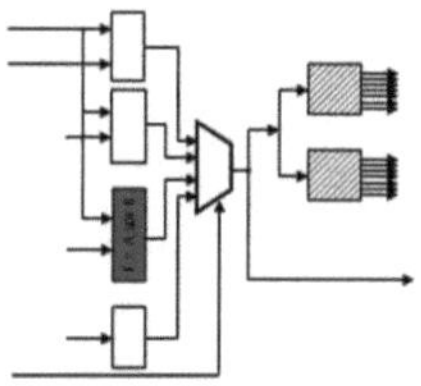

Fig. 5.19 VHDL implementation for the C3 component

```
library IEEE;
use IEEE.Std_Logic_1164.all;

entity C4 is
port (A: in std_logic_vector(7 downto 0);
      F: out std_logic_vector(7 downto 0)
     );
end C4;

architecture circuit of C4 is
begin
  F <= not A;
end circuit;
```

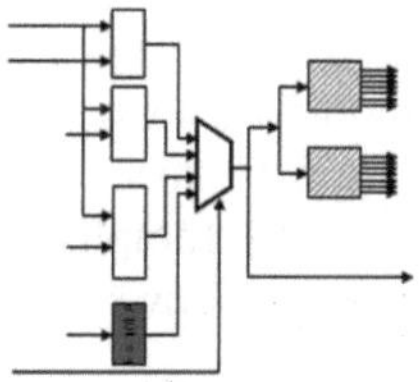

Fig. 5.20 VHDL implementation for the C4 component

```
library IEEE;
use IEEE.Std_Logic_1164.all;

entity mux4x1 is
port (w, x, y, z: in std_logic_vector(7 downto 0);
      s: in std_logic_vector(1 downto 0);
      m: out std_logic_vector(7 downto 0)
     );
end mux4x1;

architecture circuit of mux4x1 is
begin
 m <= w when s = "00" else
      x when s = "01" else
      y when s = "10" else
      z;
end circuit;
```

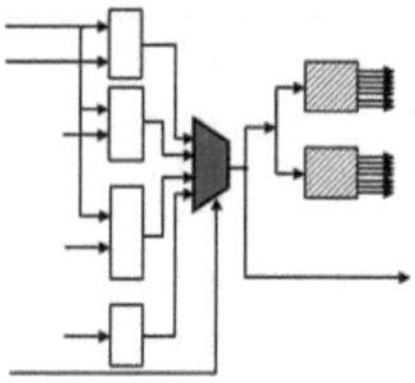

Fig. 5.21 VHDL implementation for the mux4x1 component

Fig. 5.22 Binary to 7-segments decoding table

Input	Output 6543210	Display
0000	1000000	0
0001	1111001	1
0010	0100100	2
0011	0110000	3
0100	0011001	4
0101	0010010	5
0110	0000010	6
0111	1111000	7
1000	0000000	8
...	...	9, A, b, C, d
1110	0000110	E
1111	0001110	F

included in the components declaration section of the architecture, and two copies of this component should be instantiated using *port map* statements.

The VHDL implementation listed in Fig. 5.12 decodes only four codes, "000" ('U'), "001" ('F'), "010" ('S'), and "011" ('C'). The calculator needs to displays all possible hexadecimal values, which means that the decoder receives 16 different input data, as shown in Fig. 5.22. The expected decoded outputs are shown in Fig. 5.22, but the values in the range between 9H and DH. Thus, to design the decoder, the missing codes should be defined, and the VHDL implementation in Fig. 5.12 should be adapted according. The entity for the 7-segments decoder, as shown in Fig. 5.15, has a 4 bits input signal, and a 7 bits output signal.

The tasks to be completed in this laboratory session, using Altera's Quartus II EDA tool are as follows:

- Create a project with all the components shown in Fig. 5.15;
- Create a 7-segments decoder component using the VHDL design entry editor;

Table 5.2 Truth table for the calculator

Inputs		Outputs				
$SW_{15..0}$ B A	$SW_{17..16}$ selection	$F1 = A + B$ $F2 = A$ *or* B $F3 = A$ *xor* B $F4 =$ *not* A	Simulation HEX1 HEX0	Simulation $LEDR_{6..0}$	FPGA HEX1 HEX0	FPGA $LEDR_{6..0}$
00 00	00	F1 =				
02 03	00	F1 =				
05 0A	01	F2 =				
05 FF	01	F2 =				
05 00	10	F3 =				
05 0A	10	F3 =				
00 0A	11	F4 =				
FF F5	11	F4 =				

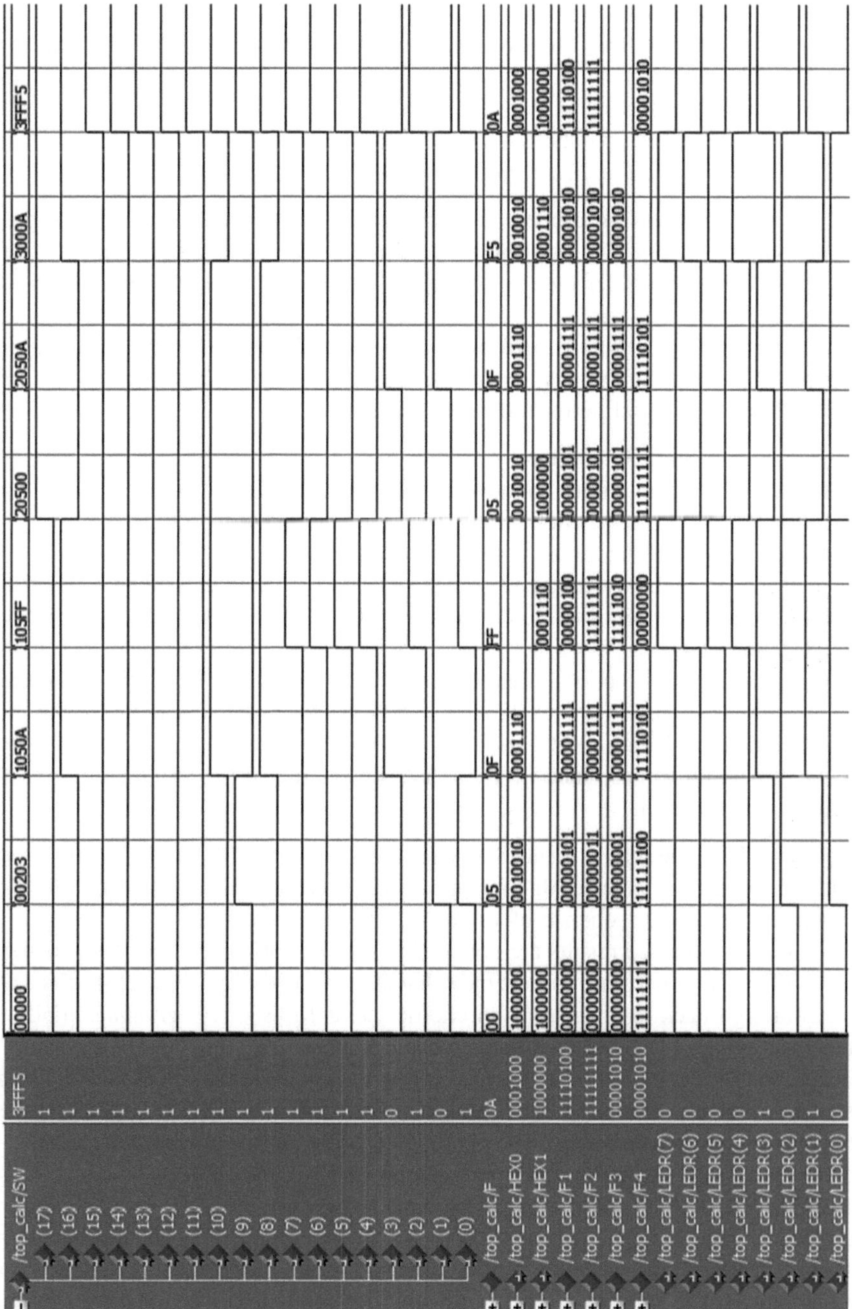

Fig. 5.23 Expected simulation results

- Modify the top-level component provided in Fig. 5.16, in order to add the new component declaration. The two instances are already created and identified by labels L6 and L7;
- Perform the synthesis;
- Perform the simulation and fix errors, if any;
- Prototype and test the circuit in the FPGA board.

The **project creation** process has been well discussed and explained in previous chapters. For the **synthesis** activity, it is essential to import the *DE2_pin_assignments.qsf* file, as discussed in Chap. 2, Step 5, and shown in Fig. 2.7.

The **simulation** activity is performed using ModelSim. To start the simulator in Quartus II, select *Tools* → *Run Simulation Tool* → *RTL Simulation.* Wait for ModelSim to open and perform the following steps:

- Simulate → Start Simulation;
- In the "Start Simulation" window, look for the "Design" tab;
- In the "work" Library, click on the "+" sign, and select the "top_calc" Entity;
- Click OK to start the simulation;
- In the Objects window drag and drop the SW vector (A, B, and the two selection bits) in the Wave window. Do the same for $LEDR_{7..0}$, $HEX0_{6..0}$, $HEX1_{6..0}$, F, F1, F2, F3 and F4.
- Using the "Force" option, as explained in Chap. 1 "Step 4", set $SW_{7..0}$ to 03H (A = 00000011), $SW_{15..8}$ to 02H (B = 00000010), and $SW_{17..16}$ to 00. Set the simulation run length to 100 ps, and press the *Run* button. Next, set $SW_{17..0}$ to the remaining values listed in Table 5.2, always running for 100 ps after each input combination.

In Fig. 5.23, the selection bits SW_{17} and SW_{16} are set to "00", "01", "10", and "11", in order to exercise all four operations. The A ($SW_{7..0}$) and B ($SW_{15..8}$) inputs are set to the sequence of values as listed in Table 3.1.

When the expected results are obtained in the simulation process, the next step is the **FPGA prototyping**. To test the calculator functionality in an actual piece of hardware, go back to Quartus II, in choose menu *Tools* → *Programmer.* Follow the instructions provided in Chap. 1, Step 5, observing the "hardware setup" instructions in Fig. 1.32 and in Fig. 1.33. Test the circuit in the FPGA board, using switches $SW_{15..0}$ to provide the A and B inputs, and switches $SW_{17..16}$ to choose the operation to be performed. The operation result is shown in binary in $LEDR_{7..0}$, and in hexadecimal on the 7-segment displays HEX0 and HEX1.

In Table 5.2, the first column has some input value examples. The second column is filled in with the selection bits ("00", "01", "10", and "11"). The third column should be filled in with the results obtained from the manual evaluation of equations F1, F2, F3 and F4. The fourth and fifth columns can be obtained straight from the simulation results, according to the waveforms in Fig. 5.23. The sixth and seventh columns should be filled in with the FPGA execution results.

Chapter 6
Sequential Circuits, Latches and Flip-Flops

The first part of this book introduces the concepts related to the design of combinational circuits in VHDL. In the second part of this book, starting from this chapter on, the design of sequential circuits in VHDL is discussed and explained. At the end of the chapter, the reader should be able:

- to understand the concepts and principles of sequential circuits;
- to understand the concepts and differences between latches and flip-flops;
- to understand the principles of registers design in VHDL;
- to design and test flip-flops, latches and registers in VHDL;
- to design and implement the proposed case study using an FPGA board.

6.1 Sequential Circuits in VHDL: The Process Statement

In the *combinational circuits* discussed in previous chapters, the results presented in their outputs depend only on their current input values. The outputs of a *sequential circuit*, on the other hand, depend also on previous values internally stored. For this reason, a sequential circuit must be designed making use of some sort of memory resources. In other words, combinational circuits can be built based on Boolean equations only, while sequential circuits must employ also storage resources as latches and flip-flops. The design of latches and flip-flops in VHDL is explored later on in this chapter.

In VHDL, as introduced in previous chapters, all signal assignments are performed in parallel. This language feature is very interesting to describe the behavior of combinational circuits. The behavior of sequential circuits is described in VHDL through the *process* statement. In Fig. 6.1 there is an example of process usage in VHDL. Important characteristics regarding the functionality of this statement are listed next:

E. A. Bezerra and D. V. Lettnin, *Synthesizable VHDL Design for FPGAs*,
DOI: 10.1007/978-3-319-02547-6_6,

Fig. 6.1 Process statement in VHDL

```
1  library ieee;
2  use ieee.std_logic_1164.all;
3  entity ProcessExample is port (
4        C: in  std_logic;
5        D: in  std_logic;
6        Q: out std_logic
7  );
8  end ProcessExample;
9  architecture behv of ProcessExample is
10       signal A, B: std_logic;
11 begin
12       A <= D;
13       Q <= B;
14       P1: process (C, A)
15       begin
16             B <= '0';
17             if (C = '1') then
18                   B <= A;
19             end if;
20       end process P1;
21 end behv;
```

- A *process* defines a sequence of commands to be performed by the circuit. This means that all the commands listed from line 16–19 in Fig. 6.1 are performed in the defined order, and one after the other.
- A *process* is cyclic, and it never ends. However, its list of commands (e.g. lines 16–19 in Fig. 6.1, and lines 17–21 in Fig. 6.2) is performed only when a change is noticed in one of its activation signals (e.g. C and D on line 14 in Fig. 6.1).
- The signals activation list of a process is known as *the sensitivity list*. A process is triggered by new "events" detected on its sensitivity list.
- After the last command of a process has been performed, the first command in the sequence is run again, but only after the next time a change is picked up in one or more signals of the sensitivity list (e.g. line 14 in Fig. 6.1).
- Some VHDL constructions can only be used inside a process as, for instance, the IF.. THEN.. ELSE shown on lines 17–19 in Fig. 6.1.
- Signals cannot be declared inside a process.
- In a process, in case of several assignments for the same signal, only the last assignment will be performed. In Fig. 6.2, B will be assigned only on line 18, and C will be assigned only on line 20.

An important VHDL process concept is that all signal attributions performed will only be valid when the execution reaches the *end process* statement. Two versions of the example in Fig. 6.2 have been implemented, and the simulation results are shown in Fig. 6.3. In Version 1, the sensitivity list of the process has only the A signal, and in Version 2 it has the A and C signals. The waveforms in Fig. 6.3 show the simulation results for Version 1 (top of the figure) and Version 2

```vhdl
library IEEE;
use IEEE.Std_Logic_1164.all;
entity TestProc is
port(
      data_in  : in  std_logic;
      data_out : out std_logic
);
end TestProc;
architecture circuit of TestProc is
   signal A, B, C, D: std_logic;
begin
   A <= data_in;
   data_out <= D;
   -- Version 1        -- Version 2
   process (A)         -- process (A, C)
   begin
     B <= A;
     B <= '0';
     C <= A and '1';
     C <= not A;
     D <= C;
   end process;
end circuit;
```

sequential statements

Fig. 6.2 Sequence of signal attributions in a process

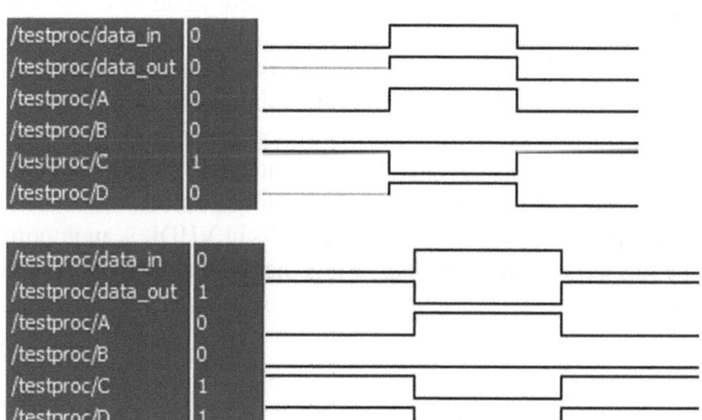

Fig. 6.3 Simulation results for versions 1 and 2 of the VHDL code listed in Fig. 6.2

(bottom of the figure). In both versions, the *data_in* input receives the values '0', '1' and '0', in this order, and when the simulation starts there is no history of previous values for any of the signals.

In Version 1, when A receives '0' from *data_in*, the process is triggered as it has signal A on its sensitivity list. Next, all the assignments are performed, sequentially. B receives '0' on line 18 in Fig. 6.2, overwriting the assignment performed on line 17. C receives '1' on line 20 (*Not A*, where A = *data_in* = '0'), overwriting the assignment performed on line 19. On line 21, D receives the value stored in C but, as shown in the top waveform in Fig. 6.3, the simulation has just started, and the simulator has no knowledge of the value of C. For this reason, the waveform shows neither '0' nor '1'. At this point, it is important to notice that the C value assigned on line 20 in Fig. 6.2 cannot be used for the D assignment on line 21, as in a process, all assignments are performed sequentially, and considering the values the signals hold when the process started.

Proceeding with the simulation of Version 1, next A receives '1' from *data_in*, and the process is triggered again as a result of this change in A ('0' to '1'). Once more, all the assignments are performed, sequentially. B receives '0' on line 18 in Fig. 6.2. C receives '0' on line 20 (*Not A*, where A = *data_in* = '1'). D receives the value stored in C on line 21, which is '1'. This is the value stored in C, when the process started, and not the value just assigned to C on line 20.

In Version 2, the only difference is the C signal added to the process sensitivity list. This tiny modification results in significant differences in the simulation results. Initially, A receives '0' from *data_in*, and the process is triggered. The assignments are performed sequentially starting by B receiving '0' on line 18 in Fig. 6.2, overwriting the assignment performed on line 17. Next, C receives '1' on line 20 (*Not A*, where A = *data_in* = '0'), overwriting the assignment performed on line 19. On line 21, D receives the value stored in C and, as there was a change in C, and as C is now on the process sensitivity list, then the process is triggered once more and D also receives '1', which is the new value of C. This simulation result is shown in the bottom waveform of Fig. 6.3.

In Fig. 6.1 the process has a label (P1 on line 14). This identification is optional, but it is very convenient when performing the simulation of a VHDL code with several processes. The sensitivity list is also optional in a process. Though, in case of processes with no sensitivity list, a *wait* statement must be included in the process body in order to hold the process from time to time.

The behavioral description of sequential circuits in VHDL is an important topic, and it will be further discussed in Chap. 9 together with a more detailed discussion of processes.

6.2 Describing a D Latch in VHDL

As mentioned before, the output of a sequential circuit depends not only on its inputs, but also on data previously stored. Consequently, sequential circuits require the implementation of storage resources. Figure 6.4 shows a basic storage element, the set-reset (SR) latch. This simple circuit manages to store data (a single bit)

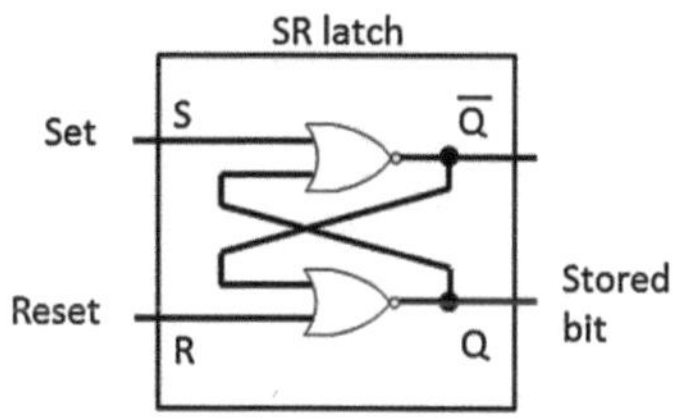

S	R	Q
0	0	hold last value
0	1	0
1	0	1
1	1	not allowed

Fig. 6.4 A basic SR latch and its truth table (adapted from [Vahid11])

```
-- Version 1 - Using process
library ieee;
use ieee.std_logic_1164.all;
entity SR_Latch_Proc is port (
     S, R : in  std_logic;
     Q, Qn: out  std_logic
);
end SR_Latch_Proc;
architecture behv of SR_Latch_Proc is
   signal auxQ, auxQn: std_logic;
begin
   process (S, R, auxQ, auxQn)
   begin
       auxQ  <= auxQn nor R;
       auxQn <= auxQ nor S;
   end process;
   Q  <= auxQ;
   Qn <= auxQn;
end behv;
```

```
-- Version 2 - Combinational circuit
library ieee;
use ieee.std_logic_1164.all;
entity SR_Latch is port (
        S, R : in  std_logic;
        Q, Qn: out  std_logic
);
end SR_Latch;
architecture behv of SR_Latch is
   signal auxQ, auxQn: std_logic;
begin
   auxQ  <= auxQn nor R;
   auxQn <= auxQ nor S;
   Q  <= auxQ;
   Qn <= auxQn;
end behv;
```

Fig. 6.5 VHDL implementation for an SR latch

through the cross-feedback loop in the pair of cross-coupled NOR gates shown in Fig. 6.4. The stored bit is available on the Q output signal.

The truth table in Fig. 6.4 describes the SR latch functionality, which is straightforward: when the S input is ‘0’ and the R input is ‘1’, the circuit outputs ‘0’ (“reset” the circuit, forcing its Q output to ‘0’); when the S input is ‘1’ and the R input is ‘0’, the circuit outputs ‘1’ (“set” the circuit, forcing its Q output to ‘1’); when both S and R inputs are equal to ‘0’, the circuit holds its current state (memory!). In all situations, one of the outputs is always the complement of the other. However, when both inputs S and R are ‘1’, the circuit can become unstable and, consequently, this is not a valid input for this circuit.

Two versions for the VHDL implementation of an SR latch are shown in Fig. 6.5. The first version, on the left, uses a process, while the second version uses just combinational logics.

The SR latch is a good option to show how a basic storage circuit works. However, the “not allowed” input suggests that extra precautions should be taken during this circuit’s usage. Basically, in most applications, the circuit around the SR latch should be designed in order to never allow a ‘1’ value to be present in both S and R inputs simultaneously.

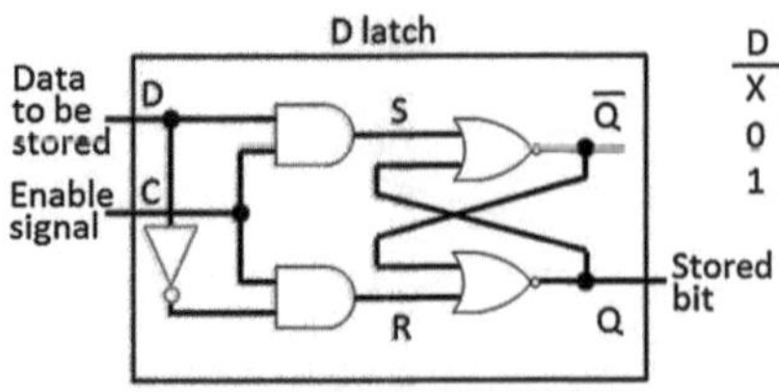

D	C	Q
X	0	hold last value
0	1	0
1	1	1

Fig. 6.6 A D latch and its truth table (adapted from [Vahid11])

A possible solution for this problem is the D latch, which is an evolution of the SR latch. A D latch is built adding two AND gates and an inverter in front of the SR latch, as shown in Fig. 6.6. With this arrangement, the "not allowed" input of the SR latch cannot happen. Now the circuit has just one data input, D. The other input, C, shown in Fig. 6.6 is used to control when a new data is stored in the memory (the D latch circuit). While the C input is '1', the input data D is stored. When C is '0', the AND gates are "closed", and changes on the D input cannot be stored in the circuit (the SR latch).

The D latch could be implemented in VHDL following the same strategy as the SR latch shown in Fig. 6.5. However, a better and more elegant way of describing a D latch is shown in Fig. 6.7. This VHDL implementation is a "behavioral" description of the D latch. The *if* statement on line 13, describes the exact behavior of the D latch truth table, i.e., when C is '1', the input data provided in D is stored. Otherwise, when C is '0', the circuit holds the value provided in D, when C was

```
1  library ieee;
2  use ieee.std_logic_1164.all;
3  entity D_latch is port (
4          C:   in std_logic;
5          D:   in std_logic;
6          Q: out std_logic
7  );
8  end D_latch;
9  architecture behv of D_latch is
10 begin
11     process(C, D)
12     begin
13         if (C = '1') then
14             Q <= D;
15         end if;
16     end process;
17 end behv;
```

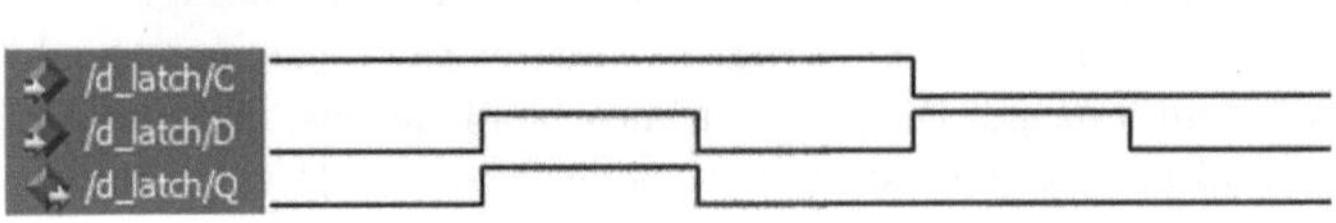

Fig. 6.7 VHDL implementation for a D latch

'1' for the last time. The waveform in Fig. 6.7 shows the D latch behavior according to C and D inputs.

The D latch is level activated, as the circuit output (Q) follows the value presented at its input (D), while the C input (which is usually connected to the clock) is active, i.e. C is kept at level '1'. This behavior is called "transparent". The advantage of the D latch is its low cost on circuitry. The drawback is that it has no precise control of the input D to output Q. For instance, if a circuit has several D latches in cascade and if C = '1' for a long period of time, the data at the D input might be propagated through many latches. Otherwise, if C = '1' for a too short period of time, it may not enable a store. The terms "long period" and "short period" are dependent on the target technology.

6.3 Describing a D Flip-Flop in VHDL

The D flip-flop is conceived aiming applications where the aforementioned situation of controlling the exact moment when the input data is stored, is an important issue. In a D flip-flop, the input data is stored just when there is a transition from '0' to '1' (or from '1' to '0') on the C input. So, the D flip-flop is said to be controlled by "edge", while the D latch is controlled by "level". In a D latch, whenever the C input is '1' (active high logic level), the D value is stored in the circuit. In a D flip-flop, the D value is stored only when the C input changes from '0' to '1' (a rising edge event). In other versions of these circuits, the D latch stores the input data when C = '0' (active low logic level), and the D flip-flop stores the input data when C changes from '1' to '0' (a falling edge event). The D Flip-flop can be implemented according to different design styles. In a typical design, two D latches are connected in a master-servant topology, as shown in Fig. 6.8.

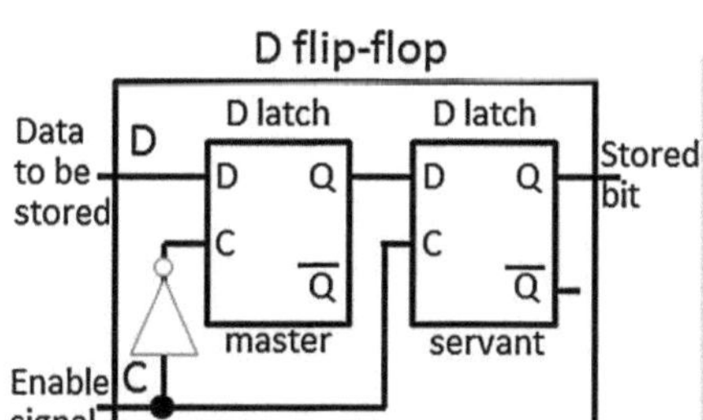

D	C	Q
X	X	hold last value
0	0->1	0
1	0->1	1

```
library ieee;
use ieee.std_logic_1164.all;
entity D_FF is port (
   C: in std_logic;
   D: in std_logic;
   Q: out std_logic
);
end D_FF;
architecture behv of D_FF is
begin
   process(C, D)
   begin
      if (C'event and C = '1') then
         Q <= D;
      end if;
   end process;
end behv;
```

Fig. 6.8 D flip-flop. Block diagram, truth table, and VHDL implementation (adapted from [Vahid11])

In Fig. 6.8, the output (Q) of the master latch is connected to the input (D) of the servant latch. The clock signal for the master latch is inverted in relation to the clock signal for the servant latch. In this sense, the master is loaded when C = '0', and the servant when C = '1'. When C changes from '0' to '1', the master is disabled and the servant is loaded with the value that was at D just before C has changed. Outside the active clock edge there is no transfer of the input value to the servant latch. As a tradeoff this architecture needs more logic gates to be implemented than the D latch. However, gate count is less of an issue nowadays.

The VHDL code listed in Fig. 6.8 is a behavioral implementation for the described D flip-flop. The hardware detection of a rising edge event on the C input is described in VHDL on line 13. A VHDL synthesis tool understands this sort of description and a physical D flip-flop is used in the final circuit, in case this D flip-flop resource is available in the target technology.

There are several variations for the D flip-flop implementation. In Fig. 6.9, a reset control signal has been added to the circuit. In this case, when the RST input is zero, the output will also be zero. Otherwise, when RST is '1', then the Q output will depend on the D and C inputs, as explained before. This D flip-flop with asynchronous reset VHDL implementation is shown on line 14. In a synchronous reset implementation, line 14 should be placed inside the *c'event* test.

Notice in Fig. 6.9, on line 16, the *elseif* VHDL statement usage. This combination of if (*RST* = '0') and *elseif* (*C'event and C* = '1'), tells the synthesis tool to generate D flip-flop with asynchronous reset. *Very important!! Any change in this syntax, may result in the synthesis tool generating another sort of circuit, which may not be the expected D flip-flop.*

Figure 6.9 shows the block diagram (symbol) of a D flip-flop with asynchronous reset. In this diagram, internal details as logic gates and latches are hidden,

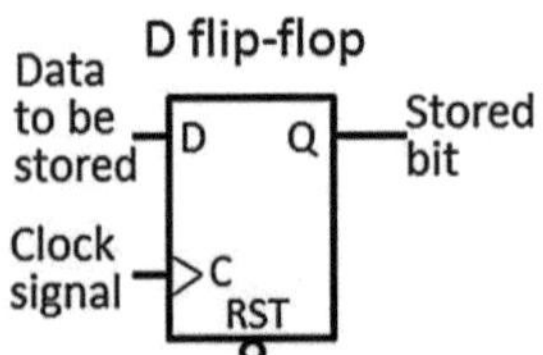

D	C	RST	Q
X	X	0	0
X	X	1	hold value
0	0->1	1	0
1	0->1	1	1

```
1  library ieee;
2  use ieee.std_logic_1164.all;
3  entity D_FF is port (
4     C  : in std_logic;
5     RST: in std_logic;
6     D  : in std_logic;
7     Q  : out std_logic
8  );
9  end D_FF;
10 architecture behv of D_FF is
11 begin
12    process(C, RST, D)
13    begin
14       if (RST = '0') then
15          Q <= '0';
16       elsif (C'event and C = '1') then
17          Q <= D;
18       end if;
19    end process;
20 end behv;
```

Fig. 6.9 D flip-flop with asynchronous reset

and only the input and output signals are shown. The small triangle on the C input represents a "clock" signal in digital systems. For the sake of simplicity, this symbol and connotation has not been used in the previous flip-flop diagram shown in Fig. 6.8. However, usually in flip-flop based designs, the C input is always connected to a clock signal in order to synchronize the data read/write operations with the remaining components of the circuit. Basically, when a component needs to store a bit in a flip-flop, if the component operations are coordinated by the same clock signal as the flip-flop, then it is possible to assure that the correct data is written. In a typical synchronous digital system, operations take place in the rising (or falling) edge of the global clock signal. It is not unusual for current digital systems to present several clock signals, and the implementation of such systems is a challenge for the designers.

Figure 6.10 shows a D flip-flop with asynchronous reset, and an *Enable* input. The VHDL implementation describes the behavior of this flip-flop where, basically, whenever the reset signal is high (RST = '1'), and there is a rising clock edge, if the enable signal is high (EN = '1'), then the input data is stored (Q <= D).

The waveform in Fig. 6.11 shows the simulation results for the D flip-flop with enable and asynchronous reset. The CLK input is fed with a clock signal, and the simulation runs for 9 clock cycles. In the first 4 clock cycles, the reset signal (RST) is kept in active-low level ('0'), and no action takes place in the flip-flop. As the reset is asynchronous, the *elseif* statement cannot be reached while RST = '0' (see line 13 in Fig. 6.10), and the changes on input signals EN and D have no effect in the circuit state. Next, in the beginning of the 5th clock cycle, the reset goes high, the enable (EN) goes low, and the D input remains high. In this case, as the enable signal is low, again the input data (D) cannot be stored in the

```
1  library ieee;
2  use ieee.std_logic_1164.all;
3  entity D_FF is port (
4     CLK, RST,
5     EN, D      : in std_logic;
6     Q          : out std_logic
7  );
8  end D_FF;
9  architecture behv of D_FF is
10 begin
11    process(CLK, RST, D, EN)
12    begin
13       if (RST = '0') then
14          Q <= '0';
15       elsif (CLK'event and CLK = '1') then
16          if EN = '1' then
17             Q <= D;
18          end if;
19       end if;
20    end process;
21 end behv;
```

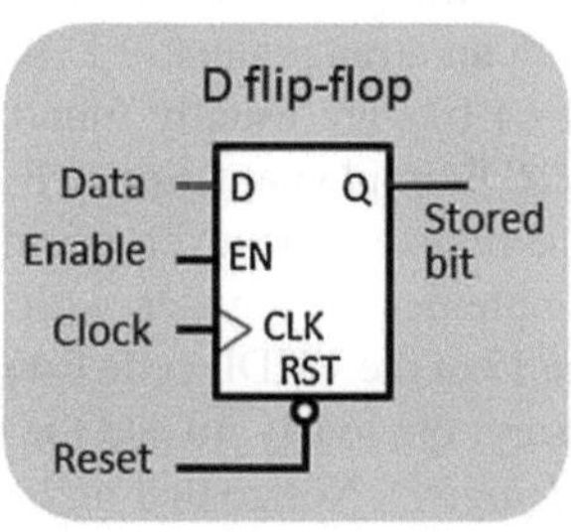

Fig. 6.10 D flip-flop with enable and asynchronous reset

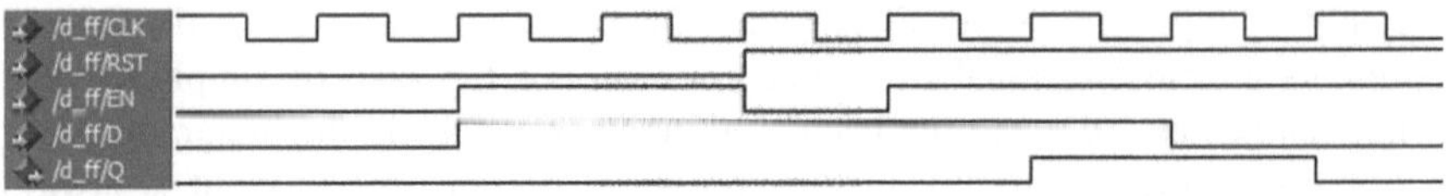

Fig. 6.11 Simulation waveform for the D flip-flop with enable and asynchronous reset

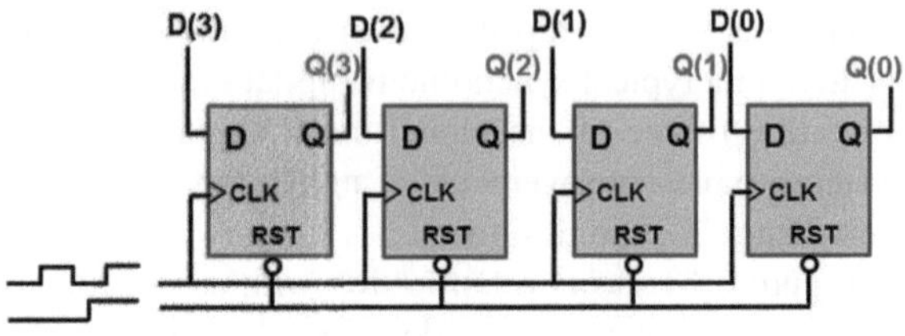

Fig. 6.12 Block diagram of a 4-bits register based on D flip-flops

flip-flop (see line 16 in Fig. 6.10). The enable signal goes high again in the beginning of the 6th clock cycle, simultaneously with the rising clock edge. In this case, there is no change in the flip-flop memory contents, as the enable signal should have been placed in an active-high level, before the rising clock edge event. Finally, in the rising edge of 7th clock cycle, the enable signal was already in active-high level, and the D input, which is '1', is stored in the flip-flop.

6.4 Implementing Registers with D Flip-Flops

A D flip-flop can be used to store a single bit of data. For circuits that perform operations with larger word sizes, registers should be used instead. A register can be built from an arrangement of D flip-flops, as can be seen in Fig. 6.12. Registers with 32 or 64 bits are usually found in commercial microprocessors.

In the VHDL implementation in Fig. 6.13, a 4-bits word present on the D input is stored in the register whenever a rising clock edge is provided in the CLK input. In order to implement a register with enable and asynchronous reset signals, just replace line 17 of the VHDL code listed in Fig. 6.13, by lines 16–18 of Fig. 6.10. It is also necessary, obviously, to add the enable signal declaration in the register entity (*EN: in std_logic;*). Notice that on line 14 of Fig. 6.10, the single bit *Q* assignment should be replaced by *Q* <="0000", as now Q is a 4 bits vector. In order to have a more generic code, the VHDL statement *Q* <= (*others* => '0'); could be used instead. In this case, all bits of Q will receive '0', no matters Q length.

Once more, it is important to notice that synthesis tools are not as smart as one expect them to be. This means that there is almost no flexibility to describe flip-flops and registers in VHDL. The designer who wants a D flip-flop or register properly synthesized, should follow the models shown in this book and in other available documents. Otherwise, the synthesis tool may not generate the hardware the designer wants.

```
library ieee;
use ieee.std_logic_1164.all;
entity D_4FF is port (
    CLK:  in std_logic;
    RST:  in std_logic;
    D   :  in std_logic_vector(3 downto 0);
    Q   : out std_logic_vector(3 downto 0)
);
end D_4FF;
architecture behv of D_4FF is
begin
    process(CLK, D)
    begin
      if RST = '0' then
          Q <= "0000";   -- or (others => '0');
      elsif (CLK'event and CLK = '1') then
          Q <= D;
      end if;
    end process;
end behv;
```

Fig. 6.13 VHDL description of a 4-bits register based on D flip-flops

6.5 Laboratory Assignment

The laboratory objectives are:

- to design 4-bits and 8-bits registers with enable and asynchronous reset;
- to add the implemented registers to the calculator developed in the last chapter.

Using as example the registers and flip-flops provided in previous sections, write the VHDL implementations for the 4-bits and 8-bit registers shown in Fig. 6.14. Also, make the necessary modifications in the top_calc component as needed to add the registers to the design.

The whole design has a total of 9 files (8 components + top), as follows:

- *c1.vhd*, provided in Fig. 5.17;
- *c2.vhd*, provided in Fig. 5.18;
- *c3.vhd*, provided in Fig. 5.19;
- *c4.vhd*, provided in Fig. 5.20;
- *mux4x1.vhd*, provided in Fig. 5.21;
- *reg4bits.vhd*, should be adapted from Figs. 6.10 and 6.13;
- *reg8bits.vhd*, should be adapted from Figs. 6.10 and 6.13;
- *decod7seg.vhd*, implemented in the previous chapter, according to Sect. 5.4 and Fig. 5.12;

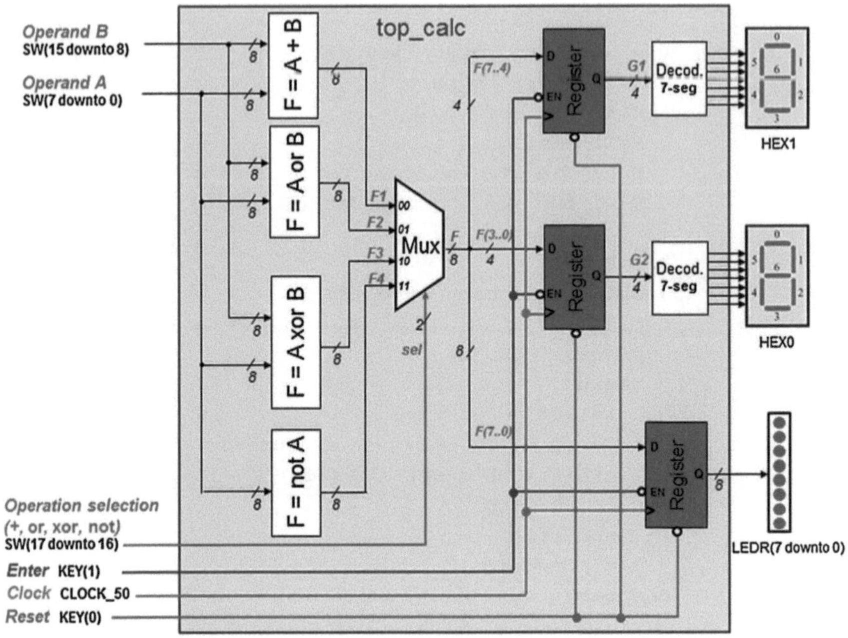

Fig. 6.14 Inserting registers in the calculator to store operation results

- *top_calc.vhd*, partially provided in Fig. 5.16. The previous chapter implementation should be adapted in order to add the new registers, and also the new input signals shown in Fig. 6.14.

The tasks to be completed in this laboratory session, using Altera's Quartus II are as follows:

- Create a project with all the components shown in Fig. 6.14;
- Using the VHDL design entry editor, create a 4-bits register and a 8-bits register, both with an enable signal, and asynchronous reset;
- Modify the top-level component developed in the previous chapter, in order to add the new components (the three registers shown in Fig. 6.14). Notice that new signals are needed to connect the register outputs, to the decoder inputs.
- Perform the synthesis;
- Perform the simulation and fix errors, if any;
- Prototype and test the circuit in the FPGA board.

The new registers used to store the calculator results are as follows:

- A 4-bits register to store the Less Significant Bits (LSB) of the operation result, to be presented in HEX0;
- A 4-bits register to store the Most Significant Bits (MSB) of the operation result, to be presented in HEX1;

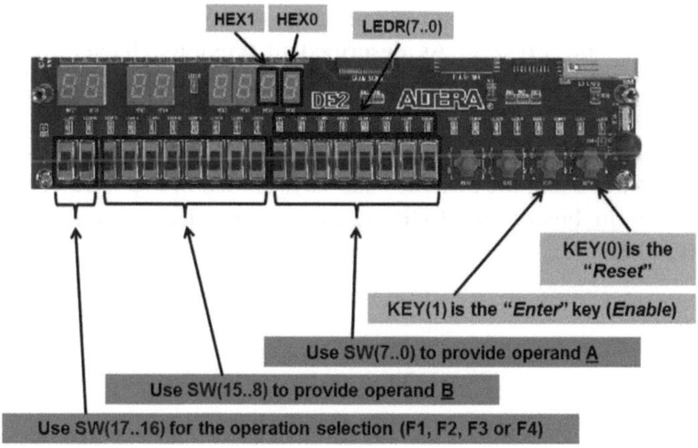

Fig. 6.15 Input and output user interface for the calculator

- An 8-bits register to store the operation result to be shown in binary on the red LEDs.

Figure 6.15 shows the user interface for the new calculator. The calculator's operation is straightforward:

- When KEY(0) is pressed (Reset), the flip-flops (registers) should be cleared. As a result, the LEDs are switched off, and the 7-segments displays (HEX0 and HEX1) should show a '0'.
- When KEY(1) is pressed (Enter), the registers are enabled for writing, allowing the operation result to be stored (memory!).

Additionally, make sure to connect the clock input to the CLOCK_50 signal provided by the DE2 board. This is a 50 MHz crystal available on the board. The top entity should be changed in order to include the new DE2 input signals:

- key: in std_logic_vector(1 downto 0); – – KEY(0) e KEY(1)
- clock_50: in std_logic; – – clock 50 MHz

Also, it is important to notice that the four push buttons available in DE2 board are active low. This means that when they are pressed, a zero ('0') is generated. In addition, there is a Schmitt Trigger based debounce circuit for the push buttons, in order to help to have just one zero pulse each time a button is pressed.

As stated before, the *project creation* process has been well discussed and explained in previous chapters. For the *synthesis* activity, it is essential to import the *DE2_pin_assignments.qsf* file, as discussed in Chap. 2, Step 5, and shown in Fig. 2.7.

The *simulation* activity is performed using ModelSim. To start the simulator in Quartus II, select *Tools→Run Simulation Tool→RTL Simulation*. Wait for ModelSim to open and perform the steps described in previous chapters.

When the expected results are obtained in the simulation process, the next step is the *FPGA prototyping*. As described in previous chapters, to download the generated bitstream to the FPGA, using Quartus II choose menu *Tools→Programmer*. Follow the instructions provided in Chap. 1, Step 5, observing the "hardware setup" instructions in Figs. 1.32 and 1.33. Test the circuit in the FPGA board, using the switches and keys shown in Fig. 6.15. The operation result is shown in binary in $LEDR_{7..0}$, and in hexadecimal on the 7-segment displays HEX0 and HEX1.

Chapter 7
Synthesis of Finite State Machines

A finite state machine (FSM) is a powerful tool used to build control circuits. Control circuits are employed in situations where a sequence of operations should be performed, according to a pre-defined execution order. The use of a FSM as a control circuit is better explored in the next chapters. This chapter introduces the basic concepts related to the design of FSMs. At the end of the chapter, the reader should be able:

- to understand the concept of FSMs;
- to understand the synthesis process of FSMs in VHDL;
- to understand the concept of counters implemented in VHDL using FSMs;
- to design and to implement in VHDL an FSM based counter.

7.1 Finite State Machines

As already discussed in Chap. 1, a computer system usually consists of a control module (i.e. control unit) and an operations execution module (i.e. datapath), as shown in Fig. 1.1. The control module is responsible for coordinating the sequence of activities to be performed by a given system. In digital systems, control signals are generated by sequential circuits. A sequential circuit passes through a series of states, and each state (every time), may provide a certain output. The outputs of each state are the signals used to control the sequence of activities executed by a target system.

In Fig. 7.1, a state diagram is used to model an FSM. The state diagram is represented by a directed graph, where the vertices (or nodes) are the FSM states, and the edges (or arcs) represent the transitions from one state to another. As shown in Fig. 7.1, an FSM has the following components:

- *States* are represented in the graph by nodes S_0, S_1, S_2 and S_3. The states are responsible, basically, for storing information regarding past changes in the

E. A. Bezerra and D. V. Lettnin, *Synthesizable VHDL Design for FPGAs*,
DOI: 10.1007/978-3-319-02547-6_7,

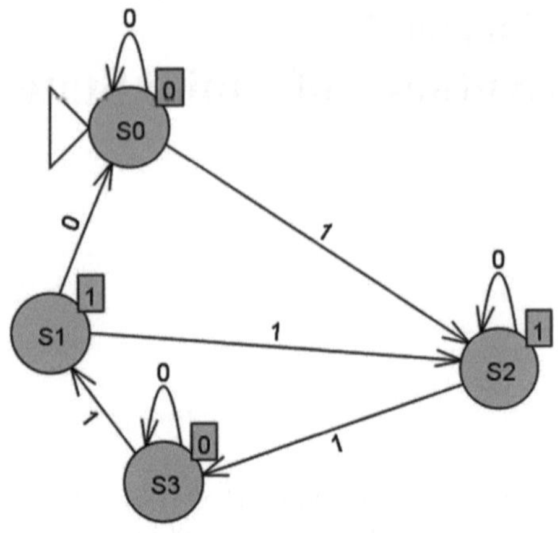

Fig. 7.1 Example of a state diagram for an FSM

inputs. In the example FSM, the triangle pointing to S_0 indicates that this is the initial state.

- *Transitions* are represented in the graph by the edges. A transition indicates a state change by means of a condition that enables the switch from a state to another.
- *Actions* are activities to be performed when the respective state is reached.

A hardware implementation of an FSM comprises a combinational circuit, and a register. As shown in Fig. 7.2, the FSM's current state is stored in the register. The combinational circuit uses the circuit's inputs and the stored "current state", in order to calculate the "next state". The combinational circuit calculates also the FSM's output.

An FSM can also be represented through a state transition table. In Table 7.1, it can be observed that when the FSM is in state S_0, a transition (change) to state S_2, takes place only when the circuit's input is '1' ("Inputs" in Fig. 7.2). While the circuit's input stays in '0', the FSM will remain in the same state S_0, as shown on the first row of Table 7.1. This action is also shown graphically in the loop with the '0' input in state S_0 in Fig. 7.1. The fourth column in Table 7.1 shows the output provided by the FSM for each state. In Fig. 7.1, the output for each state is represented by the numeric value in a square at the top right hand corner of the nodes. In Fig. 7.2, the output is represented by the signal "Outputs".

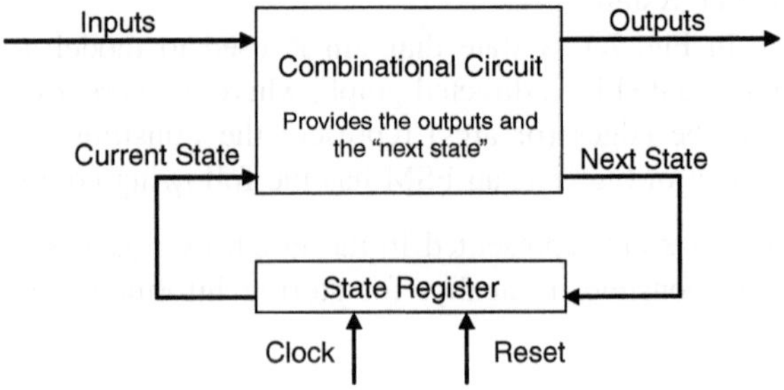

Fig. 7.2 Block diagram of an FSM hardware implementation

Table 7.1 State transition table for the FSM shown in Fig. 7.1

Current state	Input	Next state	Output
S_0	0	S_0	0
S_0	1	S_2	0
S_1	0	S_0	1
S_1	1	S_2	1
S_2	0	S_2	1
S_2	1	S_3	1
S_3	0	S_3	0
S_3	1	S_1	0

The state transition table (Table 7.1) is a direct representation of the state diagram of an FSM.

Considering that the circuit in Fig. 7.2 implements the FSM shown in Fig. 7.1, the register stores the FSM current state (Table 7.1 first column), and the combinational circuit calculates the "next state" (Table 7.1 third column) using the stored "current state" and the circuit's inputs (Table 7.1 second column). The combinational circuit calculates also the FSM's output (Table 7.1 fourth column).

A possible VHDL implementation for the FSM is shown in Fig. 7.3. In that implementation, a transition happens each time an event of a rising clock edge takes place. This event is represented in Fig. 7.1 by the graph's edges, and in Fig. 7.2 by the *clock* input.

7.2 VHDL Synthesis of Finite State Machines

An FSM is, essentially, a sequential circuit. Basically, in an FSM, from the combination of a given sequence of inputs and past states, the next states, and the respective outputs are generated sequentially.

As aforementioned in Chap. 6, a sequential circuit can be implemented in VHDL using the *process* statement. There are several ways to describe an FSM in VHDL, using 1, 2 or even 3 processes.

The FSM studied in Fig. 7.1 has been implemented in VHDL using the two processes strategy. The source code for this implementation is shown in Fig. 7.3. An important VHDL concept useful for FSMs description is the enumeration type. With the VHDL reserved word *type* the developer can create a new type. In Fig. 7.3, line 7, STATES is a new type that can hold the values S0, S1, S2 and S3. On line 8, CS (i.e. current state) and NS (i.e. next state) are two internal signals defined as STATES type. This means that these two signals can be assigned only to other signals or constants of the same type STATES. It is also important to highlight that S0, S1, S2 and S3 are not part of the VHDL language. These are just labels used in the set of possible values to be used in the new enumeration type. During the synthesis process, these labels will be translated into ones and zeros by

```
1  entity MOORE is
2      port(X, clock, reset: in std_logic;
3           Z              : out std_logic);
4  end;
5
6  architecture TwoProcesses of MOORE is
7      type STATES is (S0, S1, S2, S3);    VHDL Enumeration Type
8      signal CS, NS : STATES;
9  begin
10     process (clock, reset)     1st process: implements a register
11     begin                      that stores the current state (CS)
12         if reset= '1' then     as a function of the next state (NS)
13             CS  <= S0;
14         elsif clock'event and clock='1' then
15             CS <= NS;
16         end if;
17     end process;
18                                2nd process: implements the circuit
19     process(CS, X)             that generates the Z output,
20     begin                      according to the current state (CS)
21         case CS is             and the X input
22             when S0 =>  Z <= '0';
23                         if X='0' then NS <= S0; else NS <= S2; end if;
24             when S1 =>  Z <= '1';
25                         if X='0' then NS <= S0; else NS <= S2; end if;
26             when S2 =>  Z <= '1';
27                         if X='0' then NS <= S2; else NS <= S3; end if;
28             when S3 =>  Z <= '0';
29                         if X='0' then NS <= S3; else NS <= S1; end if;
30         end case;
31     end process;
32 end TwoProcesses;
```

Fig. 7.3 VHDL implementation of an FSM using two processes

the EDA tool, in order to generate the final hardware implementation for the abstract VHDL description. The VHDL syntax for the enumeration type is as follows:

$$\text{type indentifier is (element1, element2, element3, ...);}$$

where, "type" and "is" are VHDL reserved words, and "identifier" is the name of the new type that is being created. Following the "is" statement, there is an ordered set of values to be used in the VHDL description as the values to be assigned to the signals of type "identifier".

In Fig. 7.3, line 13, an element belonging to the set of enumeration values is directly assigned to the CS signal. In Fig. 7.3, line 15, there is an example of assignment of a signal of type STATES to another signal of the same type.

The VHDL synthesis, using two processes, of the FSM described in Fig. 7.1 is straightforward. The process P1 defines the current state (CS) for the processes P2. Process P1 is sensitive to the clock and reset signals. If an asynchronous reset

(reset = '0') occurs, CS receives the S0 value, otherwise, if a rising clock edge occurs CS is updated with the next state (NS) signal, that is, it is responsible to perform the state transition process. The process P2 updates the (NS) signal and defines new values for the output Z signal according to the CS.

In both representations, the graphical (Fig. 7.1) and the textual (Fig. 7.3) one, it is possible to notice that the circuit's outputs are defined in each of the states. As listed in Table 7.1, for states S0 and S3, the output is '0', and for states S1 and S2, the output is '1'. This type of FSM is known as a Moore machine, as the output values depend only on the CS. Another type of FSM is known as a Mealy machine, where the outputs are defined according not only to the CS, but also to the current inputs.

Using the enumeration type in VHDL designs is important also in the simulation step. As shown in Fig. 7.4, the output values (Z) are well identified for each CS state. All the transitions between states are also well identified, as the names S0, S1, S2 and S3 are a good help for the simulation visualization. In case the enumeration type had not been used in VHDL code, the CS and NS signals would have shown "00", "01", "10" and "11" instead, making it a bit more difficult to visualize all the transitions. In FSMs with more states, the situation would be even more complicated.

Another approach to describe an FSM in VHDL employs three processes instead of two. As shown in Fig. 7.5a, the P1 process updates the (CS), and sends this information to P2 and P3. In Fig. 7.5b, P2 computes the next state, but does not perform the transition. The (NS) information is sent from P2 to P1, which is the responsible for updating the (CS). P3 is the process responsible for signals (outputs) assignments and updates.

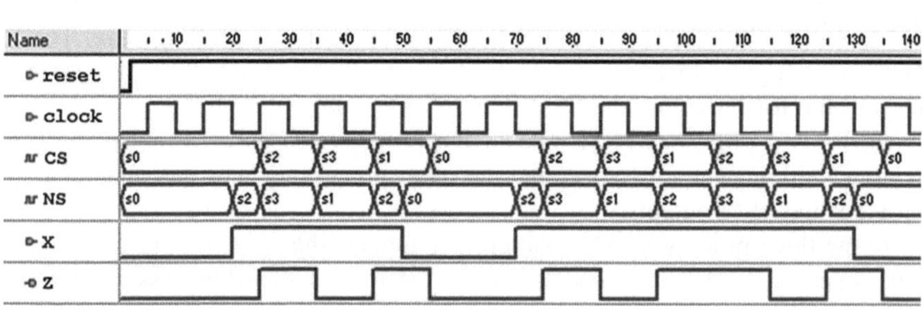

Fig. 7.4 Waveform diagram for the example FSM

Fig. 7.5 Three processes FSM diagram. **a** P1 process updates the CS; **b** P2 process defines the NS

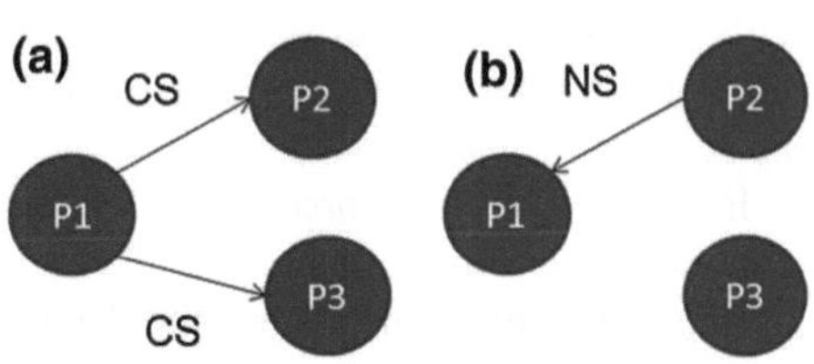

```
1  P1: process(clk)
2  begin
3      if falling_edge(clk) then
4          if rst = '0' then
5              CS <= S0;
6          else
7              CS <= NS;
8          end if;
9      end if;
10 end process;
11
12 P3: process(clk)
13 begin
14     if rising_edge(clk) then
15         case CS is
16             when S0 =>
17                 Z <= '0'
18             when S1 =>
19                 Z <= '0'
20             when S2 =>
21                 Z <= '1';
22             when others =>
23         end case;
24     end if;
25 end process;
```

```
P2: process(CS, X)
begin
    case CS is
        when S0 =>
            NS <= S1;
        when S1 =>
            if X = '1' then
                NS <= S2;
            else
                NS <= S1;
            end if;
        when S2 =>
            NS <= S1;
        when others =>
    end case;
end process;
```

Fig. 7.6 A VHDL implementation for a three processes FSM

In Fig. 7.6 there is a VHDL implementation for a three processes case study. The P1 process is triggered by the clock falling edge (i.e. clk'event and clk = '0'), performing the state transitions (line 7, CS <= NS). P2 process is triggered by CS and the X input signal. According to the input (X), P2 defines the next state of the FSM, which will be assigned to CS by P1 in the next clock falling edge. The P3 process is triggered by the rising clock edge (i.e. clk'event and clk = '1'), and it is responsible for providing the FSM outputs, according to the (CS).

In the two processes FSM, one of the processes is used to model the state register, defining the next state, and the other process updates the next state and provides the FSM outputs.

In the three processes FSM, one process models the state register defining the next state, a second process updates the next state, and a third process provides the FSM output.

In a single process FSM implementation, all these activities are performed by the same process. Figure 7.7 shows a VHDL implementation for the Fig. 7.1 FSM, using only one process. In this case, when the reset signal is '1', the initial state is set to S0. When the reset signal is '0', and there is a rising clock edge event, then the Z output is provided according to the (CS), and the (NS) is defined according to the X input.

In the two and the three processes approaches, one of the processes provides combinatorial outputs, as there is no clock signal on its sensitivity list. This can be an interesting solution, in case the designer needs to have combinatorial outputs.

```
entity MOORE is
    port(X, clock, reset: in std_logic;
         Z              : out std_logic);
end;

architecture OneProcess of MOORE is
    type STATES is (S0, S1, S2, S3);
    signal CS: STATES;
begin
    process(clock, reset)
    begin
        if reset= '1' then
            CS  <= S0;
        elsif clock'event and clock='1' then
            case CS is
                when S0 => Z <= '0';
                            if X='0' then CS <= S0; else CS <= S2; end if;
                when S1 => Z <= '1';
                            if X='0' then CS <= S0; else CS <= S2; end if;
                when S2 => Z <= '1';
                            if X='0' then CS <= S2; else CS <= S3; end if;
                when S3 => Z <= '0';
                            if X='0' then CS <= S3; else CS <= S1; end if;
            end case;
        end if;
    end process;
end OneProcess;
```

Fig. 7.7 VHDL implementation of an FSM using one process

In the one process implementation, there is no combinatorial output. On the other hand, the one process approach, historically, is a better choice regarding faster simulation times, and also it is easier to debug as there is no combinatorial processes involved. However, designers usually find the two processes approach more readable, as it is an appropriate behavioral description of the FSM classical hardware architecture shown in Fig. 7.2.

7.3 FSM Case Study: Designing a Counter

A counter is a good example of a sequential circuit, and it has interesting features suitable for FSM modeling. A counter has an output that provides expected values for each of its states, according to the input clock pulses. The output values are provided in a sequence that can be ascending, descending, or out of order. In any case, when the last value of the counting is reached, the sequence can be restarted, according to the design requirements. There are several types of counters, and among the most used ones there are the up counters, and the down counters. In Fig. 7.8 there is the VHDL description of a ring counter, performing an up counting.

```
1  library ieee;
2  use ieee.std_logic_1164.all;
3
4  entity ring_counter is
5      port (CLK    : in std_logic;
6            RST    : in std_logic;
7            Z      : out std_logic_vector(7 downto 0)
8            );
9  end ring_counter;
10
11 architecture beh of ring_counter is
12     signal aux : std_logic_vector(7 downto 0);
13 begin
14     process(CLK, RST)
15     begin
16         if (RST = '0') then
17                 aux <= "00000001";
18         elsif (CLK'event and CLK = '1') then
19             aux(7 downto 1) <= aux(6 downto 0);
20             aux(0) <= aux(7);
21         end if;
22     end process;
23     Z <= aux;
24 end beh;
```

Fig. 7.8 A ring counter described in VHDL

The ring counter is initialized on line 17, and at each rising clock edge, the counter value is rotated to the left. On line 20, the most significant bit is copied to the less significant bit position, closing the "ring".

The simulation results for this counter are shown in Fig. 7.9. After the reset (RST) signal goes up, the ring counter is load with the start value "00000001". The bit in the less significant position holds a '1', which is rotated passing through all the eight bit positions, and going back to the original position after eight clock pulses. The pattern provided on the Z output by this ring counter is also known as the "one hot" code. As shown in Fig. 7.9, at each clock pulse only one of the eight output bits is '1' ("one hot"). This is a circular up counter as the provided Z outputs, in decimal notation, are: 1, 2, 4, 8, 16, 32, 64, 128, 1, 2, 4, ….

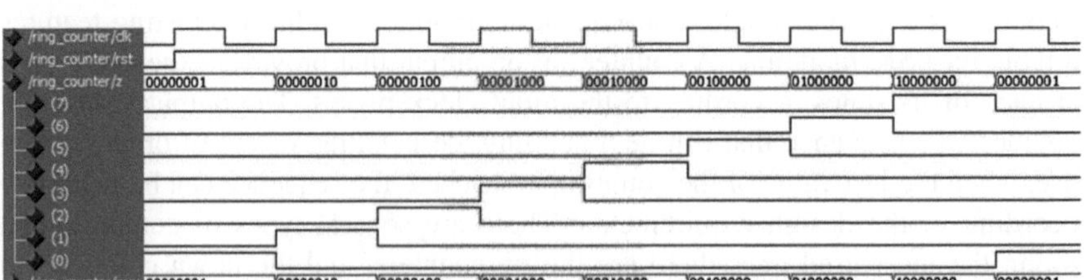

Fig. 7.9 Simulation results for the ring counter

Another example of counter is shown in Fig. 7.10. In this example, a counter is used to generate a delay. This sort of circuit is very useful, for instance, in applications where data need to be shown in a display. In Fig. 6.14, the calculation result stored in the registers is shown in the 7-segment displays each time the KEY(1) push-button (Enter) is pressed. A similar application may require the stored data to be shown on the displays periodically as, for example, every one second.

The FSM in Fig. 7.10 is used to generate a 1 s delay, considering an input clock of 27 MHz. The S1 state is included just to indicate the place where the 7-segment code could be inserted. Usually, after the data are sent to the display (in S1, S2, …, states), the FSM flow is diverged to the D1 state, which is the initial state of the delay routine. In this state, the delay counter is set to an initial value (zero in this example). Next, in the following rising clock edge, the counter (delay) is incremented (D2 state). In the next state (D3), there is a test inorder to check if the final value for the counting has been reached. If the test results in true, then 1 s has passed, and the FSM diverges back to the application (S1 state). Otherwise, if the test results in false, the FSM needs to keep counting on each rising clock edge, and the FSM goes to the D4 state. In this example, the final test value was calculated considering the 27 MHz input clock (i.e. 27 million pulses per second), and the FSM has to count 8,388,608 times in order to waste, approximately 1/3 s of time. As shown in Fig. 7.10, the FSM needs three clock pulses to perform a complete round, i.e. increment in D2, test in D3, and do nothing in D4. The D4 state has been included in this FSM with the only goal of spending one extra clock pulse. Therefore, 3 clock pulses times 1/3 s is equal to the required 1 s delay.

```
process(CLK, RST)          -- Considering a 27MHz clock, the process is triggered
begin                      -- 27 million times per second.
  if RST = '0' then
    CS <= S0;
  elsif CLK'event and CLK = '1' then
    case CS is
      when S1 =>  ...                   -- Do something (e.g. write to LCD, ...).
      when D1 =>                        -- This state starts the delay counting (1s).
        delay <= ( others => '0' );
        CS <= D2;
      when D2 =>                        -- State that generates the chosen delay
        delay <= delay + 1;             -- "delay" was set to zero in D1.
        CS <= D3;
      when D3 =>                    -- Tests if the final counting has been reached
        if delay >= x"800000" then  -- 8,388,608 / 27,000,000 = 0.3 * 3 = 1 s.
          CS <= S1;   -- When the upper limit is reached, leaves the delay
        else          -- loop, and goes back to the application.
          CS <= D4;   -- Remains in the counting loop to generate the delay.
        end if;
      when D4 =>      -- Continues the delay counting.
        CS <= D2;
    end case;
  end if;
end process;
```

Fig. 7.10 A counter used for delay generation

7.4 Laboratory Assignment

In previous chapters, several components have been glued together in order to build a basic calculator. In the next chapters, an FSM will be used in order to control the calculator operations.

In this chapter, a counter is used as a case study. Counters represent an important type of digital circuits, and FSMs can be used to model their behavior and functionality.

The laboratory objectives are:

- to design an FSM based digital circuit;
- to design a counter in VHDL.

7.4.1 Laboratory Session

The tasks to be performed in this laboratory session are as follows:

- Design and implement in VHDL a one process FSM, for the generation of the ASCII characters 'A' to 'Z'.
- The characters should be presented on the green LEDs (binary format) and on two 7-segments displays (hexadecimal format), as shown in Fig. 7.11.

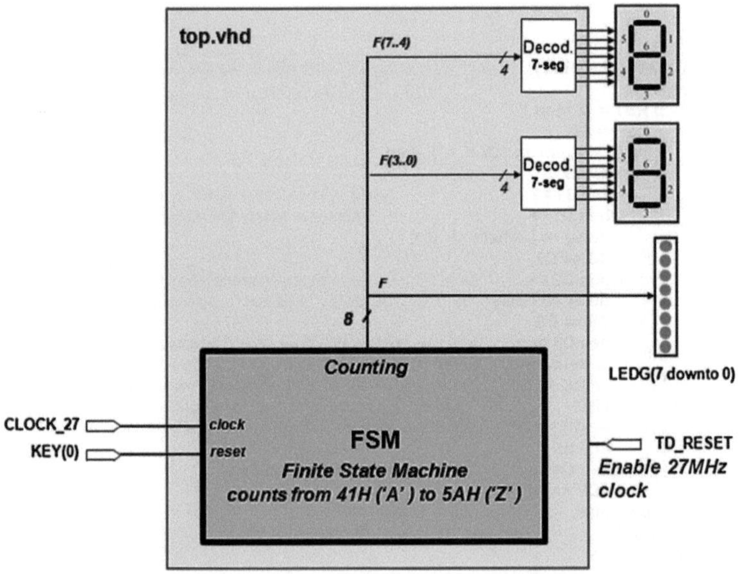

Fig. 7.11 'A' to 'Z' ASCII counter block diagram

- To write in the 7-segment displays, it is necessary to use two instances of the *decod7seg.vhd* component, described in Chap. 5 (see Fig. 5.22).
- The FSM has an asynchronous reset, KEY(0) button, used to initialize a counter with the first character in the sequente ('A' = 41H).
- On the rising clock edge (27 MHz), the counter should be incremented, generating the next ASCII table character ('B', 'C', 'D', …).
- The FSM should have a small number of states, but sufficient to increment the counter and check whether the final counting ('Z' = 5AH) has been reached.
- When the final counting is reached, the FSM should be reset to the initial state (beginning of the counting).
- In order to activate the DE2 27 MHz clock signal, the TD_RESET output has to be set to '1'.
- In Fig. 7.12 it is shown a suggestion of VHDL description for the top component, without the 7-segment decoders. In order to include the 7-segment decoders, create an F signal in the top file to connect the FSM component to the 7-segment decoders and the LEDG output.

Note Alternatively, instead of the 27 MHz clock, the rising clock edge can be generated by a push button (KEY). Warning! Beware of the debounce problem, since a single press in a mechanical switch may result in multiple presses in a digital circuit. A clock signal is a much more precise way to control the counting sequence.

In this laboratory session, all the usual activities should be performed: project creation; VHDL synthesis; VHDL simulation; and FPGA prototyping.

For the FPGA prototyping, it is important to include in the FSM some sort of the delay procedure as, for instance, the one shown in Fig. 7.10. Without a delay routine, the user will not be able to watch the counting presented on the LEDs and displays. However, for the simulation, the delay routine should be set to a smaller final counting value, otherwise the process would take too long. For instance, a 20 s simulation time in an i7 quad core processor (hyper threading, so 8 cores),

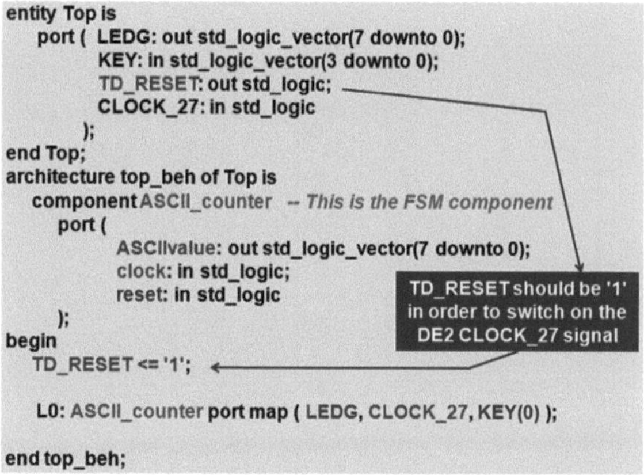

```
entity Top is
    port ( LEDG: out std_logic_vector(7 downto 0);
           KEY: in std_logic_vector(3 downto 0);
           TD_RESET: out std_logic;
           CLOCK_27: in std_logic
         );
end Top;
architecture top_beh of Top is
   component ASCII_counter -- This is the FSM component
      port (
              ASCIIvalue: out std_logic_vector(7 downto 0);
              clock: in std_logic;
              reset: in std_logic
      );
begin
   TD_RESET <= '1';

   L0: ASCII_counter port map ( LEDG, CLOCK_27, KEY(0) );

end top_beh;
```


Fig. 7.12 Suggestion of top file for the ASCII counter design without the 7-segment decoders

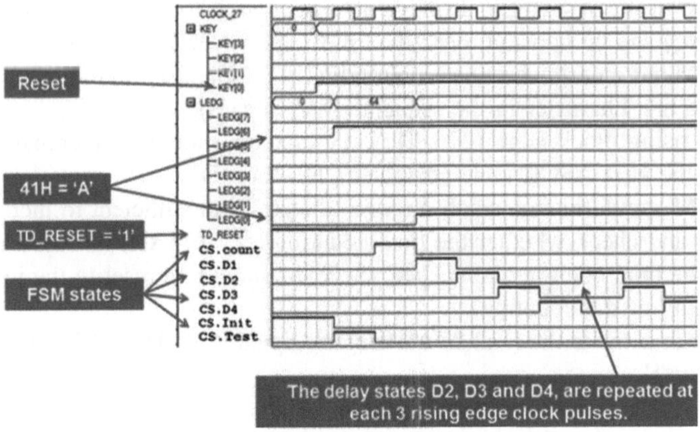

Fig. 7.13 Simulation results showing the delay states of the FSM

running at 2.93 GHz and with 8 GB of RAM, took 19 h and 28 min to be performed.

Figure 7.13 shows the ASCII counter simulation waveform. In the CS.Init state, after the KEY(0) button is released, the initial counting value is set to 41H, which is the 'A' character in the ASCII table (see LEDG). In the CS.Test state, the FSM checks if the upper counting limit 5AH ('Z') has been reached. Next, the delay states are performed, starting by the CS.D1 state, where the delay counting signal is initialized. The CS.D2, CS.D3, and CS.D4 states are performed 8,388,608, as discussed before in Fig. 7.10.

Chapter 8
Using Finite State Machines as Controllers

In the previous chapter, FSMs have been used in the design of counters. This is a good example of FSM usage, where each state of the FSM provides as output a counting value. The present chapter introduces the design of FSMs targeting one of their main applications, which is the control of a sequence of events. At the end of the chapter, the reader should be able:

- to understand the basic concepts of datapath and control unit;
- to design the FSM of a control unit;
- to implement a control module for the case study (calculator);
- to simulate and test the case study—calculator with control module;
- to prototype the case study using an FPGA board.

8.1 Designing an FSM Based Control Unit

As discussed in Chap. 1, and shown in Fig. 1.1, digital systems in general have a module responsible for performing the processing, and another module used to control the sequence of operations. Figure 8.1 shows three examples of typical digital systems where there is a control unit and a datapath. In each design there is a control unit sending a sequence of "commands" to the operational module.

In Fig. 8.1, FSMs can be used in the design of the control circuits. For instance, the Vending Machine controller has to perform all the necessary steps in order to deliver (or not) the selected product to the user. The controller's FSM should be designed considering that the operational part of the system has to perform the activities in a sequential order (i.e. payment selection; product selection; delivery product; ...). The FSMs for the remaining controllers in Fig. 8.1 should be designed following similar principles, always considering a very specific sequence of events to be performed.

Strategies for describing FSMs in VHDL have been discussed in Chap. 7. However, before writing the VHDL implementation for an FSM, the target circuit

E. A. Bezerra and D. V. Lettnin, *Synthesizable VHDL Design for FPGAs*,

DOI: 10.1007/978-3-319-02547-6_8,

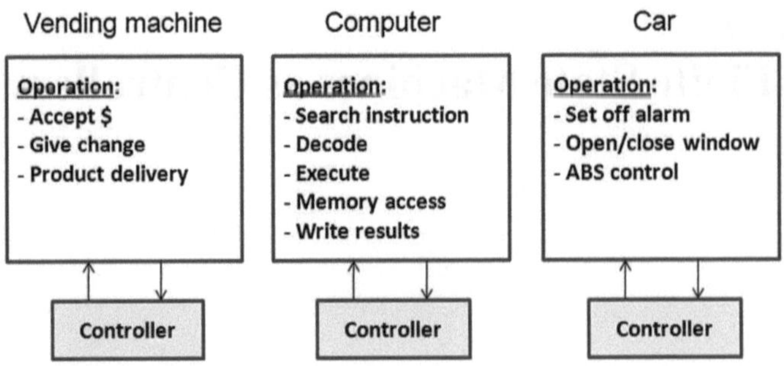

Fig. 8.1 Simulation results showing the delay states of the FSM

has to be properly designed. In Chap. 7, the FSM for the proposed counter case study has been written in VHDL in an empirical way, without following any design methodology.

In the present chapter, a controller for a new case study is implemented in VHDL, but this time a proper design methodology is followed.

The methodology adopted for the design of a controller in VHDL has the following steps:

STEP 1 Write a system description as complete as possible (requirements);

STEP 2 Prepare a graphical representation for the FSM, making as many versions as needed;

STEP 3 Write down a state transition table for the FSM, listing the inputs and outputs, including all the states (current and next);

STEP 4 Describe in VHDL the modeled FSM behavior, using the graph and the truth table as a guide, and following the examples provided in Chap. 7.

In the classical FSM controller design methodology used in digital systems in general (not in HDL based designs), the following additional steps are required:

STEP 5 Define the target circuit architecture which is, usually, represented by the block diagram in Fig. 1.3;

STEP 6 Define a unique identifier (coding strategy) for each state;

STEP 7 Make the necessary changes in the truth table, in order to have only binary numbers (replace the state symbolic identifiers by the binary codes);

STEP 8 From the truth table, obtain the Boolean equations, and design the circuit for the combinational component of Fig. 1.3.

It is important to notice that these additional steps are not used in a VHDL based design, as they are automatically performed by the synthesis tool. The synthesis tool not only extracts the required combinational logic from the VHDL FSM description, but also performs very efficient optimizations.

Next, a case study is presented and implemented in VHDL in order to demonstrate the FSM design methodology. The additional steps discussed are also performed for the case study, demonstrating the classical approach for the design of a FSM based digital system controller.

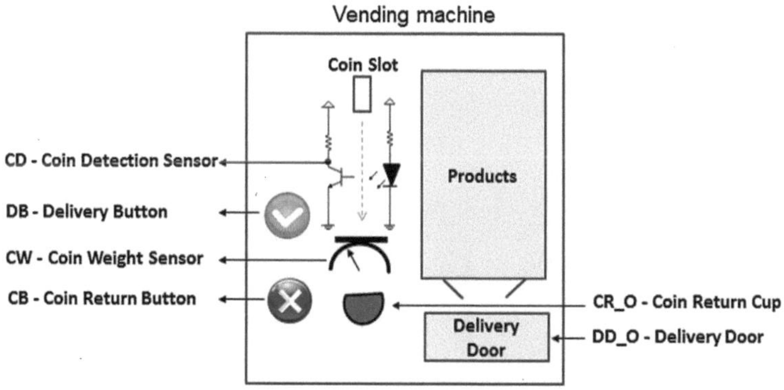

Fig. 8.2 Block diagram of the vending machine

8.2 Case Study: Designing a Vending Machine Controller

In this case study you are asked to implement a digital controller for the vending machine shown in Fig. 8.2.

STEP 1 Textual description.

The vending machine has the following features (textual description):

- The vending machine sells just one type of product, and each product costs $0.25;
- Only $0.25 coins are accepted;
- A valid $0.25 coin weights 5.670 g;
- Invalid coins are rejected automatically, and the control circuit does not need to do anything about that;
- There is a slot to insert $0.25 coins in the vending machine;
- There is a coin detection circuit that outputs '1' every time a $0.25 coin is inserted in the slot, or '0' otherwise;
- There is a digital scale that outputs '1' when a weight of 5.670 g (or more) is detected;
- There is a "delivery button" that outputs '1' every time it is pushed;
- There is a "coin return button" that outputs '1' every time it is pushed;
- There is a "coin return cup" to collect coins ejected by the "coin return button" (and also rejected coins);
- There is a door where the user can collect their product, after have pushed the "delivery button";
- When the vending machine runs out of products, it shuts itself, and the control circuit does not need to do anything about that.

The textual description together with the block diagram, comprises the Step 1 of the controller design methodology introduced in the previous section.

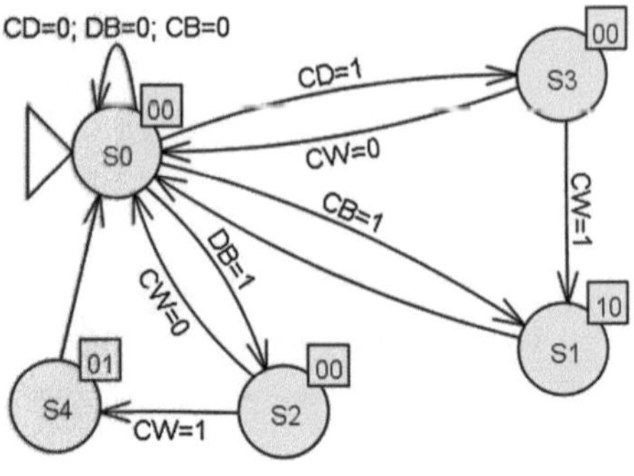

Fig. 8.3 FSM graph for the vending machine controller

STEP 2 FSM graphical representation.

In Step 2, a graphical representation for the FSM is built, and shown in Fig. 8.3. As shown in Fig. 8.2, the following list of abbreviations has been adopted for the design:

CD	Coin Detection Sensor (input)
DB	Delivery Button (input)
CW	Coin Weight Sensor (input)
CB	Coin Return Button (input)
CS	Current State (input)
NS	Next State (output)
CR_O	Coin Return Cup (output)
DD_O	Delivery Door (output)

In the graph diagram shown in Fig. 8.3, it is possible to observe that:

- S0 is the initial state.
- The FSM remains in S0 waiting for a coin be inserted in the slot (CD = 1), or the delivery button is pushed (DB = 1), or the coin return button is pushed (CB = 1).
- The output signals are represented by the two bits inside the small box on the right top of each state. The most significant bit is the "coin return" (CR_O) output, and the less significant bit is the "delivery door" (DD_O) output.
- The "coin return" (CR_O) and the "delivery door" (DD_O) outputs are equal to 0 in all states but in S1 and S4. In S1, CR_O = 1, which means that a \$0.25 coin is returned. In S4, DD_O = 1, in order to deliver a product to the customer.
- When a coin is inserted in the slot, the FSM goes to S3 state. If there was a coin already in the vending machine, the next state is S1, where the coin is returned to the customer (CR_O = 1). Otherwise, the \$0.25 coin is placed in the digital scale, and the FSM goes back to S0.
- When the delivery button is pushed, in the S2 state, the FSM goes to the S4 state to deliver a product only if the coin weight sensor indicates that there is a valid payment in the vending machine (CW = 1), otherwise, the FSM goes back to S0 state (CW = 0).

Fig. 8.4 State transition table for the vending machine

Inputs					Outputs		
CD	DB	CW	CB	CS	NS	CR_O	SD_O
0	0	X	0	S0	S0	0	0
0	0	X	1	S0	S1	0	0
0	1	X	0	S0	S2	0	0
1	0	X	0	S0	S3	0	0
X	X	X	X	S1	S0	1	0
X	X	0	X	S2	S0	0	0
X	X	1	X	S2	S4	0	0
X	X	0	X	S3	S0	0	0
X	X	1	X	S3	S1	0	0
X	X	X	X	S4	S0	0	1

STEP 3 State transition table.

In Step 3, a state transition table is built according to the FSM diagram. This sort of table is very similar to a truth table, and it has all the circuit's inputs and outputs. The main difference is the information regarding the "current state" and the "next state", even though they are also inputs and outputs for the circuit. Figure 8.4 shows the state transition table for the vending machine, where there are several "don't care" inputs. The table has been built considering that, for these inputs, the circuit has the same behavior when the respective input is high or low. For instance, on the first line of the table, it does not make any difference in the outputs if the CW input is '1' or '0'. This means that disregarding the CW input value, the outputs for this line will be NS = S0, CR_O = '0' and DD_O = '0'. The don't care inputs are very important in order to reduce the number of table lines.

STEP 4 VHDL description.

In this step, the FSM behavior is described in VHDL, as shown in Fig. 8.5. The two processes approach shown in Fig. 7.3 has been used in this implementation. The first process is used to model the state register, defining the next state, and the second process provides the controller outputs, and generates the next state, according to the input values.

IMPORTANT!!! Steps 5 through 8 are not used in VHDL based designs! They are implemented by synthesis tools, which are carefully designed in order to generate the best circuit according to the VHDL description provided in Step 4.

Steps 5 through 8 are included next just as an example of how this type of controller circuit can be manually implemented using logic gates and a register, with no synthesis tool hardware inference and automatic circuit optimizations.

As stated before, steps 1–4 are used by VHDL developers in the design of FSM based controllers, while steps 5–8 are used by synthesis tool developers. A synthesis tool developer is responsible for implementing all the algorithms used to obtain the most appropriate coding strategy for the FSM sates, the best optimization for the combinational circuit (Boolean equations), among other performance enhancement features.

```vhdl
entity VendingMachine is
    port(clock, reset  : in std_logic;
         CD, DB, CW, CB: in std_logic;
         CR_O, DD_O    : out std_logic);
end;

architecture behv of VendingMachine is
    type STATES is (S0, S1, S2, S3, S4);
    signal CS, NS : STATES;
begin
    process (clock, reset)
    begin
        if reset= '0' then
            CS <= S0;
        elsif rising_edge(clock) then
            CS <= NS;
        end if;
    end process;

    process(CS, CD, DB, CW, CB)
    begin
        case CS is
            when S0 => CR_O <= '0'; DD_O <= '0';     -- wait for an event
                       if CB='1' then NS <= S1; elsif DB='1' then NS <= S2;
                       elsif CD='1' then NS <= S3; else NS <= S0;
            when S1 => CR_O <= '1'; DD_O <= '0';     -- return a $0.25 coin
                       NS <= S0;
            when S2 => CR_O <= '0'; DD_O <= '0';     -- Is there a payment?
                       if CW='0' then NS <= S0; else NS <= S4;
            when S3 => CR_O <= '0'; DD_O <= '0';     -- Is it a valid coin?
                       if CW='0' then NS <= S0; else NS <= S1;
            when S4 => CR_O <= '0'; DD_O <= '1';     -- Product delivery
                       NS <= S0;
        end case;
    end process;
end behv;
```

Fig. 8.5 The vending machine controller described in VHDL

STEP 5 FSM controller architecture.

The block diagram of the proposed controller is shown in Fig. 8.6. All the input signals, including the "next state" are located on the left side of the diagram. The two output signals, and the "current state" information are located on the right side of the diagram. The register used to store the current state value is located in the bottom of the diagram. The combinational circuit is responsible for defining the outputs (DD_O and CR_O) and the next state, which will be stored as the FSM's current state (CS). The combinational circuit design will be provided in Step 8.

STEP 6 States coding.

During the synthesis process, all labels used to identify the FSM states are replaced by a binary code according to the available resources in the target hardware. For instance, FPGAs typically have a large amount of D flip-flops available and, in this case, the synthesis tools usually choose the "one-hot" coding strategy to identify the FSM states. In this strategy, for a group of bits that identify a state, only one bit is 1, and all the remaining are 0. The one-hot strategy uses

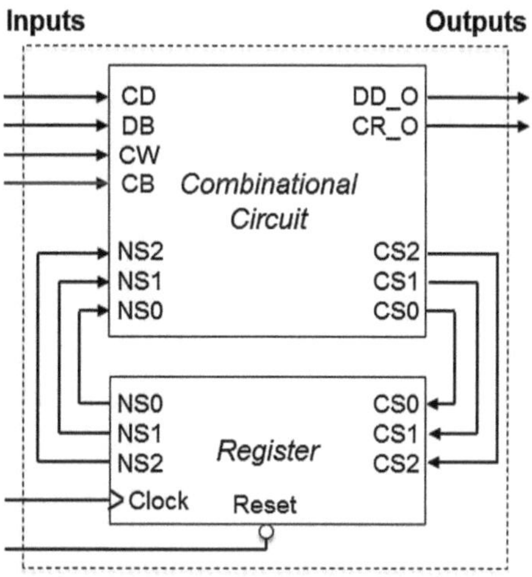

Fig. 8.6 Block diagram of the architecture for the vending machine controller

Fig. 8.7 Coding strategies for FSM states

One-hot	Binary	Gray code
00000001	000	000
00000010	001	001
00000100	010	011
00001000	011	010
00010000	100	110
00100000	101	111
01000000	110	101
10000000	111	100

more storage resources, but on the other hand it does not need any special decoder circuits. An FSM's state is identified in the group of bits that represent a state, according to the position of the bit '1'. For instance, considering an FSM with four states, S0, S1, S2, and S3, in the one-hot strategy, these states are coded, respectively, as 0001, 0010, 0100, and 1000. Other options for coding the states of an FSM include the binary and the gray code. Figure 8.7 shows the first eight codes for each one of these strategies.

STEP 7 Edit the state transition table, and replace the symbolic identifiers by their binary coding representation (Fig. 8.8).

STEP 8 Obtain the Boolean equations from the state transition table.

Using the sum of products method, the following Boolean equations are obtained, which can be further optimized using methods as Boolean equations simplification, Karnahgk maps (in cases where there are up to five variables), or the Quine–McCluskey algorithm (prime implicants method):

Fig. 8.8 State transition table, using binary codes instead of symbolic identifiers

Inputs					Outputs		
CD	DB	CW	CB	$CS_{2..0}$	$NS_{2..0}$	CR_O	SD_O
0	0	X	0	000	000	0	0
0	0	X	1	000	001	0	0
0	1	X	0	000	010	0	0
1	0	X	0	000	011	0	0
X	X	X	X	001	000	1	0
X	X	0	X	010	000	0	0
X	X	1	X	010	100	0	0
X	X	0	X	011	000	0	0
X	X	1	X	011	001	0	0
X	X	X	X	100	000	0	1

```
DD_O = CS2 CS1' CS0'
CR_O = CS2' CS1' CS0
NS0 = CS2' CS1' CS0' CD' DB' CB +
      CS2' CS1' CS0' CD DB' CB' + CS2' CS1 CS0 CW
NS1 = CS2' CS1' CS0' CD' DB CB' +
      CS2' CS1' CS0' CD DB' CB'
NS2 = CS2' CS1 CS0' CW
```

Figure 8.9 shows the block diagram of the vending machine connected to proposed FSM based controller.

In Fig. 8.9, the Boolean equations should be placed in the "Combinational Circuit" box, and the register box is implemented using three D flip-flops. The controller inputs CB, CW, DB and CD are provided by the vending machine hardware. The controller outputs DD_O and CR_O are used to command the vending machine product delivery, and coin return.

8.3 Laboratory Assignment

The laboratory objectives are:

- to understand the use of FSMs as a control module for the operations flow of a combinational circuit;
- to design a VHDL based FSM to be used as a controller for the calculator implemented in previous chapters.

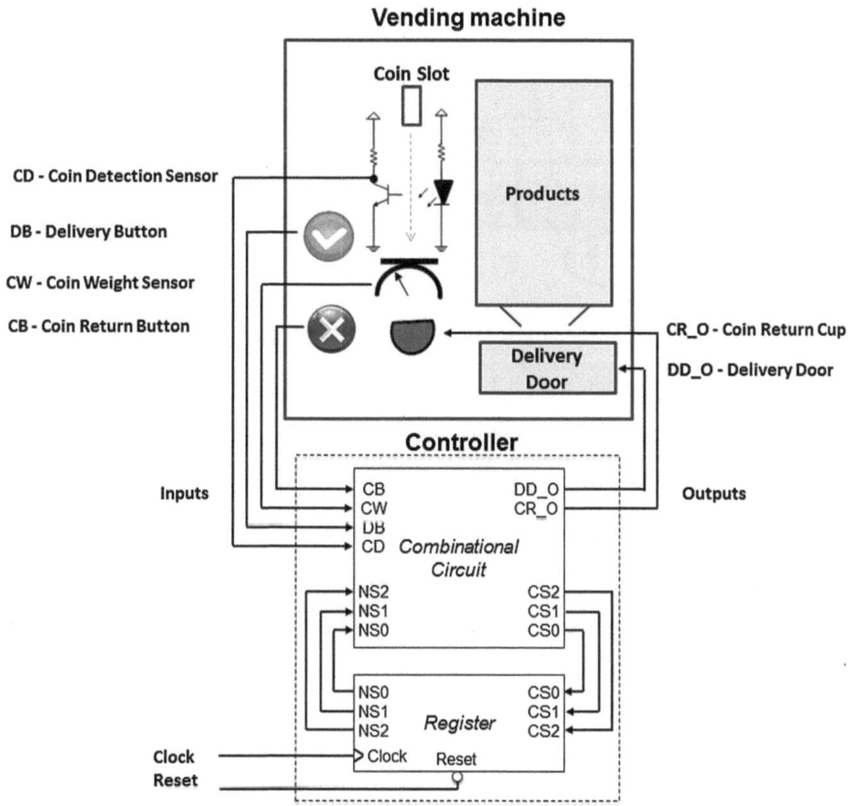

Fig. 8.9 Block diagram of the designed FSM controller, showing the input and output connections

8.3.1 *Problem Definition: Calculator with Reduced Data Input Signals*

The calculator designed in previous chapter has the following inputs:

- $SW_{7..0}$ switches for operand A; and
- $SW_{15..8}$ switches for operand B.

To decrease the amount of switches in the board layout, in this laboratory session the VHDL designer will have to use just one set of switches, $SW_{7..0}$, to input both operands A and B. In this new version, switches $SW_{17..16}$ are still used for the selection of the chosen operation. Figure 8.10 shows how the data input is performed in the DE2 board, considering the two calculator versions.

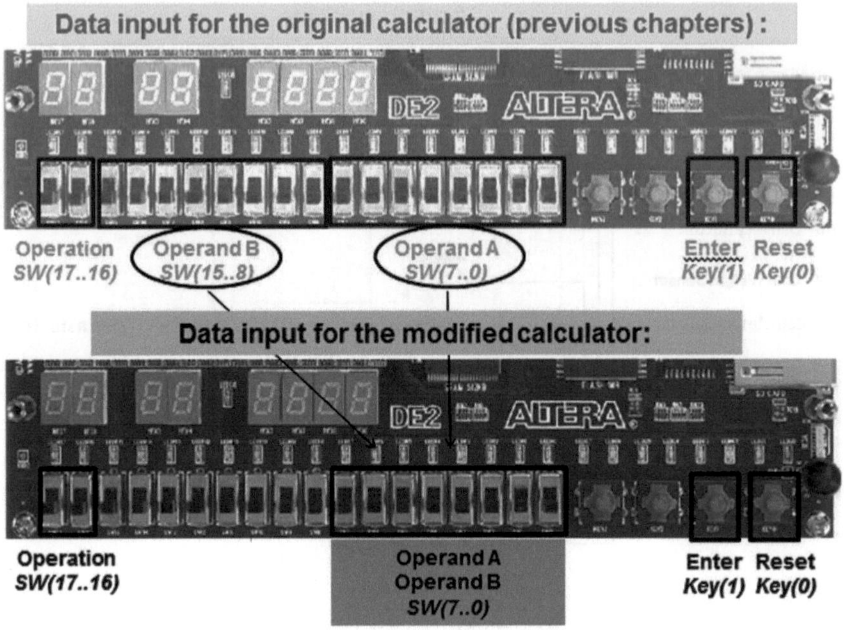

Fig. 8.10 Data input for the calculator: previous and new version

8.3.2 Laboratory Session

The tasks to be performed in this laboratory session are as follows:

- Add an input register to the calculator to hold the first operand;
- Design an FSM to control the sequence of operations in the new calculator;
- Add the FSM to the calculator design, simulate, and test the design in the FPGA board.

8.3.3 Adding a New Input Register to the Calculator Design

In Fig. 8.11, a new register is added to the calculator design. The register is located in the upper left hand corner, and it is used to store the first operand for operations that employ two operands.

Some considerations:

- There is no need to make changes in the calculator components used in previous chapters;
- In the "top_calc.vhd" file, the new register is added just by including a new *port map* VHDL statement in the architecture. Notice that there are already two

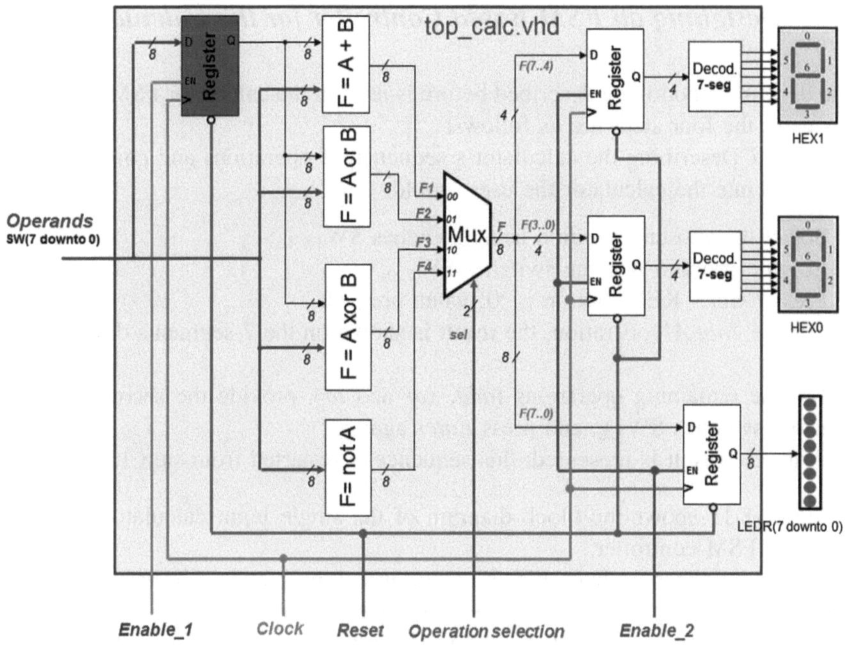

Fig. 8.11 Inserting an input register to hold the first operand

register components declared in the architecture, a four bits and an eight bits one;

- Some of the port map declarations should be changed in order to connect the new register to the remaining components, and also to the inputs;
- Another important remark is that all register enable signals, the multiplexor selector, among others, have been disconnected, as a controller FSM will be added to control the circuit operation sequence.

Next, the FSM controller is partially designed, according to the methodology described before. In Step 1, a textual description is provided for the calculator controller. In Step 2, a graphical representation for the FSM is built, according to the textual description discussed in Step 1. The proposed graphical representation is incomplete, as there are some states where all the possible outputs are not provided. Several graphical representations should be made, until one of the versions can be considered the "final" one. Step 3 is not used in this design, as the VHDL description can be extracted from the graphical view. In Step 4, a partial VHDL description for the FSM is provided. The reader should use the information in steps 3 and 4, in order to complete calculator's controller design.

8.3.4 Designing an FSM Based Controller for the Calculator

The design methodology described before is used in the calculator FSM design. In this case, the four steps are as follows:

STEP 1 Describing the calculator's sequence of operations and control flow. To operate the calculator the user should:

1. Select the chosen operation using switches $SW_{17..16}$;
2. Provide operand A using switches $SW_{7..0}$;
3. Press *Enter*—KEY_1 button is '0' when pressed;
4. For the *"not A"* operation, the result is shown on the 7-segments displays and LEDs;
5. For the remaining operations (*add*, *xor* and *or*), provide the second operand using switches $SW_{7..0}$, and press *Enter* again;
6. After the result is presented, this sequence is restarted from step 1.

Figure 8.12 shows the block diagram of the single input calculator, with the proposed FSM controller.

As can be observed from Fig. 8.12, the new top_calc.vhd file has 12 inferred components (using port map). These 12 components are inferred from nine components described in VHDL, and saved in the respective files as listed next:

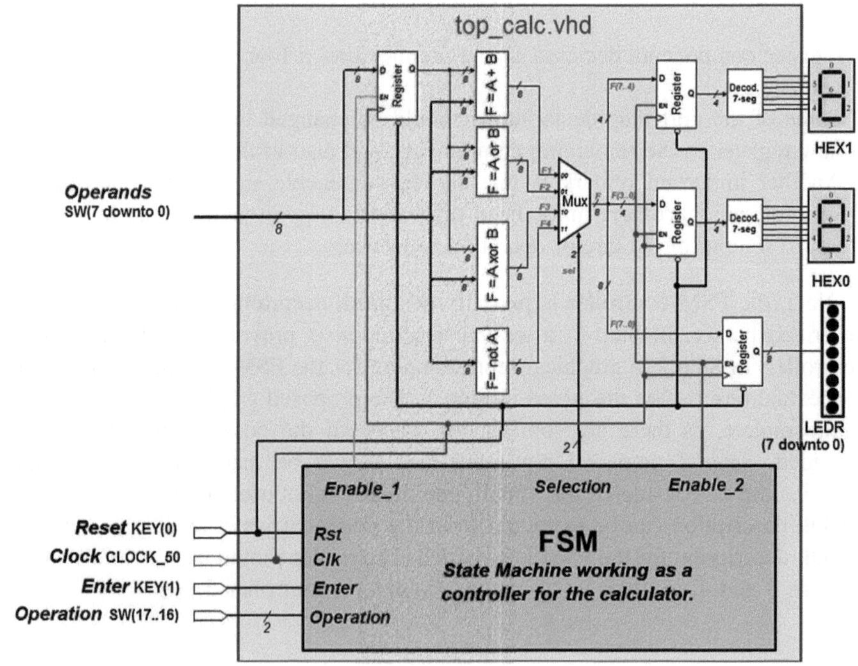

Fig. 8.12 Adding an FSM to control the calculator sequence of operations

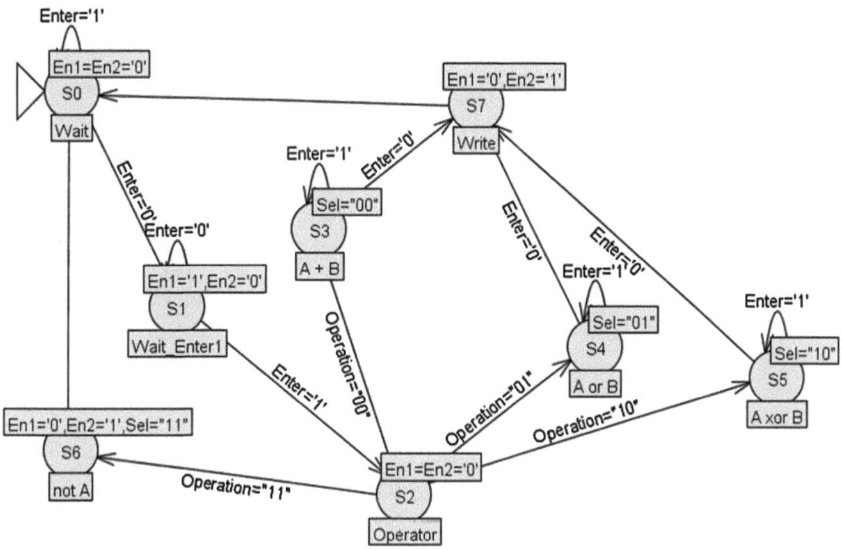

Fig. 8.13 FSM controller graphical view—incomplete version

Inputs				Outputs			
Reset	Enter	Operation	CS	NS	Enable_1	Selection	Enable_2
0	X	XX	S0	S0	0	00	0
1	1	XX	S0	S0	0	00	0
1	0	XX	S0	S1	0	00	0
1	0	XX	S1	S1	1	00	0
1	1	XX	S1	S2	1	00	0
1	X	00	S2	S3	0	00	0
1	X	01	S2	S4	0	00	0
1	X	10	S2	S5	0	00	0
1	X	11	S2	S6	0	00	0
1	0	XX	S3	S7	0	00	0
1	1	XX	S3	S3	0	00	0
1	0	XX	S4	S7	0	01	0
1	1	XX	S4	S4	0	01	0
1	0	XX	S5	S7	0	10	0
1	1	XX	S5	S5	0	10	0
1	X	XX	S6	S0	0	11	1
1	X	XX	S7	S0	0	00	1

Fig. 8.14 State transition table for the proposed FSM

```
library  ieee;
use  ieee.std_logic_1164.all;
entity FSMctrl is
  port ( Clk, Rst, Enter : in std_logic;
         Operation: in std_logic_vector(1 downto 0);
         Selection: out std_logic_vector(1 downto 0);
         Enable_1, Enable_2: out std_logic);
end FSMctrl;
architecture FSM_beh of FSMctrl is
  type states is (S0, S1, S2, S3, S4, S5, S6);
  signal CS, NS: states;
  signal clock, reset: std_logic;
begin
  clock <= Clk;    reset <= Rst;
process (clock, reset)
begin
  if reset = '0' then
    EA <= S0;
  elsif clock'event and clock = '1' then
    EA <= PE;
  end if;
end process;
process (EA, Enter)
begin
  case EA is
    when S0 =>
      if Enter = '1' then PE <= S0; else PE <= S1;
      end if;
      Enable_1 <= '0'; Enable_2 <= '0';
    when S1 =>      -- ... to be done
    when S2 =>      -- Operator
      Enable_1 <= '0'; Enable_2 <= '0';
      if Operation = "00" then
        PE <= S3;  -- add
      elsif Operation = "01" then
        PE <= S4;  -- OR
      elsif        -- ... to be done
   end case;
end process;
end FSM_beh;
```

Fig. 8.15 Partial VHDL description for the proposed FSM

- 2 × 8 bits registers—file *reg8bits.vhd*;
- 2 × 4 bits registers—file *reg4bits.vhd*;
- 1 × adder component—file *c1.vhd*;
- 1 × OR component—file *c2.vhd*;
- 1 × XOR component—file *c3.vhd*;

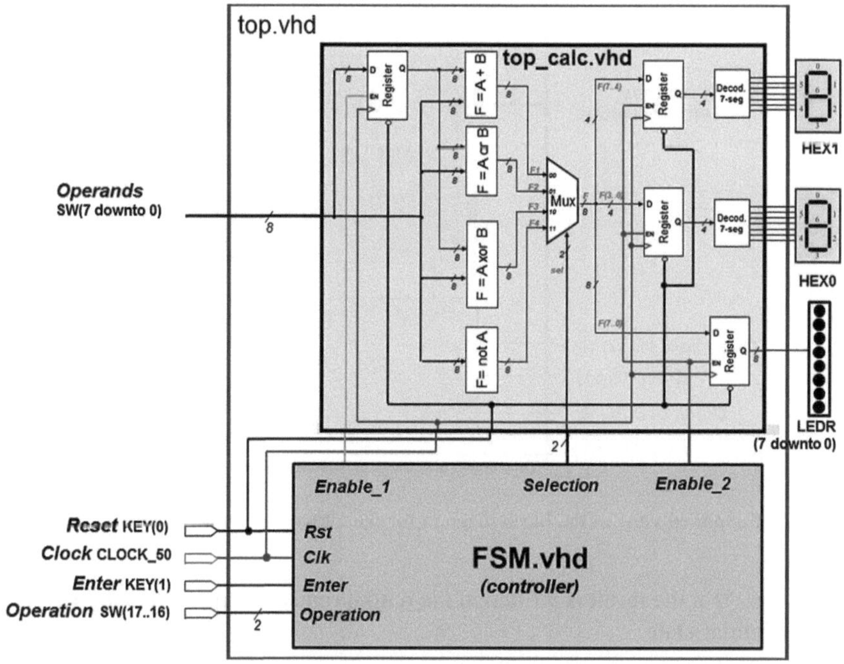

Fig. 8.16 Block diagram for the alternative implementation

- 1 × NOT component—file *c4.vhd*;
- 1 × 4 input multiplexer—file *mux4x1.vhd*;
- 2 × 7-segment decoders—file *decod7seg.vhd*;
- 1 × FSM controller—file *FSMctrl.vhd.*

STEP 2 FSM graphical representation.
The graphical view in Fig. 8.13 was built according to the description in step 1. The FSM has the following functionality:

- In the initial state "Wait", FSM outputs are disabled (*Enable_1* = '0', *Enable_2* = '0'), ensuring that there are no activities in the calculator, while the user does not provide the input data;
- When *Enter* is '0' (Key_1 is pressed), the FSM goes to the next state, and stays waiting until *Enter* goes back to '1' (push button is released);
- In the *Operator* state, according to the operation, there is a transition to the appropriate next state;
- For the "*Not A*" operation (only one operand), the result is stored in the output registers (*Enable_2* = *'1'*) and the FSM goes back to the initial state ("Wait");
- For the remaining operations, the second operator is read in an additional state—*Enable_2* = *'0'* in states *S3*, *S4*, and *S5*, followed by *Enable_2* = *'1'* in the *Write* state. In these operations, when Enter is pressed for the second time

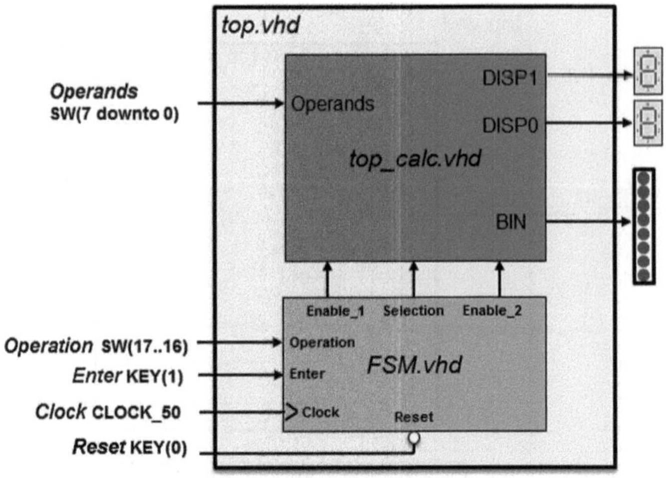

Fig. 8.17 Simplified view of the block diagram for the alternative implementation

(Key_1 = '0'), the result is written in the output registers, and the FSM goes back to the initial state.

STEP 3 FSM state transition table.

The state transition table can be obtained straight from the FSM graphical view, and it can be used to better visualize the inputs and outputs in each state (Fig. 8.14).

STEP 4 VHDL description for the modeled FSM behavior (Fig. 8.15).

Another option is to create a new top file, to connect together the FSM and the calculator components. Figure 8.16 shows this alternative implementation, where all calculator components are connected together in a top file called "top_-calc.vhd", and the FSM implementation is in the "FSM.vhd" file. Figure 8.17 shows a simplified view of this alternative implementation.

It is important to notice that in this alternative implementation it may be interesting to rename all signals listed in the "top_calc.vhd" entity (component interface), in order to avoid using DE2 labels SW, LEDR, HEX0, and HEX1. These labels are used in the entity of the new "top.vhd" file. In Fig. 8.17, the labels Operands, BIN, DISP0 and DISP1 were used instead.

Chapter 9
More on Processes and Registers

The behavioral description of sequential circuits in VHDL has been discussed in previous chapters. In Chaps. 7 and 8, the VHDL process statement has been used to describe FSMs. In Chap. 6, the process statement is introduced, and used to describe the behavior of latches, flip-flops and registers. This chapter discusses the use of processes in the implementation of both, combinational and sequential circuits. At the end of the chapter, the reader should be able:

- to understand the difference between implicit and explicit processes;
- to design combinational and sequential circuits using implicit and explicit processes;
- to implement shift registers in VHDL.

9.1 Implicit and Explicit Processes

Implicit and explicit processes have been used so far in VHDL examples all over this book. In implicit processes, the reserved word "process" is not used. Examples of implicit process include:

- Simple signal assignments, where any event in the source side of the assignment will trigger the process as, for instance, *SW*, *A*, *B*, and *C* next:
  - LEDR <= SW;
  - F <= (A and B) or C;
- Signal assignments using selection (*with/select*) and conditional (*when*) statements;
- Instantiated components (*component/portmap*).

Explicit processes are easily identified in a VHDL code by the usage of the reserved word "process". The two VHDL codes in Figs. 9.1, 9.2 when synthesized, will generate the same circuit. The code in the left hand side has three "implicit" processes performing their signal assignment operations in parallel. The

E. A. Bezerra and D. V. Lettnin, *Synthesizable VHDL Design for FPGAs*,
DOI: 10.1007/978-3-319-02547-6_9,

code in the right hand side performs the same functionality as the code on the left, but the signal assignment operations are placed inside three "explicit" processes. As shown in Fig. 9.3, both implicit and explicit processes implementations listed in Fig. 9.1 generate their Y output in parallel to the E and F assignments.

The VHDL code in Fig. 9.2 has the same three assignments as the codes in Fig. 9.1, but the circuit functionality is slightly different. As shown in Fig. 9.4, as the assignments appear in the same process, the Y output is not updated simultaneously to the E and F assignments. The Y output is updated next time the process is triggered and, consequently, it provides previous values of E and F. In Fig. 9.4, when the simulation starts, A = B = C = D = '0', and the Y output is undefined. Next, when A = B = '1', the Y output is '0', representing the previous circuit output when all input signals were '0'.

In Fig. 9.5, a multiplexer with priority is implemented using implicit (left hand side) and explicit (right hand side) processes. In VHDL, the IF statement can be used only in explicit processes. The synthesized multiplexer has a priority, as when *sel(0)* is '0', the *F* output is assigned the value in the *A* input, no matters the *sel(1)* value. The *C* input has the lowest assignment priority.

In Fig. 9.6, the selection mechanism is used in order to implement a no priority multiplexer. VHDL implicit processes use the *with/select/when* construction to synthesize no priority multiplexers. In explicit process, this mechanism is synthesized by using the *case/when* construction. In Fig. 9.6, as well as in Fig. 9.5, both the implicit and the explicit process implementation, generate the same piece of hardware.

```vhdl
library ieee;
use ieee.std_logic_1164.all;
entity Implicit is
port (
        A, B, C, D: in std_logic;
        Y          : out std_logic
);
end Implicit;
architecture Arch of Implicit is
   signal E, F: std_logic;
begin
        E <= A and B;
        F <= C and D;
        Y <= E or F;
end Arch;
```

```vhdl
library ieee;
use ieee.std_logic_1164.all;
entity Explicit_v1 is
port (
        A, B, C, D : in std_logic;
        Y          : out std_logic
);
end Explicit_v1;
architecture Arch of Explicit_v1 is
    signal E, F: std_logic;
begin
    process (A, B)
    begin
        E <= A and B;
    end process;
    process (C, D)
    begin
        F <= C and D;
    end process;
    process (E, F)
    begin
        Y <= E or F;
    end process;
end Arch;
```

Fig. 9.1 Implicit and explicit processes in VHDL

```
library ieee;
use ieee.std_logic_1164.all;
entity Explicit_v2 is
port (
        A, B, C, D : in std_logic;
        Y     : out std_logic
);
end Explicit_v2;
architecture Arch of Explicit_v2 is
   signal E, F: std_logic;
begin
    process (A, B, C, D)
    begin
        E <= A and B;
        F <= C and D;
        Y <= E or F;
    end process;
end Arch;
```

Fig. 9.2 Sequential signal assignments in an explicit process

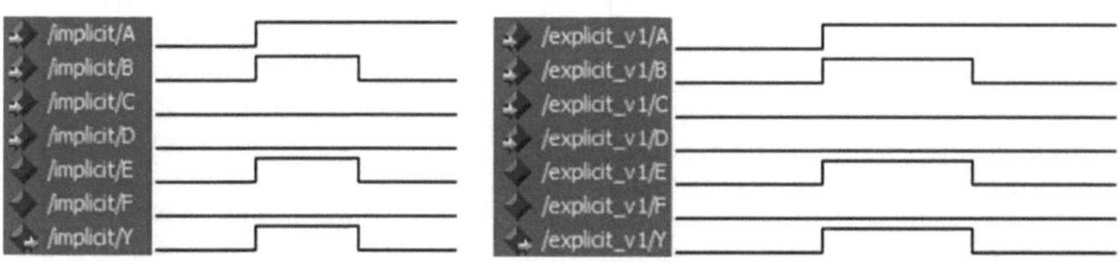

Fig. 9.3 Simulation waveforms for the VHDL codes listed in Fig. 9.1, showing their equivalence

Fig. 9.4 Simulation waveform for the VHDL code listed in Fig. 9.2

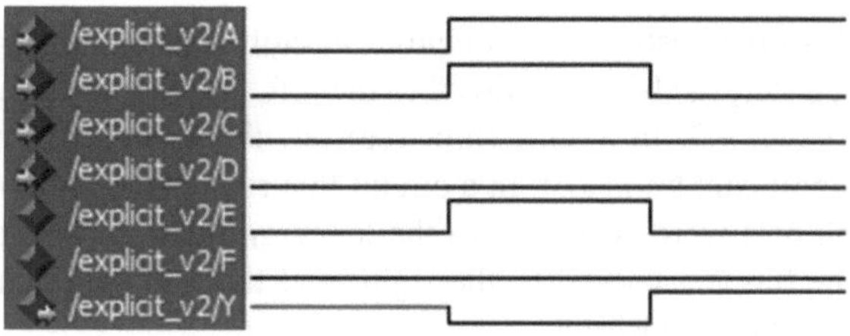

```
architecture implicit of y is
begin

   F <= A when sel(0) = '0' else
        B when sel(1) = '0' else
        C;

end implicit;
```

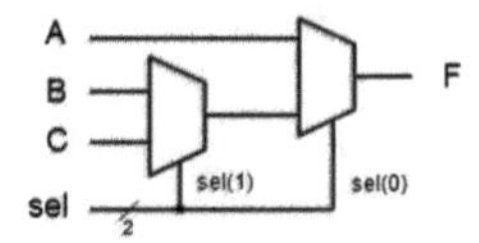


```
architecture explicit of y is
begin

  process (A, B, C, sel)
  begin
    if sel(0) = '0' then
      F <= A;
    elsif sel(1) = '0' then
      F <= B;
    else
      F <= C;
    end if;
  end process;

end explicit;
```

Fig. 9.5 Conditional assignment with priority using the "comparison" mechanism, implemented with implicit and explicit processes

```
architecture implicit of y is
begin
   with sel select
   F <= A when "00",
        B when "01",
        C when "10",
        D when others;
end implicit;
```

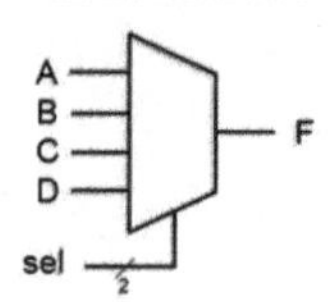


```
architecture explicit of y is
begin
  process (sel, A, B, C, D)
  begin
    case sel is
      when "00" =>
                 F <= A;
      when "01" =>
                 F <= B;
      when "10" =>
                 F <= C;
      when others =>
                 F <= D;
    end case;
  end process;
end explicit;
```

Fig. 9.6 Conditional assignment with no priority using the "selection" mechanism, implemented with implicit and explicit processes

9.2 Designing a Shift Register

A shift register is a sequential circuit used to shift bits in a word. In a shift left register, the bits are moved to the left, and the most significant bit is lost, as shown in Fig. 9.7. In a shift right register, the bits are moved to the right and the less significant bit is lost. In a circuit that rotates the bits to the left, the most significant bit of a word is moved to the less significant bit position. In a rotate right circuit, the bits are moved to the right, and the less significant bit is moved to the most significant position.

The VHDL behavioral description of registers discussed in previous chapters can be easily modified in order to implement shift and rotate registers. It is important to notice that registers are implemented in VHDL using explicit processes.

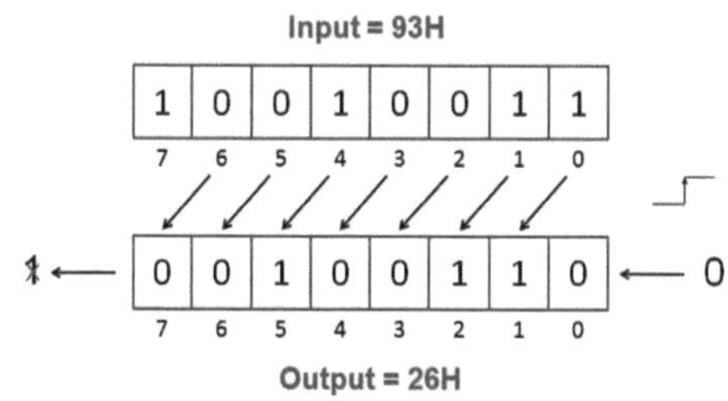

Fig. 9.7 Example of shift *left operation*

However, instead of reusing and adapting VHDL code from previous chapters, in this case study a template provided by Quartus II is used for the implementation of a shift register. To insert templates in a design, the target VHDL file must be opened in Quartus II text editor. Next, select the menu *Edit → Insert Template…* and *VHDL → Shift Registers → Basic Shift Registers*, as shown in Fig. 9.8.

In the proposed case study, an N bits input word is shifted left by one bit at each rising clock edge, and send to an output signal. This functionality is implemented by a shift register with enable and asynchronous reset. The original shift register template provided by Quartus II is changed, in order to implement the proposed functionality. The list of changes are as follows:

- A reset signal has been added.
- The "array" type was removed.
- The new input data is N bits long (instead of a single bit).
- After shifting, the less significant bit receives '0', and not the input bit.

The modified version of the shift register template is listed in Fig. 9.9, and it has the following features:

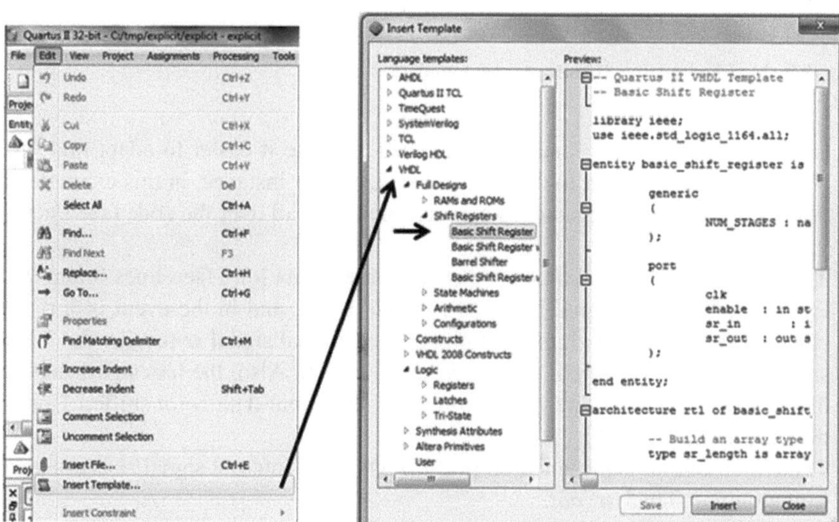

Fig. 9.8 Using quartus II to insert VHDL templates in a design

```
1  library ieee;
2  use ieee.std_logic_1164.all;
3  entity basic_shift_register is
4      generic(N : natural := 8);
5      port (clk     : in std_logic;
6            enable : in std_logic;
7            reset  : in std_logic;
8            sr_in  : in std_logic_vector((N - 1) downto 0);
9            sr_out : out std_logic_vector((N - 1) downto 0)
10     );
11 end entity;
12 architecture rtl of basic_shift_register is
13     signal sr: std_logic_vector ((N-1) downto 0);
14 begin
15     process (clk, reset)
16     begin
17         if (reset = '0') then
18             sr <= (others => '0');
19         elsif (rising_edge(clk)) then
20             if (enable = '1') then
21                 -- Shift left 1 bit each clock
22                 sr((N-1) downto 1) <= sr_in((N-2) downto 0);
23                 sr(0) <= '0';
24             end if;
25         end if;
26     end process;
27     sr_out <= sr;
28 end rtl;
```

Fig. 9.9 Modified quartus II shift register template

- On line 4, the "generic" statement is used to make it easier to adapt this shift register implementation to different word sizes. For instance, in this example an 8 bits shift register is created, as N is replaced by 8 all over the code (see lines 8, 9, 13, and 22).
- The input (*sr_in*) and output (*sr_out*) data are 8 bits long (see lines 8 and 9).
- Each time the *reset* signal is not '0', *enable* is '1', and in the event of a rising clock edge, the 7 most significant bits of the internal signal *sr* receive the 7 less significant bits of the input data *sr_in* (see line 22). Also, the less significant bit of *sr* receives '0' (see line 23). The result is the input data *sr_in* shifted 1 bit to the left, as shown in Fig. 9.7.
- The circuit output is provided on line 27, when the internal signal *sr* is assigned to the output signal *sr_out*.

The synthesized shift register component is shown in Fig. 9.10.

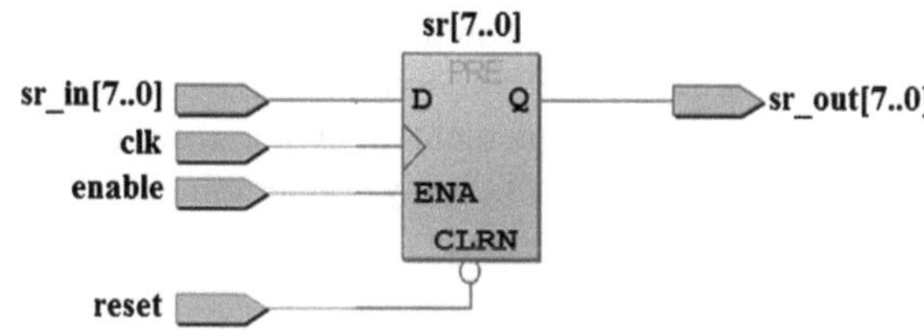

Fig. 9.10 Shift register synthesized by quartus II from VHDL listed in Fig. 9.9

9.3 Laboratory Assignment

The laboratory objectives are:

- to fix the concepts of shift register design in VHDL;
- to insert multiplication and division by 2 operations to the calculator case study.

Multiplication and division by 2 are simple operations to be performed in the base 2 numeral system. To multiply a binary number by 2, just shift the number 1 bit to the left and fill the less significant bit with '0'. The division by 2 is performed in a similar way, but shifting the number 1 bit to the right, and inserting a '0' in the most significant bit position. The operation represented in Fig. 9.7 is a

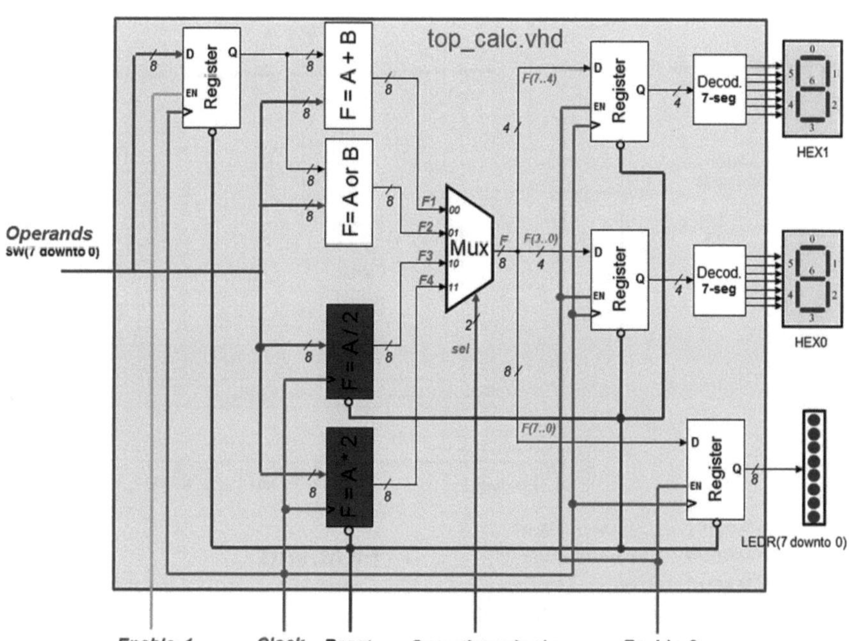

Fig. 9.11 Inserting the divide and multiply components to the calculator

multiply by 2, which resulted in an overflow as the target register is only 8 bits long, and the result is 9 bits long.

The component implemented in the shift register case study, and listed in Fig. 9.9, performs a multiply by 2 operation.

9.3.1 Laboratory Session

The tasks to be performed in this laboratory session are as follows:

- Using a shift register, design a component to perform division by 2 operations (*shift right*).
- Using a shift register, design a circuit to perform multiplication by 2 operations (*shift left*).
- Perform the simulation of both components, before inserting them in the calculator.
- Make the appropriate changes in the calculator implemented in previous chapters, in order to include the multiplication and division components:

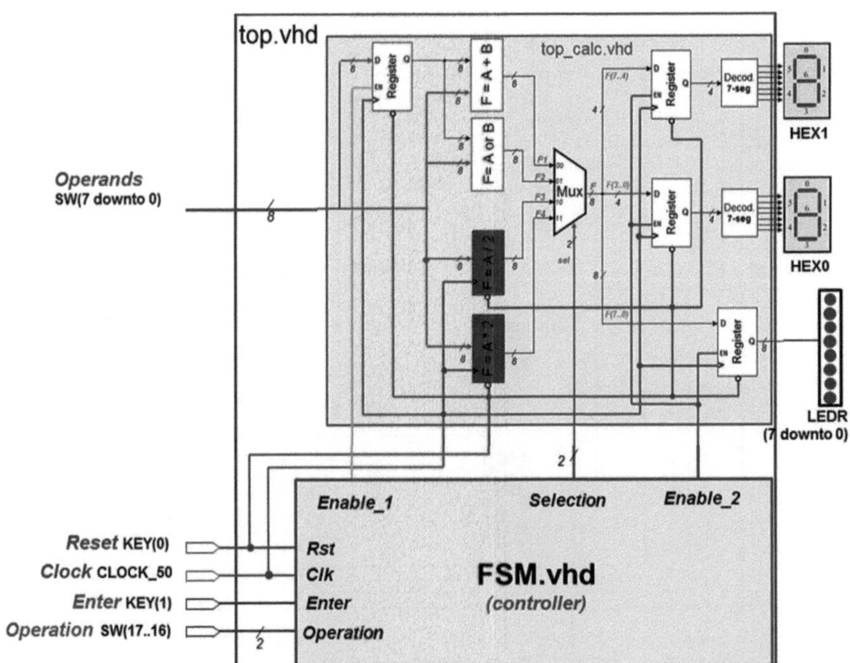

Fig. 9.12 The complete calculator design with the new components and the controller FSM

- The new multiplication and division components, as well as the not operation component, have also just one input. The FSM must be changed in order to deal with this situation.
- Replace the XOR component by the new divide by 2 component.
- Replace the NOT component by the new multiply by 2 component.

Figure 9.11 shows the block diagram of the new version for the calculator. The new components need a clock and a reset signals, but there is no need for an enable input. For instance, the component described in Fig. 9.9 shifts its input value 1 bit left. Thus, when the user provides a value to be multiplied by 2, and selects the multiplication operation, the mux output will be the input value shifted 1 bit left. At each rising clock edge, the multiplication component shifts its input 1 bit left and, while the input value remains the same, the output value will be also remain unchanged (but shifted 1 bit left in relation to the input value).

So, the shift register listed in Fig. 9.9 can be used to implement the multiply by 2 component, but without lines 20 and 24 as the enable input is not used in Fig. 9.11.

The division by 2 component can also be easily implemented from Fig. 9.9, but now the input value must be shifted to the right, and a '0' should be inserted in the most significant position bit.

Figure 9.12 shows the full block diagram of the calculator, including the controller FSM. As there is no need for enable signals in the new shift registers, the FSM can be used with minor changes. Basically, the FSM should be changed in order to manage two 1-input components, instead of just one 1-input component (NOT in the previous calculator version).

Chapter 10
Arithmetic Circuits

In previous chapters, the '+' VHDL operator is used in the adder design. In this chapter the adder component is conceived using logic gates, at the structural level. At the end of the chapter, the reader should be able:

- to understand the design process of adders using structural VHDL;
- to simulate the calculator case study using the new adder;
- to prototype the calculator case study in an FPGA board.

10.1 Half-Adder, Full-Adder, Ripple-Carry Adder

The adder is a fundamental circuit in digital systems. In an *n* bits adder, the main problem is the delay needed for the carry propagation. The carry propagation is a concern also in a multiplier circuit. The concepts behind the design of adders and multipliers are employed in the solution of several similar digital system problems. In Fig. 10.1, a four bits value ($A_{3..0}$) is added to another four bits value ($B_{3..0}$) resulting in a four bits summation ($S_{3..0}$). All bit position sum operations result also in a carry ($C_{3..0}$).

Two different components can be used in the design of this four bits adder. The half-adder component, shown in Fig. 10.2, has two inputs (A_0 and B_0) and two outputs (S_0 and C_1). The full-adder component, shown in Fig. 10.3, has three inputs (C_i, A_i and B_i) and two outputs (S_i and C_{i+1}). The half-adder component is used to perform the least significant bits (LSB) addition, and the full-adder component for the remaining bits.

In an FPGA based implementation, as there are no actual logic gates available in the device, an *n* bits adder can be implemented using only full-adders with no major resource usage penalties. The adder functionality will be implemented using the FPGA's look-up tables (LUTs), and the only concern is the amount of input signals, as there is no difference between a two input function performing a single AND operation (C_1 in the half-adder), or three ANDs and two ORs (C_{i+1} in the full-adder). Both functions will employ exactly the same amount of FPGA resources (LUTs).

E. A. Bezerra and D. V. Lettnin, *Synthesizable VHDL Design for FPGAs*,

DOI: 10.1007/978-3-319-02547-6_10,

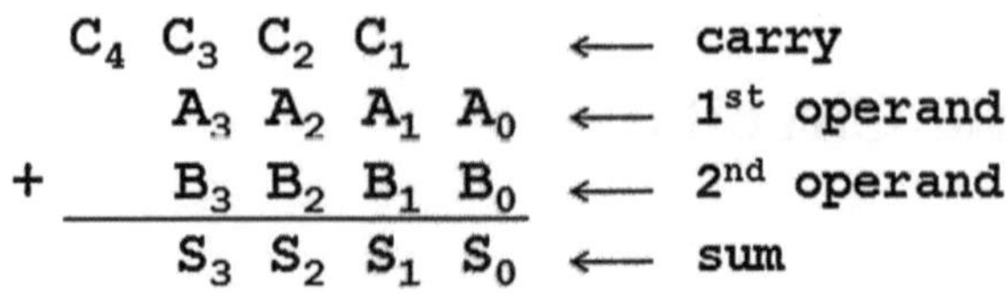

Fig. 10.1 A four bits sum operation

A_0	B_0	S_0	C_1
0	0	0	0
0	1	1	0
1	0	1	0
1	1	0	1

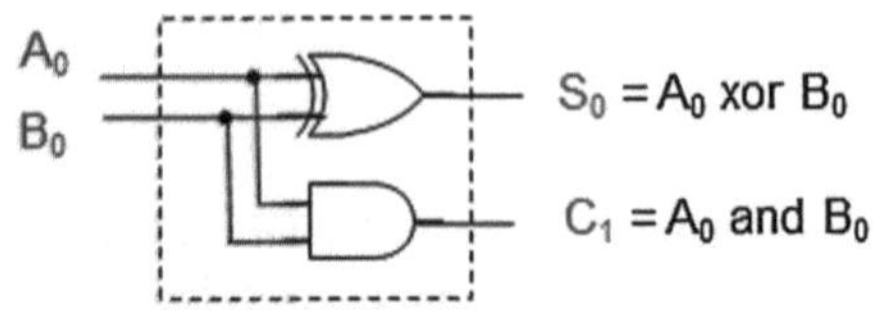

Fig. 10.2 Half-adder (HA) component. Truth table, schematic, and logic equations

C_i	A_i	B_i	S_i	C_{i+1}
0	0	0	0	0
0	0	1	1	0
0	1	0	1	0
0	1	1	0	1
1	0	0	1	0
1	0	1	0	1
1	1	0	0	1
1	1	1	1	1

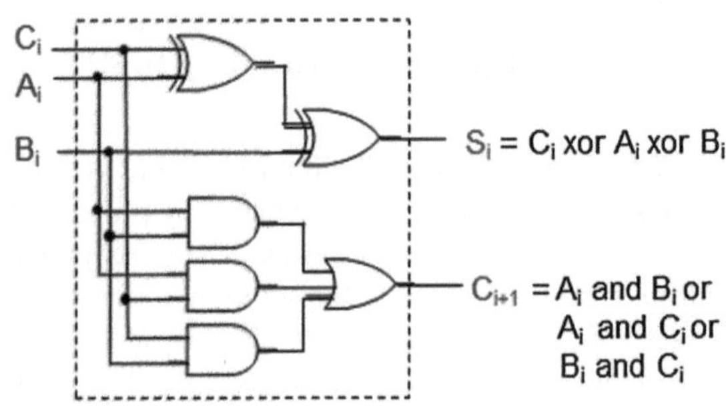

Fig. 10.3 Full-adder (FA) component. Truth table, schematic, and logic equations

There are several topologies available for implementing parallel adders. Figure 10.4 shows a four bits ripple-carry adder (RCA), which can be used to implement the operation described in Fig. 10.1. In this circuit, full-adders (FA) are used in all bit positions, including the LSB, where a half-adder could have been used. Having an FA in the LSB position is interesting also in order to have a component that can be used as a building block for larger adders or subtractors.

As shown in Fig. 10.1, each bit of the result (S_i) can only be obtained after the previous adder stage has finished its carry (C_{i-1}) calculation. Consequently, an RCA component provides a result ($S_{n..0}$) only after all carry bits ($C_{n..1}$) are stable.

When working with signed values, an RCA component can be adapted in order to implement an add/subtract circuit. In Fig. 10.5, four multiplexers and four

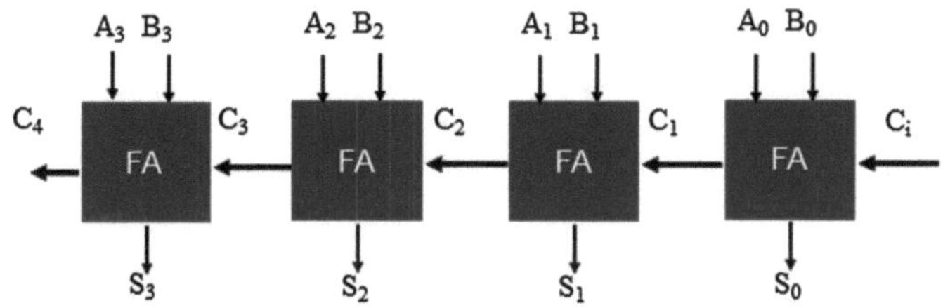

Fig. 10.4 A four bits ripple-carry adder (RCA)

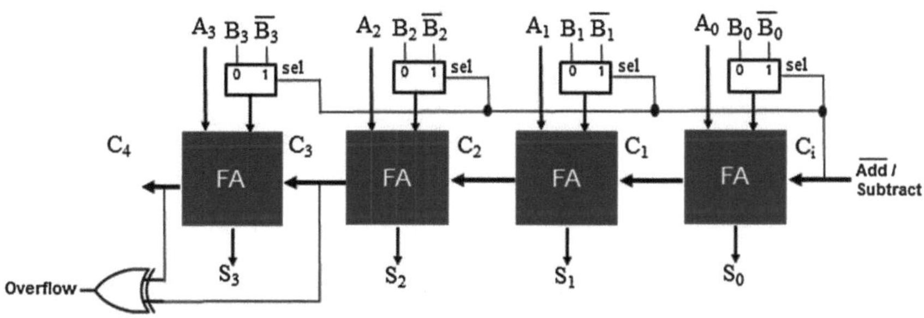

Fig. 10.5 Add/subtract circuit based on an RCA component

inverters are used in the B inputs of the RCA. When C_i is '1' the circuit works as a subtractor. In this case, the circuit performs an add operation between A and the two's complement of B:

$$
\begin{aligned}
S &= A + B' + 1 \\
S &= A + (-B) \\
S &= A - B
\end{aligned}
$$

The RCA delay to provide a valid (stable) result depends on the input data. In order to have a better understanding of the carry propagation delay problem, a *time unit* is used as a performance figure. A *time unit* is the period required by a component to evaluate one level of logic gates. For instance, the half-adder component shown in Fig. 10.2 has just one level of logic gates, and all operations in this component are evaluated in one time unit. The full-adder shown in Fig. 10.3 has two levels of logic gates, and the operations are evaluated in two time units.

Considering A = 0101 and B = 1010 (with no carries), the circuit in Fig. 10.4 provides a valid result in two time units (2t). In the worst case scenario, 8t is the maximum delay acceptable for a four bits RCA. Larger adders would result in unacceptable delay figures. An 8 bits RCA should have a maximum delay of 16t, a 32 bits RCA the delay should be 64t, and so on.

Proposed solutions to decrease the delay usually perform the carry calculation in parallel, instead of the sequential approach employed in the RCA topology.

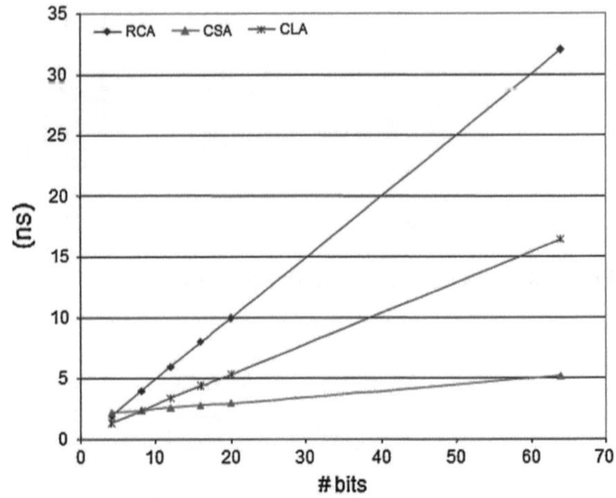

Fig. 10.6 Carry propagation delay for the three adders

The *Carry Lookahead Adder* (CLA) uses the functions *propagate* and *generate* in order to perform the parallel calculation of the carry. CLAs are faster than RCAs for four bits operands max. For longer length operands, the CLA carry generation logic complexity make it slower than the RCA. The *Carry Select Adder* (CSA) uses two FAs for each bit operation and, as a consequence, the carry generation is faster than the RCA and CLA, when considering larger operands.

Figure 10.6 shows a perform analysis between the adders. CSA and CLA present similar performance figures when for operands up to 10 bits long. For larger operands, the CSA circuit is the fastest adder. For instance, for 64 bits long operands, RCA takes more than 30 ns to provide a result, CLA takes more than 15 ns, and CSA provides the result in just 5 ns.

The drawback is that faster adders require more hardware resources to be implemented. In Fig. 10.7 it is possible to observe that the synthesized CLA takes almost twice the area than the RCA. On the other hand, CLA is about 10 % faster than RCA.

RCA is the slowest adder, but it has the less complex implementation. Two VHDL implementations of the RCA are discussed next, and compared to the

Fig. 10.7 Synthesis results summary for an RCA and a CLA implementation

Ripple-Carry Adder (RCA)

# bits	Area (logic gates)	Max delay (ns)
4	13	15.647
8	33	19.058

Carry Lookahead Adder (CLA)

# bits	Area (logic gates)	Max delay (ns)
4	23	14.268
8	62	17.484

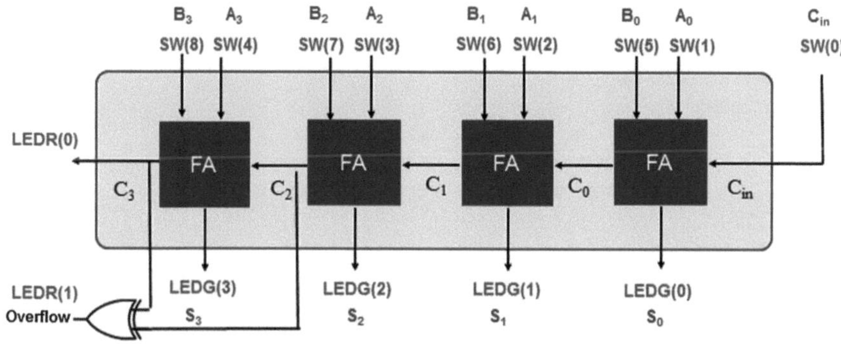

Fig. 10.8 Block diagram of the four bits RCA circuit implemented in VHDL

synthesis tool built-in adder alternative. Figure 10.8 shows the block diagram of the four bits RCA case study. This circuit considers two's complement signed numbers. The red LEDs 0 and 1 are used to indicate carry out and overflow, respectively. The green LEDs 0–3 provide sum results. The inputs are provided by the SW switches, including A and B operands, and also the carry in. Internal signals are used to connect all carries between FAs.

In the first implementation, the A and B internal signals are used just to make the code more readable. On lines 13 and 14 in Fig. 10.9, the FPGA board switches SW4..1 and SW8..5 are assigned to the internal signals A3..0 and B3..0, respectively.

```
1   LIBRARY ieee;
2   USE ieee.std_logic_1164.all;
3   ENTITY RCA IS
4      PORT (
5         SW : IN STD_LOGIC_VECTOR(17 DOWNTO 0);
6         LEDR : OUT STD_LOGIC_VECTOR(17 DOWNTO 0);
7         LEDG : OUT STD_LOGIC_VECTOR(7 DOWNTO 0)
8      );
9   END RCA;
10  ARCHITECTURE stru OF RCA IS
11     signal carry, A, B: std_logic_vector(3 downto 0);
12  BEGIN
13     A <= SW(4 downto 1);
14     B <= SW(8 downto 5);
15     LEDG(0)  <= ((A(0) xor B(0)) xor SW(0));
16     carry(0) <=  (A(0) and B(0)) or (A(0) and SW(0)) or (B(0) and SW(0));
17     LEDG(1)  <= ((A(1) xor B(1)) xor carry(0));
18     carry(1) <=  (A(1) and B(1)) or (A(1) and carry (0)) or (B(1) and carry(0));
19     LEDG(2)  <= ((A(2) xor B(2)) xor carry(1));
20     carry(2) <=  (A(2) and B(2)) or (A(2) and carry(1)) or (B(2) and carry(1));
21     LEDG(3)  <= ((A(3) xor B(3)) xor carry(2));
22     carry(3) <=  (A(3) and B(3)) or (A(3) and carry(2)) or (B(3) and carry(2));
23     LEDR(0)  <= carry(3);                -- carry out
24     LEDR(1)  <= carry(3) xor carry(2);   -- overflow
25  END stru;
```

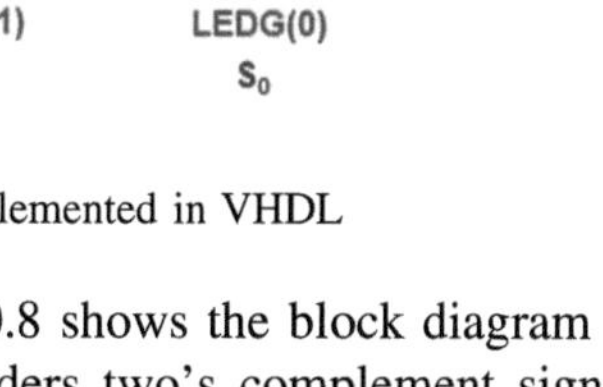

Fig. 10.9 Solution I—RCA implementation using logic equations

```
1  LIBRARY ieee;
2  USE ieee.std_logic_1164.all;
3  ENTITY RCA IS
4      PORT (
5          SW : IN STD_LOGIC_VECTOR(17 DOWNTO 0);
6          LEDR : OUT STD_LOGIC_VECTOR(17 DOWNTO 0);
7          LEDG : OUT STD_LOGIC_VECTOR(7 DOWNTO 0)
8      );
9  END RCA;
10 ARCHITECTURE stru OF RCA IS
11     signal carry, A, B: std_logic_vector(3 downto 0);
12     component FA
13         port (a, b, c: in std_logic;
14               sum, carry: out std_logic);
15     end component;
16 BEGIN
17     A <= SW(4 downto 1);
18     B <= SW(8 downto 5);
19     FA0: FA port map (A(0), B(0), SW(0),    LEDG(0), carry(0));
20     FA1: FA port map (A(1), B(1), carry(0), LEDG(1), carry(1));
21     FA2: FA port map (A(2), B(2), carry(1), LEDG(2), carry(2));
22     FA3: FA port map (A(3), B(3), carry(2), LEDG(3), carry(3));
23     LEDR(0)  <= carry(3);                -- carry out
24     LEDR(1)  <= carry(3) xor carry(2);   -- overflow
25 END stru;
26 -- FA component
27 LIBRARY ieee;
28 USE ieee.std_logic_1164.all;
29 ENTITY FA IS
30     PORT (a, b, c: in std_logic;
31           sum, carry: out std_logic);
32 END FA;
33 ARCHITECTURE FA_stru OF FCA IS
34 BEGIN
35     sum   <= (a xor b) xor c;
36     carry <= b when ((a xor b) = '0') else c;
37 END FA_stru;
```

Fig. 10.10 Solution II—RCA implementation using component/port map

In the second implementation, the FA circuit is written as a VHDL component, making the RCA design more organized and easier to follow. The FA entity/ architecture start on line 27 in Fig. 10.10. The four FA instances are created and interconnected on lines 19 through 22.

The VHDL coding style adopted in Solution II results in a more organized circuit also when using the Netlist Viewer option available in Quartus II tool. In Fig. 10.11, the four FA components used for the RCA implementation are explicitly shown in the Solution II netlist view. In order to see the FA logic gates

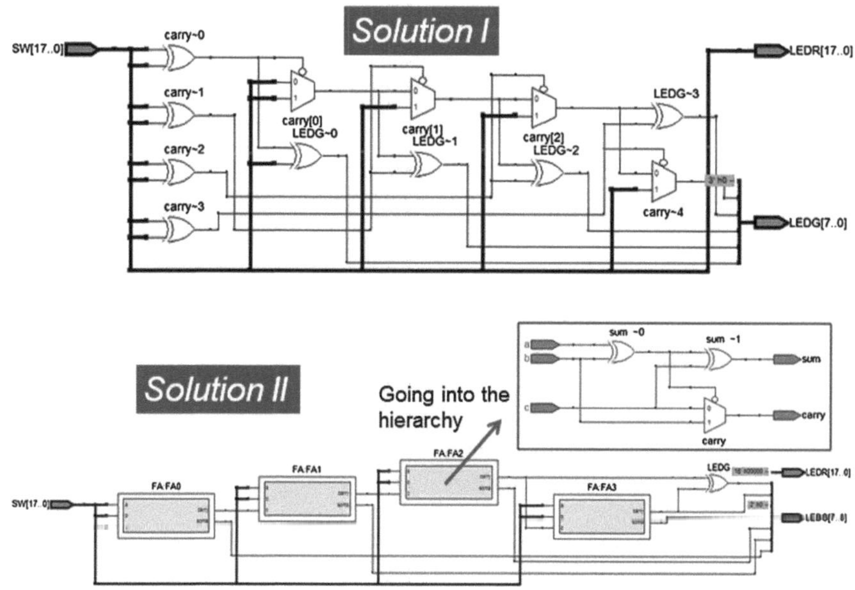

Fig. 10.11 Quartus II netlist view for the RCA circuit

implementation, the designer has to double-click in one of the FA components. On the other hand, in Solution I circuit view, the adder functionality understanding is not as straightforward as in Solution II.

In previous chapters, the adder used in the calculator design was implemented using the VHDL '+' operator. This is a much easier way to implement an adder, but the drawback is the loss of circuit's observability and controllability. In Fig. 10.12, the adder is written in a higher abstraction level, and the designer does not know which topology the synthesis tool will select for the implementation. For instance, there is no way to force the synthesis tool to implement the circuit using an RCA topology. In this example, with the lower observability degree, it is not possible to extract the overflow and carry out data.

In Fig. 10.13 there is a comparison between the synthesis results for the RCA implementation listed in Fig. 10.10 (left hand side), and the adder implementation using the '+' operator listed in Fig. 10.12 (right hand side). As expected, the '+' operator alternative, resulted in less FPGA resources usage (5 logic elements), as the synthesis tool will select the best algorithm available for the adder implementation.

However, in applications where a higher degree of controllability/observability is required, the RCA implementations described before as Solutions I and II are a better alternative for an adder design than the '+' VHDL operator.

The *carry out* and *overflow* signals observed in Solutions I and II are also known as *flags*. A flag is a strategy widely used in digital systems to indicate the status or a condition happened in the last arithmetic or logic operation. In the

```
LIBRARY ieee;
USE ieee.std_logic_1164.all;
use IEEE.std_logic_unsigned.all;
ENTITY RCA IS
    PORT (
        SW : IN STD_LOGIC_VECTOR(17 DOWNTO 0);
        LEDR : OUT STD_LOGIC_VECTOR(17 DOWNTO 0);
        LEDG : OUT STD_LOGIC_VECTOR(7 DOWNTO 0)
    );
END RCA;
ARCHITECTURE RCA_beh OF RCA IS
    signal carry, A, B: std_logic_vector(3 downto 0);
BEGIN
    A <= SW(4 downto 1);
    B <= SW(8 downto 5);
    LEDG(3 downto 0) <= A + B + ("000" & SW(0));
    -- LEDR(0)  <= carry(3);                      -- carry out
    -- LEDR(1)  <= carry(3) xor carry(2);         -- overflow
END stru;
```

Fig. 10.12 Adder implementation using the VHDL '+' operator

Family	Cyclone II	Family	Cyclone II
Device	EP2C35F672C6	Device	EP2C35F672C6
Timing Models	Final	Timing Models	Final
Met timing requirements	Yes	Met timing requirements	Yes
Total logic elements	10 / 33,216 (< 1 %)	Total logic elements	5 / 33,216 (< 1 %)
Total combinational functions	10 / 33,216 (< 1 %)	Total combinational functions	5 / 33,216 (< 1 %)
Dedicated logic registers	0 / 33,216 (0 %)	Dedicated logic registers	0 / 33,216 (0 %)

Fig. 10.13 Quartus II synthesis summary for Solution II and the '+' operator alternative

discussed examples, two flags have been implemented. One of them is used to indicate the event of a carry out in the operation, and the other one to indicate that an overflow has happen in the last operation. In this case, the user (or another circuit) can use this information in order to decide whether or not to use the result provided by the adder.

10.2 Laboratory Assignment

The laboratory objectives are:

- to understand the design of a structural RCA component in VHDL;
- to replace the adder used in the calculator by the new component.

10.2.1 Laboratory Session

The tasks to be performed in this laboratory session are as follows:

- Using the VHDL implementation of Solution II listed in Fig. 10.10, make the necessary modifications in order to have an 8-bits RCA component.
- The new component should have two 8-bits inputs, a 1-bit *carry in* input, an 8-bits sum output, and four 1-bit flags output.
- The flags should indicate the following events happened in the last operation performed by the adder: a *carry out*, an *overflow*, a *negative result*, and a *zero result*.
- Assume that the calculator makes operations in two's complement signed numbers.
- The *negative* flag can be obtained straight from the result (most significant bit).
- The *zero* flag should be on when all bits of the result are equal to zero.
- The four flags can be connected to the green LEDs, as some of the red LEDs are already in use by the calculator.

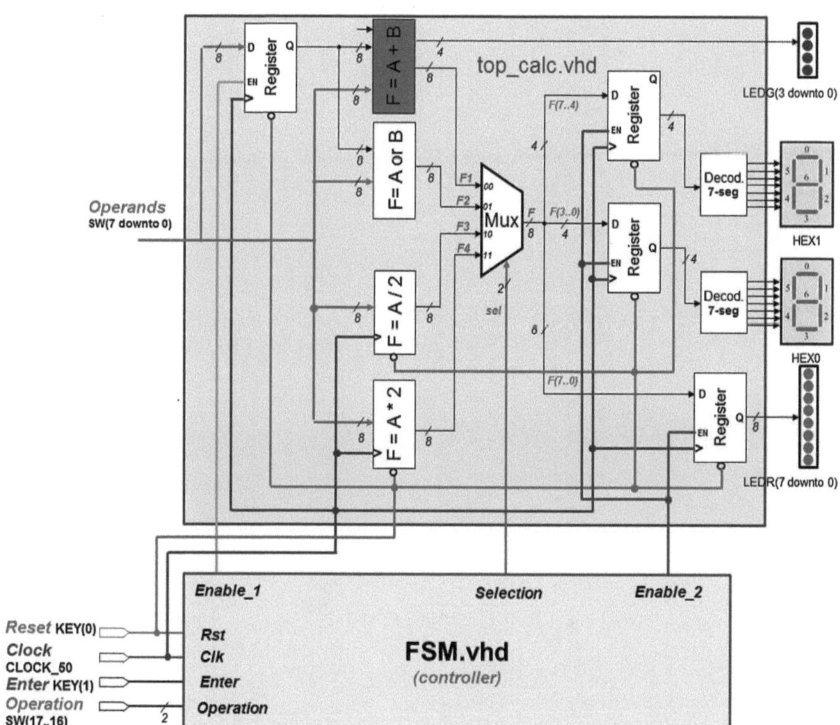

Fig. 10.14 The final calculator design, with the new RCA component

- It could be interesting to perform a timing simulation, instead of the functional simulation, in order to observe the delay needed by the RCA component to propagate the internal carry signals.
- Replace the C1 component used in previous chapters, by the new RCA component.
- The new design should be carefully designed, in order to avoid major changes in the remaining calculator components.
- Perform a final simulation in the design as a whole, this time with the new component added to the calculator.

Figure 10.14 shows the calculator design with the new RCA component, replacing the C1 component used in previous chapters. The new adder has an additional 4-bits output connected to the green LEDs (flags), and also a 1-bit input that is not connected (*carry in* signal).

Chapter 11
Writing Synthesizable VHDL Code for FPGAs

This chapter describes strategies for writing VHDL code for FPGA based designs. The strategies are presented in a way to show the stages of an application conception, considering a designer with some software development skills, but still a beginner in the hardware design language field. It starts presenting a high-level abstraction approach (software like description), but inadequate for the implementation of FPGA designs. The main mistakes made in all stages of the project are analyzed, and hardware design strategies are used in order to refine the design towards a working and efficient VHDL code.

11.1 Synthesis and Simulation

VHDL is a very complete language, but only a subset of it can be used to write synthesizable code. During the synthesis process, the VHDL source code is translated into a circuit for a target technology such as an FPGA or an ASIC, using the appropriate library components. Since VHDL offers similar resources found in typical software programming languages as, for instance, pointers and file management structures, consequently most of the features available in the language cannot be used in the synthesis processes for hardware generation. These "software-like" features have been thought targeting the development of simulation models. Considering all the available facilities, the possibility of developing hardware in a similar way to software appeals to designers with a software background. Usually, VHDL books describe the language as a whole, and it can be a bit confusing for beginners which may have the impression that systems can be implemented in VHDL in the same way as in languages like Java or C. In general, VHDL books discuss aspects of the language as, for instance, procedures, functions, data types, sequential and concurrent assignments, attributes, delta delays, but, in the examples presented of how to write a synthesizable VHDL code, only a small subset of the language is used.

The IEEE Std 1076™-2008—"IEEE Standard VHDL Language" defines the language as a whole, including simulation and synthesis semantics. The IEEE Std

E. A. Bezerra and D. V. Lettnin, *Synthesizable VHDL Design for FPGAs*,
DOI: 10.1007/978-3-319-02547-6_11,

1076.6^{TM}-2004—"IEEE Standard for VHDL register transfer level (RTL) Synthesis", describes only VHDL synthesis semantics. IEEE Std 1076.6 states that the synthesis tools should be compliant to the standard, but may have additional features. The standard covers the synthesis semantics to model, among others, level-sensitive and edge-sensitive logic. An important contribution of the standard is regarding code interoperability, as it defines in details all constructs that should be supported and also what should not be supported by the synthesis tools. The idea is to mitigate errors and mismatches between the simulation and synthesis results. This is a relatively new standard, and up to now each synthesis tool available still adopts its own VHDL synthesis semantics. The vendors are more interested in high quality synthesis results than VHDL code portability. Certainly, an acceptable level of compliance with the standard is found in all major synthesis tools, but it is still far from the ideal situation where a VHDL code written for a specific synthesis tool could be direct ported to another vendor's tool.

The situation is completely different when considering VHDL for simulation. The available tools seem to follow the IEEE 1076 standard, and usually there are no major problems when porting a VHDL simulation code from one vendor to another. VHDL simulation is a well-established subject, and its study is not one of the targets of this book. The strategies for the VHDL-based design discussed in this book target the VHDL synthesis topic.

11.2 VHDL Semantics for Synthesis

When asked to initialize a RAM memory (an array) with a sequence of values, a hardware designer learner with some sort of software background could easily come up with the VHDL code shown in Fig. 11.1.

This is an acceptable VHDL implementation to initialize a RAM memory, but it is suitable for a simulation model, and a synthesis tool would not generate a circuit with the expected functionality. This code was implemented without taking consideration of any synthesis criteria, with VHDL used like a programming language in a level of abstraction similar to the C language. The source code was written in this way because the designer had a general knowledge of the VHDL language, but was not familiar with VHDL for synthesis restrictions. Some of the problems in Fig. 11.1 cause synthesis errors, whilst others cause warnings that must be fixed in order to avoid discrepancies in the simulation before and after synthesis. The most dangerous situations are those where there are neither errors nor warnings, but in spite of this the synthesis tool creates hardware with behavior different from the expected one. The *for loop* construction, for instance, may result in a circuit having 256 adders storing a value sequentially. Another mistake made in this piece of VHDL code, is the use of two *wait* statements in the same process. This is not

```
entity RampUpMEM is
    port (clock, reset: in std_logic;
          stop: in std_logic;
          addr: out std_logic_vector(7 downto 0);
          ...
          );
end RampUpMEM;

architecture bhv of RampUpMEM is
    type RAM_TYP is array(255 downto 0) of std_logic_vector(7 downto 0)
    signal my_mem: RAM_TYP;
    signal stop1, stop2: std_logic;
begin
    process
        variable A: std_logic_vector(7 downto 0) := "00000000";
    begin
        for i in 0 to 255 loop
            my_mem(i) <= A   ;
            A := A + 1;
            wait for 10 ns;
        end loop;
        wait on stop, stop1, stop2;
        if rising_edge(clock) then
    ...
```

Fig. 11.1 Partial VHDL code for a RAM initialization application

acceptable in a VHDL for synthesis coding. Next, this and other problems are discussed and the proper VHDL semantics for synthesis is presented.

Variables and Signals

Variables and signals are both valid in VHDL for synthesis and for simulation. Signals are more suitable for hardware behavior description, as they do not change until the end of a process cycle. Variables, on the other hand, are updated instantaneously, in the same moment a value is assigned to them, and this behavior is not observed in actual hardware which is subjected to signal propagation delays. Despite that, synthesis tools allow the use of variables, but very specific rules and conventions must be followed, otherwise the generated circuit is not guarantee to function as the expected. For instance, in order to generate any sort of hardware from a variable, it is compulsory to assign the final variable result to a signal. Variables are very useful in simulation activities, as there is no delay associated to values attribution, making the simulation process much faster when comparing to the use of signals.

Wait Statements

In VHDL for synthesis, a process can have only one *wait* statement, and just in processes with no sensitivity list. The *wait* should be the first or the last statement of a process, and no timing information can be used in the *wait*. In Fig. 11.1 the VHDL semantic used is fine for simulation, but it has several problems regarding the use of *wait*, and a synthesis process will certainly fail. The process has more than one *wait* statement, and on line 20 there is even timing information (10 ns). The synthesis tool would not be able to create a circuit to generate such delay. Wait for time statements are used in testbenches, for delay generation. A synthesizable delay circuit could be designed employing timers, or counters embedded in state machines.

Memory Inference

In an FPGA design the array contents of an application can be stored in actual memory blocks, in LUTs, or in available flip-flops. Most FPGAs have a small amount of embedded and distributed memory blocks that can be freely used by the designer. FPGA vendors usually provide specific tools for memory generation. The designer can use these tools to achieve optimal memory usage figures, or they can write their own synthesizable VHDL code describing the memory behavior. The problem with the former is that very specific synthesis rules should be followed, otherwise the tool may not understand the VHDL code, and decide for allocating LUTs or flip-flops instead of available FPGA memory resources. Also, a VHDL description that generates proper memory in a synthesis tool, may not work in the same way in another tool. This sort of synthesizable VHDL code is not as portable as it should be and, unfortunately, for efficient memory generation the best alternative is still the IP creation using the vendor's tools for the target FPGA. In Fig. 11.1, the type declaration on line 10 and the signal definition on line 11 can be used by a synthesis tool to infer memory using the available memory blocks in FPGAs in general. However, the synthesis process will fail to infer memory, as the array access in the process body is not performed in an expected way.

The VHDL code in Fig. 11.2 is suitable for memory inference using Quartus II, and it worked as expected for the target Cyclone II FPGA resulting in 2,048 memory bits usage. The same VHDL code has been used to infer memory for a Xilinx FPGA, using Synplify from Synopsys, and it also worked as expected. However, when using the same VHDL code with Synplify to infer memory for the Microsemi's ProASIC3e FPGA, the inferred memory presented access problems. This is a nice way to describe a RAM memory in VHDL for synthesis, but some tools may use, for instance, flip-flops instead of the available RAM in the target device.

Additionally, Quartus II provides the following facilities for memory designs:

- There are VHDL examples of synthesizable RAM and ROM components available in the menu *Edit* → *Insert Template …* → *VHDL* → *Full Designs* → *RAMs and ROMs*.
- An IP core can be created using *Tools* → *MegaWizard Plug-In Manager* → *Create a new custom megafunction variation* → *Installed Plug-Ins* → *Memory Compiler*.

It is important to emphasize that the component generated using the second alternative can be used only in Quartus II based designs, and in this case there is no portability.

```
entity RAM is
    port (en, clk : in std_logic;
          data_in : in std_logic_vector(7 downto 0);
          addr : in std_logic_vector(7 downto 0);
          data_out : out std_logic_vector(7 downto 0)
          );
end RAM;
architecture bhv of RAM is
   type mem_type is array (255 downto 0) of std_logic_vector (7 downto 0);
   signal mem : mem_type;
begin
   process(clk, en, addr)
   begin
      if (rising_edge(clk)) then
         if (en = '1') then
            mem(conv_integer(addr)) <= data_in;
         end if;
      end if;
   end process;
   data_out <= mem(conv_integer(addr));
end bhv;
```

Fig. 11.2 RAM description suggested in synthesis tool documentation for VHDL RAM inference

Loop Statements in VHDL for Synthesis

The *for* loop in Fig. 11.1 is used, during simulation, to fill in the *my_mem* array with a numeric sequence from 0 to 255. As discussed before the VHDL code in Fig. 11.1 is not synthesizable for several reasons, and from the *for* loop construction the synthesis tool would not be able to produce and initialize a memory component. In order to generate the described loop behavior, the synthesis tool would have to unroll the *for* loop, and create a circuit to initialize all the 256 array positions simultaneously, in just one clock pulse. This is the expected behavior for the *for* loop listed in the process in Fig. 11.1. In VHDL for synthesis, the behavior of loop constructions (i.e. for, while, loop) are implicitly performed by processes. As described before, a process is cyclic, and it runs forever. Considering this behavior, a given algorithm can be modeled in VHDL, using clocked processes with test statements (i.e. if, case) to describe the behavior of all repetitions. A VHDL code with only test statements is easier to be synthesized than a code written using explicit loop constructions. The drawback is that the VHDL implementation with no explicit loop statements becomes more distant from the original algorithm, and may be more complex to test/debug.

A synthesizable version of the RAM initialization code shown in Fig. 11.1, can be obtained using the RAM component listed in Fig. 11.2 and the FSM listed in Fig. 11.3.

Therefore, in order to implement a synthesizable version for the *for* loop (Fig. 11.1, lines 17–21), in Fig. 11.3 a three process FSM has been used. In the first process the next state (PE) is assigned. The second process is responsible for defining the next state, and also for performing the "end condition" test for the

```
1  library ieee;
2  use ieee.std_logic_1164.all;
3  use ieee.std_logic_arith.all;
4  use ieee.std_logic_unsigned.all;
5
6  entity top is
7    port (clk, reset: in std_logic;
8          outp: out std_logic_vector(7 downto 0)
9    );
10 end top;
11 architecture top_beh of top is
12   type states is (S0, S1, S2);
13   signal EA, PE: states;
14   signal address, dt_out,
15          dt_in: std_logic_vector(7 downto 0);
16 component RAM
17   port (en, clk: in std_logic;
18         data_in, addr: in std_logic_vector(7 downto 0);
19         data_out: out std_logic_vector(7 downto 0)
20         );
21 end component;
22 begin
23   R0: RAM port map ('1', clk, dt_in, address, dt_out);
24   process(clk)
25   begin
26     if clk'event and clk='0' then
27       if reset='0' then
28         EA <= S0;
29       else
30         EA <= PE;
31       end if;
32     end if;
33   end process;
34   process( EA )
35   begin
36     case EA is
37       when S0 =>
38         PE <= S1;
39       when S1 =>
40         PE <= S2;
41       when S2 =>
42         if address = 255 then
43           PE <= S0;
44         else
45           PE <= S1;
46         end if;
47       when others =>
48     end case;
49   end process;
50   process(clk)
51   begin
52     if clk'event and clk='1' then
53       case EA is
54         when S0 =>
55           address <= "00000000";
56         when S1 =>
57           address <= address + '1';
58         when S2 =>
59           dt_in <= address;
60         when others =>
61       end case;
62     end if;
63   end process;
64   outp <= dt_out;
65 end top_beh;
```

Fig. 11.3 RAM initialization with no explicit loop statements

"*for*" loop (see line 42). The third process defines the "*for*" loop start value (see line 55), the memory address increment (see line 57), and the output (see line 18 in Fig. 11.1, and line 59 in Fig. 11.3). From Fig. 11.1 observation, it is straightforward that the fill in memory algorithm has an O(n) cost, as it is supposed to write a value to a memory position in each clock cycle. However, the same conclusion is not so direct when observing the three processes FSM in Fig. 11.3. This is a drawback when writing VHDL for synthesis code, that is, code readability and debugging. The IEEE Std 1076.6^{TM}-2004—"IEEE Standard for VHDL register transfer level (RTL) Synthesis", is a compulsory reading for good synthesis results. For instance, the asynchronous reset strategy used in previous chapters, states that a process must have only one clock signal, and its clock edge must be in the last *elseif* condition. No sequential statements are allowed before or after the *if* statement.

A *for* loop can be used in a synthesizable VHDL code, but the designer has to follow very specific rules. In Fig. 11.4 a for loop is used in a VHDL function.

The synthesis tool unrolls the *for* loop described in the function, and generates a hardware with 256 adders. This means that in order to perform the equation on line 23, the circuit will need one adder for each loop repetition.

```
library ieee;
use ieee.numeric_std.all;
use ieee.std_logic_1164.all;
use ieee.std_logic_unsigned.all ;
use ieee.std_logic_arith.all ;

entity count_ones is
port (
  data_in: in std_logic_vector (255 downto 0);
  cnt_out: out std_logic_vector (7 downto 0)
);
end entity count_ones;

architecture bhv of count_ones is

function count_1 (signal x: std_logic_vector)
      return std_logic_vector is
      variable total_1: integer;
begin
  total_1 := 0;
  for i in 0 to 255 loop
      if x(i) = '1' then
         total_1 := total_1 + 1;
      end if;
  end loop;
  return std_logic_vector(to_unsigned(total_1, 8));
end function count_1;

begin
 cnt_out <= count_1 (data_in);
end architecture bhv;
```

Fig. 11.4 A synthesizable *for* loop statement

11.3 HDLGen: Automatic Generation of Synthesizable VHDL

A good design approach is to split the circuit in components, and simulate/synthesize each component separately. Following this approach, when a component shows an adequate behavior, then it can be included in the whole design. The ideal situation is the one where each component is a file containing only one entity/architecture pair, with the architecture split into combinational and sequential parts, and with the sequential part having no more than two processes (ideally one). In the case of inexperienced VHDL designers, it is suggested that they first execute the synthesis and then the simulation of the design, otherwise much time is spent implementing and simulating code that may not be synthesized.

A tool has been developed targeting designers more familiar with software algorithms than hardware implementations. As shown in Fig. 11.5, using HDLGen

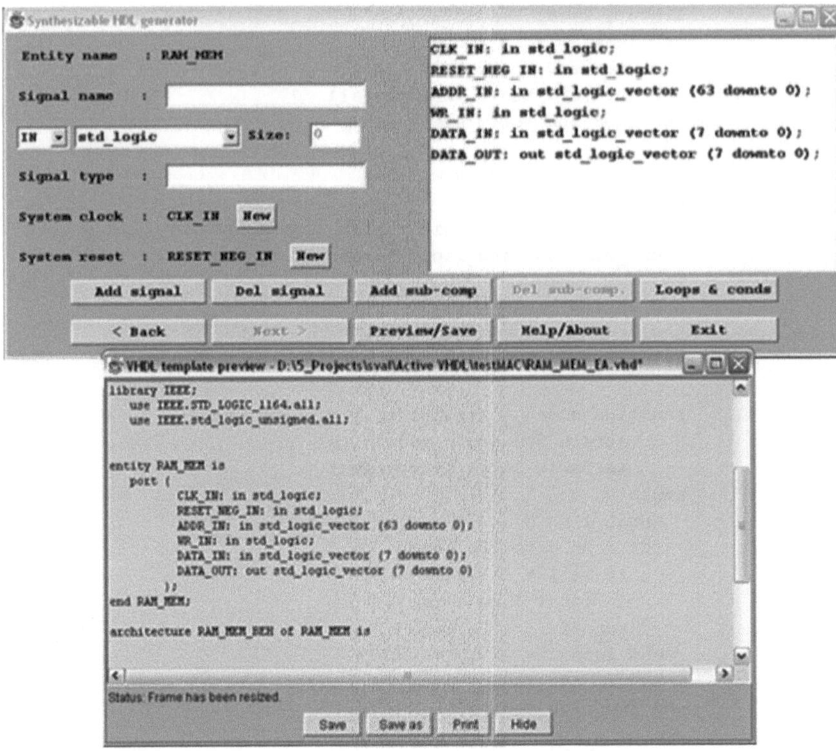

Fig. 11.5 HDLGen main window, and generated template file

the designer provides the description of the system's interface (signals), and the tool generates a template VHDL source code.

HDLGen was written in Java and it is based on strategies, methodologies and guidelines described in this book. HDLGen is free and can be downloaded off the author's website. This tool can be used by beginners in the development of synthesizable VHDL code, as a guide to avoid simple but, sometimes, expensive mistakes.

For instance, as discussed before, an experienced programmer when using VHDL for synthesis for the first time, may use a *for* statement for searching in an array which will not be understood by the synthesis tool. Using HDLGen, the programmer selects the "loop generation" option, provides the loop condition and then a state machine representing the loop constructor is generated in VHDL. More experienced developers may use HDLGen to link several components of a system, which facilitates the hierarchical view and project management. Fig. 11.5 shows HDLGen's main window (top) and the template file generated by the user

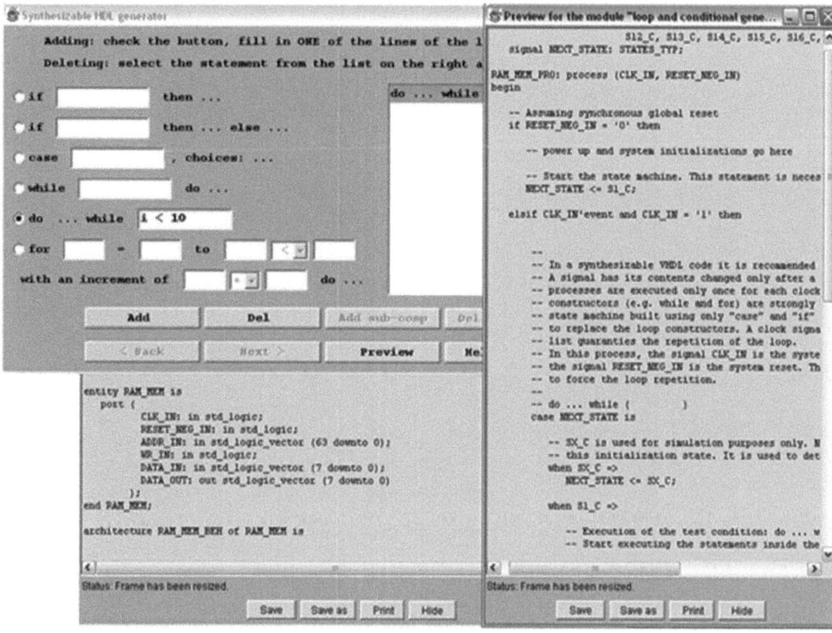

Fig. 11.6 HDLGen FSM generation for a loop statement

(bottom window). The “Loops & conds” button in the main window starts the HDLGen module responsible for thc generation of synthesizable VHDL code for loops and conditional constructions. This module is shown in Fig. 11.6 (top left window), where the tool is used to generate an FSM in VHDL for the Do … While (I < 10) loop entered in the example. This loop unrolling feature can be used as an aid to some of the problems discussed before.

Bibliography

Vahid F (2011) Digital design with RTL design, VHDL, and verilog, 2nd edn. John Wiley and Sons Publishers, New York

IEEE Standard 1076-2008–IEEE Standard VHDL Language, IEEE Computer Society, 2008

IEEE Standard 1076.6-2004—IEEE Standard for VHDL Register Transfer Level (RTL) Synthesis, IEEE Computer Society, 2004

Ashenden PJ (2008) The designer's guide to VHDL, 3rd edn (systems on silicon). Morgan Kaufmann, Burlington

Katz R, Borrielo G (2005) Contemporary logic design, 2nd edn. Prentice Hall, New Jersey

Tocci RJ (2010) Digital systems: principles and applications, 11th edn. Prentice Hall, New Jersey

Kilts S (2007) Advanced FPGA design: architecture, implementation, and optimization. Wiley-IEEE Press, Hoboken

Pedroni V (2010) Circuit design and simulation with VHDL, 2nd edn. MIT Press, Cambridge

Mano M, Kime C (1999) Logic and computer design fundamentals, 2nd edn. Prentice Hall PTR, New Jersey

E. A. Bezerra and D. V. Lettnin, *Synthesizable VHDL Design for FPGAs*,
DOI: 10.1007/978-3-319-02547-6,

MIX
Papier aus verantwortungsvollen Quellen
Paper from responsible sources
FSC® C105338

If you have any concerns about our products,
you can contact us on
ProductSafety@springernature.com

In case Publisher is established outside the EU,
the EU authorized representative is:
Springer Nature Customer Service Center GmbH
Europaplatz 3, 69115 Heidelberg, Germany

Printed by Libri Plureos GmbH
in Hamburg, Germany

VOLUME ONE HUNDRED AND ONE

ADVANCES IN VIRUS RESEARCH

Environmental Virology and Virus Ecology

Edited by

CAROLYN M. MALMSTROM

Department of Plant Biology and Graduate Program in Ecology, Evolution, and Behavioral Biology, Michigan State University, East Lansing, MI, United States

ELSEVIER

ACADEMIC PRESS

An imprint of Elsevier

Academic Press is an imprint of Elsevier
50 Hampshire Street, 5th Floor, Cambridge, MA 02139, United States
525 B Street, Suite 1650, San Diego, CA 92101, United States
The Boulevard, Langford Lane, Kidlington, Oxford OX5 1GB, United Kingdom
125 London Wall, London, EC2Y 5AS, United Kingdom

First edition 2018

Notices
Knowledge and best practice in this field are constantly changing. As new research and experience broaden our understanding, changes in research methods, professional practices, or medical treatment may become necessary.

Practitioners and researchers must always rely on their own experience and knowledge in evaluating and using any information, methods, compounds, or experiments described herein. In using such information or methods they should be mindful of their own safety and the safety of others, including parties for whom they have a professional responsibility.

To the fullest extent of the law, neither the Publisher nor the authors, contributors, or editors, assume any liability for any injury and/or damage to persons or property as a matter of products liability, negligence or otherwise, or from any use or operation of any methods, products, instructions, or ideas contained in the material herein.

ISBN: 978-0-12-814415-2
ISSN: 0065-3527

Publisher: Zoe Kruze
Acquisition Editor: Ashlie M. Jackman
Editorial Project Manager: Joanna Collett
Production Project Manager: James Selvam
Cover Designer: Mark Rogers

Typeset by SPi Global, India

CONTENTS

CONTRIBUTORS

Antonio Alcamí
Centro de Biología Molecular Severo Ochoa (Consejo Superior de Investigaciones Científicas—Universidad Autónoma de Madrid), Madrid, Spain

Pauline Bernardo
Department of Plant Pathology, Ohio State University, OARDC, Wooster, OH, United States; CIRAD, UMR BGPI; BGPI, Univ. Montpellier, CIRAD, INRA, Montpellier SupAgro, Montpellier, France

Mark J.F. Brown
School of Biological Sciences, Royal Holloway University of London, Egham, United Kingdom

Jesse L. Brunner
School of Biological Sciences, Washington State University, Pullman, WA, United States

Quentin Chesnais
Department of Entomology, University of California, Riverside, Riverside, CA, United States

Sohini Claverie
CIRAD, UMR PVBMT, F-97410 St Pierre, La Réunion, France

Denis Filloux
CIRAD, UMR BGPI; BGPI, Univ. Montpellier, CIRAD, INRA, Montpellier SupAgro, Montpellier, France

Aurora Fraile
Centro de Biotecnología y Genómica de Plantas UPM-INIA, and E.T.S.I. Agrícola, Alimentaria y de Biosistemas, Campus de Montegancedo, Universidad Politécnica de Madrid, Madrid, Spain

Serge Galzi
CIRAD, UMR BGPI; BGPI, Univ. Montpellier, CIRAD, INRA, Montpellier SupAgro, Montpellier, France

Fernando García-Arenal
Centro de Biotecnología y Genómica de Plantas UPM-INIA, and E.T.S.I. Agrícola, Alimentaria y de Biosistemas, Campus de Montegancedo, Universidad Politécnica de Madrid, Madrid, Spain

Ion Gutiérrez-Aguirre
National Institute of Biology, Ljubljana, Slovenia

Gordon W. Harkins
South African Medical Research Council Bioinformatics Unit, South African National Bioinformatics Institute, University of the Western Cape, South Africa

Penelope Hartnady
Computational Biology Division, Department of Integrative Biomedical Sciences, Institute of Infectious Diseases and Molecular Medicine, University of Cape Town, Observatory, South Africa

Roger A.C. Jones
Institute of Agriculture, Faculty of Science, University of Western Australia, Crawley; Department of Primary Industries and Regional Development, South Perth, WA, Australia

Simona Kraberger
The Biodesign Center for Fundamental and Applied Microbiomics, Center for Evolution and Medicine, School of Life Sciences, Arizona State University, Tempe, AZ, USA

Denis Kutnjak
National Institute of Biology, Ljubljana, Slovenia

Pierre Lefeuvre
CIRAD, UMR PVBMT, F-97410 St Pierre, La Réunion, France

Jean-Michel Lett
CIRAD, UMR PVBMT, F-97410 St Pierre, La Réunion, France

Darren P. Martin
Computational Biology Division, Department of Integrative Biomedical Sciences, Institute of Infectious Diseases and Molecular Medicine, University of Cape Town, Observatory, South Africa

Kerry E. Mauck
Department of Entomology, University of California, Riverside, Riverside, CA, United States

Michael J. McLeish
Centro de Biotecnología y Genómica de Plantas UPM-INIA, and E.T.S.I. Agrícola, Alimentaria y de Biosistemas, Campus de Montegancedo, Universidad Politécnica de Madrid, Madrid, Spain

Dino P. McMahon
Institute of Biology, Freie Universität Berlin; Department for Materials and Environment, BAM Federal Institute for Materials Research and Testing, Berlin, Germany

Nataša Mehle
National Institute of Biology, Ljubljana, Slovenia

Robert J. Paxton
Institute for Biology, Martin-Luther-Universität Halle-Wittenberg, Halle (Saale); German Centre for integrative Biodiversity Research (iDiv), Halle-Jena-Leipzig, Leipzig, Germany

Alberto Rastrojo
Centro de Biología Molecular Severo Ochoa (Consejo Superior de Investigaciones Científicas—Universidad Autónoma de Madrid), Madrid, Spain

Maja Ravnikar
National Institute of Biology, Ljubljana, Slovenia

Philippe Roumagnac
CIRAD, UMR BGPI; BGPI, Univ. Montpellier, CIRAD, INRA, Montpellier SupAgro, Montpellier, France

Lori R. Shapiro
Department of Applied Ecology, North Carolina State University, Raleigh, NC, United States

Arvind Varsani
The Biodesign Center for Fundamental and Applied Microbiomics, Center for Evolution and Medicine, School of Life Sciences, Arizona State University, Tempe, AZ, USA; Structural Biology Research Unit, Department of Clinical Laboratory Sciences, University of Cape Town, Rondebosch, Cape Town, South Africa

Karen D. Weynberg
School of Chemistry & Molecular Biosciences, University of Queensland, Brisbane, QLD, Australia

Lena Wilfert
Centre for Ecology and Conservation, University of Exeter, Penryn, United Kingdom

Christian M. Yarber
School of Biological Sciences, Washington State University, Pullman, WA, United States

PREFACE

Tiny but powerful, viruses are arguably some of the most elegant components of the web of life. Throughout the life sciences, viruses have served as important tools that have enabled innumerable discoveries in cell and molecular biology. Yet, in the environmental and ecological sciences, appreciation of viruses lags. This volume aims to address this problem by highlighting cutting-edge topics in environmental virology and virus ecology. The authors of these chapters are leading researchers who are probing the nature and dynamics of virus ecology across a broad range of study systems and environmental contexts.

Chapters 1–3 highlight ongoing explorations of virus diversity and prevalence in oceans, polar systems, and terrestrial plants. Wehnberg (Chapter 1) reviews the diversity of viruses in coastal waters, open oceans, and coral reefs, and their important influence on marine ecology and biogeochemical cycling. In Chapter 2, Rastrojo and Alcamí present intriguing evidence of high viral diversity in even extreme polar environments. Claverie et al. (Chapter 3) demonstrate the use of geometagenomics to uncover patterns of virus infection in terrestrial vegetation and reveal previously unknown viral species. These chapters illustrate significant findings made to date and underscore the vast potential for new discoveries as the hunt for novel viruses expands across our planet.

Chapters 4–7 examine aspects of virus transmission, a process critical for virus proliferation and one that is strongly controlled by environmental and ecological processes. Mehle et al. (Chapter 4) discuss important aspects of water-mediated transmission of viruses that infect plants, animals, and humans. Infection that can be spread broadly by surface waters, including irrigation streams, poses substantial threats in numerous systems. Brunner and Yarber (Chapter 5) focus on the particular case of ranaviruses, which are leading causes of mortality in frogs, and use modeling methods to evaluate the significance of virus transmission via water and infected carcasses to ranavirus epidemiology. Jones (Chapter 6) highlights novel and neglected pathways for transmission of plant and insect viruses, including via pollen and parasitic plants. Mauck et al. (Chapter 7) synthesize empirical literature to evaluate the extent to which plant viruses manipulate their hosts and vectors to promote their own transmission in agricultural and natural plant communities. These chapters illuminate complex pathways for virus proliferation in the environment while highlighting research directions that will improve our ability to predict and manage virus spread.

Chapters 8 and 9 extend consideration of viruses to community ecology. Despite the ubiquity of viruses across Earth's landscapes and waters (see Chapters 1–3), very little is known about how viruses fit into broader networks of species interactions, particularly in terrestrial systems. To address these unknowns, McMahon et al. (Chapter 8) set forth a molecule-to-ecology approach for examining networks of interactions among viruses and bees, while McLeish et al. (Chapter 9) explore host-range evolution among plant viruses within the framework of complex plant communities.

Taken together, these chapters illuminate the fascinating ways that viruses interact with diverse cellular life forms and persist in the face of incredible environmental challenges. Evidence to date suggests that viruses are likely present in every habitat on earth, but we have long overlooked their presence. This volume represents an important step toward tackling the vast, but challenging, opportunities for new discoveries in the fields of environmental virology and virus ecology.

Carolyn M. Malmstrom

In memory of my father, Edward James Malmstrom, Jr

CHAPTER ONE

Viruses in Marine Ecosystems: From Open Waters to Coral Reefs

Karen D. Weynberg[1]
School of Chemistry & Molecular Biosciences, University of Queensland, Brisbane, QLD, Australia
[1]Corresponding author: e-mail address: k.weynberg@uq.edu.au

Contents

Abstract

Viruses infect all kingdoms of marine life from bacteria to whales. Viruses in the world's oceans play important roles in the mortality of phytoplankton, and as drivers of evolution and biogeochemical cycling. They shape host population abundance and distribution and can lead to the termination of algal blooms. As discoveries about this huge reservoir of genetic and biological diversity grow, our understanding of the major influences viruses exert in the global marine environment continues to expand. This chapter discusses the key discoveries that have been made to date about marine viruses and the current direction of this field of research.

Advances in Virus Research, Volume 101
ISSN 0065-3527
https://doi.org/10.1016/bs.aivir.2018.02.001

1. INTRODUCTION

Viruses are the most abundant and diverse biological entities in the world's oceans and their role in the marine environment has been the subject of extensive investigation over recent years. It is estimated, from techniques such as epifluorescence microscopy (Hennes and Suttle, 1995; Noble and Fuhrman, 1998), that there are 10^{30} viruses in the oceans with an average count of 10^7 viruses per milliliter of surface seawater (Suttle, 2005). Viruses outnumber prokaryotes by an order of magnitude and cause an average of 10^{28} new infections of microbial hosts each day (Suttle, 2005, 2007).

1.1 The Marine Environment

Marine viruses are found within a complex, fluid environment that is home to organisms from all kingdoms of life. Approximately 71% of the globe is covered by ocean that contains 97% of the planet's water. Open ocean, or "pelagic," environments comprise the entire column of water beyond coastal boundaries and the edges of continental shelves and above the seabed or "benthic" zone. Open oceans are typically oligotrophic environments that are nutrient poor (Table 1) and are estimated to contain 10^4–10^5 virus-like particles per mL (VLPs mL^{-1}) (Alonso et al., 2001; Suttle, 2005, 2007). Coastal waters lie between land masses and the open ocean; they are some of the most biologically productive environments because they are enriched

Table 1 General Comparison of Three Marine Environments

Environmental Parameter	**Open Oceans**	**Coastal Waters**	**Coral Reefs**
Water depth	Deep	Moderate	Shallow
Nutrient availability	Variable: typically Fe and N depleted	Often high nutrient availability from terrestrial run-off	Low resources
Species richness	Low–moderate	Moderate	Very high
Key viral roles	Regulate phytoplankton dynamics, biogeochemical cycling	Similar to open oceans plus algal bloom control	Similar to open oceans plus pathogen impacts on keystone organisms including corals …

by nutrients from terrestrial run-off and from upwelling currents from the deep ocean. In these richer waters, concentrations of VLPs reach higher levels (10^6–10^7 VLPs mL^{-1}) than in the open oceans (Jiang and Paul, 1996; Weinbauer et al., 1993). Coral reefs are three-dimensional shallow water structures that are usually dominated by stony corals of the Scleractinia (Dubinsky, 1990). They are often located in oligotrophic waters, but can occur in a range of nutrient conditions (D'Angelo and Wiedenmann, 2014).

Much about life in the oceans is still unknown. Current estimates predict that there are approximately 2.2 ± 0.18 million species in the marine environment but that 91% of ocean species have likely yet to be discovered (Mora et al., 2011). There is particularly little known about microbes, including viruses. For example, recent work just revealed an entire new family of marine viruses, the *Autolykiviridae*, that had been hitherto completely overlooked (Kauffman et al., 2018). Among marine environments, coral reefs are the equivalent of terrestrial rainforests in terms of biodiversity and productivity. Coral reefs occupy less than 1% of the global ocean, yet support as much as 25% of known marine species (Dubinsky, 1990), or approximately 830,000 species, excluding fungi (Fisher et al., 2015).

Microbes represent more than 90% of the living biomass in the oceans and are responsible for half of the planet's primary productivity (Field et al., 1998; Worden, 2006). Phytoplankton turn over quickly, with the entire global population replaced on average weekly (Field et al., 1998). Prokaryotes dominate open-ocean systems and therefore much of the research effort on marine viruses focuses on bacteriophages (phage) that infect bacterial hosts. Among these phages are cyanophages that infect photosynthetic cyanobacteria; the most abundant cyanophages in marine waters infect the cyanobacterial genera *Synechococcus* and *Prochlorococcus* (Sullivan et al., 2003; Suttle, 2002). Marine viruses also infect algae—aquatic eukaryotic plants that range in size from unicellular microalgae to multicellular, complex macroalgae. Some viruses even infect tiny photosynthetic picoeukaryotes (PPEs) that are unicellular eukaryotes less than 3 μm in size and important primary producers in the marine environment alongside cyanobacteria.

1.2 The Beginning of Marine Virology

The tremendous abundance and diversity of viruses in the marine environment were initially recognized in the 1980s, when microscopic counts of virus particles first revealed millions of viruses within just a milliliter of seawater (Bergh et al., 1989). The stage was set for this finding by the earlier discovery in the 1970s that marine bacteria are diverse and abundant, with initial reports enumerating more than 1 million bacteria per milliliter in marine environments (Hobbie et al., 1977; Pomeroy, 1974). The revelation of marine bacterial abundance spurred further unveiling of the critical importance of microbial processes in the global oceans (Azam et al., 1983; Fuhrman et al., 1989).

The first virus found to infect a marine eukaryotic organism was reported in the 1970s in a unicellular alga—*Micromonas pusilla* (a prasinophyte, family Mamiellaceae)—that is a PPE dominant in some ocean waters (Mayer and Taylor, 1979). This finding raised the first suggestion that viruses could play significant roles in influencing the ecology of important marine primary producers (Waters and Chan, 1982). In the early 1980s, studies of viruses infecting freshwater *Chlorella* algal symbionts ultimately resulted in taxonomic classification of the *Phycodnaviridae*, a family of algal viruses (reviewed in Van Etten et al., 1983, 2002). Within the *Phycodnaviridae*, the phaeoviruses that infect marine filamentous brown algae of the Ectocarpales were further identified (reviewed in Müller and Knippers, 2011). Since then, techniques to examine natural viral populations in the laboratory and in the field have advanced to enable a broader study of marine viral ecology, including the influence of viruses on host population abundance and distribution, evolution, and horizontal gene transfer (HGT). As methodologies for investigating viruses in the marine environment improve, our knowledge of the roles and significance of viral infection widens.

This chapter describes the ecology of viruses in coastal waters, open oceans, and coral reefs. Its primary focus is on the best-studied marine viruses: those that infect unicellular eukaryotic phytoplankton and cyanophages, which are the most abundant and ecologically relevant organisms in the oceans. Only a little is known about viruses that infect members of higher marine trophic levels. For example, a few studies have found evidence of viruses in copepods (Dunlap et al., 2013) and described how zooplankton can act as vectors for algal viruses (Frada et al., 2014). Other studies have reported fish or marine mammalian viruses, but have focused on particular disease profiles (Crane and Hyatt, 2011; Fereidouni et al., 2016) that are detrimental to fisheries or aquaculture, or to populations of native mammals, such as seals (Fereidouni et al., 2016).

1.3 Techniques for Identifying Marine Viruses

A central challenge to studying marine viruses is their nanoscale size, which ranges from approximately 20 to 200 nm (Brum et al., 2013; King et al., 2011).

The initial discovery of high marine viral abundance was the result of using transmission electron microscopy (TEM) to count the number of VLPs in natural seawater samples that had been ultracentrifuged onto a grid and subsequently viewed (Bergh et al., 1989). TEM has also been used to visualize virus particles in coral reef samples (Davy and Patten, 2007; Lawrence et al., 2014; Patten et al., 2008; Pollock et al., 2014; Weynberg et al., 2017a; Wilson et al., 2005a). Additional methods have been used to directly detect and quantify viruses in seawater and coral samples, including flow cytometry (Brussaard, 2004a; Patten et al., 2006; Payet et al., 2014; Pollock et al., 2014) and epifluorescence microscopy using fluorescent stains such as SYBR Green (Noble and Fuhrman, 1998) and Yo-Pro (Hennes and Suttle, 1995). Enumeration of viruses in aquatic samples using epifluorescence microscopy is rapid (30 min), reliable, and less costly then TEM (Patel et al., 2007). Recently, it was found that fluorescent counts may also detect DNA within membrane-bound extracellular vesicles that are of similar size to viruses (Biller et al., 2017). However, it was concluded that the quantity of such vesicles in seawater samples was typically an order of magnitude lower than that of viruses and thus unlikely to overly skew counts performed via epifluorescence microscopy.

Molecular techniques, such as restriction fragment length polymorphism and denaturing gradient gel electrophoresis (DGGE) (described in the Manual of Aquatic Virology or MAVE; Wilhelm et al., 2010), have also been employed. Traditional approaches to studying marine viruses have required the host–virus system be established in the laboratory, where host monocultures are used for the isolation and propagation of viruses. This can be limiting as many host species are unable to grow under laboratory conditions, with up to 99% of prokaryotes being termed viable but nonculturable (Yamamoto, 2000)—although capable of metabolic activity these species cannot grow in culture media. Over the last decade or so, next-generation sequencing (NGS) approaches—including metagenomics, transcriptomics, and proteomics—have removed the prerequisite of isolating the host–virus system in culture. Despite technical challenges and limitations in comprehensive database and analytical resources, NGS has helped to unveil a previously unseen expanse of viral diversity and distribution in the global oceans (as highlighted by the Global Ocean Virome (GOV) dataset; Roux et al., 2016) and in coral reef ecosystems, where new findings have been revealed with both metagenomics (Correa et al., 2012, 2016; Laffy et al., 2016; Marhaver et al., 2008; Vega Thurber et al., 2008; Weynberg et al., 2014, 2017a) and transcriptomics (Bruwer et al., 2017; Levin et al., 2016; Wood-Charlson et al., 2015). It is now envisaged that combining emerging techniques such as single-cell genomics (Allen et al., 2011), microfluidic digital PCR (Tadmor et al., 2011), and phageFISH (fluorescent in situ hybridization) (Allers et al., 2013) with large microbial bioinformatics datasets will enable further insights into viral diversity and function across marine spatiotemporal ranges (Brum and Sullivan, 2015).

1.4 Morphology of Common Marine Viruses

Marine viruses are extremely diverse, both genetically and morphologically (King et al., 2011). Marine cyanophages belong to the *Caudovirales*, an order of tailed dsDNA bacteriophages that dominate global waters. The *Caudovirales* infect cyanobacteria, heterotrophic bacteria (King et al., 2011), and archaea. These cyanophages possess tail structures that protrude from the capsid containing the viral genome. The tail structures allow these viruses to recognize, attach to, and penetrate host cells. A cyanophage can use fibers on the end of its tail to anchor itself by attaching to surface receptors on a host cell—known as "adsorption"—prior to penetrating the host cell wall and/or membrane and injecting its nucleic acids into the cell interior (for an online animation of this process, see https://bilbo.bio.purdue.edu/~viruswww/Rossmann_home/movies/movies.php). The *Caudovirales* include three virus families: the *Myoviridae*, which possess contractile tails that act like a syringe by drilling down to penetrate and inject the phage genome into the host cell; the *Siphoviridae* with long noncontractile tails; and the *Podoviridae*, with short noncontractile tails (Fig. 1A–E) (for reviews of different viral tail structures, see Davidson et al., 2012; Leiman and Shneider, 2012). Members of all three families infect both unicellular and filamentous cyanobacteria (Suttle, 2002).

Other marine virus morphologies include filamentous particles, rods, and icosahedral capsids (King et al., 2011) (icosahedral capsid morphologies are shown in Fig. 2A–F). The *Phycodnaviridae* are a family of dsDNA viruses that infect marine and freshwater eukaryotic algae. These viruses typically have icosahedral capsids, without tails, and very large genomes (>100 kb). Some virus particles, including some bacteriophages, are further enclosed within a lipid envelope (Ackermann, 2006). Most members of the *Phycodnaviridae* do not have an external lipid envelope but possess an internal one that plays a critical role in the release of the viral nucleoprotein core following adsorption of the virus to the cell surface (Maier and Muller, 1998; Wolf et al., 1998). Some of the few exceptions within the *Phycodnaviridae* are certain coccolithoviruses, including EhV-86, which infect the coccolithophorid

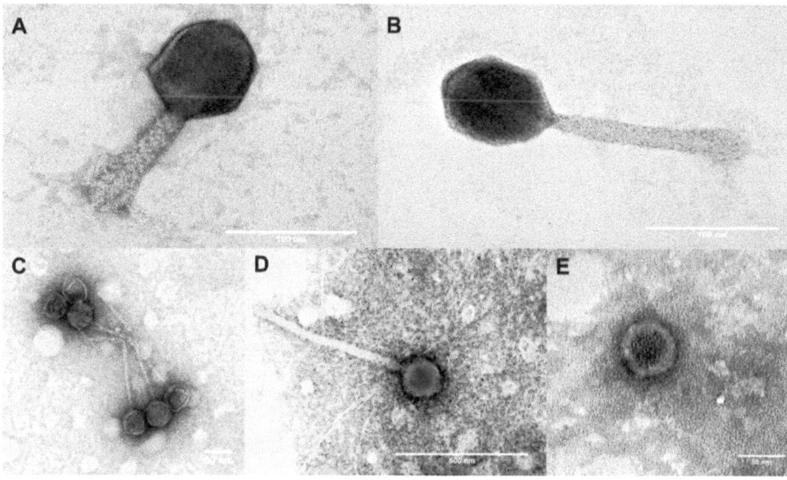

Fig. 1 Transmission electron micrographs of marine bacteriophages (order *Caudovirales*) (A and B) myoviruses (contractile tails), (C and D) siphoviruses (noncontractile viruses), and (E) podovirus (short noncontractile tail) sampled from seawater collected in the Great Barrier Reef.

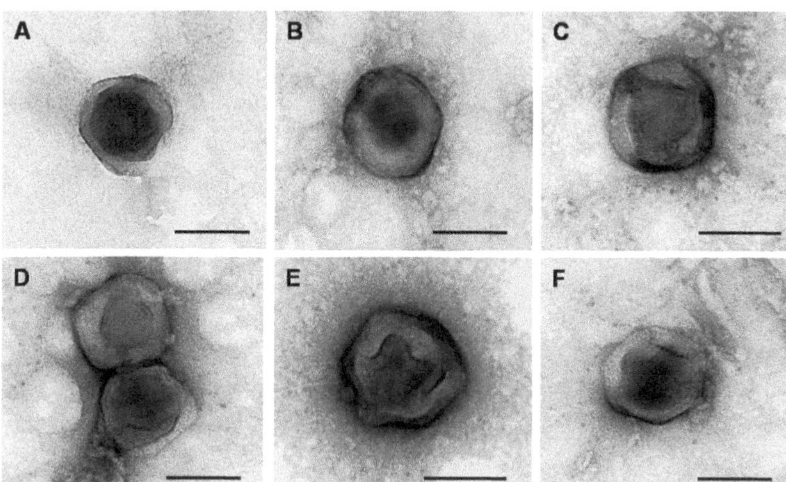

Fig. 2 (A–F) Transmission electron micrographs of icosahedral viruses (*Phycodnaviridae*) that infect marine eukaryotic algae from surface waters (<5 m) in the Western English channel. Scale bars = 100 nm.

Emiliania huxleyi. These coccolithoviruses are surrounded by an external lipid envelope that is used in a lipid fusion or endocytotic process to enter host cells, an infection strategy not seen in other algal viruses but is more akin to viral infections seen in animal hosts (Mackinder et al., 2009).

2. LYTIC INFECTION AND LYSOGENY: KEY ASPECTS OF MARINE VIRUS ECOLOGY

Viruses that infect marine bacteria and algae usually propagate via two primary processes: a lytic cycle or a lysogenic cycle. Other strategies (e.g., pseudolysogeny, chronic infection) can occur but will not be discussed here (reviewed in Clokie et al., 2011).

2.1 Lytic Cycles

In a lytic cycle, the virus introduces its genome into a host cell and initiates replication by hijacking the host's cellular machinery to make new copies of the virus. Once infection is complete, the newly replicated and assembled virus particles are released through lysis of the host cell into the surrounding waters. Viral infection resulting in lysis is ecologically significant because it causes high mortality of microbial hosts, thus structuring populations and communities and imposing strong selective pressures. Cellular lysis by viruses is further significant because it also releases energy and nutrients into surrounding waters, which other organisms can then utilize. Viruses therefore divert organic matter from lower levels of the food chain back into the particulate and dissolved organic matter pools (POM and DOM, respectively), cutting off support to higher trophic levels (Fig. 3). This process in marine microbial food webs is known as the "viral shunt" (Wilhelm and Suttle, 1999), which is estimated to channel around 25% of the carbon fixed during photosynthesis (Suttle, 2005; Wilhelm and Suttle, 1999) by marine phytoplankton. The viral shunt substantially influences biogeochemical cycling (Fuhrman, 1999) and performs a crucial recycling role in low resource systems, such as oligotrophic waters in the open ocean and coral reefs. A recent study modeled the influence of viruses on the flow of energy and nutrients through microbial food webs (Weitz et al., 2015). Viral-mediated cell lysis releases labile DOM into surrounding seawater that is easily assimilated by microbes. This results in greater nutrient availability in the euphotic zone and lowers the net export of carbon to the deep sea (Haaber and Middelboe, 2009; Short, 2012).

Phage-induced mortality has intriguing effects on marine microbial communities (Weinbauer, 2004; Wommack and Colwell, 2000). The "Kill the Winner" (KtW) model (Thingstad, 2000; Winter et al., 2010), based on classical Lotka–Volterra predator–prey dynamics, describes how viral infection

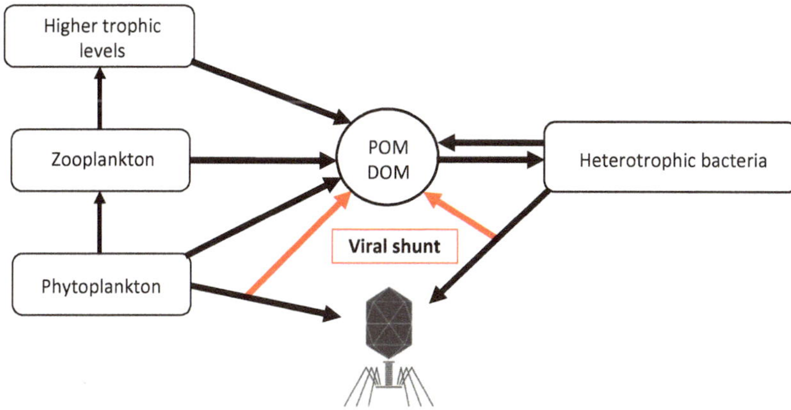

Fig. 3 Simplified diagram of the marine microbial food web and viral shunt. In the traditional food chain, grazers (zooplankton) feed on phytoplankton and are in turn consumed by higher predators, such as fish. In the traditional food chain, a good portion of nutrients and energy are passed onto higher trophic levels. In contrast, the viral shunt short circuits this chain and diverts nutrients and energy to other prokaryotes instead. By lysing marine microbes (primary producers, heterotrophic bacteria, archaea, and protists), viral infection releases dissolved organic matter (DOM) that other microbes (not bigger consumers) can utilize. This microbial recycling pathway keeps resources in the upper surface waters of the oceans and prevents carbon from sinking out of the dynamic euphotic zone to depth.

helps maintain microbial diversity by controlling species that begin to dominate, the so-called "winners." KtW postulates that viruses maintain host diversity via a frequency-dependent effect in which a particular host increases in relative abundance but consequently becomes more susceptible to viral attack. Within marine bacterial communities, there are competition specialists (*r*-strategists) and defense specialists (*k*-strategists), and both of these directly compete with each other for the same pool of often limited resources (Winter et al., 2010). Predators (phages) control the abundance of the most active *r*-strategists, allowing the slower growing defense specialists to access resources and persist (Winter et al., 2010).

2.2 Lysogeny

In lysogeny, a virus accesses a host cell but instead of immediately beginning the replication process leading to lysis, enters into a stable state of existence with the host. Phages capable of lysogeny are known as temperate phage or prophage. Lysogeny is commonly characterized by insertion of the viral genome into the host genome or other cell replicon (e.g., plasmid). The viral genome is replicated with the host genome until a trigger event, such as an environmental stressor, causes the virus to dissociate from the host genome and enter into a lytic mode of infection—a process called "induction." The capacity to suddenly switch from lysogeny to lysis means prophages represent "molecular time bombs" from the perspective of the bacterial host (Paul, 2008).

Lysogeny protects a virus from environmental factors (e.g., inactivation by UV sunlight or proteolytic digestion) that may damage the viral capsid or nucleic acid while on occasion conferring "immunity" to the host via gene expression that prevents coinfection by other viruses (Jiang and Paul, 1996). Such intimate association between viral and host genomes during lysogeny can enable HGT and may result in significant evolutionary outcomes. Studies using chemical induction with mitomycin C have found that as many as half of all culturable bacterial isolates from a range of marine environments carry prophages (Jiang and Paul, 1994, 1998; McDaniel et al., 2006), and that some bacterial isolates contain more than one prophage (Leitet and Riemann, 2006). Indeed, prophages have been experimentally induced from bacterial isolates, even when bioinformatics data gave no prior indication of their presence (Zhao et al., 2010). This discrepancy may be due to insufficient representation in databases for genomic annotation, the high diversity of phage genomes, or the presence of phage within cells as temperate plasmid-like elements not integrated into the host genome (Mobberley et al., 2008; Oakey et al., 2002). The choice to enter a lysogenic cycle, as opposed to a lytic one, may be due to factors such as seasonality or low host cell densities. Triggers that lead to a switch from lysogeny to lysis may include environmental damage to the host or its genome or, conversely, a peak in host growth and fitness that provides optimal conditions for viral replication and eventual lysis.

Lysogenic viral infections have been described in certain eukaryotic algal species and best characterized in the brown macroalgal order Ectocarpales (reviewed in Van Etten et al., 2002). Ectocarpoid host species are infected by phaeoviruses (*Phycodnaviridae*) that integrate into the host genome during the reproductive stages of the host life cycle and are passed to all daughter diploid cells during development. Infection is cryptic but sterilizing; obvious signs of viral replication and infection are evident only when the reproductive organs of the adult alga become deformed and release viruses in place of gametes or spores (Müller and Knippers, 2011). More recently, phaeoviruses have been observed in another marine macroalgal order, the Laminarales (kelp) (McKeown et al., 2017). To date, only these macroalgal species have been demonstrated to be infected by lysogenic viruses, and by displaying lysogeny, phaeoviruses are unique within the *Phycodnaviridae*.

2.3 Viruses of Eukaryotic Marine Algae

Marine algal viruses of the *Phycodnaviridae* are members of a collective called the nuclear cytoplasmic large dsDNA viruses (NCLDVs). The NCLDVs share a common ancestor and are comprised of 10 virus families that share a set of core genes (Table 2) (Iyer et al., 2001; Koonin and Yutin, 2010). Members of the NCLDVs infect eukaryotic hosts, possess large dsDNA genomes greater than 100 kilobases (kbp) and usually perform replication in the host cytoplasm, although in certain cases replication begins in the nucleus to be completed in the cytoplasm later in infection. It has been proposed that the NCLDVs be classified as a new taxonomic order, the "*Megavirales*" (Colson et al., 2013), although this has yet to be widely adopted in the literature. PPEs are less abundant than cyanobacteria but are important primary producers responsible for significant carbon fixation in global oceans. For example, PPEs contributed 76% of net carbon fixation at a Pacific Ocean coastal station (Worden et al., 2004) and up to 44% of CO_2 fixed in the sub- and tropical northeast Atlantic (Jardillier et al., 2010). Hence, the viruses infecting these eukaryotic unicellular algae are of great importance in marine microbial webs.

Marine viral infection has been studied in a wide range of eukaryotic algal host genera (Table 3) (King et al., 2011). These include dinoflagellates and diatoms (Nagasaki, 2008), brown algae (Müller and Knippers, 2011), and chlorophytes (Weynberg et al., 2017b). Algal viruses exhibit a range of genome types as exemplified by dsDNA phycodnaviruses (Van Etten et al., 2002), a dsRNA virus that infects *Micromonas* (Attoui et al., 2006), ssRNA viruses that infect *Heterocapsa* (Nagasaki, 2008), and *Heterosigma* (Nagasaki et al., 1994; Tarutani et al., 2000) species, and ssDNA viruses that infect *Chaetoceros* (King et al., 2011; Nagasaki, 2008). The diversity of algal viruses is therefore high and reflects a divergent evolutionary past. Because viruses, unlike prokaryotes and eukaryotes, do not share a universal gene, early molecular studies used PCR to target conserved genes such as the DNA polymerase gene (*polB*) (Chen and Suttle, 1995) and major capsid gene (*mcp*) (Larsen et al., 2008) to detect particular algal viruses within certain virus families in the natural environment, without the need for cultivation in the lab. Studies using the *polB* PCR primers helped to reveal the wide distribution of algal viruses across oceans (Short and Suttle, 2002) as well as in freshwater environments (Short and Short, 2008).

Table 2 Presence of NCLDV Core Genes (Groups I, II, and III) in Various NCLDV Genomes

	I												II												III						
Family Species	Vaccinia Virus D5-Type ATPase	DNA Polymerase (*otv208f*)[a]	Vaccinia Virus A32-Type ATPase (*otv92r*)	Vaccinia Virus A18-Type Helicase (*otv64f*)	Capsid Protein (*otv065f/71f/075f/093f/094f/165f/207r/211r*)	Thioloxidoreductase	Vaccinia Virus D6R-Type Helicase	Serine/Threonine Protein Kinase (*otv144r*)	VLTF2-Like Transcription Factor	TFII-Like Transcription Factor (*otv153f*)	MuT-Like NTP Pyrophosphohydrolase	Myristyolated Virion Protein A	Proliferating Cell Nuclear Antigen (*otv107f*)	Ribonucleotide Reductase, Large Subunit (*otv133r*)	Ribonucleotide Reductase, Small Subunit (*otv151r*)	Thymidylate Kinase (*otv196f*)	dUTPase (*otv199f*)	A494R-Like Uncharacterized Protein	RuvC-Like Holliday Junction Resolvase	BroA-Like	Capping Enzyme (*otv87f/156f*)	ATP-dependent Ligase (*otv181r*)	RNA Polymerase, Subunit 1	RNA Polymerase, Subunit 2	Thioredoxin/Glutaredoxin	Serine/Threonine Phosphatase	BIR Domain	Virion-Associated Membrane Protein	Topoisomerase II (*otv212r*)	SW1/SNF2 Family Helicase (*otv129f*)	RNA Polymerase, Subunit 10
Phycodnaviridae																															
OtV-1		✓	✓	✓	✓			✓		✓			✓	✓	✓	✓	✓				✓	✓							✓	✓	
OtV-2		✓	✓	✓	✓			✓		✓			✓	✓	✓	✓	✓				✓	✓							✓	✓	
OsV5		✓	✓	✓	✓			✓		✓			✓	✓	✓	✓	✓				✓	✓							✓	✓	
OlV-1		✓	✓	✓	✓			✓		✓			✓	✓	✓	✓	✓				✓	✓							✓	✓	
MpV-1		✓	✓	✓	✓			✓		✓			✓	✓	✓	✓	✓				✓	✓							✓	✓	
BpV-1		✓	✓	✓	✓			✓		✓			✓	✓	✓	✓	✓				✓	✓							✓	✓	
BpV-2		✓	✓	✓	✓			✓		✓			✓	✓	✓	✓	✓				✓	✓							✓	✓	
EhV-86	✓	✓	✓	✓	✓	✓	✓	✓	✓	✓	✓		✓	✓	✓	✓	✓	✓			✓	✓	✓	✓	✓		✓		✓		✓
PBCV-1	✓	✓	✓	✓	✓	✓	✓	✓	✓	✓	✓	✓	✓	✓	✓	✓	✓	✓			✓	✓			✓	✓			✓	✓	

EsV-1	✓	✓	✓	✓	✓	✓	✓	✓	✓				✓	✓	✓			✓	✓	✓					✓						
Mimiviridae																															
Mimivirus	✓	✓	✓	✓	✓	✓	✓	✓	✓	✓	✓	✓	✓	✓	✓			✓		✓	✓	✓	✓	✓	✓	✓	✓	✓	✓		
Iridoviridae																															
LCDV	✓	✓	✓		✓	✓	✓	✓	✓	✓	✓	✓	✓	✓	✓	✓		✓					✓	✓				✓		✓	
IIV-6	✓	✓	✓	✓	✓	✓	✓	✓	✓	✓	✓	✓	✓	✓	✓	✓	✓	✓	✓	✓			✓	✓	✓	✓	✓	✓	✓	✓	✓
Asfarviridae																															
ASFV	✓	✓	✓	✓	✓	✓	✓	✓	✓	✓	✓	✓	✓	✓	✓	✓	✓	✓			✓	✓	✓	✓			✓	✓	✓		✓
Poxviridae																															
VACV	✓	✓	✓	✓	✓	✓	✓	✓	✓	✓	✓	✓	✓	✓	✓	✓	✓		✓		✓	✓	✓	✓	✓	✓		✓			✓
MOCV	✓	✓	✓	✓	✓	✓	✓	✓	✓	✓	✓	✓	✓				✓		✓		✓		✓	✓	✓			✓			
AMEV	✓	✓	✓	✓	✓	✓	✓	✓	✓	✓	✓	✓					✓		✓	✓	✓		✓	✓	✓	✓	✓	✓			
MSEV	✓	✓	✓	✓	✓	✓	✓	✓	✓	✓	✓	✓							✓	✓	✓		✓	✓	✓		✓	✓		✓	

[a]Gene ID for OtV-1.

ASFV, African swine fever virus; *NTP*, nucleoside triphosphate; *dUTPase*, deoxyuridine triphosphatase; *ATP*, adenosine triphosphate.

Source: Based on data from this study and Moreau, H., et al., 2010. Marine prasinovirus genomes show low evolutionary divergence and acquisition of protein metabolism genes by horizontal gene transfer. J. Virol. 84, 12555–12563, https://doi.org/10.1128/jvi.01123-10, Iyer, L.M., Aravind, L., Koonin, E.V., 2001. Virology common origin of four diverse families of large eukaryotic DNA viruses. J. Virol. 75, 11720–11734. https://doi.org/10.1128/jvi.75.23.11720-11734.2001, and Raoult, D., et al., 2004. The 1.2-megabase genome sequence of mimivirus. *Science* 306, 1344–1350, https://doi.org/10.1126/science.1101485.

Table 3 Summary of Major Marine Algal Virus Groups and Their Hosts

Host Class	Host Genus	Virus Genus/Clade*	Examples of Associated Virus
Dinophyceae	*Heterocapsa*	*Dinornavirus*	HcRNAV
		Dinodnavirus	HcDNAV
Prymnesiophyceae	*Emiliania*	*Coccolithovirus*	EhV
	Phaeocystis	*Megaviridae**	PgV
	Chrysochromulina	*Prymnesiovirus*	CeV
Phaeophyceae	*Ectocarpus*	*Phaeovirus*	EsV
	Feldmannia		FsV
	Pilayella		
	Hincksia		
	Myriotrichia		
Raphidophyceae	*Heterosigma*	*Raphidovirus*	HaV
Pelagophyceae	*Aureococcus*	*Megaviridae**	AaV
Bacillariophyceae	*Chaetoceros*	*Bacilladnavirus*	CsetDNAV
	Rhizosolenia		RsRNAV
Prasinophyceae	*Ostreococcus*	*Prasinovirus*	OtV
	Micromonas		MpV
	Bathycoccus		BpV
Chlorodendrophyceae	*Tetraselmis*		TvV

*Proposed but not yet recognized by the ICTV.

2.4 Host Range of Marine Viruses

Specialist viruses have narrow host ranges, infecting only a certain species or even only certain strains within a species. Generalist viruses display wider host ranges and can infect more than one strain or species. Studies of marine phage found isolates capable of infecting only a few or as many as 20 different host strains, with a wide range of difference in sensitivity of different hosts to the same phage (Holmfeldt et al., 2007). Host range analysis of cultured marine phage isolates have revealed that some phages have extremely narrow host ranges and infect only a single host strain (Rohwer et al., 2000). Others are able to infect a number of strains of the same host species

(Wichels et al., 2002), closely related species (Comeau et al., 2006), or unrelated species (Chiura et al., 2009). The ability to infect several different hosts, known as "polyvalency," signifies an important survival strategy in the natural environment where host numbers may be low and hosts may be extremely diverse. Studies in the field, such as that conducted by De Corte and coworkers, have revealed as bacterial diversity increases with depth, viral diversity decreases indicating a broader host range may be at play (De Corte et al., 2010). Future similar studies are needed to elucidate further the range of host specificities in the marine environment.

2.5 Host Resistance to Viral Infection

Resistance to infection by cyanophages and eukaryotic viruses has been reported in marine cyanobacteria and algae. Viral resistance in bacteria can arise due to spontaneous mutations in host genes encoding cell surface receptors that prevent phage adsorbing to host cells, as well as other processes that halt intracellular viral production, including restriction modification systems. Resistance can also be attributed to an adaptive immune response in bacteria known as CRISPR-Cas systems (clustered regularly interspaced short palindromic repeats—CRISPR-associated system) (Barrangou et al., 2007), whereby bacteria retain viral elements in their genomes from previous infections that enables them to recognize and "snip" incoming viral genomes during early infection, thus stopping viral replication. The recent discovery of CRISPR-Cas in approximately 45% of all known bacteria and up to 90% of archaea (Hille and Charpentier, 2016) has yet to be properly explored in the marine environment where it has received scant attention (Wietz et al., 2014).

Marine cyanobacteria display resistance to cooccurring cyanophages. Resistance in the cyanobacterium *Synechococcus* is attributed to rapid and extensive coevolution between host and virus that results in high diversity of both the host and the virus (Marston et al., 2012). Spontaneous resistance has been reported in the cyanobacterium *Prochlorococcus*. Interestingly, observations of evolving resistant *Prochlorococcus* indicate they can improve growth rate and reduce resistance range over generations (Avrani and Lindell, 2015). The evolving arms race between *Prochlorococcus* and its phages helps explain the observed genetic diversity of the bacterial host, with varying susceptibility and resistance to viral infection, resulting in rapidly growing *Prochlorococcus* with a range of resistance to phages (Schwartz and Lindell, 2017).

Resistance to viral infection in marine eukaryotes can vary. *Heterosigma akashiwo* is a harmful bloom-forming algal species that is infected by different strains of the virus HaV. Some *H. akashiwo* hosts exhibit complete resistance to viral infection while others are completely susceptible or show susceptibility only to a subset of strains (Nagasaki, 2008; Tomaru et al., 2004). Two strains of the ssRNA virus HcRNAV that infect the red-tide causing alga *Heterocapsa circularisquama* infect different host strains, demonstrating high host specificity. These examples illustrate how the diversity of viral interactions with hosts may allow different strains to coexist within a shared environment. The occurrence of resistance mutations can limit the continuous arms race between host and phage by presenting a genetic barrier for phage that lends an evolutionary advantage to the host. Host resistance to viruses can exact a pleiotropic fitness cost, such as reduced growth rate or increased susceptibility to infection by other viral strains (Hall et al., 2011). The PPE prasinophyte species, *Ostreococcus tauri*, is infected by prasinoviruses (prasinoviruses are a genus of algal virus within the *Phycodnaviridae* family; King et al., 2011) and has been shown to respond to viral infection with three resistance types—susceptible cells, resistant cells, and resistant producers that do not lyse but instead slowly release viruses (Yau et al., 2016). Two chromosomes in *O. tauri* appear to play a role in viral resistance and are described as the outlier chromosomes—big and small—with the small outlier chromosome showing hypervariability (Yau et al., 2016). Similar chromosomes have been identified in other prasinophytes, namely *Bathycoccus prasinos* (Moreau et al., 2012) and *M. pusilla* (Worden et al., 2009). The cost of resistance to viruses in *O. tauri* is not entirely clear, as studies assessing growth rate, cell size, and chlorophyll content (Heath et al., 2017) were not unequivocal, with no apparent fitness cost observed (Heath and Collins, 2016). The role of the outlier chromosomes in *O. tauri* may explain this but further viral resistance experiments in this particular host–virus system are required.

2.6 HGT

Interesting outcomes of HGT between viruses and marine microbial hosts have been described, including the presence of auxiliary metabolic genes (AMGs) in viral genomes. One of the earliest reports was of a S-PM2 cyanophage, that acquired core photosystem II (PSII) genes from the genome of its *Synechococcus* host (Mann et al., 2003), an interesting phenomenon subsequently also reported in cyanophages infecting *Prochlorococcus*

(Lindell et al., 2004, 2005). These genes are homologous to the *psbA* and *psbD* genes that encode the D1 and D2 proteins involved in photosynthesis. The virus retains these genes, which are virally encoded during infection and ensure that the virus delays the onset of damage and reduced activity within host photosynthesis centers, thus prolonging photosynthesis to provide energy for viral replication (Puxty et al., 2015). From an early stage in infection, infected cyanobacteria are limited in their ability to fix CO_2 but photosynthetic electron transport remains unaffected (Puxty et al., 2016). Cyanophages thus use acquired photosynthesis genes to support phage replication.

On a wider scale, this phenomenon may have implications for global warming if marine CO_2 fixation is reduced by viral infection. *Synechococcus* and *Prochlorococcus* contribute as much as 10% of global photosynthesis (Field et al., 1998). Cyanophages that encode genes involved in photosynthetic electron transport and central carbon metabolism thus have potential to influence global primary productivity. As many as 88% of cyanophage genomes in a single survey were seen to contain photosynthesis genes (Sullivan et al., 2006), and these genes likewise have been found in sipho-, myo-, and podoviruses. As these genes are not confined to a single family, they serve as useful genetic markers to examine viral diversity and distribution (Sandaa and Larsen, 2006; Sullivan et al., 2006; Zeidner et al., 2005).

Another example of HGT between a virus and its eukaryotic host is that of the acquisition by the coccolithovirus EhV-86 of an entire metabolic pathway from its coccolithophorid host *E. huxleyi* (Monier et al., 2009). The EhV-86 genome was found to encode genes involved in de novo sphingolipid biosynthesis (Wilson et al., 2005b), never previously reported in a viral genome. Sphingolipids are essential components of eukaryotic cellular membranes and are also crucial in cellular signaling, for example, in apoptosis (Aguilera-Romero et al., 2014). Earlier studies postulated that a metabolic rewiring of the host's pathway resulted in downregulation of the host's genes and concomitant upregulation of the virus' corresponding copies (Rosenwasser et al., 2014). Most recently, a functional analysis of the virally encoded rate-limiting enzyme serine palmitoyltransferase, which forms a committed step in the pathway, was seen to lead to the production of a unique group of viral-specific glycosphingolipids (Ziv et al., 2016) identified as essential for infectivity and assembly of virions in the host.

An ammonium transporter protein encoded by the prasinovirus, OtV-6, that infects the PPE *Ostreococcus* has been identified as being derived from the prasinophyte host (Monier et al., 2017). During viral infection, the virally encoded transporter gene is transcribed enabling the host to increase substrate affinity and to access a range of nitrogen sources. Nitrogen can be a limiting factor in phytoplankton growth and the discovery of this viral gene demonstrates how viruses can influence nutrient uptake by host cells. Phosphate is also a significant limiting factor in the oceans for marine phytoplankton growth and is a key substrate in viral replication, too (Monier et al., 2012; Wilson et al., 1996). Viruses infecting phytoplankton have also been identified as encoding phosphate transporter genes to increase uptake by the host cell during infection (Monier et al., 2012). Such manipulation of a host cell metabolism by a virus is a fascinating and ecologically important feature of viral infection.

3. VIRUSES IN COASTAL WATERS

As noted, viral abundance is typically higher in coastal waters than in open oceans. Abundance is predominantly highest in the euphotic zone, where hosts are present in greatest numbers, and decreases exponentially with depth; however, several studies have reported unexpected viral abundance and production in subsurface sites (Boehme et al., 1993; Hara et al., 1996; Wommack and Colwell, 2000), perhaps explained by virus absorption to sediment particles, lower viral decay rates, and poorly understood microbial population dynamics (Engelhardt et al., 2014). In coastal waters, cyanophages of the genetically diverse virus family *Myoviridae* that infect *Synechococcus* species can reach concentrations greater than $10^5\,mL^{-1}$ although values may fluctuate seasonally (McDaniel et al., 2006) and with salinity, temperature, and host abundance (Suttle and Chan, 1994; Wilson et al., 1993). The higher incidence of encounter between *Synechococcus* and cyanophages in near-shore waters results in selection for phage resistance in hosts; as encounters decline further offshore, host communities exhibit lower resistance (Suttle, 2002). Viral abundance in coastal waters varies seasonally (Bergh et al., 1989; Jiang and Paul, 1994) and is typically higher in summer and autumn than in winter (Wommack and Colwell, 2000). An extended time series in the Sargasso Sea revealed that viral abundance followed seasonally recurrent patterns related to water-column stability and the distribution of bacterial hosts (Parsons et al., 2012). Further multiyear time series are needed to more closely examine such fluctuations.

The best-studied examples of eukaryotic unicellular phytoplankton in coastal waters are the prasinophytes, including *M. pusilla* and *O. tauri* (reviewed in Weynberg et al., 2017b). Metagenomic studies have demonstrated prasinoviruses are highly prevalent in the aquatic environment (Lopez-Bueno et al., 2009). In marine systems, algal species such as *Micromonas* and *Ostreococcus* undergo transient blooms (O'Kelly et al., 2003) rather than massive large-scale blooms that suffer sudden crashes, often attributed to viral infection in so-called "bloom and bust" scenarios (Schroeder et al., 2003). It has thus been proposed that *Ostreococcus* and its associated viruses could serve as a highly suitable model for host–virus interaction studies (Weynberg et al., 2017b). Prasinoviruses infecting *O. tauri* (OtVs) have been detected and isolated from coastal and lagoon sites, whereas viruses infecting *O. lucimarinus* (OlVs) are found in more widespread locations, including oligotrophic waters of the Atlantic and Pacific Oceans (Bellec et al., 2010a, b; Derelle et al., 2015). Another genus of prasinoviruses, the *Micromonas* viruses (MVs), also occurs over a wide geographical scale. A study conducted in the Gulf of Mexico revealed that MV abundances fell from $10^5\,mL^{-1}$ in January to $10^3\,mL^{-1}$ in April, with rising seawater temperature (Cottrell and Suttle, 1995). These changes in viral abundance may reflect declines in host abundances, as host strains appear to grow poorly under higher temperatures. A similar pattern in both seasonal (Sahlsten, 1998) and vertical (Sahlsten and Karlson, 1998) distributions was reported for MVs in Scandinavian coastal waters.

3.1 Algal Bloom Dynamics and Viruses

Marine algal blooms are seasonally dependent events in which in a particular algal species expands to dominate an area and then sometimes suddenly diminishes. As a bloom develops, viral infection begins to increase following the KtW model; in this framework, frequency-dependent selection on abundant "winner" hosts that support high viral densities ultimately control the "winner's" population density (Bratbak et al., 1996; Castberg et al., 2001; Jacquet et al., 2002). The greatest support for this model is evident in blooms of certain algal species, such as *E. huxleyi* (Martinez et al., 2012; Pagarete et al., 2011), *H. akashiwo* (Nagasaki, 2008; Tarutani et al., 2000; Tomaru et al., 2004, 2008), and *Phaeocystis globosa* (Baudoux et al., 2006; Brussaard et al., 2007). For viruses to replicate within a dynamically complex environment such as an algal bloom, they need to display successful strategies such as short latent periods and large burst sizes, as well as means to out-compete grazers and survive other selection pressures (Brussaard, 2004b).

The coccolithophorid *E. huxleyi* is one of the best studied in terms of bloom development, progression, and termination due to viral infection. The collapse of blooms in mesocosm experiments coupled with the observation of large VLPs within cells and in surrounding seawater (Bratbak et al., 1993) has led to further research into the role of viruses in algal bloom control. Field studies in a Norwegian fjord (Bratbak et al., 1995), during a North Sea research cruise (Bratbak et al., 1996) and in the western English channel (Wilson et al., 2002) detailed how the collapse of *E. huxleyi* blooms was accompanied by increased presence in the water column of VLPs with morphology (icosahedral capsids of approximately 170 nm diameter) similar to that of the mesocosm viruses as determined by TEM, indicating similar viruses. Since this discovery, numerous coccolithoviruses have been isolated and characterized and whole genomes have been sequenced (Nissimov et al., 2012). Different strains of the coccolithovirus EhV have been distinguished using PCR targeting of the major capsid protein (Schroeder et al., 2003).

Although some isolates share similar characteristics such as host range, studies using the *mcp* gene as a marker indicated that genotypic diversity of coccolithoviruses can be high (Highfield et al., 2014; Martinez et al., 2007; Schroeder et al., 2003). In mesocosm experiments examining *E. huxleyi* blooms, differences in the *mcp* viral marker gene were identified with DGGE. DGGE revealed that a wide diversity of viral genotypes were present at bloom start but that only a few, which remained until the end, were responsible for bloom termination (Schroeder et al., 2003). Monitoring of a natural bloom in the English channel found that no single genotype dominated, suggesting that a complex assemblage of viruses, not just a single dominant genotype, can kill a bloom (Highfield et al., 2014). An additional study, which also used DGGE and marker sequencing, found that the genotypes of coccolithoviruses varied across space and with depth in the English channel, despite reports that the host population was genetically stable (Highfield et al., 2014). The authors concluded that host–virus interactions within *E. huxleyi* blooms are complex and likely influenced by environmental factors. For example, mesocosm experiments conducted in coastal waters revealed that viral infection of *E. huxleyi* is limited in phosphate-deplete conditions but unhindered in nutrient-rich conditions; phosphorus is essential for efficient marine viral replication (Maat and Brussaard, 2016; Wilson et al., 1996, 1998).

Viral infection of algae can influence large-scale phenomena including nutrient and biogeochemical cycling in the oceans, as noted earlier. Moreover, viral lysis of *E. huxleyi* blooms may potentially influence the weather and release of greenhouse gases. When *E. huxleyi* cells expire they release dimethylsulfoniopropionate, which is a major source of dimethyl sulfide (DMS), an important greenhouse gas (Bratbak et al., 1995). DMS is highly volatile and diffuses into the atmosphere from the water column. Through processes such as photooxidation, DMS can be converted to sulfate aerosols that serve as cloud condensation nuclei and in turn, increase cloud albedo (Charlson et al., 1987). Recently, 3°C increase in temperature was shown to lead to host resistance against infection by EhV, suggesting the potential for feedbacks between climate change and EhV influence on carbon cycling and DMS production (Kendrick et al., 2014).

The unicellular marine alga *H. akashiwo* forms toxic algal blooms in coastal waters and can cause large fish kills in temperate and subarctic regions along the Pacific Rim. Early reports of VLPs associated with the termination of *H. akashiwo* blooms were made in waters off the coast of Japan (Nagasaki et al., 1994). Since then, the effects of viruses on the abundance and physiology of *H. akashiwo* have been assessed. *H. akashiwo* virus (HaV) has been sequenced (Ogura et al., 2016) and found to be a dsDNA virus belonging to the *Phycodnaviridae* (Tarutani et al., 2000; Tomaru et al., 2004, 2008). Interestingly, HaV is not the only characterized virus that infects *H. akashiwo*. A ssRNA virus, *H. akashiwo* RNA virus (HaRNAV), assigned to the family *Marnaviridae* of the *Picornavirales* (King et al., 2011), also infects the same host (Lawrence et al., 2006). Analysis of the viral infection cycle revealed that the dsDNA HaV had shorter latent periods than HaRNAV but smaller burst sizes, thus indicating the need to consider species-specific traits when modeling host–virus interactions. Viruses have been found to at least partly regulate bloom formation in several other host taxa, including *Chrysochromulina brevifilum* (Suttle and Chan, 1995) and *Phaeocystis pouchetii* (Baudoux et al., 2006; Brussaard et al., 2007), which cause blooms in the North Atlantic and North Sea coastal waters, and *Aureococcus anophagefferens* (Moniruzzaman et al., 2016), which is a photosynthetic protist that causes brown tides along the US East coast.

4. VIRUSES IN THE OPEN OCEAN

Large-scale spatial studies of viral distribution in the Pacific and Southern Oceans have used flow cytometry to demonstrate that viral abundance in the surface 200 m of the water column is high in tropical and subtropical regions but lower in Antarctic waters (De Corte et al., 2016; Yang et al., 2010). Abundances of picoeukaryotes and cyanobacteria, namely

Synechococcus and *Prochlorococcus*, accounted for almost two-thirds of the variability in viral abundance, confirming the importance of cyanophages and prasinoviruses in the marine environment. A virus "hot spot" was identified in the mid-latitude region of the North Pacific supporting the notion that large-scale distribution patterns of viruses are affected by host distributions and physical processes (Yang et al., 2010). Cultivation-independent approaches, including metagenomics, have revealed that the Southern Ocean encircling Antarctica is dominated by lysogenic viruses (Brum et al., 2017) that are termed "seasonal time bombs" because they can switch to a lytic cycle as their bacterial host production increases. The dominance of prophages (temperate viruses) in the Southern Ocean gives marine viruses and their hosts in this region a distinct genetic structure as compared to those at lower latitudes. Only little is known about viruses in the Arctic Ocean and their interactions with algal hosts. For example, a recent study characterized prasinoviruses that infect *Micromonas* in the Arctic and found that rising temperatures shortened latent times and increased burst sizes during viral infection, suggesting that climate change may influence viral dynamics and PPE community structure (Maat et al., 2017).

Global ocean-going research surveys, such as the Global Ocean Survey (GOS) (Rusch et al., 2007) and the TARA Oceans survey (Hingamp et al., 2013), have been ambitious efforts to expand knowledge of the world's seas, including their microbial and viral inhabitants. The GOS was the first global study to assess surface ocean microbial communities (Nealson and Venter, 2007) and employed extensive sequencing efforts that helped to demonstrate the abundance, diversity, and genetic influence of viruses in the marine environment (Williamson et al., 2008). Most recently, the French-led TARA Oceans survey has set out to assess the taxonomic and functional diversity in ocean surface waters of organisms from viruses to fish. This global expedition has revealed more about the extent of virus diversity, not least that environmental factors, including salinity, temperature, and oxygen concentrations, shape viral community structure (Brum et al., 2013), and has enabled the mapping of giant viruses in the seas (Hingamp et al., 2013). Virome data collected have identified virally encoded AMGs in surface waters that appear to be involved in sulfur and nitrogen cycling (Roux et al., 2016). As described earlier, AMGs expressed by viruses during infection steer central host metabolic pathways, such as photosynthesis and nutrient acquisition, toward facilitating viral replication. Furthermore, virome data analyses from the TARA expeditions have helped to establish a global oceanic virome dataset for dsDNA viruses that form an invaluable resource for assessing viral diversity and influence on host populations across a wide geographical expanse (Brum et al., 2015).

Bacteriophages were most abundant in the GOS survey followed by giant dsDNA viruses of the proposed family *Megaviridae* (Monier et al., 2008). The *Megaviridae* includes *Mimivirus*, *Mamavirus*, *Megavirus*, and other marine viruses such as *P. pouchetii* virus (PpV) and *Chrysochromulina ericina* virus (CeV) (Colson et al., 2013). Intriguingly, the *Megaviridae* identified in the GOS appeared to be associated with oomycetes, filamentous microorganisms that resemble fungi but are actually phylogenetically placed with stramenopiles (Hingamp et al., 2013). It has been postulated that the *Megaviridae* and oomycetes may share a common host or form direct virus/host interaction. Further studies will be needed to elucidate their relationship.

The Pacific Ocean Virome (POV) (Hurwitz and Sullivan, 2013) is another significantly important curated dsDNA virome dataset that comprises 32 quantitatively representative viromes collected from different depths and seasons in transects from coastal waters to open-ocean waters in the Pacific Ocean. The POV study demonstrated that richness of viruses in the Pacific decreased from deeper to surface waters, from winter to summer, and, in surface layers, with distance from the shore. In other work, genomic data from the Malaspina 2010–11 Circumnavigation Expedition, which assessed pelagic processes along the Indian, Pacific, and Atlantic Oceans (Duarte, 2015), along with data collected during the Tara expedition, were used to develop a global virome map of dsDNA viruses sampled in surface and deep-ocean waters (Roux et al., 2016). These expeditions utilized metagenomics approaches to triple the number of known marine viral populations. The authors found that 38 of 867 viral clusters were locally or globally abundant and represented almost half of the viral populations in any given GOV. Such large datasets are pivotal in expanding our knowledge of marine viruses on a global scale and are invaluable in unveiling uncultivated novel viruses and generating more robust marine ecosystem models.

A remaining challenge is to extend such established metagenomic pipelines to include ssDNA and RNA viromes. RNA viruses, particularly picorna-like viruses, have been speculated to be highly abundant in both coastal and open ocean waters, infecting eukaryotes, and outnumbering DNA viruses (Culley et al., 2003, 2006, 2014). However, difficulties in quantifying RNA viruses, due in part to the smaller size of their genomes, which is below the limit of detection using fluorescent methods, have to date prevented extensive confirmation of their abundance and ecological contributions (Steward et al., 2013).

5. VIRUSES ASSOCIATED WITH CORAL REEFS

Most work on marine viruses has concentrated on temperate coastal waters and the open ocean. It has only been during the last 10–15 years that viruses associated with tropical coral reefs have been investigated. Corals are animal invertebrates that form complex associations with bacteria, protists, fungi, archaea, and an important symbiotic relationship with a dinoflagellate alga, *Symbiodinium*. The alga resides within the coral's tissues and can provide as much as 95% of the coral's food via photosynthetic exudate (Muscatine and Weis, 1992). Coral bleaching is the phenomenon that results when this symbiosis breaks down.

Viruses can infect all organisms within this coral holobiont (Rohwer et al., 2002). Unlike in open oceans, where prokaryotes and bacteriophages dominate, coral reefs are home to a variety of both eukaryotic viruses and phages, including several NCLDV families, *Retroviridae*, *Herpesvirales*, and *Caudovirales* (Vega Thurber et al., 2017; Weynberg et al., 2017c). Sponges are important filter feeders within coral reefs and can filter thousands of liters of seawater daily. Despite early reports of VLPs in sponge tissue (Vacelet and Gallissian, 1978) and viral predation by sponges as a modeled nutrient flow pathway (Hadas et al., 2006), research into viral interactions with sponges is lacking. A bioinformatic pipeline, HoloVir (Laffy et al., 2016), is now available for analyzing metagenomic data derived from the viral fraction of coral reefs and recently a sponge virome has been described and compared with coral viromes.

Microscopy of coral tissue has revealed the presence of viruses in both tissue and the mucus layer that forms over living coral tissue (Hanh et al., 2015; Patten et al., 2008; Pollock et al., 2014). The coral photosymbiont has been established in laboratory cultures. TEM examination of cultured *Symbiodinium* cells has revealed VLPs in the nuclei and other cellular compartments (Lawrence et al., 2014; Weynberg et al., 2017a; Wilson et al., 2001). Through a transcriptomic analysis of *Symbiodinium* under temperature stress, a novel viral genome was assembled that shares some homology with a characterized ssRNA dinoflagellate virus that infects the dinoflagellate *H. circularisquama* (Levin et al., 2016). A molecular assay has recently been developed, which will enable the detection of this dinornavirus in the field (Montalvo-Proano et al., 2017) and potentially provide future insights into the role of this virus in coral ecology, including bleaching events.

Metagenomics studies on coral reef viromes have revealed that the most dominant families of viruses in corals, as in coastal waters and open oceans, are the *Siphoviridae*, *Myoviridae*, and *Podoviridae*, members of the tailed bacteriophage virus order, the *Caudovirales* (Vega Thurber et al., 2008; Weynberg et al., 2014). Viral communities are seen to shift between healthy, diseased, and bleached states in corals but a core virome has emerged from sequencing studies. A metaanalysis of coral sequencing studies has revealed approximately 60 virus families associated with corals (Wood-Charlson et al., 2015). The core virome is comprised of 9–12 virus families with dsDNA genomes, ssDNA genomes, and retrotranscribing RNA viruses (reviewed in Vega Thurber et al., 2017). Dominant eukaryotic viruses belonging to the *Phycodnaviridae* and members of the giant virus family *Mimiviridae* are abundant in corals; additional NCLDV families that associate with corals include *Poxviridae*, *Ascoviridae*, and *Iridioviridae*. Signatures of Herpes-like viruses (dsDNA) have been reported in corals too (Vega Thurber et al., 2008) but these are described as atypical in terms of their observed location inside cells and restricted homology to known herpes virus sequences (Correa et al., 2016). Members of the *Circoviridae* (ssDNA genomes), and ssRNA retroviruses are also prevalent in corals (Correa et al., 2012, 2016; Weynberg et al., 2014).

The "Piggyback the Winner" model is based on data collected within coral reefs in the Pacific and Atlantic Oceans and predicts, in contrast to the "Kill-the-Winner" model, that viruses switch to lysogeny when microbial host abundance is high (Knowles et al., 2016). Model testing included assessment of 24 coral viromes, which found that marker genes indicative of lysogenic viruses increased with microbial host abundance, resulting in a significant decrease in virus-to-microbe ratios in seawater. This model posits that as host microbial numbers rise, viruses switch to lysogeny, thus piggybacking on the success of their microbial host's growth and population increase. Viruses remain dormant within host cells and are replicated only during host cell division, thereby slowly increasing their numbers before eventually switching to lysis and killing their hosts. This confers advantages on the virus as during the lysogenic phase there is no need to compete for host cells or to evade the immune response of the host.

5.1 Coral Health, Disease, Bleaching, and Viruses

Corals continually produce a mucus layer of polysaccharides and glycoproteins onto their outer surfaces as a protective buffer against the external environment (Brown and Bythell, 2005; Bythell and Wild, 2011). This surface mucus layer is estimated to contain $0.1–3 \times 10^7\,\text{mL}^{-1}$ viruses and $2–5 \times 10^6$ bacteria mL^{-1}—a 10–100-fold greater concentration than in the surrounding seawater (Hanh et al., 2015; Leruste et al., 2012; Sweet and Bythell, 2017). Bacteriophages are known to be part of the core coral virome and likely perform both negative and positive roles in terms of coral health and disease, as they can target both beneficial and pathogenic bacterial hosts associated with the coral holobiont. Between 20 and 30 coral diseases have been described (Bourne et al., 2009; Harvell et al., 2007; Willis et al., 2004) but, to date, no disease profile has been clearly attributed to a viral infection. It is highly unlikely that viruses are not involved in at least one of these diseases but this has yet to be proven.

The best-characterized coral disease is black band disease (BBD), which is caused by a multiconsortium of bacteria that form a black mat over living coral tissue. The mat moves at a rate of 2–4 mm per day (Sato et al., 2016) and creates a high sulfide anoxic environment that leads the underlying coral tissue to disintegrate and die (Bourne et al., 2011, 2013). The dominant species in the BBD mat is a filamentous cyanobacterium that harbors lysogenic prophages that may add virulence to the bacterial host (Buerger et al., 2016). Viral-enhanced virulence has been likewise hypothesized to contribute to another coral disease (white syndrome), caused by the bacterium *Vibrio coralliilyticus* (Ben-Haim et al., 2003). Thus, the virulence of BBD mats and white syndrome may be increased by viral infection in a manner like that of *V. cholerae*, in which the bacterial pathogen that causes cholera outbreaks shows increased virulence due to the presence of the CTX prophage (Waldor and Mekalanos, 1996; Weynberg et al., 2015).

The presence of a CRISPR-Cas system in the BBD dominant cyanobacterium genome indicates an adaptive immune response against viral infection (Buerger et al., 2016), suggesting that there is a dynamic "arms race" between virus and cyanobacterial host that shapes the virulence of coral disease. Importantly, phages in the coral holobiont can exert top-down control on pathogenic and virulent bacteria, thus counteracting these disease effects and promoting coral health. The bacteriophage adherence immunity model (Barr et al., 2013) has demonstrated how phage within mucus layers, as those found in corals, can adhere to bacteria in ways that decrease microbial colonization, drive bacterial lysis, and control pathogenicity. Viral control of harmful bacteria raises the possibility of phage therapy for diseased corals, whereby phages could be delivered to prevent, treat, and mitigate pathogenesis caused by bacteria. A small-scale pilot study using this approach was reported in the Red Sea (Atad et al., 2012) but further controlled experiments are needed. The nascent but growing field of coral reef virology holds much promise for future understanding of these ecologically important areas of the world's oceans.

6. SUMMARY

Viruses are the most abundant and diverse biological entities in the world's oceans with as many as 10 million viruses in just a single drop of seawater. Viral infection of cyanobacteria and marine algae has major ecological and environmental effects on phenomena across scales, from the dynamics of host–virus interactions to global biogeochemical cycling. Marine viruses play key roles in HGT and evolution, as well as population and community structure. Viral lysis of host cells contributes to the DOM pool, increasing the availability of surface water nutrients for microbial consumption and diverting them from higher trophic levels. Viral lysis of important primary producers in the marine environment has major implications for nutrient and energy cycling globally. From early anecdotal reports of the presence of viruses in marine samples, the field of marine virology has led to intensive investigations over the past 30 years, which have demonstrated that cyanophages and algal viruses are ubiquitous and highly important members of the microbial assemblage and oceanic ecosystems. Approaches such as metagenomics have vastly expanded our understanding and identification of viruses in the marine environment and has helped reveal links to previously unidentified and uncultivated hosts. Future challenges include the development of robust ecological models based on detailed studies of species-specific host–virus interactions in marine ecosystems. In the future, marine virology will likely produce many new groundbreaking discoveries about the "dark matter" of the world's oceans.

REFERENCES

Ackermann, H.W., 2006. In: Calendar, R. (Ed.), The Bacteriophages. Oxford University Press, New York, pp. 8–16.

Aguilera-Romero, A., Gehin, C., Riezman, H., 2014. Sphingolipid homeostasis in the web of metabolic routes. BBA—Mol. Cell Biol. L. 1841, 647–656. https://doi.org/10.1016/j.bbalip.2013.10.014.

Allen, L.Z., et al., 2011. Single virus genomics: a new tool for virus discovery. PLoS One 6, e17722, https://doi.org/10.1371/journal.pone.0017722.

Allers, E., et al., 2013. Single-cell and population level viral infection dynamics revealed by phageFISH, a method to visualize intracellular and free viruses. Environ. Microbiol. 15, 2306–2318. https://doi.org/10.1111/1462-2920.12100.

Alonso, M.C., Jimenez-Gomez, F., Rodriguez, J., Borrego, J.J., 2001. Distribution of virus-like particles in an oligotrophic marine environment (Alboran Sea, western Mediterranean). Microb. Ecol. 42, 407–415. https://doi.org/10.1007/s00248-001-0015-y.

Atad, I., Zvuloni, A., Loya, Y., Rosenberg, E., 2012. Phage therapy of the white plague-like disease of Favia favus in the Red Sea. Coral Reefs 31, 665–670. https://doi.org/10.1007/s00338-012-0900-5.

Attoui, H., et al., 2006. Micromonas pusilla reovirus: a new member of the family Reoviridae assigned to a novel proposed genus (Mimoreovirus). J. Gen. Virol. 87, 1375–1383. https://doi.org/10.1099/vir.0.81584-0.

Avrani, S., Lindell, D., 2015. Convergent evolution toward an improved growth rate and a reduced resistance range in Prochlorococcus strains resistant to phage. Proc. Natl. Acad. Sci. U. S. A. 112, E2191–E2200. https://doi.org/10.1073/pnas.1420347112.

Azam, F., et al., 1983. The ecological role of water-column microbes in the sea. Mar. Ecol. Prog. Ser. 10, 257–263. https://doi.org/10.3354/meps010257.

Barr, J.J., et al., 2013. Bacteriophage adhering to mucus provide a non-host-derived immunity. Proc. Natl. Acad. Sci. 110, 10771–10776. https://doi.org/10.1073/pnas.1305923110.

Barrangou, R., et al., 2007. CRISPR provides acquired resistance against viruses in prokaryotes. Science 315, 1709.

Baudoux, A.C., Noordeloos, A.A.M., Veldhuis, M.J.W., Brussaard, C.P.D., 2006. Virally induced mortality of Phaeocystis globosa during two spring blooms in temperate coastal waters. Aquat. Microb. Ecol. 44, 207–217. https://doi.org/10.3354/ame044207.

Bellec, L., Grimsley, N., Derelle, E., Moreau, H., Desdevises, Y., 2010a. Abundance, spatial distribution and genetic diversity of Ostreococcus tauri viruses in two different environments. Environ. Microbiol. Rep. 2, 313–321. https://doi.org/10.1111/j.1758-2229.2010.00138.x.

Bellec, L., Grimsley, N., Desdevises, Y., 2010b. Isolation of prasinoviruses of the green unicellular algae Ostreococcus spp. on a worldwide geographical scale. Appl. Environ. Microbiol. 76, 96–101. https://doi.org/10.1128/aem.01799-09.

Ben-Haim, Y., et al., 2003. Vibrio coralliilyticus sp nov., a temperature-dependent pathogen of the coral Pocillopora damicornis. Int. J. Syst. Evol. Microbiol. 53, 309–315. https://doi.org/10.1099/ijs.0.02402-0.

Bergh, O., Borsheim, K.Y., Bratbak, G., Heldal, M., 1989. High abundance of viruses found in aquatic environments. Nature 340, 467–468. https://doi.org/10.1038/340467a0.

Biller, S.J., et al., 2017. Membrane vesicles in sea water: heterogeneous DNA content and implications for viral abundance estimates. ISME J. 11, 394–404. https://doi.org/10.1038/ismej.2016.134.

Boehme, J., et al., 1993. Viruses, bacterioplankton, and phytoplankton in the southeastern Gulf of Mexico—distribution and contribution to oceanic DNA pools. Mar. Ecol. Prog. Ser. 97, 1–10. https://doi.org/10.3354/meps097001.

Bourne, D.G., et al., 2009. Microbial disease and the coral holobiont. Trends Microbiol. 17, 554–562. https://doi.org/10.1016/j.tim.2009.09.004.

Bourne, D.G., Muirhead, A., Sato, Y., 2011. Changes in sulfate-reducing bacterial populations during the onset of black band disease. ISME J. 5, 559–564. https://doi.org/10.1038/ismej.2010.143.

Bourne, D.G., van der Zee, M.J.J., Botte, E.S., Sato, Y., 2013. Sulfur-oxidizing bacterial populations within cyanobacterial dominated coral disease lesions. Environ. Microbiol. Rep. 5, 518–524. https://doi.org/10.1111/1758-2229.12055.

Bratbak, G., Egge, J.K., Heldal, M., 1993. Viral mortality of the marine alga Emiliania huxleyi (Haptophyceae) and termination of algal blooms. Mar. Ecol. Prog. Ser. 93, 39–48. https://doi.org/10.3354/meps093039.

Bratbak, G., et al., 1995. Viral activity in relation to Emiliania huxleyi blooms: a mechanism of DMSP release? Mar. Ecol. Prog. Ser. 128, 133–142. https://doi.org/10.3354/meps128133.
Bratbak, G., Wilson, W., Heldal, M., 1996. Viral control of Emiliania huxleyi blooms? J. Mar. Syst. 9, 75–81. https://doi.org/10.1016/0924-7963(96)00018-8.
Brown, B.E., Bythell, J.C., 2005. Perspectives on mucus secretion in reef corals. Mar. Ecol. Prog. Ser. 296, 291–309. https://doi.org/10.3354/meps296291.
Brum, J.R., Sullivan, M.B., 2015. Rising to the challenge: accelerated pace of discovery transforms marine virology. Nat. Rev. Microbiol. 13, 147. https://doi.org/10.1038/nrmicro3404.
Brum, J.R., Schenck, R.O., Sullivan, M.B., 2013. Global morphological analysis of marine viruses shows minimal regional variation and dominance of non-tailed viruses. ISME J. 7, 1738–1751. https://doi.org/10.1038/ismej.2013.67.
Brum, J.R., et al., 2015. Patterns and ecological drivers of ocean viral communities. Science 348. https://doi.org/10.1126/science.1261498.
Brum, J.R., Hurwitz, B.L., Schofield, O., Ducklow, H.W., Sullivan, M.B., 2017. Seasonal time bombs: dominant temperate viruses affect Southern Ocean microbial dynamics. ISME J. 11, 588.
Brussaard, C.P.D., 2004a. Optimization of procedures for counting viruses by flow cytometry. Appl. Environ. Microbiol. 70, 1506–1513. https://doi.org/10.1128/aem.70.3.1506-1513.2004.
Brussaard, C.P., 2004b. Viral control of phytoplankton populations—a review. J. Eukaryot. Microbiol. 51, 125–138.
Brussaard, C.P.D., Bratbak, G., Baudoux, A.C., Ruardij, P., 2007. Phaeocystis and its interaction with viruses. Biogeochemistry 83, 201–215. https://doi.org/10.1007/s10533-007-9096-0.
Bruwer, J.D., Agrawal, S., Liew, Y.J., Aranda, M., Voolstra, C.R., 2017. Association of coral algal symbionts with a diverse viral community responsive to heat shock. BMC Microbiol. 17. https://doi.org/10.1186/s12866-017-1084-5.
Buerger, P., Wood-Charlson, E.M., Weynberg, K.D., Willis, B.L., van Oppen, M.J.H., 2016. CRISPR-Cas defense system and potential prophages in cyanobacteria associated with the coral black band disease. Front. Microbiol. 7. https://doi.org/10.3389/fmicb.2016.02077.
Bythell, J.C., Wild, C., 2011. Biology and ecology of coral mucus release. J. Exp. Mar. Biol. Ecol. 408, 88–93. https://doi.org/10.1016/j.jembe.2011.07.028.
Castberg, T., et al., 2001. Microbial population dynamics and diversity during a bloom of the marine coccolithophorid Emiliania huxleyi (Haptophyta). Mar. Ecol. Prog. Ser. 221, 39–46. https://doi.org/10.3354/meps221039.
Charlson, R.J., Lovelock, J.E., Andreae, M.O., Warren, S.G., 1987. Oceanic phytoplankton, atmospheric sulfur, cloud albedo and climate. Nature 326, 655–661. https://doi.org/10.1038/326655a0.
Chen, F., Suttle, C.A., 1995. Amplification of DNA polymerase gene fragments from viruses infecting microalgae. Appl. Environ. Microbiol. 61, 1274–1278.
Chiura, H.X., Uchiyama, N., Kogure, K., 2009. Broad-host range gene transporter particles produced by Aliivibrio fischeri. Microbes Environ. 24, 322–329. https://doi.org/10.1264/jsme2.ME09153.
Clokie, M.R.J., Millard, A.D., Letarov, A.V., Heaphy, S., 2011. Phages in nature. Bacteriophage 1, 31–45. https://doi.org/10.4161/bact.1.1.14942.
Colson, P., et al., 2013. "Megavirales", a proposed new order for eukaryotic nucleocytoplasmic large DNA viruses. Arch. Virol. 158, 2517–2521. https://doi.org/10.1007/s00705-013-1768-6.

Comeau, A.M., Chan, A.M., Suttle, C.A., 2006. Genetic richness of vibriophages isolated in a coastal environment. Environ. Microbiol. 8, 1164–1176. https://doi.org/10.1111/j.1462-2920.2006.01006.x.

Correa, A.M.S., Welsh, R.M., Vega Thurber, R.L., 2012. Unique nucleocytoplasmic dsDNA and +ssRNA viruses are associated with the dinoflagellate endosymbionts of corals. ISME J 7, 13–27. http://www.nature.com/ismej/journal/v7/n1/suppinfo/ismej201275s1.html.

Correa, A.M.S., et al., 2016. Viral outbreak in corals associated with an in situ bleaching event: a typical herpes-like viruses and a new megavirus infecting symbiodinium. Front. Microbiol. 7. https://doi.org/10.3389/fmicb.2016.00127.

Cottrell, M.T., Suttle, C.A., 1995. Dynamics of a lytic virus infecting the photosynthetic marine picoflagellate Micromonas pusilla. Limnol. Oceanogr. 40, 730–739.

Crane, M., Hyatt, A., 2011. Viruses of fish: an overview of significant pathogens. Virus 3, 2025–2046. https://doi.org/10.3390/v3112025.

Culley, A.I., Lang, A.S., Suttle, C.A., 2003. High diversity of unknown picorna-like viruses in the sea. Nature 424, 1054–1057. https://doi.org/10.1038/nature01886.

Culley, A.I., Lang, A.S., Suttle, C.A., 2006. Metagenomic analysis of coastal RNA virus communities. Science 312, 1795–1798. https://doi.org/10.1126/science.1127404.

Culley, A.I., et al., 2014. The characterization of RNA viruses in tropical seawater using targeted PCR and metagenomics. mBio 5. https://doi.org/10.1128/mBio.01210-14.

D'Angelo, C., Wiedenmann, J., 2014. Impacts of nutrient enrichment on coral reefs: new perspectives and implications for coastal management and reef survival. Curr. Opin. Environ. Sustain. 7, 82–93. https://doi.org/10.1016/j.cosust.2013.11.029.

Davidson, A.R., Cardarelli, L., Pell, L.G., Radford, D.R., Maxwell, K.L., 2012. In: Rossmann, M.G., Rao, V.B. (Eds.), Viral Molecular Machines. Springer, USA, pp. 115–142.

Davy, J.E., Patten, N.L., 2007. Morphological diversity of virus-like particles within the surface microlayer of scleractinian corals. Aquat. Microb. Ecol. 47, 37–44. https://doi.org/10.3354/ame047037.

De Corte, D., et al., 2010. Links between viral and prokaryotic communities throughout the water column in the (sub)tropical Atlantic Ocean. ISME J. 4, 1431–1442. https://doi.org/10.1038/ismej.2010.65.

De Corte, D., Sintes, E., Yokokawa, T., Lekunberri, I., Herndl, G.J., 2016. Large-scale distribution of microbial and viral populations in the South Atlantic Ocean. Environ. Microbiol. Rep. 8, 305–315. https://doi.org/10.1111/1758-2229.12381.

Derelle, E., et al., 2015. Diversity of viruses infecting the green microalga ostreococcus lucimarinus. J. Virol. 89, 5812–5821. https://doi.org/10.1128/jvi.00246-15.

Duarte, C.M., 2015. Seafaring in the 21st century: the Malaspina 2010 circumnavigation expedition. Limnol. Oceanogr. Bull. 24, 11–14.

Dubinsky, Z., 1990. Coral Reefs. Elsevier, Amsterdam, New York.

Dunlap, D.S., et al., 2013. Molecular and microscopic evidence of viruses in marine copepods. Proc. Natl. Acad. Sci. 110, 1375.

Engelhardt, T., Kallmeyer, J., Cypionka, H., Engelen, B., 2014. High virus-to-cell ratios indicate ongoing production of viruses in deep subsurface sediments. ISME J. 8, 1503–1509. https://doi.org/10.1038/ismej.2013.245.

Fereidouni, S., Munoz, O., Von Dobschuetz, S., De Nardi, M., 2016. Influenza virus infection of marine mammals. EcoHealth 13, 161–170. https://doi.org/10.1007/s10393-014-0968-1.

Field, C.B., Behrenfeld, M.J., Randerson, J.T., Falkowski, P., 1998. Primary production of the biosphere: integrating terrestrial and oceanic components. Science 281, 237–240. https://doi.org/10.1126/science.281.5374.237.

Fisher, R. et al. Species richness on coral reefs and the pursuit of convergent global estimates. Curr. Biol. 25, 2015, 500–505, https://doi.org/10.1016/j.cub.2014.12.022.

Frada, M.J., et al., 2014. Zooplankton may serve as transmission vectors for viruses infecting algal blooms in the ocean. Curr. Biol. 24, 2592–2597. https://doi.org/10.1016/j.cub.2014.09.031.
Fuhrman, J.A., 1999. Marine viruses and their biogeochemical and ecological effects. Nature 399, 541. https://doi.org/10.1038/21119.
Fuhrman, J.A., Sleeter, T.D., Carlson, C.A., Proctor, L.M., 1989. Dominance of bacterial biomass in the Sargasso Sea and its ecological implications. Mar. Ecol. Prog. Ser. 57, 207–217. https://doi.org/10.3354/meps057207.
Haaber, J., Middelboe, M., 2009. Viral lysis of Phaeocystis pouchetii: implications for algal population dynamics and heterotrophic C, N and P cycling. ISME J 3, 430–441.
Hadas, E., Marie, D., Shpigel, M., Ilan, M., 2006. Virus predation by sponges is a new nutrient-flow pathway in coral reef food webs. Limnol. Oceanogr. 51, 1548–1550.
Hall, A.R., Scanlan, P.D., Morgan, A.D., Buckling, A., 2011. Host–parasite coevolutionary arms races give way to fluctuating selection. Ecol. Lett. 14, 635–642. https://doi.org/10.1111/j.1461-0248.2011.01624.x.
Hanh, N.-K., et al., 2015. Coral mucus is a hot spot for viral infections. Appl. Environ. Microbiol. 81, 5773–5783. https://doi.org/10.1128/aem.00542-15.
Hara, S., Koike, I., Terauchi, K., Kamiya, H., Tanoue, E., 1996. Abundance of viruses in deep oceanic waters. Mar. Ecol. Prog. Ser. 145, 269–277.
Harvell, D., et al., 2007. Coral disease, environmental drivers and the balance between coral and microbial associates. Oceanography 20, 172–195.
Heath, S.E., Collins, S., 2016. Mode of resistance to viral lysis affects host growth across multiple environments in the marine picoeukaryote Ostreococcus tauri. Environ. Microbiol. 18, 4628–4639. https://doi.org/10.1111/1462-2920.13586.
Heath, S.E., Knox, K., Vale, P.F., Collins, S., 2017. Virus resistance is not costly in a marine alga evolving under multiple environmental stressors. Viruses-Basel 9. https://doi.org/10.3390/v9030039.
Hennes, K.P., Suttle, C.A., 1995. Direct counts of viruses in natural waters and laboratory cultures by epifluorescence microscopy. Limnol. Oceanogr. 40, 1050–1055. https://doi.org/10.4319/lo.1995.40.6.1050.
Highfield, A., Evans, C., Walne, A., Miller, P.I., Schroeder, D.C., 2014. How many coccolithovirus genotypes does it take to terminate an Emiliania huxleyi bloom? Virology 466, 138–145. https://doi.org/10.1016/j.virol.2014.07.017.
Hille, F., Charpentier, E., 2016. CRISPR-Cas: biology, mechanisms and relevance. Philosophical Transactions of the Royal Society B: Biological Sciences 371, 20150496. https://doi.org/10.1098/rstb.2015.0496.
Hingamp, P., et al., 2013. Exploring nucleo-cytoplasmic large DNA viruses in Tara Oceans microbial metagenomes. ISME J. 7 (9), 1678–1695. https://doi.org/10.1038/ismej.2013.59.
Hobbie, J.E., Daley, R.J., Jasper, S., 1977. Use of nuclepore filters for counting bacteria by fluorescence microscopy. Appl. Environ. Microbiol. 33, 1225–1228.
Holmfeldt, K., Middelboe, M., Nybroe, O., Riemann, L., 2007. Large variabilities in host strain susceptibility and phage host range govern interactions between lytic marine phages and their Flavobacterium hosts. Appl. Environ. Microbiol. 73, 6730–6739. https://doi.org/10.1128/aem.01399-07.
Hurwitz, B.L., Sullivan, M.B., 2013. The pacific ocean virome (pov): a marine viral metagenomic dataset and associated protein clusters for quantitative viral ecology. PLoS One 8. https://doi.org/10.1371/journal.pone.0057355.
Iyer, L.M., Aravind, L., Koonin, E.V., 2001. Common origin of four diverse families of large eukaryotic DNA viruses. J. Virol. 75, 11720–11734. https://doi.org/10.1128/jvi.75.23.11720-11734.2001.
Jacquet, S., et al., 2002. Flow cytometric analysis of an Emiliana huxleyi bloom terminated by viral infection. Aquat. Microb. Ecol. 27, 111–124. https://doi.org/10.3354/ame027111.

Jardillier, L., Zubkov, M.V., Pearman, J., Scanlan, D.J., 2010. Significant CO_2 fixation by small prymnesiophytes in the subtropical and tropical Northeast Atlantic Ocean. ISME J. 4, 1180–1192. https://doi.org/10.1038/ismej.2010.36.

Jiang, S.C., Paul, J.H., 1994. Seasonal and diel abundance of viruses and occurrence of lysogeny/bacteriocinogeny in the marine environment. Mar. Ecol. Prog. Ser. 104, 163–172. https://doi.org/10.3354/meps104163.

Jiang, S.C., Paul, J.H., 1996. Occurrence of lysogenic bacteria in marine microbial communities as determined by prophage induction. Mar. Ecol. Prog. Ser. 142, 27–38. https://doi.org/10.3354/meps142027.

Jiang, S.C., Paul, J.H., 1998. Significance of lysogeny in the marine environment: studies with isolates and a model of lysogenic phage production. Microb. Ecol. 35, 235–243. https://doi.org/10.1007/s002489900079.

Kauffman, K.M., et al., 2018. A major lineage of non-tailed dsDNA viruses as unrecognized killers of marine bacteria. Nature https://doi.org/10.1038/nature25474. https://www.nature.com/articles/nature25474#supplementary-information.

Kendrick, B.J., et al., 2014. Temperature-induced viral resistance in Emiliania huxleyi (Prymnesiophyceae). PLoS One 9, e112134, https://doi.org/10.1371/journal.pone.0112134.

King, A.M.Q., Adams, M.J., Carstens, E.B., Lefkowitz, E.J., 2011. Virus Taxonomy: Ninth Report of the International Committee on Taxonomy of Viruses. Elsevier Academic Press.

Knowles, B., et al., 2016. Lytic to temperate switching of viral communities. Nature 531, 466.

Koonin, E.V., Yutin, N., 2010. Origin and evolution of eukaryotic large nucleo-cytoplasmic DNA viruses. Intervirology 53, 284–292. https://doi.org/10.1159/000312913.

Laffy, P.W., et al., 2016. HoloVir: a workflow for investigating the diversity and function of viruses in invertebrate holobionts. Front. Microbiol. 7. https://doi.org/10.3389/fmicb.2016.00822.

Larsen, J.B., Larsen, A., Bratbak, G., Sandaa, R.A., 2008. Phylogenetic analysis of members of the Phycodnaviridae virus family, using amplified fragments of the major capsid protein gene. Appl. Environ. Microbiol. 74, 3048–3057. https://doi.org/10.7128/aem.02548-07.

Lawrence, J.E., Brussaard, C.P.D., Suttle, C.A., 2006. Virus-specific responses of Heterosigma akashiwo to infection. Appl. Environ. Microbiol. 72, 7829–7834. https://doi.org/10.1128/aem.01207-06.

Lawrence, S.A., Wilson, W.H., Davy, J.E., Davy, S.K., 2014. Latent virus-like infections are present in a diverse range of Symbiodinium spp. (Dinophyta). J. Phycol. 50, 977–997. https://doi.org/10.1111/jpy.12242.

Leiman, P.G., Shneider, M.M., 2012. In: Michael, G., Rossmann, , Rao, V.B. (Eds.), Viral Molecular Machines. Springer, USA, pp. 93–114.

Leitet C, Riemann B & A., H. Plasmids and prophages in Baltic Sea bacterioplankton isolates. J. Mar. Biol. Assoc. U. K. 86, 567–575 (2006).

Leruste, A., Bouvier, T., Bettarel, Y., 2012. Enumerating viruses in coral mucus. Appl. Environ. Microbiol. 78, 6377–6379. https://doi.org/10.1128/aem.01141-12.

Levin, R.A., Voolstra, C.R., Weynberg, K.D., Oppen, M.J.H., 2016. Evidence for a role of viruses in the thermal sensitivity of coral photosymbionts. ISME J 11, https://doi.org/10.1038/ismej.2016.154.

Lindell, D., et al., 2004. Transfer of photosynthesis genes to and from Prochlorococcus viruses. Proc. Natl. Acad. Sci. U. S. A. 101, 11013–11018. https://doi.org/10.1073/pnas.0401526101.

Lindell, D., Jaffe, J.D., Johnson, Z.I., Church, G.M., Chisholm, S.W., 2005. Photosynthesis genes in marine viruses yield proteins during host infection. Nature 438, 86–89. https://doi.org/10.1038/nature04111.

Lopez-Bueno, A., et al., 2009. High diversity of the viral community from an Antarctic lake. Science 326, 858–861. https://doi.org/10.1126/science.1179287.
Maat, D.S., Brussaard, C.P.D., 2016. Both phosphorus- and nitrogen limitation constrain viral proliferation in marine phytoplankton. Aquat. Microb. Ecol. 77, 87–97.
Maat, D.S., et al., 2017. Characterization and temperature dependence of arctic micromonas polaris viruses. Viruses 9, 134. https://doi.org/10.3390/v9060134.
Mackinder, L.C.M., et al., 2009. A unicellular algal virus, Emiliania huxleyi virus 86, exploits an animal-like infection strategy. J. Gen. Virol. 90, 2306–2316. https://doi.org/10.1099/vir.0.011635-0.
Maier, I., Muller, D.G., 1998. Virus binding to brown algal spores and gametes visualized by DAPI fluorescence microscopy. Phycologia 37, 60–63. https://doi.org/10.2216/i0031-8884-37-1-60.1.
Mann, N.H., Cook, A., Millard, A., Bailey, S., Clokie, M., 2003. Marine ecosystems: bacterial photosynthesis genes in a virus. Nature 424, 741. https://doi.org/10.1038/424741a.
Marhaver, K.L., Edwards, R.A., Rohwer, F., 2008. Viral communities associated with healthy and bleaching corals. Environ. Microbiol. 10. https://doi.org/10.1111/j.1462-2920.2008.01652.x.
Marston, M.F., et al., 2012. Rapid diversification of coevolving marine Synechococcus and a virus. Proc. Natl. Acad. Sci. U. S. A. 109, 4544–4549. https://doi.org/10.1073/pnas.1120310109.
Martinez, J.M., Schroeder, D.C., Larsen, A., Bratbak, G., Wilson, W.H., 2007. Molecular dynamics of Emiliania huxleyi and cooccurring viruses during two separate mesocosm studies. Appl. Environ. Microbiol. 73, 554–562. https://doi.org/10.1128/aem.00864-06.
Martinez, J.M., Schroeder, D.C., Wilson, W.H., 2012. Dynamics and genotypic composition of Emiliania huxleyi and their co-occurring viruses during a coccolithophore bloom in the North Sea. FEMS Microbiol. Ecol. 81, 315–323. https://doi.org/10.1111/j.1574-6941.2012.01349.x.
Mayer, J.A., Taylor, F.J.R., 1979. A virus which lyses the marine nanoflagellate Micromonas pusilla. Nature (London) 281, 299–301. https://doi.org/10.1038/281299a0.
McDaniel, L.D., DelaRosa, M., Paul, J.H., 2006. Temperate and lytic cyanophages from the Gulf of Mexico. J. Mar. Biol. Assoc. U.K. 86, 517–527. https://doi.org/10.1017/s0025315406013427.
McKeown, D.A., et al., 2017. Phaeoviruses discovered in kelp (Laminariales). ISME J. https://doi.org/10.1038/ismej.2017.130.
Mobberley, J.M., Authement, R.N., Segall, A.M., Paul, J.H., 2008. The temperate marine phage φHAP-1 of Halomonas aquamarina possesses a linear plasmid-like prophage genome. J. Virol. 82, 6618–6630.
Monier, A., Claverie, J.M., Ogata, H., 2008. Taxonomic distribution of large DNA viruses in the sea. Genome Biol 9, R106. https://doi.org/10.1186/gb-2008-9-7-r106.
Monier, A., et al., 2009. Horizontal gene transfer of an entire metabolic pathway between a eukaryotic alga and its DNA virus. Genome Res. 19, 1441–1449. https://doi.org/10.1101/gr.091686.109.
Monier, A., et al., 2012. Phosphate transporters in marine phytoplankton and their viruses: cross-domain commonalities in viral–host gene exchanges. Environ. Microbiol. 14, 162–176. https://doi.org/10.1111/j.1462-2920.2011.02576.x.
Monier, A., et al., 2017. Host-derived viral transporter protein for nitrogen uptake in infected marine phytoplankton. Proc. Natl. Acad. Sci. U. S. A. 114, E7489–E7498. https://doi.org/10.1073/pnas.1708097114.
Moniruzzaman, M., et al., 2016. Diversity and dynamics of algal Megaviridae members during a harmful brown tide caused by the pelagophyte, Aureococcus anophagefferens. FEMS Microbiol. Ecol. 92. https://doi.org/10.1093/femsec/fiw058.

Montalvo-Proano, J., Buerger, P., Weynberg, K.D., van Oppen, M.J.H., 2017. A PCR-based assay targeting the major capsid protein gene of a dinorna-like ssRNA virus that infects coral photosymbionts. Front. Microbiol. 8. https://doi.org/10.3389/fmicb.2017.01665.

Mora, C., Tittensor, D.P., Adl, S., Simpson, A.G.B., Worm, B., 2011. How many species are there on earth and in the ocean? PLoS Biol. 9, e1001127, https://doi.org/10.1371/journal.pbio.1001127.

Moreau, H., et al., 2012. Gene functionalities and genome structure in Bathycoccus prasinos reflect cellular specializations at the base of the green lineage. Genome Biol. 13, R74.

Müller, D.G., Knippers, R., 2011. In: Tidona, C., Darai, G. (Eds.), The Springer Index of Viruses. Springer, New York, pp. 1259–1263.

Muscatine, L., Weis, V.M., 1992. In: Falkowski, P.G., Woodhead, A. (Eds.), Primary Productivity in the Sea. Plenum Press, New York, pp. 257–272.

Nagasaki, K., 2008. Dinoflagellates, diatoms, and their viruses. J. Microbiol. (Seoul, Korea) 46, 235–243. https://doi.org/10.1007/s12275-008-0098-y.

Nagasaki, K., Ando, M., Itakura, S., Imai, I., Ishida, Y., 1994. Viral mortality in the final stage of Heterosigma akashiwo (Raphidophyceae) red tide. J. Plankton Res. 16, 1595–1599. https://doi.org/10.1093/plankt/16.11.1595.

Nealson, K.H., Venter, J.C., 2007. Metagenomics and the global ocean survey: what's in it for us, and why should we care? ISME J. 1, 185–187. https://doi.org/10.1038/ismej.2007.43.

Nissimov, J.I., et al., 2012. Draft genome sequence of four coccolithoviruses: Emiliania huxleyi virus EhV-88, EhV-201, EhV-207, and EhV-208. J. Virol. 86, 2896–2897. https://doi.org/10.1128/jvi.07046-11.

Noble, R.T., Fuhrman, J.A., 1998. Use of SYBR green I for rapid epifluorescence counts of marine viruses and bacteria. Aquat. Microb. Ecol. 14, 113–118. https://doi.org/10.3354/ame014113.

O'Kelly, C.J., Sieracki, M.E., Thier, E.C., Hobson, I.C., 2003. A transient bloom of ostreococcus (Chlorophyta, Prasinophyceae) in West Neck Bay, Long Island, New York. J. Phycol. 39, 850–854.

Oakey, H.J., Cullen, B.R., Owens, L.L., 2002. The complete nucleotide sequence of the Vibrio harveyi bacteriophage VHML. J. Appl. Microbiol. 93, 1089–1098.

Ogura, Y., Hayashi, T., Ueki, S., 2016. Complete genome sequence of a phycodnavirus, Heterosigma akashiwo virus strain 53. Genome Announc. 4. https://doi.org/10.1128/genomeA.01279-16.

Pagarete, A., et al., 2011. Unveiling the transcriptional features associated with coccolithovirus infection of natural Emiliania huxleyi blooms. FEMS Microbiol. Ecol. 78, 555–564. https://doi.org/10.1111/j.1574-6941.2011.01191.x.

Parsons, R.J., Breitbart, M., Lomas, M.W., Carlson, C.A., 2012. Ocean time-series reveals recurring seasonal patterns of virioplankton dynamics in the northwestern Sargasso Sea. ISME J. 6, 273–284. https://doi.org/10.1038/ismej.2011.101.

Patel, A., et al., 2007. Virus and prokaryote enumeration from planktonic aquatic environments by epifluorescence microscopy with SYBR green I. Nat. Protoc. 2, 269. https://doi.org/10.1038/nprot.2007.6.

Patten, N.L., Seymour, J.R., Mitchell, J.G., 2006. Flow cytometric analysis of virus-like particles and heterotrophic bacteria within coral-associated reef water. J. Mar. Biol. Assoc. U. K. 86, 563–566. https://doi.org/10.1017/s0025315406013476.

Patten, N.L., Harrison, P.L., Mitchell, J.G., 2008. Prevalence of virus-like particles within a staghorn scleractinian coral (Acropora muricata) from the great barrier reef. Coral Reefs 27, 569–580. https://doi.org/10.1007/s00338-008-0356-9.

Paul, J.H., 2008. Prophages in marine bacteria: dangerous molecular time bombs or the key to survival in the seas? ISME J. 2, 579–589. https://doi.org/10.1038/ismej.2008.35.

Payet, J.P., McMinds, R., Burkepile, D.E., Thurber, R.L.V., 2014. Unprecedented evidence for high viral abundance and lytic activity in coral reef waters of the South Pacific Ocean. Front. Microbiol. 5. https://doi.org/10.3389/fmicb.2014.00493.

Pollock, F.J., et al., 2014. Abundance and morphology of virus-like particles associated with the coral Acropora hyacinthus differ between healthy and white syndrome-infected states. Mar. Ecol. Prog. Ser. 510, 39–43. https://doi.org/10.3354/meps10927.

Pomeroy, L.R., 1974. The ocean's food web, a changing paradigm. Bioscience 24, 499–504.

Puxty, R.J., Millard, A.D., Evans, D.J., Scanlan, D.J., 2015. Shedding new light on viral photosynthesis. Photosynth. Res. 126, 71–97. https://doi.org/10.1007/s11120-014-0057-x.

Puxty, R.J., Millard, A.D., Evans, D.J., Scanlan, D.J., 2016. Viruses inhibit CO_2 fixation in the most abundant phototrophs on earth. Curr. Biol. 26, 1585–1589. https://doi.org/10.1016/j.cub.2016.04.036.

Rohwer, F., et al., 2000. The complete genomic sequence of the marine phage roseophage SIO1 shares homology with nonmarine phages. Limnol. Oceanogr. 45, 408–418.

Rohwer, F., Seguritan, V., Azam, F., Knowlton, N., 2002. Diversity and distribution of coral-associated bacteria. Mar. Ecol. Prog. Ser. 243, 1–10.

Rosenwasser, S., et al., 2014. Rewiring host lipid metabolism by large viruses determines the fate of Emiliania huxleyi, a bloom-forming alga in the ocean. Plant Cell 26, 2689–2707. https://doi.org/10.1105/tpc.114.125641.

Roux, S., et al., 2016. Ecogenomics and potential biogeochemical impacts of globally abundant ocean viruses. Nature 537, 689–+. https://doi.org/10.1038/nature19366.

Rusch, D.B., et al., 2007. The sorcerer II Global Ocean sampling expedition: Northwest Atlantic through eastern tropical Pacific. PLoS Biol. 5, e77, https://doi.org/10.1371/journal.pbio.0050077.

Sahlsten, E., 1998. Seasonal abundance in Skagerrak–Kattegat coastal waters and host specificity of viruses infecting the marine photosynthetic flagellate Micromonas pusilla. Aquat. Microb. Ecol. 16, 103–108. https://doi.org/10.3354/ame016103.

Sahlsten, E., Karlson, B., 1998. Vertical distribution of virus-like particles (VLP) and viruses infecting Micromonas pusilla during late summer in the southeastern Skagerrak, North Atlantic. J. Plankton Res. 20, 2207–2212. https://doi.org/10.1093/plankt/20.11.2207.

Sandaa, R.A., Larsen, A., 2006. Seasonal variations in virus–host populations in Norwegian coastal waters: focusing on the cyanophage community infecting marine Synechococcus spp. Appl. Environ. Microbiol. 72, 4610–4618. https://doi.org/10.1128/aem.00168-06.

Sato, Y., Civiello, M., Bell, S.C., Willis, B.L., Bourne, D.G., 2016. Integrated approach to understanding the onset and pathogenesis of black band disease in corals. Environ. Microbiol. 18, 752–765. https://doi.org/10.1111/1462-2920.13122.

Schroeder, D.C., Oke, J., Hall, M., Malin, G., Wilson, W.H., 2003. Virus succession observed during an Emiliania huxleyi bloom. Appl. Environ. Microbiol. 69, 2484–2490. https://doi.org/10.1128/AEM.69.5.2484-2490.2003.

Schwartz, D.A., Lindell, D., 2017. Genetic hurdles limit the arms race between Prochlorococcus and the T7-like podoviruses infecting them. ISME J. 11, 1836. https://doi.org/10.1038/ismej.2017.47. https://www.nature.com/articles/ismej201747#supplementary-information.

Short, S.M., 2012. The ecology of viruses that infect eukaryotic algae. Environ. Microbiol. 14, 2253–2271. https://doi.org/10.1111/j.1462-2920.2012.02706.x.

Short, S.M., Short, C.M., 2008. Diversity of algal viruses in various North American freshwater environments. Aquat. Microb. Ecol. 51, 13–21. https://doi.org/10.3354/ame01183.

Short, S.M., Suttle, C.A., 2002. Sequence analysis of marine virus communities reveals that groups of related algal viruses are widely distributed in nature. Appl. Environ. Microbiol. 68, 1290–1296. https://doi.org/10.1128/aem.68.3.1290-1296.2002.

Steward, G.F., et al., 2013. Are we missing half of the viruses in the ocean? ISME J. 7, 672–679. https://doi.org/10.1038/ismej.2012.121.

Sullivan, M.B., Waterbury, J.B., Chisholm, S.W., 2003. Cyanophages infecting the oceanic cyanobacterium Prochlorococcus. Nature 424, 1047–1051. https://doi.org/10.1038/nature01929.

Sullivan, M.B., et al., 2006. Prevalence and evolution of core photosystem II genes in marine cyanobacterial viruses and their hosts. PLoS Biol. 4, 1344–1357. https://doi.org/10.1371/journal.pbio.0040234.

Suttle, C.A., 2002. In: Whitton, B.A., Potts, M. (Eds.), The Ecology of Cyanobacteria: Their Diversity in Time and Space. Kluwer, pp. 563–589.

Suttle, C.A., 2005. Viruses in the sea. Nature 437, 356–361. https://doi.org/10.1038/nature04160.

Suttle, C.A., 2007. Marine viruses—major players in the global ecosystem. Nat. Rev. Microbiol. 5, 801–812. https://doi.org/10.1038/nrmicro1750.

Suttle, C.A., Chan, A.M., 1994. Dynamics and distribution of cyanophages and their effect on marine synechococcus spp. Appl. Environ. Microbiol. 60, 3167–3174.

Suttle, C.A., Chan, A.M., 1995. Viruses infecting the marine prymnesiophyte Chrysochromulina spp—isolation, preliminary characterization and natural-abundance. Mar. Ecol. Prog. Ser. 118, 275–282. https://doi.org/10.3354/meps118275.

Sweet, M., Bythell, J., 2017. The role of viruses in coral health and disease. J. Invertebr. Pathol. 147, 136–144. https://doi.org/10.1016/j.jip.2016.12.005.

Tadmor, A.D., Ottesen, E.A., Leadbetter, J.R., Phillips, R., 2011. Probing individual environmental bacteria for viruses by using microfluidic digital PCR. Science 333, 58–62. https://doi.org/10.1126/science.1200758.

Tarutani, K., Nagasaki, K., Yamaguchi, M., et al., 2000. Appl. Environ. Microbiol. 66, 4916–4920. https://doi.org/10.1128/aem.66.11.4916-4920.2000.

Thingstad, T.F., 2000. Elements of a theory for the mechanisms controlling abundance, diversity, and biogeochemical role of lytic bacterial viruses in aquatic systems. Limnol. Oceanogr. 45, 1320–1328.

Tomaru, Y., Tarutani, K., Yamaguchi, M., Nagasaki, K., 2004. Quantitative and qualitative impacts of viral infection on a Heterosigma akashiwo (Raphidophyceae) bloom in Hiroshima Bay, Japan. Aquat. Microb. Ecol. 34, 227–238. https://doi.org/10.3354/ame034227.

Tomaru, Y., Shirai, Y., Nagasaki, K., 2008. Ecology, physiology and genetics of a phycodnavirus infecting the noxious bloom-forming raphidophyte Heterosigma akashiwo. Fish. Sci. 74, 701–711. https://doi.org/10.1111/j.1444-2906.2008.01580.x.

Vacelet, J., Gallissian, M.F., 1978. Virus-like particles in cells of sponge Verongia cavernicola (Demospongiae, Dictyoceratida) and accompanying tissues changes. J. Invertebr. Pathol. 31, 246–254. https://doi.org/10.1016/0022-2011(78)90014-9.

Van Etten, J.L., Burbank, D.E., Xia, Y., Meints, R.H., 1983. Growth cycle of a virus, PBCV-1, that infects chlorella-like algae. Virology 126, 117–125.

Van Etten, J.L., Graves, M.V., Muller, D.G., Boland, W., Delaroque, N., 2002. Phycodnaviridae—large DNA algal viruses. Arch. Virol. 147, 1479–1516. https://doi.org/10.1007/s00705-002-0822-6.

Vega Thurber, R.L., et al., 2008. Metagenomic analysis indicates that stressors induce production of herpes-like viruses in the coral Porites compressa. Proc. Natl. Acad. Sci. U. S. A. 105. https://doi.org/10.1073/pnas.0808985105.

Vega Thurber, R., Payet, J.P., Thurber, A.R., Correa, A.M.S., 2017. Virus–host interactions and their roles in coral reef health and disease. Nat. Rev. Microbiol. 15. https://doi.org/10.1038/nrmicro.2016.176.

Waldor, M.K., Mekalanos, J.J., 1996. Lysogenic conversion by a filamentous phage encoding cholera toxin. Science 272, 1910–1914. https://doi.org/10.1126/science.272.5270.1910.

Waters, R.E., Chan, A.T., 1982. Micromonas pusilla virus—the virus growth-cycle and associated physiological events within the host-cells—host range mutation. J. Gen. Virol. 63, 199–206. https://doi.org/10.1099/0022-1317-63-1-199.
Weinbauer, M.G., 2004. Ecology of prokaryotic viruses. FEMS Microbiol. Rev. 28, 127–181. https://doi.org/10.1016/j.femsre.2003.08.001.
Weinbauer, M.G., Fuks, D., Peduzzi, P., 1993. Distribution of viruses and dissolved DNA along a coastal trophic gradient in the Northern Adriatic Sea. Appl. Environ. Microbiol. 59, 4074–4082.
Weitz, J.S., et al., 2015. A multitrophic model to quantify the effects of marine viruses on microbial food webs and ecosystem processes. ISME J. 9, 1352–1364. https://doi.org/10.1038/ismej.2014.220.
Weynberg, K.D., Wood-Charslon, E.M., Suttle, C., van Oppen, M.J., 2014. Generating viral metagenomes from the coral holobiont. Front. Microbiol. 5. https://doi.org/10.3389/fmicb.2014.00206.
Weynberg, K.D., Voolstra, C.R., Neave, M.J., Buerger, P., Oppen, M.J.H., 2015. From cholera to corals: viruses as drivers of virulence in a major coral bacterial pathogen. Sci. Rep. 5. https://doi.org/10.1038/srep17889.
Weynberg, K.D., et al., 2017a. Prevalent and persistent viral infection in cultures of the coral algal endosymbiont Symbiodinium. Coral Reefs 1–12. https://doi.org/10.1007/s00338-017-1568-7.
Weynberg, K.D., Allen, M.J., Wilson, W.H., 2017b. Marine prasinoviruses and their tiny plankton hosts: a review. Viruses 9, E43. https://doi.org/10.3390/v9030043.
Weynberg, K.D., et al., 2017c. Coral-associated viral communities show high levels of diversity and host auxiliary functions. PeerJ 5, e4054, https://doi.org/10.7717/peerj.4054.
Wichels, A., Gerdts, G., Schutt, C., 2002. Pseudoalteromonas spp. phages, a significant group of marine bacteriophages in the North Sea. Aquat. Microb. Ecol. 27, 233–239. https://doi.org/10.3354/ame027233.
Wietz, M., Millán-Aguiñaga, N., Jensen, P.R., 2014. CRISPR-Cas systems in the marine actinomycete Salinispora: linkages with phage defense, microdiversity and biogeography. BMC Genomics 15, 936. https://doi.org/10.1186/1471-2164-15-936.
Wilhelm, S.W., Suttle, C.A., 1999. Viruses and nutrient cycles in the sea—viruses play critical roles in the structure and function of aquatic food webs. Bioscience 49, 781–788. https://doi.org/10.2307/1313569.
Wilhelm, S.W., Weinbauer, M.G., Suttle, C.A. (Eds.), 2010. American Society of Limnology and Oceanography. American Society of Limnology and Oceanography. Waco, TX.
Williamson, S.J., et al., 2008. The sorcerer II Global Ocean sampling expedition: metagenomic characterization of viruses within aquatic microbial samples. PLoS One 3, e1456, https://doi.org/10.1371/journal.pone.0001456.
Willis, B.E., Page, C.A., Dinsdale, E.A., 2004. Coral Health and Disease. Springer-Verlag Publishing, Berlin, Heidelberg, pp. 69–104.
Wilson, W.H., Joint, I.R., Carr, N.G., Mann, N.H., 1993. Isolation and molecular characterization of 5 marine cyanophages propagated on synechococcus sp strain Wh7803. Appl. Environ. Microbiol. 59, 3736–3743.
Wilson, W.H., Carr, N.G., Mann, N.H., 1996. The effect of phosphate status on the kinetics of cyanophage infection in the oceanic cyanobacterium Synechococcus sp WH7803. J. Phycol. 32, 506–516. https://doi.org/10.1111/j.0022-3646.1996.00506.x.
Wilson, W.H., Turner, S., Mann, N.H., 1998. Population dynamics of phytoplankton and viruses in a phosphate-limited mesocosm and their effect on DMSP and DMS production. Estuar. Coast. Shelf Sci. 46, 49–59. https://doi.org/10.1006/ecss.1998.0333.
Wilson, W.H., Francis, I., Ryan, K., Davy, S.K., 2001. Temperature induction of viruses in symbiotic dinoflagellates. Aquat. Microb. Ecol. 25, 99–102. https://doi.org/10.3354/ame025099.

Wilson, W.H., et al., 2002. Isolation of viruses responsible for the demise of an Emiliania huxleyi bloom in the English Channel. J. Mar. Biol. Assoc. U. K. 82, 369–377. https://doi.org/10.1017/s002531540200560x.

Wilson, W.H., Dale, A.L., Davy, J.E., Davy, S.K., 2005a. An enemy within? Observations of virus-like particles in reef corals. Coral Reefs 24, 145–148. https://doi.org/10.1007/s00338-004-0448-0.

Wilson, W.H., et al., 2005b. Complete genome sequence and lytic phase transcription profile of a coccolithovirus. Science 309, 1090–1092. https://doi.org/10.1126/science.1113109.

Winter, C., Bouvier, T., Weinbauer, M.G., Thingstad, T.F., 2010. Trade-offs between competition and defense specialists among unicellular planktonic organisms: the "killing the winner" hypothesis revisited. Microbiol. Mol. Biol. Rev. 74, 42–57. https://doi.org/10.1128/mmbr.00034-09.

Wolf, S., Maier, I., Katsaros, C., Muller, D.G., 1998. Virus assembly in Hincksia hincksiae (Ectocarpales, Phaeophyceae) an electron and fluorescence microscopic study. Protoplasma 203, 153–167. https://doi.org/10.1007/bf01279472.

Wommack, K.E., Colwell, R.R., 2000. Virioplankton: viruses in aquatic ecosystems. Microbiol. Mol. Biol. Rev. 64, 69–+. https://doi.org/10.1128/mmbr.64.1.69-114.2000.

Wood-Charlson, E.M., Weynberg, K.D., Suttle, C., Roux, S., van Oppen, M.J.H., 2015. Metagenomic characterisation of viral communities in corals: mining biological signal from methodological noise. Environ. Microbiol. 17. https://doi.org/10.1111/1462-2920.12803.

Worden, A.Z., 2006. Picoeukaryote diversity in coastal waters of the Pacific Ocean. Aquat. Microb. Ecol. 43, 165–175. https://doi.org/10.3354/ame043165.

Worden, A.Z., Nolan, J.K., Palenik, B., 2004. Assessing the dynamics and ecology of marine picophytoplankton: the importance of the eukaryotic component. Limnol. Oceanogr. 49, 168–179. https://doi.org/10.4319/lo.2004.49.1.0168.

Worden, A.Z., et al., 2009. Green evolution and dynamic adaptations revealed by genomes of the marine picoeukaryotes micromonas. Science 324, 268–272. https://doi.org/10.1126/science.1167222.

Yamamoto, H., 2000. Viable but nonculturable state as a general phenomenon of non-spore-forming bacteria, and its modeling. J. Infect. Chemotherapy 6, 112–114. https://doi.org/10.1007/s101560000024.

Yang, Y.H., Motegi, C., Yokokawa, T., Nagata, T., 2010. Large-scale distribution patterns of virioplankton in the upper ocean. Aquat. Microb. Ecol. 60, 233–246. https://doi.org/10.3354/ame01428.

Yau, S., et al., 2016. A viral immunity chromosome in the marine picoeukaryote, *Ostreococcus tauri*. PLoS Pathog. 12, e1005965, https://doi.org/10.1371/journal.ppat.1005965.

Zeidner, G., et al., 2005. Potential photosynthesis gene recombination between Prochlorococcus and Synechococcus via viral intermediates. Environ. Microbiol. 7, 1505–1513. https://doi.org/10.1111/j.1462-2920.2005.00833.x.

Zhao, Y., et al., 2010. Searching for a "hidden" prophage in a marine bacterium. Appl. Environ. Microbiol. 76, 589–595.

Ziv, C., et al., 2016. Viral serine palmitoyltransferase induces metabolic switch in sphingolipid biosynthesis and is required for infection of a marine alga. Proc. Natl. Acad. Sci. 113, E1907–E1916. https://doi.org/10.1073/pnas.1523168113.

CHAPTER TWO

Viruses in Polar Lake and Soil Ecosystems

Alberto Rastrojo, Antonio Alcamí[1]
Centro de Biología Molecular Severo Ochoa (Consejo Superior de Investigaciones Científicas—Universidad Autónoma de Madrid), Madrid, Spain
[1]Corresponding author: e-mail address: aalcami@cbm.csic.es

Contents

Abstract

Viruses play an important role in the control of microbial communities, and it has been suggested that the influence of viruses in polar ecosystems, with low nutrients and under extreme environmental conditions, may be greater. Viral metagenomics allows the genetic characterization of complex viral communities without the need to isolate and grow viruses. Recent investigations in Antarctica and the Arctic are uncovering a great diversity of DNA viruses, including bacteriophages, circular single-stranded DNA viruses, algal-infecting phycodnaviruses, and virophages, adapted to these extreme environments. The limited sequence similarity between viruses in Antarctica and the Arctic suggests that viral communities in the two polar regions have evolved independently since the formation of the Antarctic continent, estimated to occur 25 million years ago. The community of RNA viruses in Antarctica is dominated by the order Picornavirales and their quasispecies composition suggests that higher genetic variability may correlate with viral adaptation to new environmental conditions.

Advances in Virus Research, Volume 101
ISSN 0065-3527
https://doi.org/10.1016/bs.aivir.2018.02.002

1. VIRAL METAGENOMICS IN NATURAL ECOSYSTEMS

Research in virology in the past has focused on the identification, isolation, and growth of viruses causing diseases in humans, animals, and plants. More recently, it has been realized that there is a large and diverse viral community present in biological systems, including natural ecosystems and the human body. It has been estimated that the presence of virus particles in seawater and fresh water systems is in the order of billions per liter, and the carbon mass of virus particles in the planet is equivalent to that of 75 million blue whales (Suttle, 2005). What is the identity of these viruses? Which are their hosts? What is the impact of viruses in ecosystems, infecting and killing different hosts? A better understanding of viruses in natural ecosystems is one of the major challenges of virology in the 21st century.

Next-generation sequencing (NGS) technologies, lead initially by 454-Roche and now by Illumina, have allowed the sequencing of complete viral communities in ecosystems, known as viral metagenomics, without the need to previously isolate and grow the viruses in the laboratory. Illumina technologies generate millions of sequencing reads of viral genomes, and recent technologies such as PacBio or MinION allow the sequence of long reads, even complete viral genomes in a single sequencing read (Carrasco-Ramiro et al., 2017). NGS is uncovering abundant and highly diverse viral communities in biological systems.

The initial studies on viral metagenomics were performed on seawater samples and identified thousands of viral species previously unknown (Rohwer and Thurber, 2009; Suttle, 2005) (Fig. 1A and D). In such studies, the host cells are removed by filtration (0.45 or 0.2 μm pore size), and virus particles are concentrated by tangential filtration or flocculation (Rastrojo and Alcami, 2017). Some protocols include additional steps such as ultracentrifugation through CsCl gradients or sucrose cushions. Metagenomic studies of prokaryotes and eukaryotes are normally based on sequencing of hypervariable regions of ribosomal 16S and 18S rDNA genes, respectively, and show the diversity of species within a complex community or microorganisms. Metagenomic studies consisting of the sequence of cellular genomes are limited but offer information on the function of genes present in microbial genomes, which may be specific for different environments. Because viruses do not have a common gene, viral metagenomics consists of the sequence of viral genomes extracted from semipurified preparations of virus particles. The majority of sequences from viral metagenomics are

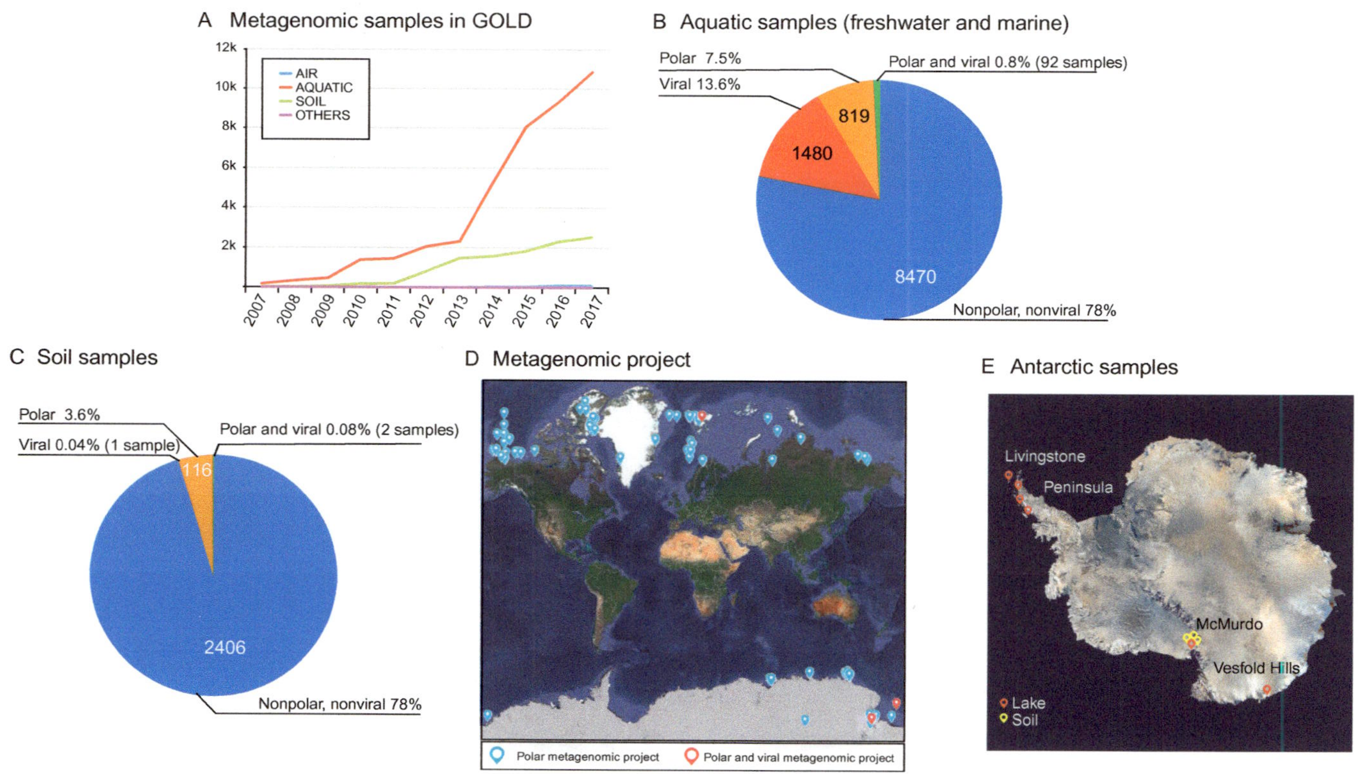

Fig. 1 Metagenomic studies of microbes in natural ecosystems. Microbial metagenomics samples sequenced over the years, as archived in Genomes OnLine Database (GOLD, https://gold.jgi.doe.gov/) (A). Distribution of aquatic (B) and soil (C) metagenomics samples. Map showing the metagenomics sampling sites, created with Google Maps and coordinates retrieved from GOLD (D). Map of Antarctica showing the location where lake and soil samples were taken (E).

not significantly related to sequences deposited in databases, illustrating that viruses in natural ecosystems are largely unknown (for additional discussion see chapter "From spatial metagenomics to molecular characterization of plant viruses: A geminivirus case study" by Claverie et al., this volume).

Metagenomics studies in environmental samples have focused mainly on cellular microorganisms, with studies on viral metagenomics being more limited. Studies on viruses in marine, freshwater, or soil samples from polar regions are scarce (Fig. 1B and C), maybe because of the need to collect large amounts of sample or technical difficulties to purify viral preparations. A better knowledge of viruses from extreme environments, such as high salt, low pH, or high temperatures, has been an area of particular interest since these viruses must have evolved to survive under adverse conditions. Here, we review metagenomics studies of viral communities that inhabit the polar regions, focusing on lake and soil ecosystems.

2. ANTARCTICA AND THE ARCTIC

Polar regions are the coldest places on Earth and experience extreme environmental conditions. However, the two regions are distinct since Antarctica is a continent surrounded by an ocean, whereas the Arctic is an ocean surrounded by continents. The environmental conditions in Antarctica are more severe than at similar latitudes in the Arctic. For example, the annual average temperature in the North Pole is −18°C whereas it is −50°C in the South Pole (Cavicchioli, 2015). The Russian Vostok Station, located in the center of the Antarctic continent, registered in 1983 the lowest temperature recorded in the planet, −89°C. In 2010, temperatures determined from satellites estimated an even lower temperature, −92.3°C, in East Antarctica. Most of the Antarctic continent, 98%, is covered by a thick layer of ice that reaches up to 4000 m (Cavicchioli, 2015).

The Antarctic continent has never been inhabited by native human populations and is considered one of the pristine natural ecosystems on Earth. Life in polar regions is limited due to the extreme conditions and is dominated by microorganisms such as protists (unicellular eukaryotes, like protozoa and algae) and bacteria. Interestingly, microbial biodiversity in polar regions is unexpectedly high, but the drivers of such diversity are not fully characterized (Anesio and Bellas, 2011; Chown et al., 2015). Our knowledge of viruses in polar regions remains similarly limited.

3. DNA VIRUSES IN ANTARCTICA

One of the most interesting questions in Antarctic research is to identify the forms of life that have been adapted to this hostile environment within a continent isolated from other continents for approximately 25 million years. The ocean surrounding Antarctica is particularly rich in nutrients and animal life. By contrast, life on the Antarctic continent itself is limited by extreme conditions: low levels of nutrients, darkness during several months every year, and very low temperatures. Under these extreme conditions only microorganisms (including viruses) are able to survive, dominating both aquatic and terrestrial ecosystems in Antarctica (Cavicchioli, 2015).

It has been proposed that viruses play a special role in the control of microbial communities under low nutrient conditions (Anesio and Bellas, 2011; Cavicchioli, 2015; Laybourn-Parry and Pearce, 2007). The lower abundance of grazers in these ecosystems suggests that viruses may be responsible for a high proportion of cellular mortality and consequently influence microbial evolution. But viruses present in extreme polar ecosystems have been poorly characterized, and a few scientific reports have described a high abundance of virus-like particles detected by fluorescence, calculated infection rates of host cells in these environments, and identified the presence of virus particles by electron microscopy (reviewed by Cavicchioli, 2015; Laybourn-Parry and Pearce, 2007).

3.1 DNA Viruses in Antarctic Lakes

The first metagenomics study of DNA viruses in Antarctica was carried out in Limnopolar Lake (Byers Peninsula, Livingston Island) (Fig. 1E). Byers Peninsula is an Antarctic Specially Protected Area of scientific interest because it is the largest ice-free area in the South Shetland Islands. The Peninsula expanded for approximately 8000 years BP as the Rotch Dome glacier retreated, leaving a unique area with numerous freshwater lakes that display some of the greatest microbial biodiversity in Antarctica (Oliva et al., 2016; Toro et al., 2007). Initial analysis by electron microscopy showed a large diversity of viral capsids in Limnopolar Lake (Lopez-Bueno et al., 2009) (Fig. 2). DNA sequencing of viral genomes with 454-Roche technology revealed a large percentage (81%–94%) of new viruses unrelated to

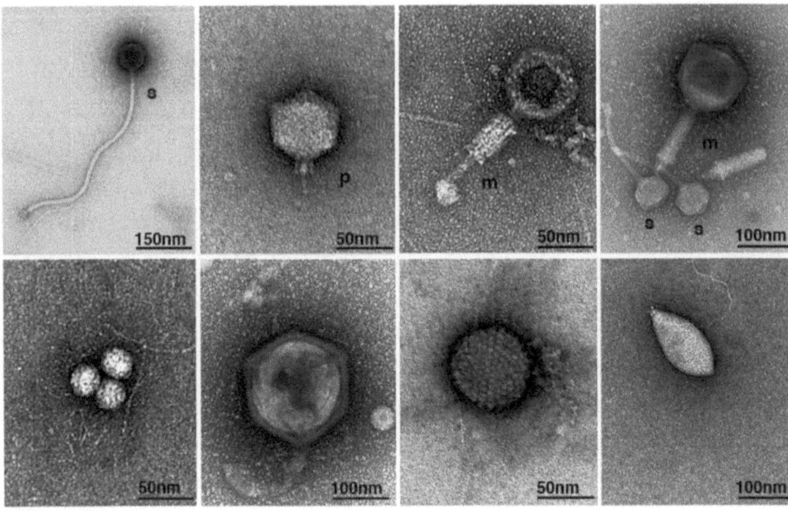

Fig. 2 Electron microscopy of viruses present in Limnopolar Lake (Byers Peninsula, Livingston Island, Antarctica). *Upper panel*: bacteriophages such as siphoviruses (s), podoviruses (p), and myoviruses (m). *Lower panels*: virus particles with spikes, and lemon shape and small size virions. *Reproduced with permission from Lopez-Bueno, A., Tamames, J., Velazquez, D., Moya, A., Quesada, A., Alcami, A., 2009. High diversity of the viral community from an Antarctic lake. Science 326, 858–861.*

viruses previously described in other aquatic environments (Lopez-Bueno et al., 2009). Phages infecting bacteria were identified, but small viruses with circular ssDNA viruses, related to circoviruses, geminiviruses, or satellites, were abundant in the samples. The abundance of circular ssDNA viruses may be overrepresented due to the preferential amplification of circular genomes when using multiple displacement amplification. However, circular ssDNA viruses are evidently abundant in these ecosystems even if a 100-fold overamplification is considered to have been introduced by this methodology and after the data were corrected for genome length. These viruses have unique genomic organizations compared to other ssDNA viruses previously described, with open reading frames transcribed unidirectionally, and are separated phylogenetically, suggesting that they may represent new viruses adapted to extreme polar environments (Lopez-Bueno et al., 2009).

Extreme ecosystems were expected to harbor low biodiversity and they could facilitate the study of the role of viruses in simple ecosystems with reduced number of species (Convey and Stevens, 2007; Convey et al., 2008). By contrast, the first viral metagenomics study in Antarctica uncovered a viral community with unexpectedly high diversity (Lopez-Bueno et al., 2009). Computer analysis estimated the presence of nearly 10,000 viral species as compared to less than 800 viral species estimated in North American lakes. This study also showed a change in the composition of the viral community between the spring, when the lake is covered with a thick ice layer, and the summer, when algae-infecting phycodnaviruses increase as a result of ice melt. An increase in light provides the necessary energy for primary production and generates organic matter that promotes algal survival during the winter (Cavicchioli, 2015; Wilkins et al., 2013). Seasonal environmental changes select for microbial populations that use different survival strategies (i.e., autotrophs vs heterotrophs). Thus, over the annual cycle, the growth of different hosts determines seasonal changes in the viral community present in the lake. Mathematical models of the influence of light cycles in polar regions on virus–host interactions have identified potential devastating effects of viruses on bacterial populations (Lauro et al., 2011). On the other hand, mathematical models predict that the interaction of virophages—viruses that parasitize other viruses—and phycodnaviruses may benefit ecosystems. A virophage that preys on phycodnaviruses that infect prasinophytes (flagellated algae) may reduce the virulence of phycodnaviruses and the overall mortality of algal hosts. This will increase the frequency of algal blooms during polar summer light periods and stimulate secondary production (Yau et al., 2011).

The initial viral metagenomics studies in Limnopolar Lake were expanded with a comparative study of the viral community in eight lakes along the Antarctic Peninsula (Fig. 1E) (Aguirre de Carcer et al., 2016). This study confirmed the presence of a high diversity of viruses in Antarctic aquatic ecosystems unrelated to known viruses, and an abundance of circular ssDNA viruses. Viral diversity did not correlate with the geographical location of the lakes, indicating that the latitudinal gradient of biodiversity described for macroorganisms and in viral metagenomics studies in oceans is not observed in these Antarctic viral communities (Angly et al., 2006).

Small circular ssDNA viruses have also been identified in a freshwater pond on the McMurdo Ice Shelf (Fig. 1E), and their low sequence similarity to ssDNA viruses sequenced in Limnopolar Lake suggests high diversity of these viruses in Antarctica (Zawar-Reza et al., 2014). Other studies on Antarctic viruses have been carried out in a pristine, meromictic (permanently stratified) lake, Ace Lake, located in the Vestfold Hills of East Antarctica (Fig. 1E) (Lauro et al., 2011; Wilkins et al., 2013). Ace Lake was formed at the end of the Quaternary, ~12,000 years ago, and was initially fresh. Seawater invaded the lake basin during an early Holocene sea level highstand (~7000 years), and the lake was reformed after subsequent sea-level fall after ~5000 years. This lake system is of special interest because the capture of marine-derived biota allows the study of the short-term evolution of species and ecosystem function. Metagenomics and metaproteomics studies have provided genetic and functional information about the microbial communities in Ace Lake (Lauro et al., 2011). The upper, oxygen-rich layer is dominated by phycodnaviruses, whereas the bottom anoxic layer is dominated by bacteriophages (siphovirus, myovirus, podovirus).

Studies in Ace Lake and the nearby Organic Lake identified the presence of virophages, associated with phycodnaviruses that may play an important role in aquatic ecosystems (Lauro et al., 2011; Yau et al., 2011). Virophages use the replicative cytoplasmic virion factory of giant viruses for their own replication (Fischer and Suttle, 2011). Metaproteomics in samples from Deep Lake, also located in the Vestfold Hills (Fig. 1E), showed the presence of several archeal viruses, consistent with a dominance of haloarchaea in this hypersaline lake, and mapped repetitions and spacers of the CRISPR defense system in haloarchaea conferring host immunity against viruses (Tschitschko et al., 2015). This study also showed that viruses may contribute to the generation of host surface protein variants, by mutating and shuffling the host genes. The virus supplying the variant host gene may benefit from reduced competition by more destructive and virulent viruses that do not efficiently infect through these surface protein variants.

3.2 DNA Viruses in Antarctic Soils

Previous studies have identified a high concentration and diversity of viruses in soil samples from different environments (Williamson et al., 2017). Besides in lake systems, a large diversity of viruses in Antarctica has been reported in soil samples. The first report characterized the viral community in the Miers Valley (McMurdo Dry Valleys, Ross Dependency, eastern Antarctica) (Fig. 1E), in samples from surface soil and communities of hypoliths (extremophile photosynthetic organisms) found on the underside of rocks (Zablocki et al., 2014). These viral communities were dominated by bacteriophages, with phycodnaviruses and mimiviruses (giant viruses that infect amoeba) being the second most-abundant taxa. Virophages were also identified in open soil samples. A recent study of the viral community in 14 soil samples from the Mackay glacier region, in Victoria Land, close to McMurdo (Fig. 1E), showed a high diversity of viruses within a soil sample and heterogeneity of viruses when comparing different sites with relative geographical proximity, with dominance of bacteriophages, phycodnaviruses, and mimiviruses (Adriaenssens et al., 2017). Soil factors, such as pH, calcium content, and site altitude, were suggested as the main drivers of viral community structure. Viral metagenomics studies have also been reported in microbial mats, multilayered structures composed of a variety of microorganisms (cianobacteria, algae, protozoa), in Byers Peninsula (Livingston Island, Fig. 1E) (Velazquez et al., 2016). High diversity of viruses, including bacteriophages, ssDNA viruses, and phycodnaviruses, was also reported in this study.

4. DNA VIRUSES IN THE ARCTIC

Viral metagenomics studies in Antarctica led to a relevant question: whether viruses inhabiting Antarctica and the Arctic, with similar environmental conditions, are related. The first metagenomics study of DNA viruses in lakes and soils of the Arctic was carried out recently, in lakes from Svalbard islands (Norway, 78°N latitude) (Aguirre de Carcer et al., 2015). This study showed that viral communities in the Antarctic and the Arctic belong to the same families, and that both have an abundance of small ssDNA circular viruses. However, the low genome sequence similarity suggested that the Arctic and Antarctic viral communities are largely unrelated to each other and have evolved independently since the Antarctic continent became isolated approximately 25 million years ago. Intriguingly, two Arctic ssDNA viral lineages were found to share a substantial nucleotide sequence similarity with Antarctic viruses, illustrating that some very limited connectivity may be possible between these two environments. Further comparison has found that polar lake viromes show little connectivity with other lakes in Europe, North America, and the Sahara (Fig. 3). Previous work found that viral diversity in the Arctic ocean is low compared to the diversity found in more temperate oceans (Angly et al., 2006). However, as described in Antarctica, high viral diversity was also found in freshwater lakes from the Arctic. Thus, the diversity gradient described for macroorganisms, diminishing toward the Poles, is not observed in viral communities from lakes (Chown et al., 2015).

Viral metagenomics studies in the Arctic have also been carried out in cryoconite holes. These are small holes in the ablation zone of glacial ice surfaces, formed as a result of the deposition of cryoconite (a mixture of inorganic and organic particles), which absorbs solar radiation and causes the ice to melt (Anesio and Laybourn-Parry, 2012; Boetius et al., 2015).

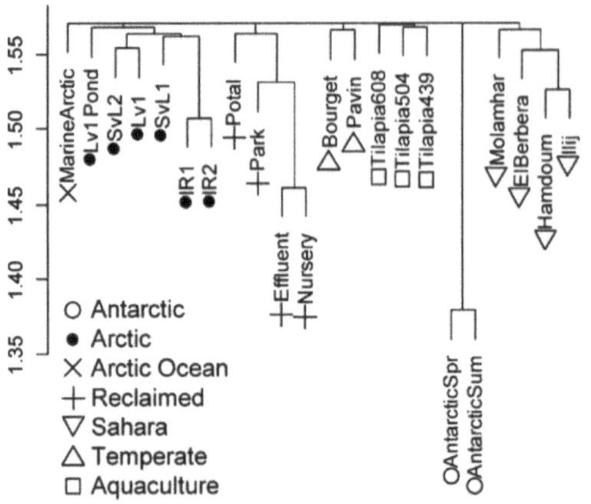

Fig. 3 Reference-independent cross-assembly (crAss) comparison of individual freshwater viromes. The dendrogram represents a hierarchical clustering (average linkage method) of *Wooters* distance values derived from the cross-assembly of freshwater viromes, using all metagenomic reads. Viromes from the same environment are represented by equal forms. The analysis evidences very low fine genetic overlap between viromes from different environments. Aquaculture and reclaimed samples were form North America, and the temperate samples are from Europe. *Reproduced with permission from Aguirre de Carcer, D., Lopez-Bueno, A., Pearce, D. A., Alcami, A., 2015. Biodiversity and distribution of polar freshwater DNA viruses. Sci. Adv. 1, e1400127.*

Phage infection rates of bacteria in cryoconite holes appear to exceed those found in temperate freshwaters, in spite of the extreme conditions of the cryoconite environment (Anesio et al., 2007; Bellas et al., 2013). Metagenomics studies in cryoconite holes in Svalbard glaciers and ice sheets in Greenland showed the abundance of bacteriophages (Caudovirales) (Bellas et al., 2015). The frequent presence of integrases in the genome sequences of these bacteriophages suggested the establishment of lysogeny, the integration of the viral genome into bacterial genomes. Lysogeny has been suggested as a bacteriophage strategy to survive polar winters in a dormant stage with no need for virus replication, changing to lytic replication in the summer when higher activity of the bacterial hosts favors replication and transmission of viruses. Interestingly, Bellas and coworkers found a phage encoding a CRISPR array that may target other phages, suggesting that a phage lysogen can confer superimmunity to its bacterial host via CRISPRs (Bellas et al., 2015). In addition, this study identified nucleocytoplasmic DNA viruses infecting eukaryotes, including algae viruses (phycodnaviruses) and virophages. It has been experimentally shown that viruses present in cryoconite holes, which may have been transported by air from long distance locations, can infect the bacterial community from a nearby lake, suggesting that viruses from glaciers have the capacity to colonize other ecological niches (Anesio et al., 2007).

As in temperate regions, viral assemblages from marine and freshwater systems in polar regions are clearly distinct, which was illustrated by the low connectivity of viruses between Arctic freshwater lakes and the Arctic Ocean (Aguirre de Carcer et al., 2015; Angly et al., 2006). This is consistent with the idea that salinity might be the most important driver of the viral community structure (Logares et al., 2009).

5. RNA VIRUSES IN ANTARCTICA

Viral metagenomics studies have uncovered a great diversity of DNA viruses in polar lakes and soils but RNA viruses have received less attention so far, in spite of evidence suggesting that RNA viruses may be more abundant than DNA viruses in oceans (Steward et al., 2013). Initial studies of RNA viromes in nonpolar seawater and freshwater showed a dominance of single positive strand viral genomes belonging to the order Picornavirales (Culley et al., 2006; Steward et al., 2013).

The only published metagenomics study of RNA viruses in Antarctica was carried out in Limnopolar Lake (Cavicchioli and Erdmann, 2015; Lopez-Bueno et al., 2015), where a large diversity of DNA viruses was previously described (Lopez-Bueno et al., 2009). Viruses belonging to the order Picornavirales dominate the RNA viral community in this lake, and it was possible to assemble the genome of several viruses, named Antarctic Picorna-like virus (APLV) 1–4. Due to the low fidelity of RNA polymerases that introduce mutations during viral replication, RNA viruses consist of a collection of closely related viral genomes known as viral quasispecies (Andino and Domingo, 2015). The high coverage of the APLV 1–4 genomes obtained by NGS allowed a detailed study of their genetic variability to determine the quasispecies composition (Fig. 4). The most abundant virus, APLV-1, showed a quasispecies of relatively low genetic variability, and its consensus sequence was conserved throughout a 4-year period, suggesting good adaptation to the hosts and environmental conditions of the lake. By contrast, the quasispecies compositions of less abundant viruses were highly variable. Interestingly, some of the less abundant viruses, such as APLV-2, were also found in microbial mats in the lake basin, where they were much less genetically variable (Fig. 4). Lopez-Bueno et al. (2015) proposed that these viruses were transferred from microbial mats to the

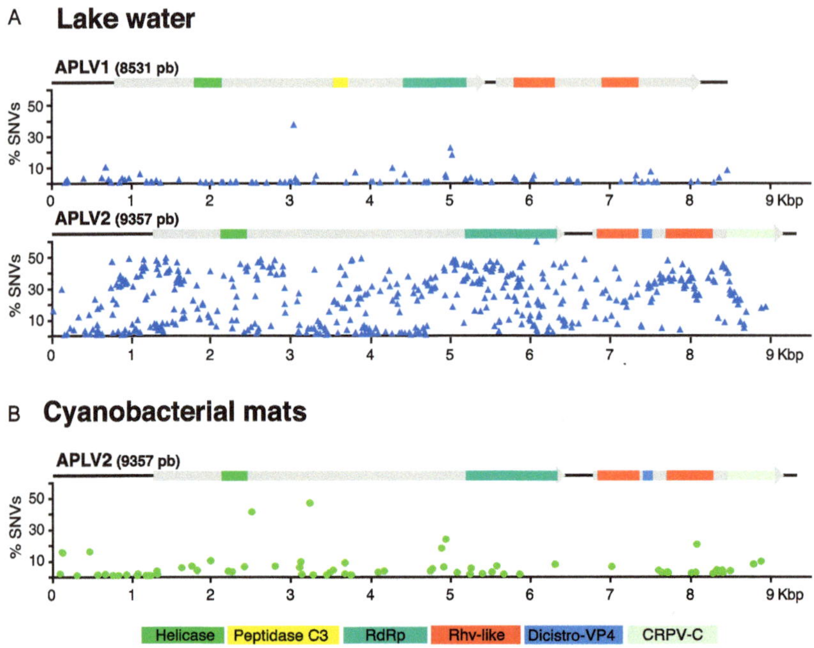

Fig. 4 Quasispecies of Antarctic picorna-like viruses (APNVs). The distribution of single-nucleotide variants (SNVs) along the APLV-1 and APLV-2 genomes is shown. Each *colored symbol* represents SNVs with frequencies higher than 1% forming the quasispecies of a virus in a particular sample. Genome organization schemes are drawn above graphs to properly locate the SNVs. The open reading frames are marked in *gray* and specific protein activities are *colored*. Quasispecies of viruses present in Lake Limnopolar water in 2010 (A) and in cyanobacterial mats (B) are represented. *Modified from Lopez-Bueno, A., Rastrojo, A., Peiro, R., Arenas, M., Alcami, A., 2015. Ecological connectivity shapes quasispecies structure of RNA viruses in an Antarctic lake. Mol. Ecol. 24, 4812–4825.*

lake by summer snow melt, and that the viruses' subsequent diversity within the lake water reflected a process by which the viruses were trying to adapt to the new environmental conditions in competition with resident viruses already better adapted to the lake conditions (Lopez-Bueno et al., 2015). This was the first RNA viral quasispecies study in natural ecosystems and suggests that genetic variability may facilitate the adaptation of RNA viruses to changes in natural environments.

The relevant role of RNA viruses in polar microbial ecosystems was illustrated in a recent study in sea samples near Palmer Station (Anvers Island, near Antarctic Peninsula) (Fig. 1E). This study estimated that RNA viruses contributed up to 65% to total viroplankton and that the majority of RNA viruses are closely related to viruses in the order Picornavirales

(Miranda et al., 2016). The authors assembled several viral genomes and proposed that these viruses infect algae and control algae blooms in the Antarctic sea water.

To our knowledge, metagenomics studies of RNA viruses in the Arctic have not been published so far.

6. FUTURE PERSPECTIVES

Initial studies of viruses in polar regions have demonstrated the existence of a great diversity of new viruses adapted to the extreme conditions of high latitudes, and a very limited connectivity between Antarctica and the Arctic, suggesting an independent evolution of these viral communities since the formation of the Antarctic continent, estimated to occur 25 million years ago.

A large system of approximately 380 lakes has been found below the Antarctic ice sheet (Wright and Siegert, 2012). An important question of Antarctic research is to find out which microorganisms—including viruses—may survive under these conditions: high atmospheric pressure, temperatures below 0°C, complete darkness, and very high oxygen concentration. Lake Vostok is the largest subglacial lake, sitting below ~3.8 km of ice under the Russian research station. Exploration of Lake Vostok has been highly controversial due to problems associated with the use of drilling fluid and kerosene to keep the bore hole open. In spite of these problems, metagenomics data have been published on microorganisms found in the accretion ice from Lake Vostok (Rogers et al., 2013; Shtarkman et al., 2013). Other projects have tried to reach subglacial lakes Ellsworth and Williams using drilling technologies that are less contaminating. The Ellsworth project failed to penetrate ~3.4 km of ice to reach the lake (Schiermeier, 2014). Access to Williams Lake was successful using hot water to pass through ~800 m of ice and metagenomics studies showed a community of proteobacteria and a range of other bacteria, and ammonia-oxidizing archaea (Christner et al., 2014; Tranter, 2014). Microorganisms appear to use energy derived from the oxidation of geochemically formed reduced sulfur and iron. The identification of viral communities inhabiting subglacial lakes and the characterization of the unique properties that allow them to adapt to these extreme environments will be of great interest, but there are no reports describing viruses in subglacial lakes so far.

The information available on viruses in polar regions is very limited, and increasing our knowledge of viral communities adapted to the extreme conditions of polar regions will be of interest. These studies will identify biological functions encoded by viral genes, which are relevant for adaptation to these extreme environments, and will help us to understand the role that these viruses play in polar microbial communities. This is one of the challenges for research in Antarctica and the Arctic in the future years.

ACKNOWLEDGMENTS

Viral metagenomics research in the author's laboratory was funded by the Spanish Polar Programme, Spanish Ministry of Economy and Competitiveness, and European Union.

REFERENCES

Adriaenssens, E.M., Kramer, R., Van Goethem, M.W., Makhalanyane, T.P., Hogg, I., Cowan, D.A., 2017. Environmental drivers of viral community composition in Antarctic soils identified by viromics. Microbiome 5, 83.

Aguirre de Carcer, D., Lopez-Bueno, A., Pearce, D.A., Alcami, A., 2015. Biodiversity and distribution of polar freshwater DNA viruses. Sci. Adv. 1, e1400127.

Aguirre de Carcer, D., Lopez-Bueno, A., Alonso-Lobo, J.M., Quesada, A., Alcami, A., 2016. Metagenomic analysis of lacustrine viral diversity along a latitudinal transect of the Antarctic Peninsula. FEMS Microbiol. Ecol. 92, fiw074.

Andino, R., Domingo, E., 2015. Viral quasispecies. Virology 479–480, 46–51.

Anesio, A.M., Bellas, C.M., 2011. Are low temperature habitats hot spots of microbial evolution driven by viruses? Trends Microbiol. 19, 52–57.

Anesio, A.M., Laybourn-Parry, J., 2012. Glaciers and ice sheets as a biome. Trends Ecol. Evol. 27, 219–225.

Anesio, A.M., Mindl, B., Laynbourn-Parry, J., Hodson, A.J., Sattler, B., 2007. Viral dynamics in cryoconite holes on a Hogh Arctic glacier (Svalbard). J. Geophys. Res. 112, G04531.

Angly, F.E., Felts, B., Breitbart, M., Salamon, P., Edwards, R.A., Carlson, C., Chan, A.M., Haynes, M., Kelley, S., Liu, H., Mahaffy, J.M., Mueller, J.E., Nulton, J., Olson, R., Parsons, R., Rayhawk, S., Suttle, C.A., Rohwer, F., 2006. The marine viromes of four oceanic regions. PLoS Biol. 4, e368.

Bellas, C.M., Anesio, A.M., Telling, J., Stibal, M., Tranter, M., Davis, S.A., 2013. Viral impacts on bacterial communities in Arctic cryoconite. Environ. Res. Lett. 8, 045021.

Bellas, C.M., Anesio, A.M., Barker, G., 2015. Analysis of virus genomes from glacial environments reveals novel virus groups with unusual host interactions. Front. Microbiol. 6, 656.

Boetius, A., Anesio, A.M., Deming, J.W., Mikucki, J.A., Rapp, J.Z., 2015. Microbial ecology of the cryosphere: sea ice and glacial habitats. Nat. Rev. Microbiol. 13, 677–690.

Carrasco-Ramiro, F., Peiro-Pastor, R., Aguado, B., 2017. Human genomics projects and precision medicine. Gene Ther. 24, 551–561.

Cavicchioli, R., 2015. Microbial ecology of Antarctic aquatic systems. Nat. Rev. Microbiol. 13, 691–706.

Cavicchioli, R., Erdmann, S., 2015. The discovery of Antarctic RNA viruses: a new game changer. Mol. Ecol. 24, 4809–4811.

Chown, S.L., Clarke, A., Fraser, C.I., Cary, S.C., Moon, K.L., McGeoch, M.A., 2015. The changing form of Antarctic biodiversity. Nature 522, 431–438.

Convey, P., Stevens, M.I., 2007. Ecology. Antarctic biodiversity. Science 317, 1877–1878.

Convey, P., Gibson, J.A., Hillenbrand, C.D., Hodgson, D.A., Pugh, P.J., Smellie, J.L., Stevens, M.I., 2008. Antarctic terrestrial life—challenging the history of the frozen continent? Biol. Rev. Camb. Philos. Soc. 83, 103–117.
Christner, B.C., Priscu, J.C., Achberger, A.M., Barbante, C., Carter, S.P., Christianson, K., Michaud, A.B., Mikucki, J.A., Mitchell, A.C., Skidmore, M.L., Vick-Majors, T.J., Wissard Science Team, 2014. A microbial ecosystem beneath the West Antarctic ice sheet. Nature 512, 310–313.
Culley, A.I., Lang, A.S., Suttle, C.A., 2006. Metagenomic analysis of coastal RNA virus communities. Science 312, 1795–1798.
Fischer, M.G., Suttle, C.A., 2011. A virophage at the origin of large DNA transposons. Science 332, 231–234.
Lauro, F.M., DeMaere, M.Z., Yau, S., Brown, M.V., Ng, C., Wilkins, D., Raftery, M.J., Gibson, J.A., Andrews-Pfannkoch, C., Lewis, M., Hoffman, J.M., Thomas, T., Cavicchioli, R., 2011. An integrative study of a meromictic lake ecosystem in Antarctica. ISME J. 5, 879–895.
Laybourn-Parry, J., Pearce, D.A., 2007. The biodiversity and ecology of Antarctic lakes: models for evolution. Philos. Trans. R. Soc. Lond. B Biol. Sci. 362, 2273–2289.
Logares, R., Brate, J., Bertilsson, S., Clasen, J.L., Shalchian-Tabrizi, K., Rengefors, K., 2009. Infrequent marine-freshwater transitions in the microbial world. Trends Microbiol. 17, 414–422.
Lopez-Bueno, A., Tamames, J., Velazquez, D., Moya, A., Quesada, A., Alcami, A., 2009. High diversity of the viral community from an Antarctic lake. Science 326, 858–861.
Lopez-Bueno, A., Rastrojo, A., Peiro, R., Arenas, M., Alcami, A., 2015. Ecological connectivity shapes quasispecies structure of RNA viruses in an Antarctic lake. Mol. Ecol. 24, 4812–4825.
Miranda, J.A., Culley, A.I., Schvarcz, C.R., Steward, G.F., 2016. RNA viruses as major contributors to Antarctic virioplankton. Environ. Microbiol. 18, 3714–3727.
Oliva, M., Antoniades, D., Giralt, S., Granados, I., Pla-Rabes, S., Toro, M., Liu, E.J., Sanjurjo, J., Vieira, G., 2016. The Holocene deglaciation of the Byers Peninsula (Livingston Island, Antarctica) based on the dating of lake sedimentary records. Geomorphology 261, 89–102.
Rastrojo, A., Alcami, A., 2017. Aquatic viral metagenomics: lights and shadows. Virus Res. 239, 87–96.
Rogers, S.O., Shtarkman, Y.M., Kocer, Z.A., Edgar, R., Veerapaneni, R.S., D'Elia, T., 2013. Ecology of subglacial lake Vostok. Biology 2, 629–650.
Rohwer, F., Thurber, R.V., 2009. Viruses manipulate the marine environment. Nature 459, 207–212.
Schiermeier, Q., 2014. Polar drilling problems revealed. Nature 505, 463.
Shtarkman, Y.M., Kocer, Z.A., Edgar, R., Veerapaneni, R.S., D'Elia, T., Morris, P.F., Rogers, S.O., 2013. Subglacial Lake Vostok (Antarctica) accretion ice contains a diverse set of sequences from aquatic, marine and sediment-inhabiting bacteria and eukarya. PLoS One 4, e67221.
Steward, G.F., Culley, A.I., Mueller, J.A., Wood-Charlson, E.M., Belcaid, M., Poisson, G., 2013. Are we missing half of the viruses in the ocean? ISME J. 7, 672–679.
Suttle, C.A., 2005. Viruses in the sea. Nature 437, 356–361.
Toro, M., Camacho, A., Rochera, C., Rico, E., Bañon, M., Fernandez-Valiente, E., Marco, E., Justel, A., Avendaño, M.C., Ariosa, Y., Vincent, W.F., Quesada, A., 2007. Limnological characteristics of the freshwater ecosystems of Byers Peninsula, Livingston Island, in maritime Antarctica. Polar Biol. 30, 635–649.
Tranter, M., 2014. Biogeochemistry: microbes eat rock under ice. Nature 512, 256–257.
Tschitschko, B., Williams, T.J., Allen, M.A., Paez-Espino, D., Kyrpides, N., Zhong, L., Raftery, M.J., Cavicchioli, R., 2015. Antarctic archaea-virus interactions: metaproteome-led analysis of invasion, evasion and adaptation. ISME J. 9, 2094–2107.

Velazquez, D., Lopez-Bueno, A., Aguirre de Carcer, D., de los Rios, A., Alcami, A., Quesada, A., 2016. Ecosystem function decays by fungal outbreaks in Antarctic microbial mats. Sci. Rep. 6, 22954.

Wilkins, D., Yau, S., Williams, T.J., Allen, M.A., Brown, M.V., DeMaere, M.Z., Lauro, F.M., Cavicchioli, R., 2013. Key microbial drivers in Antarctic aquatic environments. FEMS Microbiol. Rev. 37, 303–335.

Williamson, K.E., Fuhrmann, J.J., Wommack, K.E., Radosevich, M., 2017. Viruses in soil ecosystems: unknown quantity within an unexplored territory. Annu. Rev. Virol. 4, 201–219.

Wright, A., Siegert, M., 2012. A fourth inventory of Antarctic subglacial lakes. Antarct. Sci. 24, 659–664.

Yau, S., Lauro, F.M., DeMaere, M.Z., Brown, M.V., Thomas, T., Raftery, M.J., Andrews-Pfannkoch, C., Lewis, M., Hoffman, J.M., Gibson, J.A., Cavicchioli, R., 2011. Virophage control of Antarctic algal host-virus dynamics. Proc. Natl. Acad. Sci. U. S. A. 108, 6163–6168.

Zablocki, O., van Zyl, L., Adriaenssens, E.M., Rubagotti, E., Tuffin, M., Cary, S.C., Cowan, D., 2014. High-level diversity of tailed phages, eukaryote-associated viruses, and virophage-like elements in the metaviromes of Antarctic soils. Appl. Environ. Microbiol. 80, 6888–6897.

Zawar-Reza, P., Arguello-Astorga, G.R., Kraberger, S., Julian, L., Stainton, D., Broady, P.A., Varsani, A., 2014. Diverse small circular single-stranded DNA viruses identified in a freshwater pond on the McMurdo Ice Shelf (Antarctica). Infect. Genet. Evol. 26, 132–138.

CHAPTER THREE

From Spatial Metagenomics to Molecular Characterization of Plant Viruses: A Geminivirus Case Study

Sohini Claverie*, Pauline Bernardo[†,‡,§], Simona Kraberger[¶], Penelope Hartnady[||], Pierre Lefeuvre*, Jean-Michel Lett*, Serge Galzi[‡,§], Denis Filloux[‡,§], Gordon W. Harkins[#], Arvind Varsani[¶,], Darren P. Martin[||], Philippe Roumagnac[‡,§,1]**

*CIRAD, UMR PVBMT, F-97410 St Pierre, La Réunion, France
[†]Department of Plant Pathology, Ohio State University, OARDC, Wooster, OH, United States
[‡]CIRAD, UMR BGPI, Montpellier, France
[§]BGPI, Univ. Montpellier, CIRAD, INRA, Montpellier SupAgro, Montpellier, France
[¶]The Biodesign Center for Fundamental and Applied Microbiomics, Center for Evolution and Medicine, School of Life Sciences, Arizona State University, Tempe, AZ, USA
[||]Computational Biology Division, Department of Integrative Biomedical Sciences, Institute of Infectious Diseases and Molecular Medicine, University of Cape Town, Observatory, South Africa
[#]South African Medical Research Council Bioinformatics Unit, South African National Bioinformatics Institute, University of the Western Cape, South Africa
**Structural Biology Research Unit, Department of Clinical Laboratory Sciences, University of Cape Town, Rondebosch, Cape Town, South Africa
[1]Corresponding author: e-mail address: philippe.roumagnac@cirad.fr

Contents

Advances in Virus Research, Volume 101
ISSN 0065-3527
https://doi.org/10.1016/bs.aivir.2018.02.003

Abstract

The number of plant viruses that are known likely remains only a vanishingly small fraction of all extant plant virus species. Consequently, the distribution and population dynamics of plant viruses within even the best-studied ecosystems have only ever been studied for small groups of virus species. Even for the best studied of these groups very little is known about virus diversity at spatial scales ranging from an individual host, through individual local host populations to global host populations. To date, metagenomics studies that have assessed the collective or metagenomes of viruses at the ecosystem scale have revealed many previously unrecognized viral species. More recently, novel georeferenced metagenomics approaches have been devised that can precisely link individual sequence reads to both the plant hosts from which they were obtained, and the spatial arrangements of these hosts. Besides illuminating the diversity and the distribution of plant viruses at the ecosystem scale, application of these "geometagenomics" approaches has enabled the direct testing of hypotheses relating to the impacts of host diversity, host spatial variations, and environmental conditions on plant virus diversity and prevalence. To exemplify how such top-down approaches can provide a far deeper understanding of host–virus associations, we provide a case-study focusing on geminiviruses within two complex ecosystems containing both cultivated and uncultivated areas. Geminiviruses are a highly relevant model for studying the evolutionary and ecological aspects of viral emergence because the family *Geminiviridae* includes many of the most important crop pathogens that have emerged over the past century. In addition to revealing unprecedented degrees of geminivirus diversity within the analyzed ecosystems, the geometagenomics-based approach enabled the focused in-depth analysis of the complex evolutionary dynamics of some of the highly divergent geminivirus species that were discovered.

1. INTRODUCTION

Although viruses are the most abundant and diverse entities on earth and are arguably "the most successful form of life" (Wasik and Turner, 2013), they are still primarily regarded as agents of disease. As exemplified by other chapters of this book focusing on a range of host types, it is now apparent, however, that viruses are core components of global ecosystems within which the spectrum of virus–host interactions include parasitism and pathogenicity at one extreme and symbiosis and mutualism at the other

(Roossinck, 2011b, 2015; Suttle, 2007). The ecological roles of viruses and the benefits that they can bestow on host species have mainly been overlooked because of how difficult it is to detect and observe them. This difficulty has historically restricted most virological studies to the small minority of pathogenic virus species that are associated with diseases of cultivated plants, domesticated animals, and humans.

The beginnings of plant virology—and virology per se—date back to the late 19th century when the first virus ever discovered, later named tobacco mosaic virus (Family *Virgaviridae*, Genus *Tobamovirus*), was reported by Ivanovski in 1892 and Beijerinck in 1898 (Bos, 1999). Since these two pioneering studies, approximately 4400 other virus species have been described of which approximately 1400 infect plants (10th Report of the International Committee on Taxonomy of Viruses, ICTV). The vast majority of plant viruses are transmitted by a vector organism that has the capacity to directly inject virus particles into the cytoplasm of plant cell types that are amenable to virus replication.

While most known plant virus species have single-stranded (ss) RNA genomes (Zaitlin and Palukaitis, 2000), the biggest emerging worldwide threat to crops are arthropod-borne viruses with ssDNA genomes in the family *Geminiviridae*. Examples of important emergent diseases that are caused by geminiviruses include maize streak disease (Harkins et al., 2009b), cassava mosaic disease (Patil and Fauquet, 2009), cotton leaf curl disease (Briddon and Markham, 2000), and tomato leaf curl disease (Accotto et al., 2000).

Geminiviruses display substitution rates that are as high as those of viruses with ssRNA genomes and they are known to readily recombine both within and between species. Viruses within this genus are therefore predisposed to rapid adaptive evolution (Duffy and Holmes, 2008; Harkins et al., 2009a; Lefeuvre et al., 2010; Monjane et al., 2011). They are therefore good, and relevant, models for studying plant virus evolution and emergence.

With the increasing availability of geminivirus genome sequences, population-scale studies looking in-depth at genetic diversity within individual crop-infecting geminivirus species have recently been reported (Kraberger et al., 2017b; Pande et al., 2017). Besides revealing ample evidence of recombination, these studies have also determined that the origins of much of the recombinationally acquired genetic material are likely geminiviruses belonging to species that have never been characterized. In cases where parental viruses belong to species that have been characterized, many of these "parental" virus species are known to primarily infect uncultivated hosts. Because of their focus on individual host and virus species, however, such studies cannot directly address hypotheses relating to the potentially complex, but highly relevant, evolutionary, and ecological factors underlying the origin and spread of such recombinants.

Crucially, the development and application of sequence-nonspecific virus discovery approaches (Roossinck et al., 2015; Rosario et al., 2012b) has tremendously increased the rate at which novel highly divergent geminiviruses have been discovered within uncultivated plants, including *Euphorbia caput-medusae* (Bernardo et al., 2013), cleome (Fontenele et al., 2017), wild rice (Kraberger et al., 2017a), uncultivated grapevines (Perry et al., 2018), and *Plantago lanceolata* (Susi et al., 2017). The picture of global-geminivirus diversity that is emerging from these virus discovery studies is that the economically relevant geminivirus crop pathogens represent a minute fraction of the geminiviruses that could potentially emerge from uncultivated species to threaten crop production in the future.

In this chapter, we initially review the ongoing development of viral metagenomics-based approaches, including the most recent iterations of the georeferenced metagenomics approaches. Besides providing insights into the distribution and prevalence of plant viruses across landscapes, these approaches have enabled the quantitative estimation of the genetic diversity within global plant virus populations. We then focus on a geminivirus case study to highlight how a top-down georeferenced metagenomics approach can both illuminate the extant of geminivirus diversity within an ecosystem and reveal new host-geminivirus associations, paving the way toward a better ecosystem scale understanding of host–parasite dynamics and the emergence of new crop diseases.

2. METAGENOMICS-BASED APPROACHES: THE WITHOUT A PRIORI INVESTIGATION REVOLUTION

2.1 Pioneering Metagenomics Studies

Although the term "metagenomics" was first used by Jo Handelsman and coworkers in 1998 to describe "the genomic analysis of a microorganism's population in an environmental sample" (Handelsman et al., 1998), the conceptual and technical framework upon which metagenomics is based was first introduced more than a decade earlier. In the early 1980s, Pace and colleagues used 5S and 16S rRNA culture-independent analyses to describe microbial communities in environmental samples (Pace et al., 1985). This culture-independence readily provided reasonably unbiased views of microbial populations within environmental samples—populations in which approximately 99% of species could not be cultured (Amann et al., 1995).

The multitude of culture-independent microbe discovery and quantification techniques that were devised subsequent to the seminal work of Pace et al. (1985)—techniques that are now collectively referred to as metagenomics approaches—enabled the large-scale discovery, identification, and study of microbial taxa within thousands of the Earth's myriad environments (Alberti et al., 2017; Gilbert and Dupont, 2011; Temperton and Giovannoni, 2012).

2.2 Viral Metagenomic Studies: Virus Discovery and Dark Matter

Although metagenomics was first used to characterize bacterial, archaeal, and fungal communities, it is increasingly being employed in virology investigations (Delwart, 2007). Whereas there are universally conserved genes in bacteria, archaea, and fungi that allow the direct comparison of all members of these kingdoms, there are no genes that are conserved across all viruses that can be used as canonical phylogenetic markers.

To circumvent this problem, protocols have been developed to enable the specific but mostly sequence-independent amplification of virus sequences. In their seminal work, Allander et al. (2001) used a technique based on DNase sequence-independent single primer amplification. This technique allowed the first metagenomic study of uncultured marine viral communities (Breitbart et al., 2002) and marked the emergence of metagenomics in virology as a means of identifying new viruses (Breitbart and Rohwer, 2005). The strategy used was based on (i) the enrichment of viruses in an environmental sample using a combination of differential filtration and ultracentrifugation and (ii) the amplification of viral genomes with random primers.

A growing number of methodologies, based primarily on high-throughput sequencing technologies, are now available to interrogate the genetic diversity within virus populations without any prior knowledge of what species the populations contain. Using these approaches, novel viral communities have been studied in a wide range of environments including those within the human body (Breitbart et al., 2003; Lecuit and Eloit, 2014), in the oceans (Breitbart et al., 2004, Brum et al., 2015, Suttle, 2005; see chapter "Viruses in marine ecosystems: From open waters to coral reefs" by Weynberg), in fresh water lakes (Dayaram et al., 2016, Roux et al., 2012; see chapters "Water-mediated transmission of plant, animal, and human viruses" by Mehle et al. and "Viruses in polar lake and soil ecosystems" by Rastrojo and Alcamí), and in terrestrial ecosystems (Ng et al., 2011b; Roossinck, 2012).

Along with the discovery of hundreds of individual new viral species, the rich viral diversity that has been identified in such studies emphasizes that viruses are likely essential components within ecosystems ranging in scale from that of the human gut (Scarpellini et al., 2015; Stecher et al., 2012) to that of the planet-wide oceanic planktonic or bacterial ecosystems (Alberti et al., 2017; Sunagawa et al., 2015). Furthermore, these studies have shown that up to 60%–99% of the sequences recovered are "dark matter" because they can only tentatively be classified as being of viral origin due to their lacking any convincingly detectable homology to any nucleotide sequences with known origins that are currently deposited within large publically nucleotide sequence databases such as GenBank (Rosario and Breitbart, 2011). Although our inability to definitively classify these "dark matter" sequences can hamper ecological and evolutionary studies of viromes, their mere presence strongly suggests that global viral diversity could far exceed that which is presently known (Rosario and Breitbart, 2011).

One of the main reasons for suspecting that many of these dark matter sequences are of viral origin is that even when metagenomics-derived sequence reads do display homology to known virus sequences, the degree of similarity of these sequences to their best-matching viral homologs within the sequence databases is frequently close to the threshold of what would be considered "convincingly detectable homology." For example, a recent analysis of replication-associated protein (Rep) sequences of ssDNA viruses demonstrated the extent of how much diversity exists even within a well-known evolutionarily conserved virus protein. Phylogenetic analysis of Rep sequences from 659 circular Rep-encoding ssDNA viruses (CRESS-DNA viruses) that had been discovered in various metagenomics studies revealed that the vast majority of these clustered outside of the four CRESS-DNA virus families that are presently recognized by the International Committee for Virus Taxonomy (Simmonds et al., 2017), and that most of the viruses from which these sequences were derived are likely representative of tens, if not hundreds, of presently uncharacterized family level CRESS-DNA virus lineages.

2.3 Plant Viral Metagenomics: Looking Beyond Cropping Systems

As is true for viruses more generally, there are profound gaps in our understanding of the nature and diversity of plant viruses. Existing knowledge is biased by a focus on only a small subset of plant–virus interactions. Since the 19th century discovery of tobacco mosaic virus, most plant virus studies have been motivated by an imperative to manage the diseases of crop and ornamental plants. Nevertheless, paradoxical metagenomics-based reports on uncultivated plant species of high frequencies of viral infections but low levels of disease symptoms (Muthukumar et al., 2009; Roossinck et al., 2010), and reconstructions of the emergence histories of important crop pathogens suggesting that many of these likely originated in uncultivated hosts, have prompted a reassessment of how viral diseases of plants originate (Fargette et al., 2006). Also, the discovery that many different plant virus species can coexist and potentially interact either within an individual host, within multiple individuals of a single host species, or within different host species within the same plant community (Vayssier-Taussat et al., 2014), has seriously undermined the plausibility of the "one virus—one disease" model. An alternative model is that viral diseases are the outcome of evolutionary and ecological processes operating at the level of whole viral communities, or viromes, and that diseases are a consequence of otherwise benign viruses inflicting excessive harm on particular host plants (Vayssier-Taussat et al., 2014).

It is in fact becoming increasingly evident that the emergence of disease causing viruses involves processes operating at scales ranging from individual molecular interactions within host cells to the interactions of host and potential host organisms across landscapes. Hence, it has been posited that the long-term coevolution of plants and their viruses within resilient natural ecosystems may lead to mutual adaptation and low disease burdens (Cooper and Jones, 2006; Jones, 2009). However, the reality is probably more complex because the influence of viruses on natural hosts can range from fitness reducing (e.g., Kelley, 1994; Remold, 2002; Malmstrom et al., 2005) to fitness enhancing (Márquez et al., 2007; Xu et al., 2008). By contrast, ecosystems that have been disturbed, for example by human activities such as agriculture, can potentially foster an environment where there are no strong evolutionary constraints preventing the emergence of viruses and/or vectors that inflict considerable damage on certain host species (Fargette et al., 2006; Roossinck and Garcia-Arenal, 2015).

Given that most of the information that we presently have on plant virus diversity is derived from viruses belonging to the small number of species that cause recognizable diseases on crop and ornamental plants (Malmstrom et al., 2011; Wren et al., 2006), we still cannot fully understand why and how plant viral diseases emerge. Recent viral metagenomic studies have therefore started to look beyond cropping systems and substantial methodological advances have been made enabling the study of plant viral diversity and evolution at the scale of entire ecosystems (Roossinck et al., 2015; Stobbe and Roossinck, 2014).

2.4 Georeferenced Metagenomics Approaches: Exploring Contemporary Plant Viral Dynamics Within a Defined Spatiotemporal Framework

Ecology can be loosely defined as the study of how organisms interact with each other and with the environment. While determining the inventory of virus species that infect individual plants within an ecosystem could facilitate the inference of probable host ranges, comparing the genetic composition of virus populations in different hosts could enable the inference of viral community structure. Furthermore, comparing the species composition of viral communities and the genetic structures of virus populations before and after particular environmental perturbations, could provide insight about how viral populations respond to such perturbations.

In the context of understanding viral emergence, the scale at which viral ecology studies are particularly pertinent is that of agroecological interfaces that span cropping systems and natural environments (Alexander et al., 2014). At these interfaces, changes in environmental factors (e.g., the abundances of host and vector species) are expected to be more abrupt than in natural environments. Crucially, we suggest that the environmental heterogeneity at these interfaces will potentially leave imprints both on the community structures of virus populations in these ecosystems, and on the individual rapidly evolving genomes from which these populations are composed.

In 2010, Roossinck et al. developed a viral metagenomics approach that could precisely link individual sequence reads from bulked multiplexed sequencing reactions to information on the individual hosts from which the sequences were sampled (Roossinck et al., 2010). This paved the way toward quantitatively studying viral diversity, prevalence, and spatial distributions across landscapes. Using this approach, pioneering "ecogenomics" studies in natural environments indicated that ~70% of sampled plants in a Costa Rican forest (Roossinck et al., 2010) and ~25% of plants in an Oklahoma prairie (Muthukumar et al., 2009) harbored identifiable plant viral sequences (Fig. 1).

"Geometagenomics" is a subsequent enhancement of the "ecogenomics" approach, which has been developed to further consider the spatial arrangements of plant samples, and the precise environmental contexts of individual sampling sites (Bernardo et al., 2018). The geometagenomics experimental design involves systematically defined sampling locations within a predefined georeferenced grid that might contain, for example, 100 sampling points (or geonodes) at 500-m spacing (10 nodes × 10 nodes; Bernardo et al., 2018).

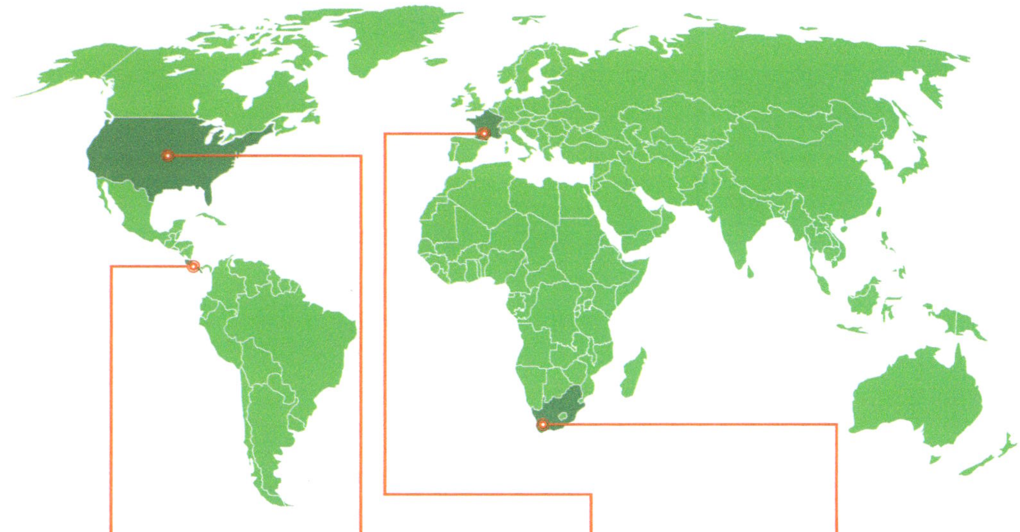

	Costa Rica (Área de Conservación Guanacaste)	USA (Tall Grass Prairie Preserve - Oklahoma)	France (Camargue)	South Africa (Cape Floristic Region)
Year	2010	2010	2010 and 2012	2010
Ecosystem	Dry forest, cloud forest, and rain forest	Tallgrass prairie	Mediterranean wetlands/crops	Fynbos/crops
Approach	Ecogenomics	Ecogenomics	Geometagenomics	Geometagenomics
Plant types	Uncultivated	Uncultivated	Cultivated vs uncultivated	Cultivated vs uncultivated
Plant diversity	High	Low	Low vs high	Low vs high
Virus prevalence	70%	25%	26%, 2010 36%, 2012	28%
Reference	Roossinck et al. (2010)	Muthukumar et al. (2009)	Bernardo et al. (2017)	Bernardo et al. (2017)

Fig. 1 Summary of the published plant virus ecogenomics and geometagenomics studies that have been conducted over the past decade.

This georeferenced approach yields geographically tagged DNA amplicons from both known and unknown virus species with both RNA and DNA genomes, and further allows viral sequences to be linked to specific host plants at a specific location and time. The a priori choice of sampling points allows the identification of reference ecosystems that would be most appropriate for determining, for example, the impacts of agriculture on viral demographics and evolution within natural endangered ecosystems, or the transmission rates of viruses between cultivated and uncultivated plants in areas where crop fields border natural vegetation.

To date, the geometagenomics approach has been applied to the study of only two ecosystems (Fig. 1): one in the Western Cape region of South Africa and the other in the Rhône River Delta region in France (Bernardo et al., 2018). The South African site included endangered evergreen shrubland (renosterveld) and beach scrub (strandveld) situated beside fertile plains planted with barley, wheat, and other crops (Bernardo et al., 2018). The French site included saline steppes and xero-halophitic meadows surrounded by areas under intensive rice, wheat, and alfalfa cultivation (Bernardo et al., 2018). The objectives of this study were to assess whether (i) plant-associated virus communities were more prevalent but less diverse in cultivated areas and (ii) viruses from particular known families were significantly associated with cultivated or uncultivated areas.

These French and South African geometagenomics studies revealed that 26%–36% of the plant samples throughout the georeferenced sampling grids in the two countries contained plant and/or fungal virus sequences and that viral prevalence in these ecosystems was similar to that observed in the Oklahoma prairie ecosystem (Muthukumar et al., 2009) (Fig. 1). Importantly, these and other estimates of plant viral prevalence should be taken as lower bounds of the actual prevalences within particular environments. This is because there are two biases that lead viral prevalence to be underestimated, both of which have impacted all plant virus ecogenomics and geometagenomics studies to date. The first bias arises because viral genomes may not be isolated with the same efficiency from all host species, and so some viruses may be missed. The second bias arises because sequenced genomic fragments from viruses that are closely related to previously characterized plant viruses are easier to identify as being of plant viral origin than are those from viruses that are only distantly related (or completely unrelated) to any previously characterized taxon. As a result of these two biases, many viruses that are present within analyzed plant samples will either not be present at high enough titers to be isolated and sequenced, or, if they are isolated and sequenced, will be so distantly related to currently known viruses that they will not be identifiable as being virus derived and will be categorized as dark matter of indeterminate origin.

In the single geometagenomics study so far published (Bernardo et al., 2018), the detected proportion of infected plant samples (a proxy for virus prevalence) was higher for cultivated host plants than for uncultivated ones. Likewise, it was found that overall virus prevalence was generally higher in cultivated crop areas than in natural vegetation areas. Furthermore, viruses from some virus families showed strong associations with either cultivated or uncultivated areas. These results are consistent with hypotheses that generally predict increased pathogen prevalence as host abundance increases (Agrawal et al., 2006; Keesing et al., 2010). It is, for example, known that the prevalence of a pathogen might be increased through nonrandom biodiversity loss caused by changes to the physical or biotic environment (Lacroix et al., 2014; Pagan et al., 2012), through increasing the absolute abundance of hosts and/or pathogen-transmitting vectors (Mitchell et al., 2002; Roche et al., 2012), or through increasing the probability that susceptible hosts and vectors will encounter each other (Allan et al., 2003).

In the endangered South African Fynbos Biome, Bernardo et al. (2018) further found that the prevalence of plant viruses was significantly higher in exotic plants than in indigenous plants. Invasive exotic plants are considered to be the most severe threat currently facing the Fynbos Biome because competition between these exotic and indigenous plants reduces the diversity of the indigenous plants (Downey and Richardson, 2016). An additional threat posed by exotic plants may be their capacity to increase pathogen loads in the indigenous species with which they compete (Borer et al., 2007; Malmstrom et al., 2005). On the other hand, disease-causing plant viruses might also, over time, cause declines in the density and distribution of some exotic species and, ultimately, facilitate the recovery of indigenous species, as is possible for pathogens more generally (Flory and Clay, 2013).

2.5 Vector-Enabled Metagenomics (VEM): Exploring the Diversity of Plant Viruses Using Their Insect Vectors

Using ecogenomics or geometagenomics to characterize virus populations at the scale of whole ecosystems will likely yield only limited information about the transmission dynamics of viruses among plants within an ecosystem. VEM has therefore been developed to complement plant-focused metagenomics studies. VEM approaches generally investigate the diversity of plant viruses within their insect vectors. With VEM, insects are sampled and, as with the ecogenomics and geometagenomics approaches, viral particles are purified and viral nucleic acids are sequenced (Ng et al., 2011a).

The VEM approach exploits the natural ability of highly mobile insect vectors to accumulate viruses from many plant species over time and space within an ecosystem (Ng et al., 2011a). The VEM approach can even be expanded to include the sampling of insect predators (such as dragonflies) that feed on plant virus transmission vectors (Rosario et al., 2012a). The VEM approach has been used on several insect vectors and insect vector predators such as leafhoppers (Fontenele et al., 2018; Kamali et al., 2017), earwigs (Godinho et al., 2017), whiteflies (Ng et al., 2011a; Rosario et al., 2014, 2015, 2016), and dragonflies (Rosario et al., 2013) to identify novel and known plant viruses. Besides enabling the discovery of dozens of previously unknown viruses, the VEM approach has prompted reevaluation of the geographical ranges of known viruses. For example, a VEM study focusing on Puerto Rican dragonflies identified the first mastrevirus genome from the Americas (Rosario et al., 2013)—a region long believed to be devoid of this important geminivirus genus. Since then, additional mastreviruses have been reported from the Americas using VEM or metagenomics-based approaches to analyze switchgrass (Agindotan et al., 2015), sugarcane (Boukari et al., 2017; Candresse et al., 2014), leafhoppers and maize (Fontenele et al., 2018).

3. DISCOVERY OF NEW VIRAL TAXA USING METAGENOMICS-BASED APPROACHES: THE GEMINIVIRUS TEST CASE

3.1 Establishing Taxonomic Groups From Metagenomics Datasets

One current limitation of viral metagenomics studies is the difficulty of assigning virus-derived sequencing reads to established taxonomic groups such as species, genera, and families. This is due in part to the fact that there is no robust means by which small genome fragments can be taxonomically classified with a high degree of accuracy. Therefore, most studies attempt to group sequences, usually using an arbitrarily defined degree of pairwise nucleotide identity, into so-called operational taxonomic units (OTUs) that might or might not reflect widely accepted taxonomic groups such as strains, species, genera, or families (Roossinck, 2012). This approach works well with commonly used pairwise sequence similarity searching approaches (e.g., BLASTn or BLASTx) and tentative classifications that are made using such approaches can be further refined using phylogenetic analyses (Bernardo et al., 2018).

In the geometagenomics study of Bernardo et al. (2018), 78% of detected plant virus OTUs shared 27%–75% identity (median = 49%) with isolates of known plant virus species, suggesting that these OTUs might represent novel species within 19 of the 22 plant virus families that are currently recognized by the ICTV (King et al., 2012; Roossinck, 2011a). Interestingly, ~81% of these apparently novel viruses were found within uncultivated plants, supporting the notion that the presently known crop-infecting plant virus species represent just a relatively small fraction of all terrestrial plant-infecting virus species (Wren et al., 2006). Interestingly, six OTUs from the Bernardo et al. (2018), study potentially represented highly divergent geminivirus species. Since many of the most important pathogens of cultivated crops that have emerged over the past 50 years have been geminiviruses, here we will look in more detail at these six OTUs and use classical molecular approaches to fully characterize them.

3.2 The Discovery and Characterization of Novel Geminiviruses

Four out of these six OTUs were from four different South African uncultivated endemic plant species (*E. caput-medusae*, *Limeum africanum*, *Exomis microphylla*, and *Polygala garcinii*). Whereas one of the two remaining OTUs that were discovered in France was identified in an uncultivated species (*Juncus maritimus*) the other was from a cultivated species (*Medicago sativa*, alfalfa).

While the first step of a top-down georeferenced metagenomics approach involves extensively sampling viral sequences from large numbers of plants each with known geographical coordinates, the second step involves the use of standard molecular approaches such as the cloning and full genome sequencing of the identified viruses. Therefore, DNA was extracted from geminivirus-infected samples from each of four species (*L. africanum*, *E. microphylla*, *P. garcinii*, and *J. maritimus*) using the DNeasy Plant Mini Kit (Qiagen, Germany). Extracted DNA was used as a template for PCR amplification of the complete genomes of the novel geminiviruses using pairs of abutting primers (designed based on assembled reads from the metagenomics step; Table 1), each of which spanned a restriction enzyme site. Amplicons from *L. africanum* were then cloned as described by Susi et al. (2017), whereas those from the other three plants were cloned as described by Bernardo et al. (2013). The cloned genomes were Sanger-sequenced using primer walking. Note that genomes recovered from *M. sativa* (alfalfa leaf curl virus, ALCV) and *E. caput-medusae* (*E. caput-medusae* latent virus, EcmLV)

Table 1 List of Abutting Primers Used for PCR Amplification of the Complete Genomes of the Four Novel Geminivirus Detected in the Western Cape Region of South Africa and the Rhône Delta Region of France

Host Plant	Virus[a]	Accession Number	Abutting Primers
Juncus maritimus	JmaV	MG001958	5′-TGGTACCAATTGAGCCGCTC-3′
			5′-GGGTACCCAAGGGCAATTTTG-3′
Polygala garcinii	PgaV	MG001959	5′-ATACTTCTAGATCAATTCGCTTCCTG-3′
			5′-CGGTCTCTAGACTGCGGTTGCACAA-3′
Exomis microphylla	EmaV	MG001960	5′-AGTCGACCTCTGGCTATCTC-3′
			5′-TGTCGACGAATGTGACCGTTA-3′
Limeum africanum	LaaV	MG001961	5′-TTCTTGAACTGAAGATAGGCCTCCTCTTC-3′
			5′-GGAGGCCCAGAAACTCCTACAGAA-3′

[a]JmaV, juncus maritimus-associated virus (2740 nts); PgaV, polygala garcinii-associated virus (2798 nts); EmaV, exomis microphylla-associated virus (2974 nts); and LaaV, limeum africanum-associated virus (2963 nts).

have already been reported on Bernardo et al. (2013) and Roumagnac et al. (2015) and were both assigned to the same new genus, *Capulavirus* (Varsani et al., 2017).

The four newly determined genome sequences were for juncus maritimus-associated virus (JmaV; MG001958; 2740 nts) from *J. maritimus*, polygala garcinii-associated virus (PgaV; MG001959; 2798 nts) from *P. garcinii*, limeum africanum-associated virus (LaaV; MG001961; 2963 nts) from *L. africanum*, and exomis microphylla-associated virus (EmaV; MG001960, 2974 nts) from *E. microphylla* (Fig. 2).

While three of these sequences have the characteristic geminivirus virion strand replication origin "TAATTATAC" nonanucleotide motif, the sequence from *E. microphylla* contains a "TAAGATTCC" nonanucleotide that has so far been found in two other geminiviruses: the eragrovirus eragrovirus curvula streak virus and the becurtovirus beet curly top Iran virus. The arrangement of open reading frames (ORFs) within the four novel geminivirus genomes are similar to those described previously for other geminiviruses (Fig. 2).

The four novel geminivirus genomes were aligned using MUSCLE (Edgar, 2004) with representative sequences from the nine geminivirus genera that are currently recognized by the ICTV and four sequences from novel geminiviruses that have not yet been assigned to genera by the ICTV. The alignment was used to infer a neighbor-joining (NJ) phylogenetic tree with a Jukes-Cantor nucleotide substitution model and 1000 bootstrap iterations in MEGA7 (Kumar et al., 2016). The NJ phylogenetic tree was midpoint rooted, and branches with <60% support were collapsed using TreeGraph2 (Stover and Muller, 2010).

The capsid protein (CP) and Rep amino acid sequences of the representative geminiviruses and the four new sequences reported here were aligned using MUSCLE (Edgar, 2004). The resulting alignments were used to infer maximum-likelihood phylogenetic trees using PHYML 3.0 (Guindon et al., 2010) using the rtREV+G and rtREV+G+I amino acid substitution models for CP and Rep, respectively, inferred to be the best-fitting models using ProtTest (Abascal et al., 2005). The trees were rooted with CP and Rep sequences of representative genomoviruses (Krupovic et al., 2016). Branches with <0.8 aLRT branch support were collapsed using TreeGraph2 (Stover and Muller, 2010). All pairwise identities (genome, CP, and Rep) were determined using SDT v1.2 (Muhire et al., 2014).

While the genome-wide pairwise nucleotide identities of the four new genomes were <60% to all other known geminiviruses (Fig. 3), phylogenetic

Fig. 2 Genome organizations of viruses belonging to the different geminivirus genera. *Abbreviations*: TYLCV, tomato yellow leaf curl virus, BGMV, bean golden mosaic virus, BCTIV, beet curly top Iran virus, EcmLV, Euphorbia caput-medusae latent virus, BCTV, beet curly top virus, GRBV, grapevine red blotch virus, ECSV, Eragrostis curvula streak virus, MSV, maize streak virus, TPCTV, tomato pseudo-curly top virus, TCTV, turnip curly top virus, CCDaV, citrus chlorotic dwarf associated virus, MMDav, mulberry mosaic dwarf associated virus, AGmV, apple geminivirus, TaGV1, tomato associated geminivirus 1, EmaV, Exomis microphylla associated virus, JmaV, Juncus maritimus associated virus, LaaV, Limeum africanum associated virus and PgaV, Polygala garcinii associated virus.

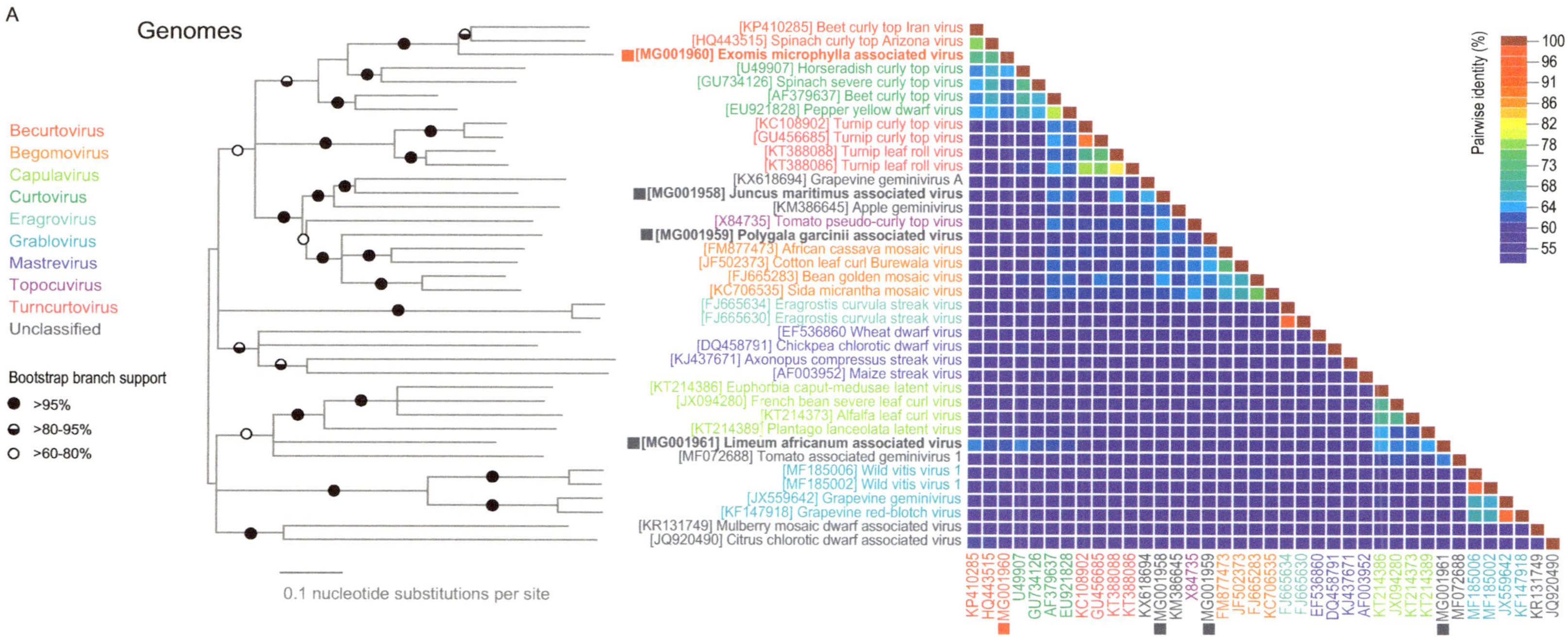

Fig. 3 Phylogenetic and nucleotide pairwise identity analyses. (A, *left*) Unrooted neighbor-joining tree inferred from aligned full-genome nucleotide sequences of representative isolates from the various geminivirus genera. (*Branch support*) *Solid dot*, >95% bootstrap support. *Half-filled dot*, 80%–95%, or *unfilled dots*, 60%–80%. *Branches with less than 60% bootstrap support* have been collapsed. The four novel geminiviruses discovered in France and South Africa are indicated with *red* and *gray squares*, to the *right*. (A, *right*) Genome-wide pairwise identities of representative isolates from various geminivirus genera as determined using SDT v1.2. (B) Maximum-likelihood phylogenetic trees, respectively, applying the rtREV + G and rtREV+G + I amino acid substitution models for CP (B, *left*) and Rep (B, *right*) of representative geminivirus genera. *Numbers associated with branches* indicate percentage aLRT support for these branches. *Both trees* are rooted with CP and Rep sequences of representative genomoviruses (not shown). *Branches with less than 80% aLRT support* have been collapsed.

(Continued)

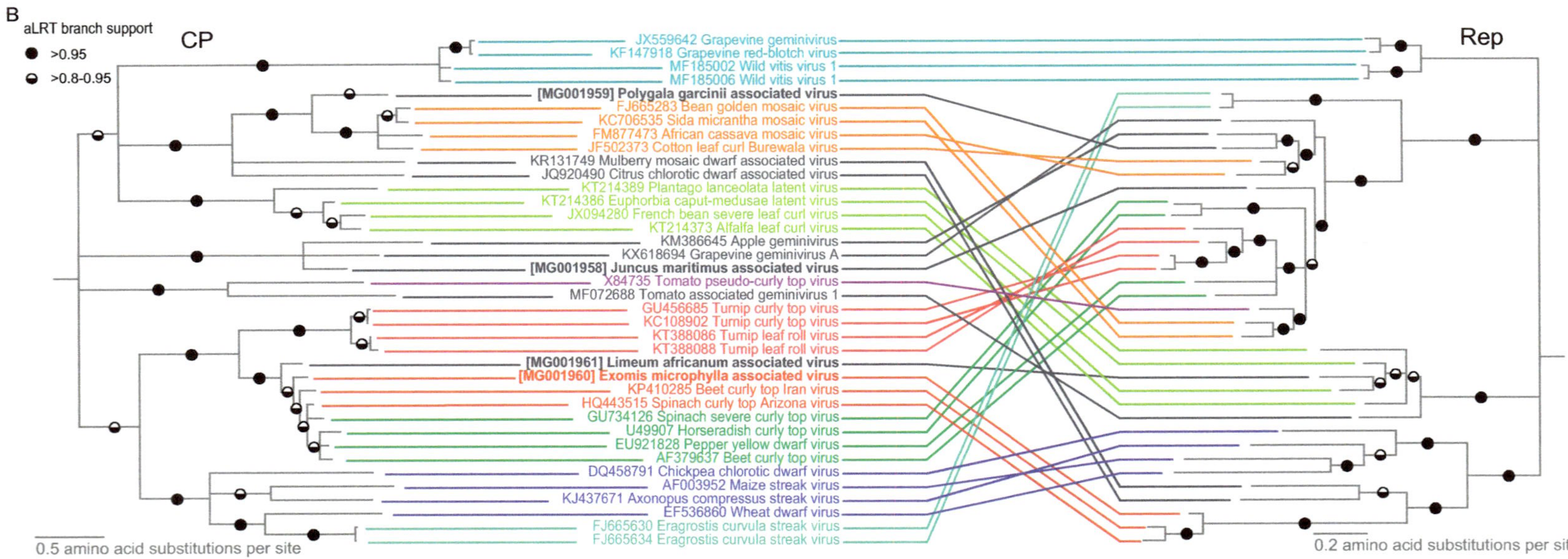
B
aLRT branch support
>0.95
>0.8-0.95
CP
Rep
JX559642 Grapevine geminivirus
KF147918 Grapevine red-blotch virus
MF185002 Wild vitis virus 1
MF185006 Wild vitis virus 1
[MG001959] Polygala garcinii associated virus
FJ665283 Bean golden mosaic virus
KC706535 Sida micrantha mosaic virus
FM877473 African cassava mosaic virus
JF502373 Cotton leaf curl Burewala virus
KR131749 Mulberry mosaic dwarf associated virus
JQ920490 Citrus chlorotic dwarf associated virus
KT214389 Plantago lanceolata latent virus
KT214386 Euphorbia caput-medusae latent virus
JX094280 French bean severe leaf curl virus
KT214373 Alfalfa leaf curl virus
KM386645 Apple geminivirus
KX618694 Grapevine geminivirus A
[MG001958] Juncus maritimus associated virus
X84735 Tomato pseudo-curly top virus
MF072688 Tomato associated geminivirus 1
GU456685 Turnip curly top virus
KC108902 Turnip curly top virus
KT388086 Turnip leaf roll virus
KT388088 Turnip leaf roll virus
[MG001961] Limeum africanum associated virus
[MG001960] Exomis microphylla associated virus
KP410285 Beet curly top Iran virus
HQ443515 Spinach curly top Arizona virus
GU734126 Spinach severe curly top virus
U49907 Horseradish curly top virus
EU921828 Pepper yellow dwarf virus
AF379637 Beet curly top virus
DQ458791 Chickpea chlorotic dwarf virus
AF003952 Maize streak virus
KJ437671 Axonopus compressus streak virus
EF536860 Wheat dwarf virus
FJ665630 Eragrostis curvula streak virus
FJ665634 Eragrostis curvula streak virus
0.5 amino acid substitutions per site
0.2 amino acid substitutions per site

Fig. 3—Cont'd

analysis of the genomes (Fig. 3) revealed that PgaV, EmaV, LaaV, and JmaV, respectively, cluster most closely with begomoviruses, becurtoviruses, capulaviruses, and an unclassified genus-level geminivirus lineage containing the two viruses, apple geminivirus, and grapevine geminivirus A (Al Rwahnih et al., 2016; Liang et al., 2015). EmaV is the first becurtovirus-like virus to be identified in Africa (Fig. 3).

PgaV and LaaV do not cluster with the sequences of any particular geminivirus genus but instead branch basal to the begomovirus and capulavirus clades, respectively. This might indicate that the full breadth of geminivirus diversity encompassed within the capulavirus and curtovirus genera is broader than previously known. If, however, it is discovered that these viruses have unique transmission vectors, it might be more appropriate to accommodate these within a new geminivirus genus (as was done for the capulaviruses; Varsani et al., 2017).

Consistent with previous reports (Briddon et al., 1996; Stanley et al., 1986), there is substantial incongruence between the phylogenies of the CP and Rep proteins (Fig. 3). It is plausible that this incongruence is a consequence of intergenus recombination events having played a part in the origin of some of the geminivirus genera (Bernardo et al., 2013; Briddon et al., 2010; Hernandez-Zepeda et al., 2013; Klute et al., 1996; Varsani et al., 2009). While both the Rep and CP sequences of PgaV cluster with begomovirus sequences and those of EmaV cluster with becurtovirus sequences, the Rep of LaaV clusters with capulaviruses and its CP clusters with becurtoviruses. Similarly, the Reps of JmaV, apple geminivirus, and grapevine geminivirus A are begomovirus-like but phylogenetically their CPs do not cluster closely with those of any other geminiviruses (Fig. 3).

The South African and French geometagenomics studies (Bernardo et al., 2018) therefore revealed an unexpected degree of geminivirus diversity within an area encompassing just ~20 km^2 of the Earth's surface. Although nine geminivirus genera are now accepted by ITCV, when the samples in these studies were first collected in 2010–12, only four geminivirus genera were known (Zerbini et al., 2017). It is noteworthy that despite six completely novel geminivirus sequences having been identified in this study, not a single previously known geminivirus species was found. The six novel geminivirus sequences have an average pairwise Hamming distance (a proxy of the identity score) of 0.396 to one another, which is almost as high as the average pairwise Hamming distances (0.405) of the genome sequences representing the breadth of previously known geminivirus diversity (Fig. 3).

3.3 Divergent Newly Discovered Geminiviruses Such as the Capulaviruses Can Reveal Much About the Ecology and Evolution of Geminiviruses

Although geminiviruses have been well studied since the 1970s (Bock et al., 1974; Duffus and Gold, 1973), the discovery of several new major geminivirus lineages using metagenomic approaches, and the subsequent biological characterization of isolates representing some of these lineages, has illuminated several gaps in our knowledge regarding the ecology and evolution of this family (Agindotan et al., 2015; Bernardo et al., 2013; Boukari et al., 2017; Candresse et al., 2014; Fontenele et al., 2017; Kraberger et al., 2017a; Kreuze et al., 2009; Rosario et al., 2013; Susi et al., 2017). For example, a more in-depth characterization of the capulaviruses discovered in the French and South African metagenomics surveys has prompted the reevaluation of our understanding of geminivirus pathogenicity. Indeed, as with other recently discovered geminiviruses found in uncultivated hosts (Kraberger et al., 2017a; Perry et al., 2018), the capulaviruses EcmLV and *P. lanceolata* latent virus (PlLV) do not cause obvious infection symptoms in the uncultivated hosts from which they were initially isolated (Bernardo et al., 2013; Susi et al., 2017). If these viruses do indeed have decreased degrees of pathogenicity this could have arisen due to transmission being hampered within natural environments (Keesing et al., 2006), such that selection has favored less pathogenic virus variants which do not shorten their host's life span and can therefore be transmitted over longer time periods. It is plausible that geminiviruses such as EcmLV and PlLV are examples of viruses that, over hundreds or thousands of years, have become well adapted to infecting particular host species within particular natural environments.

It is noteworthy, however, that EcmlV has a broad experimental host range that includes tomato and *Nicotiana benthamiana*, in which this virus causes severe symptoms (Bernardo et al., 2013). This result possibly exemplifies the role that uncultivated plant species and their viruses might play during the emergence of new diseases following the introduction of exotic crop species into an ecosystem. The number of "new encounter" situations, where introduced cultivated plants first come in contact with viruses adapted to indigenous uncultivated plants, has increased tremendously with the progressive intensification and globalization of agriculture over the past two centuries (Jones, 2009). It is plausible that, if they are brought into contact with a susceptible cultivated host species, large numbers of presently unnoticed virus species (i.e., primarily those asymptomatically infecting their natural hosts) such as EcmLV and PlLV, are at this moment poised to emerge as serious agricultural pathogens (Elena et al., 2014; Webster et al., 2007).

Unlike EcmLV and PlLV, the capulavirus discovered infecting the cultivated host, *M. sativa*, ALCV, is already potentially an emerging pathogen (Bernardo et al., 2016). ALCV causes severe disease symptoms in *M. sativa* and is transmissible by *Aphis craccivora* (Roumagnac et al., 2015)—an aphid species with a broad host range and almost worldwide distribution (CIE, 1983). The simple fact that ALCV is transmissible by an aphid rather than by any of the other previously known geminivirus transmission vectors (whiteflies, leafhoppers, and treehoppers; Zerbini et al., 2017) increases the risk that it could emerge as a serious pathogen.

The discovery that ALCV is aphid-transmitted supports the hypothesis that geminivirus coat protein sequences as a whole may be codiverging with geminivirus vector species (Rybicki, 1994). Under this evolutionary scenario, the most recent common ancestor of all the present day geminivirus vector species may have transmitted the most recent common ancestor of all the extant geminiviruses. As these vectors diversified over millions of years into aphids, treehoppers, leafhoppers, and whiteflies, the different geminivirus coat protein genes may have adapted over the same time scale to retain the capacities of their cognate viruses to be transmitted by these distinct insect lineages (Rybicki, 1994).

4. FROM METAGENOMICS TO BIOLOGICAL AND MOLECULAR CHARACTERIZATION OF PLANT VIRUSES: A CONCEPTUAL FRAMEWORK FOR IMPROVING OUR UNDERSTANDING OF VIRAL ECOLOGY AND EVOLUTION

Plant virus metagenomics studies have so far emphasized both that many more plant-infecting viruses remain undiscovered than those which are currently known, and that the geographical and host ranges of known viruses may be far broader than was previously believed. As exemplified by the geminivirus test case, a systematic top-down georeferenced metagenomics approach combining the power of the NGS-based approach for detecting and identifying viral sequences, and the reliability of classical molecular approaches for obtaining full-length genomes of newly discovered viruses, can represent a crucial step toward a better understanding of the ecology and evolution of viruses that are important to humans. In the near future, metagenomics-based approaches (including RCA, ecogenomics, geometagenomics, and VEM), should be widely applicable to the epidemiological surveillance of viral pathogens, the routine diagnosis of viral diseases, and the large-scale spatiotemporal modeling of virus molecular evolution.

However, before the full potential of plant virus metagenomics is achieved, several technical issues need to be resolved. The first issue is that, with current technology, the full virus genomes and large viral sequence contigs that are assembled during metagenomics studies may be chimeras of reads from different viral genomes and, as such, may not actually exist. This problem reduces the usefulness of metagenomics sequences in phylogenetic-based inference of molecular clock rates, positive selection, geographical dispersal patterns, and population fluctuations. The second technical issue that must be addressed is the accurate differentiation of sequences: both between exogenous and endogenous virus sources, and between sequences derived from different components of segmented virus genomes. Consequently, the characterization of viruses discovered by metagenomics approaches at present still relies on full genome cloning from source material and Sanger sequencing.

Fortunately, third-generation sequencing techniques that are capable of much longer-read lengths (>15 kilobases on average) are under development by companies such as Pacific Biosciences and Oxford Nanopore (Goodwin et al., 2016). By negating the need to assemble full genomes from fragmentary short sequence reads, these new sequencing techniques should address both of the issues mentioned above. This will pave the way toward the direct use of metagenomics datasets in determining both the roles of plant viruses in global ecosystems, and which environmental conditions might prevent or precipitate the evolution of benign viruses into crop pathogens (Green et al., 2017; Shendure et al., 2017).

ACKNOWLEDGMENTS

Thank you to Bradley White for drawing Fig. 1. D.P.M., G.W.H., and A.V. have received research grants from the National Research Foundation of South Africa. P.R. has received an EU grant FP7-PEOPLE-2013-IOF (No. PIOF-GA-2013-622571). S.C., P.L., and J.M.L. were supported by the European Union (ERDF), the Conseil Régional de La Réunion and CIRAD. S.C. was a recipient of a PhD fellowship from the Agropolis Fondation (E-Space) and CIRAD. D.P.M., G.W.H., A.V., S.C., P.L., J.M.L., and P.R. declare that they have no conflict of interest.

REFERENCES

Abascal, F., Zardoya, R., Posada, D., 2005. Prottest: selection of best-fit models of protein evolution. Bioinformatics 21, 2104–2105.

Accotto, G.P., Navas-Castillo, J., Noris, E., Moriones, E., Louro, D., 2000. Typing of tomato yellow leaf curl viruses in Europe. Eur. J. Plant Pathol. 106, 179–186.

Agindotan, B.O., Domier, L.L., Bradley, C.A., 2015. Detection and characterization of the first north American Mastrevirus in switchgrass. Arch. Virol. 160, 1313–1317.

Agrawal, A.A., Lau, J.A., Hamback, P.A., 2006. Community heterogeneity and the evolution of interactions between plants and insect herbivores. Q. Rev. Biol. 81, 349–376.

Al Rwahnih, M., Alabi, O.J., Westrick, N.M., Golino, D., Rowhani, A., 2016. Description of A novel monopartite geminivirus and its defective subviral genome in grapevine. Phytopathology. Phyto07160282r.

Alberti, A., Poulain, J., Engelen, S., Labadie, K., Romac, S., Ferrera, I., Albini, G., Aury, J.M., Belser, C., Bertrand, A., Cruaud, C., Da Silva, C., Dossat, C., Gavory, F., Gas, S., Guy, J., Haquelle, M., Jacoby, E., Jaillon, O., Lemainque, A., Pelletier, E., Samson, G., Wessner, M., Acinas, S.G., Royo-Llonch, M., Cornejo-Castillo, F.M., Logares, R., Fernandez-Gomez, B., Bowler, C., Cochrane, G., Amid, C., Ten Hoopen, P., De Vargas, C., Grimsley, N., Desgranges, E., Kandels-Lewis, S., Ogata, H., Poulton, N., Sieracki, M.E., Stepanauskas, R., Sullivan, M.B., Brum, J.R., Duhaime, M.B., Poulos, B.T., Hurwitz, B.L., Pesant, S., Karsenti, E., Wincker, P., Team, G.T., Consortium, T.O., 2017. Viral to metazoan marine plankton nucleotide sequences from the Tara oceans expedition. Sci. Data 4, 170093.

Alexander, H.M., Mauck, K.E., Whitfield, A.E., Garrett, K.A., Malmstrom, C.M., 2014. Plant-virus interactions and the agro-ecological interface. Eur. J. Plant Pathol. 138, 529–547.

Allan, B.F., Keesing, F., Ostfeld, R.S., 2003. Effect of forest fragmentation on Lyme disease risk. Conserv. Biol. 17, 267–272.

Allander, T., Emerson, S.U., Engle, R.E., Purcell, R.H., Bukh, J., 2001. A virus discovery method incorporating DNase treatment and its application to the identification of two bovine parvovirus species. Proc. Natl. Acad. Sci. U.S.A 98, 11609–11614.

Amann, R.I., Ludwig, W., Schleifer, K.H., 1995. Phylogenetic identification and in-situ detection of individual microbial-cells without cultivation. Microbiol. Rev. 59, 143–169.

Bernardo, P., Golden, M., Akram, M., Naimuddin, Nadarajan, N., Fernandez, E., Granier, M., Rebelo, A.G., Peterschmitt, M., Martin, D.P., Roumagnac, P., 2013. Identification and characterisation of a highly divergent Geminivirus: evolutionary and taxonomic implications. Virus Res. 177, 35–45.

Bernardo, P., Muhire, B., Francois, S., Deshoux, M., Hartnady, P., Farkas, K., Kraberger, S., Filloux, D., Fernandez, E., Galzi, S., Ferdinand, R., Granier, M., Marais, A., Monge Blasco, P., Candresse, T., Escriu, F., Varsani, A., Harkins, G.W., Martin, D.P., Roumagnac, P., 2016. Molecular characterization and prevalence of two Capulaviruses: alfalfa leaf curl virus from France and Euphorbia caput-medusae latent virus from South Africa. Virology 493, 142–153.

Bernardo, P., Charles-Dominique, T., Barakat, M., Ortet, P., Fernandez, E., Filloux, D., Hartnady, P., Rebelo, T.A., Cousins, S.R., Mesleard, F., Cohez, D., Yavercovski, N., Varsani, A., Harkins, G.W., Peterschmitt, M., Malmstrom, C.M., Martin, D.P., Roumagnac, P., 2018. Geometagenomics illuminates the impact of agriculture on the distribution and prevalence of plant viruses at the ecosystem scale. ISME J 12 (1), 173–184.

Bock, K.R., Guthrie, E.J., Woods, R.D., 1974. Purification of maize streak virus and its relationship to viruses associated with streak diseases of sugarcane and Panicum maximum. Ann. Appl. Biol. 77, 289–296.

Borer, E.T., Hosseini, P.R., Seabloom, E.W., Dobson, A.P., 2007. Pathogen-induced reversal of native dominance in a grassland community. Proc. Natl. Acad. Sci. U. S. A. 104, 5473–5478.

Bos, L., 1999. Beijerinck's work on tobacco mosaic virus: historical context and legacy. Philos. Trans. R. Soc. Lond. B Biol. Sci. 354, 675–685.

Boukari, W., Alcala-Briseno, R.I., Kraberger, S., Fernandez, E., Filloux, D., Daugrois, J.H., Comstock, J.C., Lett, J.M., Martin, D.P., Varsani, A., Roumagnac, P., Polston, J.E., Rott, P.C., 2017. Occurrence of a novel Mastrevirus in sugarcane germplasm collections in Florida, guadeloupe and reunion. Virol. J. 14, 146.

Breitbart, M., Rohwer, F., 2005. Here a virus, there a virus, everywhere the same virus? Trends Microbiol. 13, 278–284.

Breitbart, M., Salamon, P., Andresen, B., Mahaffy, J.M., Segall, A.M., Mead, D., Azam, F., Rohwer, F., 2002. Genomic analysis of uncultured marine viral communities. Proc. Natl. Acad. Sci. U. S. A. 99, 14250–14255.

Breitbart, M., Hewson, I., Felts, B., Mahaffy, J.M., Nulton, J., Salamon, P., Rohwer, F., 2003. Metagenomic analyses of an uncultured viral community from human feces. J. Bacteriol. 185, 6220–6223.

Breitbart, M., Felts, B., Kelley, S., Mahaffy, J.M., Nulton, J., Salamon, P., Rohwer, F., 2004. Diversity and population structure of a near-shore marine-sediment viral community. Proc. Biol. Sci. 271, 565–574.

Briddon, R.W., Markham, P.G., 2000. Cotton leaf curl virus disease. Virus Res. 71, 151–159.

Briddon, R.W., Bedford, I.D., Tsai, J.H., Markham, P.G., 1996. Analysis of the nucleotide sequence of the treehopper-transmitted Geminivirus, tomato pseudo-curly top virus, suggests a recombinant origin. Virology 219, 387–394.

Briddon, R.W., Heydarnejad, J., Khosrowfar, F., Massumi, H., Martin, D.P., Varsani, A., 2010. Turnip curly top virus, a highly divergent Geminivirus infecting turnip in Iran. Virus Res. 152, 169–175.

Brum, J.R., Ignacio-Espinoza, J.C., Roux, S., Doulcier, G., Acinas, S.G., Alberti, A., Chaffron, S., Cruaud, C., De Vargas, C., Gasol, J.M., Gorsky, G., Gregory, A.C., Guidi, L., Hingamp, P., Iudicone, D., Not, F., Ogata, H., Pesant, S., Poulos, B.T., Schwenck, S.M., Speich, S., Dimier, C., Kandels-Lewis, S., Picheral, M., Searson, S., Tara Oceans, C., Bork, P., Bowler, C., Sunagawa, S., Wincker, P., Karsenti, E., Sullivan, M.B., 2015. Ocean plankton. Patterns and ecological drivers of ocean viral communities. Science 348, 1261498.

Candresse, T., Filloux, D., Muhire, B., Julian, C., Galzi, S., Fort, G., Bernardo, P., Daugrois, J.H., Fernandez, E., Martin, D.P., Varsani, A., Roumagnac, P., 2014. Appearances can be deceptive: revealing a hidden viral infection with deep sequencing in a plant quarantine context. PLoS One 9, E102945.

CIE, 1983. Distribution Maps of Plant Pests. Cab International, Wallingford, UK.

Cooper, I., Jones, R.A., 2006. Wild plants and viruses: under-investigated ecosystems. Adv. Virus Res. 67, 1–47.

Dayaram, A., Galatowitsch, M.L., Arguello-Astorga, G.R., Van Bysterveldt, K., Kraberger, S., Stainton, D., Harding, J.S., Roumagnac, P., Martin, D.P., Lefeuvre, P., Varsani, A., 2016. Diverse circular replication-associated protein encoding viruses circulating in invertebrates within a Lake ecosystem. Infect. Genet. Evol. 39, 304–316.

Delwart, E.L., 2007. Viral metagenomics. Rev. Med. Virol. 17, 115–131.

Downey, P.O., Richardson, D.M., 2016. Alien plant invasions and native plant extinctions: a six-threshold framework. AoB Plants 8.

Duffus, J.E., Gold, A.H., 1973. Infectivity neutralization used in serological tests with partially purified beet curly top virus. Phytopathology 63, 1107–1110.

Duffy, S., Holmes, E.C., 2008. Phylogenetic evidence for rapid rates of molecular evolution in the single-stranded Dna Begomovirus tomato yellow leaf curl virus. J. Virol. 82, 957–965.

Edgar, R.C., 2004. Muscle: multiple sequence alignment with high accuracy and high throughput. Nucleic Acids Res. 32, 1792–1797.

Elena, S.F., Fraile, A., Garcia-Arenal, F., 2014. Evolution and emergence of plant viruses. Adv. Virus Res. 88, 161–191.

Fargette, D., Konate, G., Fauquet, C., Muller, E., Peterschmitt, M., Thresh, J.M., 2006. Molecular ecology and emergence of tropical plant viruses. Annu. Rev. Phytopathol. 44, 235–260.

Flory, S.L., Clay, K., 2013. Pathogen accumulation and long-term dynamics of plant invasions. J. Ecol. 101, 607–613.

Fontenele, R.S., Lamas, N.S., Lacorte, C., Lacerda, A.L.M., Varsani, A., Ribeiro, S.G., 2017. A novel Geminivirus identified in tomato and cleome plants sampled in Brazil. Virus Res. 240, 175–179.

Fontenele, R.S., Alves-Freitas, D.M.T., Silva, P.I.T., Foresti, J., Silva, P.R., Godinho, M.T., Varsani, A., Ribeiro, S.G., 2018. Discovery of the first maize-infecting Mastrevirus in the Americas using a vector-enabled metagenomics approach. Arch. Virol. 163, 263–267.

Gilbert, J.A., Dupont, C.L., 2011. Microbial metagenomics: beyond the genome. Ann. Rev. Mar. Sci. 3, 347–371.

Godinho, M.T., Paula, D.P., Varsani, A., Ribeiro, S.G., 2017. Genome sequence of cauliflower mosaic virus identified in earwigs (Doru Luteipes) through a metagenomic approach. Genome Announc. 5, e00043-17.

Goodwin, S., Mcpherson, J.D., Mccombie, W.R., 2016. Coming of age: ten years of next-generation sequencing technologies. Nat. Rev. Genet. 17, 333–351.

Green, E.D., Rubin, E.M., Olson, M.V., 2017. The future of Dna sequencing. Nature 550, 179–181.

Guindon, S., Dufayard, J.F., Lefort, V., Anisimova, M., Hordijk, W., Gascuel, O., 2010. New algorithms and methods to estimate maximum-likelihood phylogenies: assessing the performance of Phyml 3.0. Syst. Biol. 59, 307–321.

Handelsman, J., Rondon, M.R., Brady, S.F., Clardy, J., Goodman, R.M., 1998. Molecular biological access to the chemistry of unknown soil microbes: a new frontier for natural products. Chem. Biol. 5, R245–249.

Harkins, G.W., Delport, W., Duffy, S., Wood, N., Monjane, A.L., Owor, B.E., Donaldson, L., Saumtally, S., Triton, G., Briddon, R.W., Shepherd, D.N., Rybicki, E.P., Martin, D.P., Varsani, A., 2009a. Experimental evidence indicating that Mastreviruses probably did not co-diverge with their hosts. Virol. J. 6, 104.

Harkins, G.W., Martin, D.P., Duffy, S., Monjane, A.L., Shepherd, D.N., Windram, O.P., Owor, B.E., Donaldson, L., Van Antwerpen, T., Sayed, R.A., Flett, B., Ramusi, M., Rybicki, E.P., Peterschmitt, M., Varsani, A., 2009b. Dating the origins of the maize-adapted strain of maize streak virus, Msv-a. J. Gen. Virol. 90, 3066–3074.

Hernandez-Zepeda, C., Varsani, A., Brown, J.K., 2013. Intergeneric recombination between a new, spinach-infecting curtovirus and a new geminivirus belonging to the genus becurtovirus: first new world exemplar. Arch. Virol. 158, 2245–2254.

Jones, R.A.C., 2009. Plant virus emergence and evolution: origins, new encounter scenarios, factors driving emergence, effects of changing world conditions, and prospects for control. Virus Res. 141, 113–130.

Kamali, M., Heydarnejad, J., Pouramini, N., Masumi, H., Farkas, K., Kraberger, S., Varsani, A., 2017. Genome sequences of beet curly top Iran virus, oat dwarf virus, turnip curly top virus, and wheat dwarf virus identified in leafhoppers. Genome Announc. 5, e01674-16.

Keesing, F., Holt, R.D., Ostfeld, R.S., 2006. Effects of species diversity on disease risk. Ecol. Lett. 9, 485–498.

Keesing, F., Belden, L.K., Daszak, P., Dobson, A., Harvell, C.D., Holt, R.D., Hudson, P., Jolles, A., Jones, K.E., Mitchell, C.E., Myers, S.S., Bogich, T., Ostfeld, R.S., 2010. Impacts of biodiversity on the emergence and transmission of infectious diseases. Nature 468, 647–652.

Kelley, S.E., 1994. Viral pathogens and the advantage of sex inthe perennial grass Anthoxanthum odoratum. Philos. Trans. R. Soc. Lond. B Biol. Sci. 346 (1317), 295–302.

King, A.M.Q., Adams, M.J., Carstens, E.B., Lefkowitz, E.J., 2012. In: King, A.M.Q., Adams, M.J., Carstens, E.B., Lefkowitz, E.J. (Eds.), Virus taxonomy. classification and nomenclature of viruses. Ninth Report of the International committee on Taxonomy of Viruses. Elsevier, Amsterdam, Boston, Heidelberg, London, New York, Oxford, Paris, San Diego, San Francisco, Singapore, Sydney, Tokyo.

Klute, K.A., Nadler, S.A., Stenger, D.C., 1996. Horseradish curly top virus is a distinct subgroup II geminivirus species with rep and C4 genes derived from a subgroup III ancestor. J. Gen. Virol. 77 (Pt 7), 1369–1378.

Kraberger, S., Geering, A.D.W., Walters, M., Martin, D.P., Varsani, A., 2017a. Novel Mastreviruses identified in Australian wild rice. Virus Res. 238, 193–197.

Kraberger, S., Saumtally, S., Pande, D., Khoodoo, M.H.R., Dhayan, S., Dookun-Saumtally, A., Shepherd, D.N., Hartnady, P., Atkinson, R., Lakay, F.M., Hanson, B., Redhi, D., Monjane, A.L., Windram, O.P., Walters, M., Oluwafemi, S., Michel-Lett, J., Lefeuvre, P., Martin, D.P., Varsani, A., 2017b. Molecular diversity, geographic distribution and host range of monocot-infecting Mastreviruses in Africa and surrounding Islands. Virus Res. 238, 171–178.

Kreuze, J.F., Perez, A., Untiveros, M., Quispe, D., Fuentes, S., Barker, I., Simon, R., 2009. Complete viral genome sequence and discovery of novel viruses by deep sequencing of small Rnas: a generic method for diagnosis, discovery and sequencing of viruses. Virology 388, 1–7.

Krupovic, M., Ghabrial, S.A., Jiang, D., Varsani, A., 2016. Genomoviridae: a new family of widespread single-stranded DNA viruses. Arch. Virol. 161, 2633–2643.

Kumar, S., Stecher, G., Tamura, K., 2016. Mega7: molecular evolutionary genetics analysis version 7.0 for bigger datasets. Mol. Biol. Evol. 33, 1870–1874.

Lacroix, C., Jolles, A., Seabloom, E.W., Power, A.G., Mitchell, C.E., Borer, E.T., 2014. Non-random biodiversity loss underlies predictable increases in viral disease prevalence. J. R. Soc. Interface 11, 20130947.

Lecuit, M., Eloit, M., 2014. The diagnosis of infectious diseases by whole genome next generation sequencing: a new era is opening. Front. Cell. Infect. Microbiol. 4.

Lefeuvre, P., Martin, D.P., Harkins, G., Lemey, P., Gray, A.J., Meredith, S., Lakay, F., Monjane, A., Lett, J.M., Varsani, A., Heydarnejad, J., 2010. The spread of tomato yellow leaf curl virus from the middle east to the world. PLoS Pathog. 6, E1001164.

Liang, P., Navarro, B., Zhang, Z., Wang, H., Lu, M., Xiao, H., Wu, Q., Zhou, X., Di Serio, F., Li, S., 2015. Identification and characterization of a novel geminivirus with monopartite genome infecting apple trees. J. Gen. Virol. 96, 2411–2420.

Malmstrom, C.M., Mccullough, A.J., Johnson, H.A., Newton, L.A., Borer, E.T., 2005. Invasive annual grasses indirectly increase virus incidence in California native perennial bunchgrasses. Oecologia 145, 153–164.

Malmstrom, C.M., Melcher, U., Bosque-Perez, N.A., 2011. The expanding field of plant virus ecology: historical foundations, knowledge gaps, and research directions. Virus Res. 159, 84–94.

Márquez, L.M., Redman, R.S., Rodriguez, R.J., Roossinck, M.J., 2007. A virus in a fungus in a plant: three-way symbiosis required for thermal tolerance. Science 315, 513–515.

Mitchell, C.E., Tilman, D., Groth, J.V., 2002. Effects of grassland plant species diversity, abundance, and composition on foliar fungal disease. Ecology 83, 1713–1726.

Monjane, A.L., Harkins, G.W., Martin, D.P., Lemey, P., Lefeuvre, P., Shepherd, D.N., Oluwafemi, S., Simuyandi, M., Zinga, I., Komba, E.K., Lakoutene, D.P., Mandakombo, N., Mboukoulida, J., Semballa, S., Tagne, A., Tiendrebeogo, F., Erdmann, J.B., Van Antwerpen, T., Owor, B.E., Flett, B., Ramusi, M., Windram, O.P., Syed, R., Lett, J.M., Briddon, R.W., Markham, P.G., Rybicki, E.P., Varsani, A., 2011. Reconstructing the history of maize streak virus strain a dispersal to reveal diversification hot spots and its origin in southern Africa. J. Virol. 85, 9623–9636.

Muhire, B.M., Varsani, A., Martin, D.P., 2014. Sdt: a virus classification tool based on pairwise sequence alignment and identity calculation. PLoS One 9, E108277.

Muthukumar, V., Melcher, U., Pierce, M., Wiley, G.B., Roe, B.A., Palmer, M.W., Thapa, V., Ali, A., Ding, T., 2009. Non-cultivated plants of the tallgrass prairie preserve of northeastern Oklahoma frequently contain virus-like sequences in particulate fractions. Virus Res. 141, 169–173.

Ng, T.F.F., Duffy, S., Polston, J.E., Bixby, E., Vallad, G.E., Breitbart, M., 2011a. Exploring the diversity of plant DNA viruses and their satellites using vector-enabled metagenomics on whiteflies. PLoS One 6, E19050.

Ng, T.F.F., Willner, D.L., Lim, Y.W., Schmieder, R., Chau, B., Nilsson, C., Anthony, S., Ruan, Y.J., Rohwer, F., Breitbart, M., 2011b. Broad surveys of DNA viral diversity obtained through viral metagenomics of mosquitoes. PLoS One 6, e20579.

Pace, N.R., Stahl, D.A., Lane, D.J., Olsen, G.J., 1985. Analyzing natural microbial populations by rRNA sequences. ASM News 51, 4–12.

Pagan, I., Gonzalez-Jara, P., Moreno-Letelier, A., Rodelo-Urrego, M., Fraile, A., Pinero, D., Garcia-Arenal, F., 2012. Effect of biodiversity changes in disease risk: exploring disease emergence in a plant-virus system. PLoS Pathog. 8, E1002796.

Pande, D., Madzokere, E., Hartnady, P., Kraberger, S., Hadfield, J., Rosario, K., Jaschke, A., Monjane, A.L., Owor, B.E., Dida, M.M., Shepherd, D.N., Martin, D.P., Varsani, A., Harkins, G.W., 2017. The role of Kenya in the trans-African spread of maize streak virus strain a. Virus Res. 232, 69–76.

Patil, B.L., Fauquet, C.M., 2009. Cassava mosaic geminiviruses: actual knowledge and perspectives. Mol. Plant Pathol. 10, 685–701.

Perry, K.L., Mclane, H., Thompson, J.R., Fuchs, M., 2018. A novel grablovirus from non-cultivated grapevine (vitis Sp.) in north America. Arch. Virol. 163, 259–262.

Remold, S.K., 2002. Unapparent virus infection and host fitness in three weedy grass species. J. Ecol. 90 (6), 967–977.

Roche, B., Dobson, A.P., Guegan, J.F., Rohani, P., 2012. Linking community and disease ecology: the impact of biodiversity on pathogen transmission. Philos. Trans. R. Soc., B 367, 2807–2813.

Roossinck, M.J., 2011a. The big unknown: plant virus biodiversity. Curr. Opin. Virol. 1, 63–67.

Roossinck, M.J., 2011b. The good viruses: viral mutualistic symbioses. Nat. Rev. Microbiol. 9, 99–108.

Roossinck, M.J., 2012. Plant virus metagenomics: biodiversity and ecology. Annu. Rev. Genet. 46, 359–369.

Roossinck, M.J., 2015. Move over, bacteria! Viruses make their mark as mutualistic microbial symbionts. J. Virol. 89, 6532–6535.

Roossinck, M.J., Garcia-Arenal, F., 2015. Ecosystem simplification, biodiversity loss and plant virus emergence. Curr. Opin. Virol. 10c, 56–62.

Roossinck, M.J., Saha, P., Wiley, G., Quan, J., White, J., Lai, H., Chavarria, F., Shen, G., Roe, B., 2010. Ecogenomics: using massively parallel pyrosequencing to understand virus ecology. Mol. Ecol. 19, 81–88.

Roossinck, M.J., Martin, D.P., Roumagnac, P., 2015. Plant virus metagenomics: advances in virus discovery. Phytopathology 105, 716–727.

Rosario, K., Breitbart, M., 2011. Exploring the viral world through metagenomics. Curr. Opin. Virol. 1, 1–9.

Rosario, K., Dayaram, A., Marinov, M., Ware, J., Kraberger, S., Stainton, D., Breitbart, M., Varsani, A., 2012a. Diverse circular ssdna viruses discovered in dragonflies (Odonata: Epiprocta). J. Gen. Virol. 93, 2668–2681.

Rosario, K., Duffy, S., Breitbart, M., 2012b. A field guide to eukaryotic circular single-stranded Dna viruses: insights gained from metagenomics. Arch. Virol. 157, 1851–1871.

Rosario, K., Padilla-Rodriguez, M., Kraberger, S., Stainton, D., Martin, D.P., Breitbart, M., Varsani, A., 2013. Discovery of a novel Mastrevirus and Alphasatellite-like circular DNA in dragonflies (Epiprocta) from Puerto Rico. Virus Res. 171, 231–237.

Rosario, K., Capobianco, H., Ng, T.F., Breitbart, M., Polston, J.E., 2014. Rna viral metagenome of whiteflies leads to the discovery and characterization of a whitefly-transmitted Carlavirus in North America. PLoS One 9, E86748.

Rosario, K., Seah, Y.M., Marr, C., Varsani, A., Kraberger, S., Stainton, D., Moriones, E., Polston, J.E., Duffy, S., Breitbart, M., 2015. Vector-enabled metagenomic (Vem) surveys using whiteflies (Aleyrodidae) reveal novel Begomovirus species in the new and oldworlds. Virus 7, 5553–5570.

Rosario, K., Marr, C., Varsani, A., Kraberger, S., Stainton, D., Moriones, E., Polston, J.E., Breitbart, M., 2016. Begomovirus-associated satellite DNA diversity captured through vector-enabled metagenomic (Vem) surveys using whiteflies (Aleyrodidae). Virus 8, 36.

Roumagnac, P., Granier, M., Bernardo, P., Deshoux, M., Ferdinand, R., Galzi, S., Fernandez, E., Julian, C., Abt, I., Filloux, D., Mesleard, F., Varsani, A., Blanc, S., Martin, D.P., Peterschmitt, M., 2015. Alfalfa leaf curl virus: an aphid-transmitted geminivirus. J. Virol. 89, 9683–9688.

Roux, S., Enault, F., Robin, A., Ravet, V., Personnic, S., Theil, S., Colombet, J., Sime-Ngando, T., Debroas, D., 2012. Assessing the diversity and specificity of two freshwater viral communities through metagenomics. PLoS One 7.

Rybicki, E.P., 1994. A phylogenetic and evolutionary justification for 3 genera of Geminiviridae. Arch. Virol. 139, 49–77.

Scarpellini, E., Ianiro, G., Attili, F., Bassanelli, C., De Santis, A., Gasbarrini, A., 2015. The human gut microbiota and virome: potential therapeutic implications. Dig. Liver Dis. 47 (12), 1007.

Shendure, J., Balasubramanian, S., Church, G.M., Gilbert, W., Rogers, J., Schloss, J.A., Waterston, R.H., 2017. DNA sequencing at 40: past, present and future. Nature 550, 345–353.

Simmonds, P., Adams, M.J., Benko, M., Breitbart, M., Brister, J.R., Carstens, E.B., Davison, A.J., Delwart, E., Gorbalenya, A.E., Harrach, B., Hull, R., King, A.M., Koonin, E.V., Krupovic, M., Kuhn, J.H., Lefkowitz, E.J., Nibert, M.L., Orton, R., Roossinck, M.J., Sabanadzovic, S., Sullivan, M.B., Suttle, C.A., Tesh, R.B., Van Der Vlugt, R.A., Varsani, A., Zerbini, F.M., 2017. Consensus statement: virus taxonomy in the age of metagenomics. Nat. Rev. Microbiol. 15, 161–168.

Stanley, J., Markham, P.G., Callis, R.J., Pinner, M.S., 1986. The nucleotide sequence of an infectious clone of the geminivirus beet curly top virus. EMBO J. 5, 1761–1767.

Stecher, B., Denzler, R., Maier, L., Bernet, F., Sanders, M.J., Pickard, D.J., Barthel, M., Westendorf, A.M., Krogfelt, K.A., Walker, A.W., Ackermann, M., Dobrindt, U., Thomson, N.R., Hardt, W.D., 2012. Gut inflammation can boost horizontal gene transfer between pathogenic and commensal Enterobacteriaceae. Proc. Natl. Acad. Sci. U. S. A. 109, 1269–1274.

Stobbe, A.H., Roossinck, M.J., 2014. Plant virus metagenomics: what we know and why we need to know more. Front. Plant Sci. 5, 150.

Stover, B.C., Muller, K.F., 2010. Treegraph 2: combining and visualizing evidence from different phylogenetic analyses. BMC Bioinf. 11, 7.

Sunagawa, S., Coelho, L.P., Chaffron, S., Kultima, J.R., Labadie, K., Salazar, G., Djahanschiri, B., Zeller, G., Mende, D.R., Alberti, A., Cornejo-Castillo, F.M., Costea, P.I., Cruaud, C., D'ovidio, F., Engelen, S., Ferrera, I., Gasol, J.M., Guidi, L., Hildebrand, F., Kokoszka, F., Lepoivre, C., Lima-Mendez, G., Poulain, J., Poulos, B.T., Royo-Llonch, M., Sarmento, H., Vieira-Silva, S., Dimier, C., Picheral, M., Searson, S., Kandels-Lewis, S., Tara Oceans, C., Bowler, C., De Vargas, C., Gorsky, G., Grimsley, N., Hingamp, P., Iudicone, D., Jaillon, O., Not, F., Ogata, H., Pesant, S., Speich, S., Stemmann, L., Sullivan, M.B., Weissenbach, J., Wincker, P., Karsenti, E., Raes, J., Acinas, S.G., Bork, P., 2015. Ocean Plankton. Structure and function of the global ocean microbiome. Science 348, 1261359.

Susi, H., Laine, A.L., Filloux, D., Kraberger, S., Farkas, K., Bernardo, P., Frilander, M.J., Martin, D.P., Varsani, A., Roumagnac, P., 2017. Genome sequences of a capulavirus infecting *Plantago lanceolata* in the Aland Archipelago of Finland. Arch. Virol. 162, 2041–2045.

Suttle, C.A., 2005. Viruses in the sea. Nature 437, 356–361.

Suttle, C.A., 2007. Marine viruses—major players in the global ecosystem. Nat. Rev. Microbiol. 5, 801–812.

Temperton, B., Giovannoni, S.J., 2012. Metagenomics: microbial diversity through a scratched lens. Curr. Opin. Microbiol. 15, 605–612.

Varsani, A., Shepherd, D.N., Dent, K., Monjane, A.L., Rybicki, E.P., Martin, D.P., 2009. A highly divergent south African geminivirus species illuminates the ancient evolutionary history of this family. Virol. J. 6, 36.

Varsani, A., Roumagnac, P., Fuchs, M., Navas-Castillo, J., Moriones, E., Idris, A., Briddon, R.W., Rivera-Bustamante, R., Zerbini, F.M., Martin, D.P., 2017. Capulavirus and Grablovirus: two new genera in the family *Geminiviridae*. Arch. Virol. 162 (6), 1819–1831.

Vayssier-Taussat, M., Albina, E., Citti, C., Cosson, J.F., Jacques, M.A., Lebrun, M.H., Le Loir, Y., Ogliastro, M., Petit, M.A., Roumagnac, P., Candresse, T., 2014. Shifting the paradigm from pathogens to Pathobiome: new concepts in the light of meta-omics. Front. Cell. Infect. Microbiol. 4, 29.

Wasik, B.R., Turner, P.E., 2013. On the biological success of viruses. Annu. Rev. Microbiol. 67, 519–541.

Webster, C.G., Coutts, B.A., Jones, R.A.C., Jones, M.G.K., Wylie, S.J., 2007. Virus impact at the Interface of an ancient ecosystem and a recent agroecosystem: studies on three legume-infecting Potyviruses in the southwest Australian floristic region. Plant Pathol. 56, 729–742.

Wren, J.D., Roossinck, M.J., Nelson, R.S., Scheets, K., Palmer, M.W., Melcher, U., 2006. Plant virus biodiversity and ecology. PLoS Biol. 4, E80.

Xu, P., Chen, F., Mannas, J.P., Feldman, T., Sumner, L.W., Roossinck, M.J., 2008. Virus infection improves drought tolerance. New Phytol. 180 (4), 911–921.

Zaitlin, M., Palukaitis, P., 2000. Advances in understanding plant viruses and virus diseases. Annu. Rev. Phytopathol. 38, 117–143.

Zerbini, F.M., Briddon, R.W., Idris, A., Martin, D.P., Moriones, E., Navas-Castillo, J., Rivera-Bustamante, R., Roumagnac, P., Varsani, A., Ictv Report, C., 2017. Ictv virus taxonomy profile: Geminiviridae. J. Gen. Virol. 98, 131–133.

CHAPTER FOUR

Water-Mediated Transmission of Plant, Animal, and Human Viruses

Nataša Mehle, Ion Gutiérrez-Aguirre, Denis Kutnjak, Maja Ravnikar[1]
National Institute of Biology, Ljubljana, Slovenia
[1]Corresponding author: e-mail address: maja.ravnikar@nib.si

Contents

Abstract

Viruses represent the most abundant and diverse of the biological entities in environmental waters, including the seas and probably also freshwater systems. They are important players in ecological networks in waters and influence global biochemical cycling and community composition dynamics. Among the many diverse viruses from terrestrial environments found in environmental waters, some are plant, animal, and/or human pathogens. The majority of pathogenic viral species found in waters are very

Advances in Virus Research, Volume 101
ISSN 0065-3527
https://doi.org/10.1016/bs.aivir.2018.02.004

stable and can survive outside host cells for long periods. The occurrence of such viruses in environmental waters has raised concerns because of the confirmation of the infectivity of waterborne viruses even at very low concentrations. This chapter focuses mainly on the survival of human, animal, and plant pathogenic viruses in aqueous environments, the possibility of their water-mediated transmission, the ecological implications of viruses in water, the methods adapted for detecting such viruses, and how to minimize the risk of viruses spreading through water.

1. PRESENCE OF INFECTIVE VIRUSES IN ENVIRONMENTAL WATERS

Viruses have the obligatory need of a host for their multiplication. Nevertheless, their nucleic acids can be detected outside the host organisms using metagenomic analyses, such as in wastewaters (Aw et al., 2014; Cantalupo et al., 2011), reclaimed waters (Rosario et al., 2009a), freshwater lakes (Djikeng et al., 2009), polar freshwater lakes (see chapter "Viruses in polar lake and soil ecosystems" by Rastrojo and Alcamí, in this volume; Aguirre de Cárcer et al., 2015; López-Bueno et al., 2009), river waters (Fernandez-Cassi et al., 2017), and seawater (Angly et al., 2006; Breitbart et al., 2002; Culley et al., 2006). Among the viral nucleic acids detected in aqueous ecosystems, some are known plant, animal, and/or human pathogens. Moreover, infectious viruses of terrestrial hosts have been isolated from various types of water in many locations around the world, including in well developed, industrialized countries (for reviews, see e.g., Fong and Lipp, 2005; Mehle and Ravnikar, 2012; Xagoraraki et al., 2014). These viruses are listed in Table 1, including infective plant pathogenic viruses that cause diseases in agricultural crops, along with a heterogeneous group of infectious enteric viruses (viruses that primarily infect the intestinal tract through ingestion of fecally contaminated food or water) that are associated with infections and diseases in humans and animals. In several studies, seasonal changes in the concentrations of some viruses have been reported. As an example, the highest concentrations of human noroviruses (noncultivable viruses from family *Caliciviridae*, genus *Norovirus*) in sewage are found in the winter, which is an epidemic period (Haramoto et al., 2006; Iwai et al., 2009; Kitajima et al., 2012). The occurrence of pathogenic viruses in environmental waters raises concerns regarding the possibility of plant, animal, and human exposure, and waterborne infections.

Table 1 Infectious Plant, Animal, and Human Viruses' Found in Environmental Waters

Virus Family	Virus Genus	Virus (Acronym)	Origin of Water Sample	Reference
Plant Pathogenic Viruses				
Alphaflexiviridae	*Potexvirus*	*Potato virus X* (PVX)	Ponds, creeks in forest area	Büttner and Nienhaus (1989)
Bromoviridae	*Cucumovirus*	*Cucumber mosaic virus* (CMV)	River	Piazzolla et al. (1986)
Tombusviridae	*Carmovirus*	*Carnation mottle virus* (CarMV)	River	Koenig and Lesemann (1985)
	Dianthovirus	*Carnation ringspot virus* (CRSV)	Canal near sewage plant; ditches, streams, drainage canals in agricultural areas; stream with no direct contact with fields	Koenig et al. (1988, 1989) and Yi et al. (1992)
	Necrovirus	*Tobacco necrosis virus* (TNV)	Ditches, streams, drainage canals in agricultural areas; rivers, lakes	Koenig et al. (1989), Tomlinson et al. (1983), and Yi et al. (1992)
	Tombusvirus	*Carnation Italian ringspot virus* (CIRV)	Ponds, creeks in forest area	Büttner and Nienhaus (1989) and Büttner et al. (1987)
		Grapevine Algerian latent virus (GALV)	Ditches, streams in agricultural areas	Yi et al. (1992)
		Petunia asteroid mosaic virus (PeAMV)	Ditches, drainage canals in agricultural area; seawater	Fuchs et al. (1996), Koenig (1988), and Koenig et al. (1989)
		Tomato bushy stunt virus (TBSV)	Ponds, creeks in forest area; rivers	Büttner and Nienhaus (1989), Koenig and Lesemann (1985), Tomlinson and Faithfull (1984), and Tomlinson et al. (1983)

Continued

Table 1 Infectious Plant, Animal, and Human Viruses' Found in Environmental Waters—cont'd

Virus Family	Virus Genus	Virus (Acronym)	Origin of Water Sample	Reference
Virgaviridae	*Tobamovirus*	*Tobacco mosaic virus* (TMV)	Rivers, forest ditches, and brook	Koenig and Lesemann (1985), Piazzolla et al. (1986), Pleše et al. (1996), and Tošić and Tošić (1984)
		Tomato mosaic virus (ToMV)	Ponds, creeks, brooks, streams, lakes in forest area; rivers, gravel pit, clouds, and fog	Boben et al. (2007), Büttner and Nienhaus (1989), Castello et al. (1995), Jacobi and Castello (1991), and Pleše et al. (1996)
Human and/or Animal Pathogenic Enteric Viruses				
Adenoviridae	*Mastadenovirus*	Human adenoviruses	Wastewater, seawater, creek; recreational water (defined here and below to include bathing water and other waters, such as lakes, used for recreational activities); water for animal consumption, and drinking water	Aslan et al. (2011), Fongaro et al. (2013), Krikelis et al. (1985), Lee and Kim (2002), Rigotto et al. (2010), Staggemeier et al. (2015), and Xagoraraki et al. (2007)
		Other or nonidentified adenoviruses	Wastewater, treated sewage, rivers; drinking water	Lee and Kim (2002), Lee et al. (2005), Sedmak et al. (2005), and Tani et al. (1992, 1995)
Picornaviridae	*Enterovirus*	Enteroviruses or nonidentified viruses from genus *Enterovirus*	Wastewater, treated sewage, rivers, seawater; shellfish harvesting water, recreational water, and drinking water	Aslan et al. (2011), Borchardt et al. (2004), Gerba et al. (1979), Green and Lewis (1999), Katayama et al. (2002), Lee et al. (2005), Lipp et al. (2001a, b), Lodder and de Roda Husman (2005), Powell et al. (2003), and Sedmak et al. (2005)

		Coxsackieviruses	Wastewater, treated sewage, river, seawater; recreational water, drinking water	Costan-Longares et al. (2008), Deetz et al. (1984), Goyal et al. (1984), Krikelis et al. (1985), Lee and Kim (2002), Rose et al. (1987), Shieh et al. (1997), and Tani et al. (1992, 1995)
		Echoviruses	Wastewater, treated sewage, river; recreational water, drinking water	Costan-Longares et al. (2008), Krikelis et al. (1985), Lee and Kim (2002), Rose et al. (1987), Shieh et al. (1997), and Tani et al. (1992, 1995)
		Polioviruses	Wastewater, treated sewage, river; recreational water, drinking water	Costan-Longares et al. (2008), Krikelis et al. (1985), Lee and Kim (2002), Rose et al. (1987), and Tani et al. (1992, 1995)
	Hepatovirus	*Hepatitis A virus* (HAV)	Drinking water	Borchardt et al. (2004)
Reoviridae	*Rotavirus*	Rotaviruses	Wastewater, river, creek; recreational water, drinking water	Deetz et al. (1984), Li et al. (2010, 2011), Rigotto et al. (2010), Rose et al. (1987), and Rutjes et al. (2009)
	Other genuses of family *Reoviridae*	Reoviruses	Wastewater, treated sewage, and river	Lodder and de Roda Husman (2005), Sedmak et al. (2005), and Tani et al. (1992, 1995)

2. ORIGINS OF VIRUSES IN WATERS

Plant viruses can be released into aqueous environments from injured or decaying plant residues, and from infected roots of plants growing near to water or in hydroponic systems (Koenig, 1988; Krczal et al., 1995; Mehle et al., 2014; Pares et al., 1992; Schwarz et al., 2010; Yarwood, 1960). Additionally, virus-containing seeds, virus-transmitting nematodes, zoospores, or resting sporangia of fungi, can be washed away by running soil water or drainage water (Campbell, 1996; Koenig, 1986; Matthews, 1992). It was suggested that, as many European forests stand on what was once agricultural land, viruses isolated from forest trees, soils, and waters from these area might have originated from infected crop residues (Fillhart et al., 1998). Viruses isolated in forest ecosystems with no prior history of agriculture might be indigenous or introduced by other mechanisms (Fillhart et al., 1998).

Human and animal waste has also been defined as possible sources of enteric and plant viruses in water. Many viruses can be excreted in feces and urine (Carter, 2005; Fong and Lipp, 2005; Melnick, 1984), including human enteric viruses that are associated with gastroenteritis (Clark and McKendrick, 2004; Gerba et al., 1996; Middleton, 1996), and plant pathogenic viruses (Zhang et al., 2006). If the concentrations of viruses initially present are too high, even extensive treatments of wastewaters before injection into the soil or discharging into coastal waters will not remove all potentially infectious virus particles (Goyal et al., 1984; Irving and Smith, 1981; Yates et al., 1985). Additionally, during major storms, sewage treatment plants might not be able to treat all of the incoming sewage waters because their collection systems also collect the storm-water run-off. Thus, incompletely treated or untreated sewage can be discharged into surface waters (Donovan et al., 2008). Rainfall also results in greater percolation of water into the soil, which results in higher groundwater levels (Keswick and Gerba, 1980). These in turn reduce the unsaturated soil distances between septic tank drains and the groundwater (Robertson and Edberg, 1997). Human activities can also favor the spread of viruses, such as by application of dung and liquid manure to fields, whereby the washout of the soil surface layer can transfer viruses into environmental waters (Keswick and Gerba, 1980). Furthermore, penetration of viruses via water to depths as great as 67 m and over horizontal migration as far as 408 m in glacial till has been reported, with migration of 1600 m reported for fractured limestone (Keswick and Gerba, 1980; Robertson and Edberg, 1997). Viruses that are introduced into the subsurface by a variety of routes might therefore migrate into the potable groundwater.

3. SURVIVAL OF PLANT AND ENTERIC VIRUSES IN THE AQUEOUS ENVIRONMENT

The potential for viruses to spread through water depends to a large extent on their persistence within the aqueous environment. Viruses cannot replicate outside of the host organism. Virus survival outside the host organism is affected by a combination of biological (virus structure, presence of microorganisms, organic matter), physical (temperature, ultraviolet—UV), and chemical (pH, salts, antiviral substances) factors (Carter, 2005; Keswick and Gerba, 1980; Rzeżutka and Cook, 2004; Vasickova et al., 2010; Xagoraraki et al., 2014).

In the following lines, terms such as "survival," "stability," and "inactivation" will be used. With survival we mean the time that the virus remains infective, in e.g., water media. Stability indicates the ability of the virus to keep its vital structural and genomic features intact or functional. Inactivation is the loss of the infective ability of the virus due to disruption of its structural or genomic components.

The stability of some plant viruses is extremely high. *Tomato bushy stunt virus* (TBSV; family *Tombusviridae*; genus *Tombusvirus*) and *Pepper mild mottle virus* (PMMoV; family *Virgaviridae*; genus: *Tobamovirus*) have been shown to retain their infectivity after ingestion and subsequent exposure at 37°C to the enzymes of the human alimentary tract, the bile salts, and the wide fluctuations in pH (from ~3.0 to 8.8) (Tomlinson et al., 1982; Zhang et al., 2006). Zhang et al. (2006) demonstrated that PMMoV can survive standard food processing, such as in the manufacture of hot chili sauce and powdered chili. *Pepino mosaic virus* (PepMV; family *Flexiviridae*; genus: *Potexvirus*) can survive in plant debris and on contaminated surfaces, where it can remain infectious for several weeks (Van der Vlugt, 2009). Some plant viruses can survive in seeds or in resting fungal spores (Campbell, 1996; Matthews, 1992). *Melon necrotic spot virus* (MNSV; family *Tombusviridae*; genus: *Carmovirus*) can attach to the outer covering of the aquatic zoospores of the chytrid fungus *Olpidium bornovanus* and can remain viable in the soil for several years (Campbell, 1996).

Enteric viruses are characterized by their stability both in the gastrointestinal tract and in the environment. They can be excreted through human and animal feces and resist environmental contaminants (e.g., in soil and water) for long periods (Staggemeier et al., 2015). Rotaviruses (RoV; family *Reoviridae*) and several other enteric viruses that enter the sewer systems can survive wastewater treatments, with the consequent contamination of the receiving waters (Gerba et al., 1996; Irving and Smith, 1981). Where such waters are then treated for use as drinking water, the contaminating viruses might survive this treatment too (Carter, 2005). Therefore, it has been suggested that enteric viruses are important indicators of fecal contamination (Carducci et al., 2008; De Oliveira et al., 2012; Hundesa et al., 2006; Jiang et al., 2001).

3.1 Influence of Temperature and Solution on Virus Survival

There are little data available on the survival of plant viruses in water. The half-life of *Tobacco necrosis virus* (TNV; family *Tombusviridae*; genus: *Necrovirus*) in water was estimated as 46–80 h (Yarwood, 1960). *Tomato mosaic virus* (ToMV; family *Virgaviridae*; genus: *Tobamovirus*) and PMMoV were shown to remain infectious in nutrient solution for at least 6 months, independent of the storage medium, and either stored in a glasshouse or at 4°C (Pares et al., 1992), with the consequent risk for plants grown in hydroponics systems, which typically use such nutrient solutions. Mehle et al. (2014) showed that PepMV can remained infectious in water at $20 \pm 4\,°C$ for up to 3 weeks, and the *Potato virus Y* NTN strain (PVY^{NTN}; family *Potyviridae*; genus: *Potyvirus*) for up to 1 week. When PVY^{NTN} was stored at 4°C, it survived in the water much longer (up to 10 weeks) (Mehle et al., 2014). Like plant viruses, enteric viruses have also been reported to survive longer at lower temperatures (Yates et al., 1985). Raphael et al. (1985) showed that in treated tap water at 4°C, there was no significant drop in rotavirus titer even after 64 days, whereas at 20°C the titer in the treated tap water was reduced by about $2\log_{10}$ over the same period. Poliovirus 1 was not detected after 300 days in artificially contaminated mineral water stored at room temperature, whereas it was still infectious after 1 year when contaminated mineral water was stored at 4°C (Biziagos et al., 1988). Gantzer et al. (1998) showed that in seawater, it took 671 days to inactivate 90% of Poliovirus 1 (family *Picornaviridae*; genus: *Enterovirus*) and *Hepatitis A virus* (HAV; family *Picornaviridae*; genus: *Hepatovirus*) at 4°C, but only 25 days at 25°C. Wetz et al. (2004) reported that at 22°C, 9.1 days were required for a 90% reduction in poliovirus in filtered seawater, while 3.4 days were required for a similar reduction at 30°C.

High temperatures can damage the virus capsid and the nucleic acids. For example, Biziagos et al. (1988) reported that in artificially contaminated mineral water stored at room temperature, *Hepatitis A virus* was still infectious whereas Poliovirus 1 was not detected after 300 days. In a comparative study by Cannon et al. (2006), *Feline calicivirus* (FCV; family *Caliciviridae*; genus: *Vesivirus*) was shown to be more stable than *Murine norovirus* 1 (MNV 1; family *Caliciviridae*; genus: *Norovirus*) at 56°C and in solution at room temperature.

Viral persistence in waters can be strongly related to predation by flagellates, extracellular proteases, nucleases, and other enzymes and factors (Fong and Lipp, 2005). Wetz et al. (2004) showed that poliovirus survived much longer in filtered seawater compared to unfiltered seawater at both 22 and 30°C. Similarly, poliovirus RNA was more stable in filtered seawater than in unfiltered seawater at both 4 and 23°C (Tsai et al., 1995). Finally, the sampling season (e.g., summer vs winter) was shown to have a greater effect on the survival of poliovirus than the incubation temperature (Skraber et al., 2004). Seasonal influences to the concentration of some viruses were also observed by Lipp et al. (2001b) and Tani et al. (1995).

3.2 The influence of UV Radiation on Virus Survival

In water, virus survival depends not only on temperature but also on exposure to UV light. Fujioka and Yoneyama (2002) reported that several enteroviruses in seawater were inactivated much more rapidly in the presence of sunlight than in the dark. Johnson et al. (1997) observed 90% inactivation of polioviruses in marine water after 24 h of incubation in the dark, while there was 99.9% inactivation if the water was exposed to sunlight. Viruses with double-stranded genomes are suggested to be less susceptible to UV inactivation as those with single-stranded genomes, since only one strand of the nucleic acid may be damaged during UV exposure, and their undamaged strand of the nucleic acid can serve as a template for repair by host enzymes (Thurston-Enriquez et al., 2003). For example, the doses of UV required to achieve 99% inactivation of adenovirus type 40 (double-stranded DNA virus from family *Adenoviridae*) in water was 109 mJ/cm^2, while only 16 mJ/cm^2 was required to achieve the same inactivation rate of single-stranded RNA *Feline calicivirus* (Thurston-Enriquez et al., 2003).

3.3 Influence of Virus Adsorption to Particles on Virus Survival

Viruses have often been found attached to suspended solids. Rao et al. (1984) showed that 72% of 18 analyzed samples of suspended solids of seawater were positive for enteroviruses, whereas only 14% of 35 seawater samples without solids yielded the virus. Aggregation with and/or adsorption to solid materials such as clay and silicates, or to organic debris, bacteria, or algae, greatly increase the survival times of viruses in water, by offering protection from enzymes and other degrading factors, and from UV inactivation (Koenig, 1986; Labelle and Gerba, 1980; Meschke and Sobsey, 1998; Piazzolla et al., 1986). Rao et al. (1984) showed that when associated with solids, poliovirus, and rotavirus retain their infectious quality for 19 days, while the same viruses remained infectious for only 9 days when freely suspended in seawater. Studies by Helmi et al. (2008) and Lehtola et al. (2007) have revealed that viral particles can penetrate biofilms and thereby obtain protection against environmental stress. Biofilms are typically composed of bacterial cells attached to each other and to a living or nonliving surface, together with a particular extracellular matrix, and are formed in response to different environmental factors. During biofilm erosion or sloughing, protected immobilized viral particles can be released anew into the environment; thus, biofilms can be considered a potential secondary source of viral contamination of water.

The strength of adsorption of viruses to inorganic or organic matter is expected to vary across the different viruses, such as through differences in the surface charge of the virions, and also across the different adsorbing materials and environmental conditions, such as pH, salt content, and presence of other materials. It has been shown that enteric viruses are destabilized, and subsequently inactivated, in water that is low (or depleted) in salt ions (e.g., Mg^{2+}) (Vasickova et al., 2010). In contrast, increased concentrations of salts (NaCl) might be virucidal for several other kinds of viruses (Vasickova et al., 2010). Pancorbo et al. (1987) showed that the rotavirus infectivity reduction rate significantly correlated with water pH; i.e., as pH increased, infectivity decreased more rapidly. In their study, the infectivity of rotavirus declined more rapidly in polluted creek water at pH 7.8 than in nonpolluted lake water at pH 6.0, although the polluted creek water also showed higher suspended solids (42 mg/L) than the nonpolluted lake water (10 mg/L). They assumed that the water pH has an influence on viral aggregation and thereby the rotavirus susceptibility to virucidal factors in the water. In addition, pH was also shown to have direct effects on the reduction of infectivity; however, the susceptibility of viruses to pH differs dramatically between different viruses. In a comparative study, Cannon et al. (2006) demonstrated that murine norovirus 1 was stable across the entire pH range tested (i.e., pH 2–10), with less than 1 log reduction in infectivity at pH 2, whereas *Feline calicivirus* was inactivated rapidly at pH 3–9.

In summary, the differences in the survival observed among viruses are linked to their different virus properties, e.g., nature of the genome (double-stranded or single-stranded genome), structure of virus coat protein and differences in the surface charge of the virions.

4. INFECTION OF HOSTS WITH CONTAMINATED WATER

Enormous quantities of water are used for irrigation and other human activities (e.g., clean water for drinking, recreation). In several areas worldwide, there is a shortage of clean water, and thus the demand for recycling of wastewaters is very high (Thebo et al., 2017). Recycled wastewaters can help to reduce the use of clean water. However, such waters represent an important source of a wide variety of pathogen microorganisms, including viruses. As humans, animals, and plants are in constant contact with water, there is continual high risk of infection (Stewart-Wade, 2011; Toze, 2006).

The presence of viruses in waters can have epidemiological importance if the viruses can enter a host. Waterborne transmission of some plant viruses has been confirmed (Mehle and Ravnikar, 2012), although most probably this is not the most efficient route of virus transmission. Waterborne plant viruses infect plants, from which the viruses can then rapidly and effectively spread to neighboring plants, either mechanically or via different vectors (Mehle et al., 2014). Infection of plants with waterborne viruses can occur through the roots or upper parts of plants when using infected water for irrigation (reviewed by Koenig, 1986; Mandahar, 1990; Mehle and Ravnikar, 2012). *Lettuce big-vein associated virus* (LBVaV; family *Rhabdoviridae*; genus *Varicosavirus*), *Mirafiori lettuce big-vein virus* (MLBVV; family *Ophioviridae*; genus *Ophiovirus*), TNV, *Cucumber necrosis virus* (CNV; family *Tombusviridae*; genus *Tombusvirus*), *Cucumber leaf spot virus* (CLSV; family *Tombusviridae*), MNSV and PepMV can infect plants through the roots with the aid of vectors, such as the aquatic zoospores of fungi (Alfaro-Fernandez et al., 2010; Campbell, 1996; Jones, 2004; Navarro et al., 2005; Stanghellini and Rasmussen, 1994; Tomlinson and Faithfull, 1980; Tomlinson and Thomas, 1986). For PepMV, PVY, ToMV, *Pelargonium flower break virus* (PFBV; family *Tombusviridae*; genus *Alphacarmovirus*), however, vectorless transmission via the roots is the most likely route or one of the possible route (Krczal et al., 1995; Mehle et al., 2014; Pares et al., 1992; Schwarz et al., 2010). Virus infection of roots can occur even under sterile conditions and is more effective in sand than in liquid media, which suggests that the viruses enter the plants through damaged root hairs or small wounds that appear during root growth. Infection with waterborne viruses can also occur during plant germination (Teakle and Morris, 1981).

In humans and animals, viruses are responsible for a significant proportion of cases of gastroenteritis and other illnesses related to contaminated water (Borchardt et al., 2003; Reynolds et al., 2008). Enteric viruses can be transmitted by oral ingestion, with many contamination routes known, such as drinking water (Beer et al., 2015; Crabtree et al., 1997; Kukkula et al., 1997); shellfish grown in contaminated waters near treatment plant outfall (Croci et al., 1999; Green and Lewis, 1999); vegetables or fruit irrigated or processed using contaminated water (Calder et al., 2003; Carter, 2005); and surface waters ingested during recreational activities (Crabtree et al., 1997; Gerba et al., 1996; Papapetropoulou and Vantarakis, 1998). Enteric viruses transmitted by the fecal-to-oral route are predominantly transmitted by routes other than water. The most important mechanisms involve personal contact and transfer of viruses in droplets or inhalation of viruses in aerosols, or with airborne particles (Carter, 2005). Of note, water can also transmit viruses to humans and other animals via inhalation (e.g., showering) or contact with the skin and eyes (e.g., swimming), which can cause respiratory and ocular infections (Gall et al., 2015; reviewed by Griffin et al., 2003).

5. METHODS FOR DETECTION OF VIRUSES IN WATER

Viruses in water are usually present in concentrations that are too low for detection using conventional methods. Despite these relatively low concentrations, they can still pose significant health risks, as often only very low titers are required for infection (Fong and Lipp, 2005; Mehle and Ravnikar, 2012). Virological investigations of water samples are therefore nearly always multistage processes that involve sampling, concentration, removal of inhibitors, and virus isolation in cell culture or on test plants, and/or identification by molecular or serological methods.

5.1 Virus Concentration Methods

A concentration step is an important requisite when viruses are monitored in water samples (Cashdollar and Wymer, 2013; Fong and Lipp, 2005; Ikner et al., 2012). Depending on the complexity of the sample (e.g., waste water vs drinking water) and the expected degree of viral contamination, the required starting volumes can range from 1 to 1000 L or more. In contrast, the volumes after concentration should be in the range of 10 mL or lower before they are subjected to the detection methods. To accomplish this, a two-step procedure is generally used. In the initial primary concentration step, the volume is reduced to the 100 mL range, and then in the secondary concentration step, the volume is further reduced to 1–10 mL. An ideal concentration method needs to be cost effective, rapid, easy-to-apply, deployable in the field, highly efficient for viral recoveries, efficient independent of virus type, and able to remove inhibitors for the downstream methods. Unfortunately, there is no 100% ideal concentration method, as each has its pros and cons, as reviewed by Cashdollar and Wymer (2013) and Ikner et al. (2012). However, good results may be achieved by several approaches.

The most commonly used primary concentration methods include adsorption–elution to electropositive (1MDS Virosorb, Nanoceram, Z Plus, glass wool) or electronegative (nitrocellulose) media, and application of different filtration strategies (e.g., tangential filtration, vortex filtration) (Cai et al., 2015; De Keuckelaere et al., 2013; Jones et al., 2014; McMinn and Korajkic, 2015). Adsorption–elution relies on the electrostatic binding of viruses to charged filters, with the viruses then eluted, while the filtration strategies rely on virus size and the consequent cut-off of the filters used. Other methods that are used in both primary and secondary concentration steps include precipitation–flocculation (aluminum hydroxide, polyethylene glycol, and skimmed milk), ultracentrifugation, and ultrafiltration (Fong and Lipp, 2005; Ikner et al., 2012). Normally, each laboratory uses its own standard methods that are the best adapted to its particular needs.

One new method that is being increasingly used for concentration of different viruses is based on convective interaction media (CIM) monolithic chromatographic supports (Fig. 1). CIM monoliths are a new generation of chromatographic materials, which are manufactured as columns of different sizes. Owing to their particular structure and properties, these allow the binding and elution of viruses without the disadvantages of classic chromatography (Kramberger et al., 2015). The availability of different monolith chemistries allows the binding and concentration of viruses through different interactions, such as electrostatic (quaternary amine monoliths for viruses in fresh water; Boben et al., 2007; Gutiérrez-Aguirre et al., 2009, 2011; Kovač et al., 2009; Kramberger et al., 2004; Steyer et al., 2015) and hydrophobic (butyl monoliths for viruses in saline water; Balasubramanian et al., 2016). Monoliths have been used to concentrate plant and human viruses

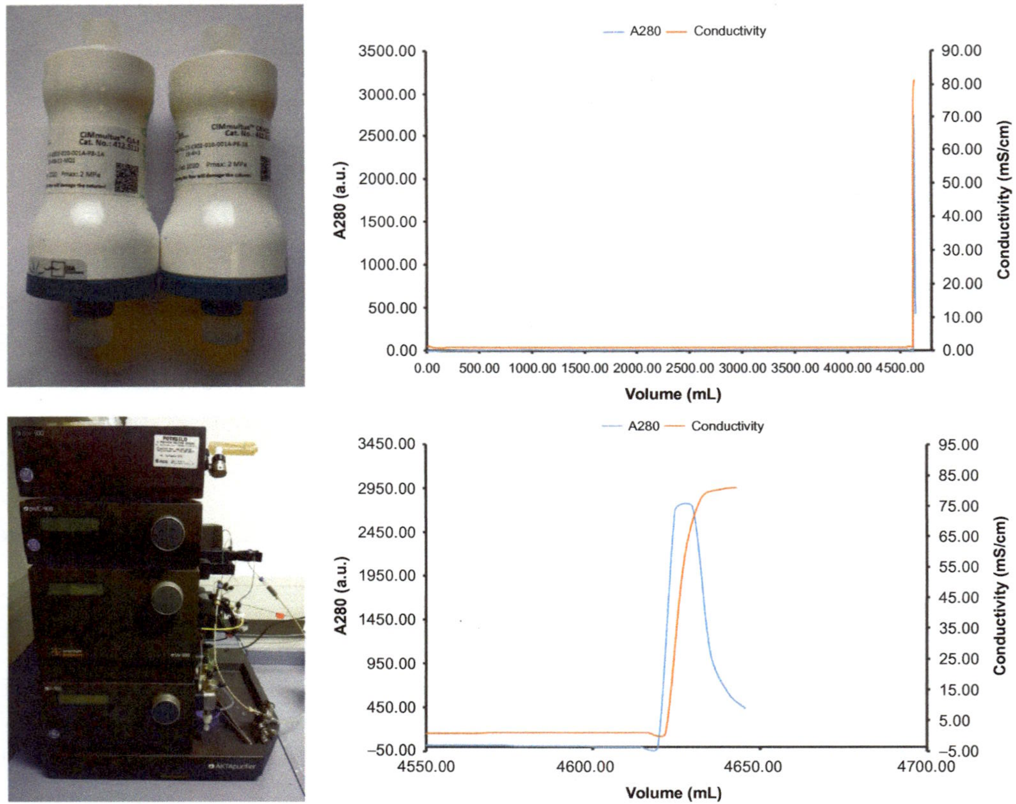

Fig. 1 Concentration of viruses with convective interaction media (CIM) monolithic columns. *Upper left panel*, 8 mL bed volume CIM columns (BIA Separations, Slovenia) of QA (ionic exchange) and C4 chemistries (hydrophobic interaction). *Lower left panel*, example of a typical FPLC system (AKTA, GE Healthcare, Sweden) used to concentrate viruses in the lab. Modular pumps and detectors can be used for concentrations in the field. *Upper right panel*, chromatogram, showing the UV and conductivity signals after loading cca 4500 mL of water in a CIM QA 8 mL column, with subsequent elution highlighted in the *lower right panel*. When the salt (conductivity) is increased, the bound viruses are eluted in cca 10 mL volume.

with different structures, as icosahedral to rod-like (Boben et al., 2007; Gutiérrez-Aguirre et al., 2009, 2011; Kramberger et al., 2004). Monoliths have been shown to concentrate five different human enteric viruses from treated wastewaters in a single step, with recoveries for Rotavirus of up to 98% (Steyer et al., 2015), and they can also be deployed in the field (Balasubramanian et al., 2016; Gutiérrez-Aguirre et al., 2011). Recently, monoliths were also used as the virus concentration step to enrich the virus fraction in wastewater samples that were used for viral metagenomic studies (Gutiérrez-Aguirre et al., 2018). For all of these reasons, commercially available CIM monoliths (BIA Separations, Ljubljana, Slovenia) appear to be a promising new tool to add to the pool of already available virus concentration methods.

5.2 Virus Detection Methods

In the past, the presence of viruses in water samples after concentration was monitored mainly by mechanical inoculation of test plant leaves (Büttner et al., 1987; Büttner and Nienhaus, 1989; Koenig et al., 1988, 1989, 2004; Koenig and Lesemann, 1985; Piazzolla et al., 1986; Tomlinson et al., 1983) or by inoculation of cell cultures (Farrah et al., 1977; Gerba et al., 1979; Goyal et al., 1984; Lipp et al., 2001a, b; Rao et al., 1984). Infected test plants develop visible symptoms and infected cell cultures undergo morphological changes that can be seen by eye or under an optical microscope, which are known as cytopathic effects. These changes do not allow specific identification of the viruses, and therefore further characterization is needed. With cell cultures, the number of virus particles that can form plaques per unit volume can be measured (i.e., the "plaque forming units"), and therefore this method has been also used to estimate virus concentrations in water samples (Rao et al., 1984; Tani et al., 1995).

Individual viruses can be examined using electron microscopy (Jacobi and Castello, 1991; Steinmann, 1981; Tošić and Tošić, 1984). Electron microscopy shows directly and in considerable detail the virus morphology and inner particle structure, and thus helps in virus recognition. Enzyme-linked immunosorbent assays are the most widely used methods in virology, while other serological methods have also been applied to the detection of viruses in concentrated water samples (Boben et al., 2007; Castello et al., 1995; Goyal et al., 1984; Steinmann, 1981). However, these methods are no longer commonly used for diagnosis of water samples, due to their low sensitivity.

Molecular techniques have proven to be very suitable for the detection and identification of viruses. Among these, the methods that amplify the nucleic acids are today the most widely used, such as those based on polymerase chain reaction (PCR). PCR-based assays are sensitive, but also rapid and specific, if a well-designed assay is developed. Conventional PCR or reverse transcriptase (RT)-PCR has been used in a number of water-quality monitoring studies (e.g., Griffin et al., 1999; Pina et al., 1998; Reynolds et al., 1995; Soule et al., 2000). Unknown viruses that have been found in environmental waters have also been characterized using RT-PCR and sequencing (Koenig et al., 2004). Gosalvez et al. (2003) developed RT-PCR for detection of *Melon necrotic spot virus* in concentrated water samples. The method was shown to be much more sensitive (detection limit, $10\,ng\,mL^{-1}$) than enzyme-linked immunosorbent assays (detection limit, $160\,\mu g\,mL^{-1}$).

Conventional PCR has been modified to improve the specificity, sensitivity, and efficiency, and also to quantify the number of viruses detected in water samples. Some variations in conventional PCR include antigen- and immunomagnetic-capture PCR, multiplex PCR, nested PCR, real-time PCR (qPCR), and digital PCR (dPCR). Much of this work has focused on human pathogens. For example, antigen- and immunomagnetic-capture RT-PCR were developed for detection of enteroviruses, *Hepatitis A virus* and rotaviruses in highly inhibitory environmental samples, such as sewage (Casas and Suñén, 2002; Graff et al., 1993). The application of multiplex PCR-based assays (where several sets of primers and/or probes against several targets are included in a single PCR) might save time and costs, as several different viruses can be detected in a single assay. Triplex RT-PCR was developed and used for simultaneous detection of poliovirus, *Hepatitis A virus*, and rotavirus in sewage and seawater (Tsai et al., 1994). The development of a multiplex PCR-based assay, however, is not easy, and requires careful optimization of the reaction mixtures and the PCR conditions. Nested PCR assays increase the sensitivity and specificity of PCR through the use of an internal primer set. Nested PCR has been used for detection of adenoviruses in concentrated water samples from swimming pools (Papapetropoulou and Vantarakis, 1998). Seminested RT-PCR followed by restriction fragment length polymorphism analysis were used for the detection and characterization of rotaviruses in raw sewage and treated effluent samples, here without the use of a preliminary concentration step (Dubois et al., 1997).

More recently, there has been increased use of real-time or quantitative PCR (qPCR); this is a very sensitive quantitative assay in which the target sequences are simultaneously amplified and quantified. The procedure is suitable for high-throughput analysis and is less time consuming comparing to conventional PCR, as agarose gel electrophoresis is not required. The entire analysis can be done within a closed system, which reduces the potential for contamination. Quantitative reverse transcription PCR (RT-qPCR) has been successfully introduced for environmental monitoring of enteroviruses (EV; family *Picornaviridae*; genus *Enterovirus*) in sewage and sludge, and in marine samples (Donaldson et al., 2002; Monpoeho et al., 2000; Nordgren et al., 2009). The RT-qPCR developed by Donaldson et al. (2002) for detection of enteroviruses has a detection limit of 9.3 viral particles mL^{-1} seawater. In addition, RT-qPCR has been shown to be useful for surveying enteric viruses in concentrates from large (Haramoto et al., 2004) and relatively small volumes of freshwater samples (1 L) (Fuhrman et al., 2005). RT-qPCR has also been used to quantify plant pathogenic viruses, such as PMMoV in concentrated samples from seawater (Rosario et al., 2009b) and river water (Hamza et al., 2011). Boben et al. (2007) developed an RT-qPCR method for to detect and quantify ToMV in irrigation waters. This enabled environmental water samples to be defined as ToMV positive at concentrations as low as 4.2×10^{-10} mg mL^{-1}, without any concentration step.

Quantification using qPCR depends on the use of reference materials with known concentrations, which are rarely commercially available, especially in the field of plant virology. In addition, quantification accuracy is highly dependent on the amplification efficiencies between the sample and the reference material (Cankar et al., 2006). One of the factors that can significantly influence amplification efficiency is the presence of inhibitors of PCR amplification, which are indeed commonly found in aquatic samples (Griffin et al., 2003; Rački et al., 2014a). Digital PCR (dPCR) has emerged as a promising technology to overcome these limitations of qPCR (Gutiérrez-Aguirre et al., 2015). In the dPCR the reaction mixture is divided into hundreds to millions of partitions. After end point amplification, positive and negative compartments are counted and the absolute concentration of target copies in the initial sample is derived by applying Poisson's distribution without the need for any standard. Among the dPCR technologies, digital droplet reverse transcription PCR (RT-ddPCR) has already been shown to be a valuable tool for detection and quantification of PMMoV (Rački et al., 2014a) and rotaviruses (Rački et al., 2014b) at the particularly low concentrations usually found in water samples.

The possibility of simplifying the whole procedure of molecular diagnostics has also been explored, with the aim of providing a simple way to confirm the presence of viruses in the field. CIM technology in combination with RT-qPCR detection using the portable SmartCycler Real-Time PCR Thermocycler (Cepheid, CA, USA) has been adapted for detection of rotaviruses in tap water and other environmental water samples (Gutiérrez-Aguirre et al., 2011). As an attractive alternative to PCR-based methods, isothermal amplification-based techniques have been reported, which do not require the temperature cycler. Among these, nucleic acid sequence-based amplification has been successfully used for detection of enteric viruses in concentrated water samples (Rutjes et al., 2005), and recently it has been shown that loop-mediated isothermal amplification technology can be used for rapid in-the-field detection of PepMV in concentrated water samples (Mehle et al., 2017). Loop-mediated isothermal amplification provides target amplification without the need for expensive thermocyclers, and without the need to extract the nucleic acids (Francois et al., 2011; Kogovšek et al., 2015).

The molecular methods described above for viral disease diagnosis are, in general, designed to detect one or a few virus species at the same time. Microarrays are an established strategy for detection of millions of sequences in a single assay, and these have also been used for the detection and genotyping of waterborne viruses (Brinkman and Fout, 2009; Lemarchand et al., 2004). Nowadays, a metagenomics diagnostic technique that uses next-generation sequencing (NGS) is rapidly emerging as an important, and the most promising, universal technology in virus diagnostics. When combined with bioinformatics, this technology revolutionized the discovery and diagnoses of viruses.

6. METAGENOMICS APPROACHES FOR THE STUDY OF VIRAL COMMUNITIES IN WATER

Across the many different environmental waters, most extensively in seawater (Suttle, 2016) but also in freshwater systems (Aguirre de Cárcer et al., 2015; Djikeng et al., 2009; López-Bueno et al., 2009), viruses have been shown to be the most abundant and diverse biological entities. Viruses have important roles in the ecological networks in these waters, by influencing global biochemical cycling and the composition dynamics the biological communities (Fuhrman, 1999). However, their small size, dilution in water, nonculturability, and high genomic diversity impose specific challenges for the characterization of the whole of the viral communities (i.e., the virome) in water samples (Edwards and Rohwer, 2005). Owing to their high genomic diversity, gene-based generic molecular identification and detection methods are not feasible for viruses, which rules out amplification and sequencing of specific highly conserved genes (such as 16S rRNA for bacteria or 18S rRNA for eukaryotes). Truly generic methods that allow "blind" identification of all of the nucleic acids in investigated samples (i.e., the metagenome) are needed to capture the complete diversity of the virome in such samples. In this section, some basic principles and historical outline of the use of metagenomics approaches for the analysis of water samples will be presented, finally highlighting the findings of plant, animal, and human viruses. More detailed reviews about viruses in freshwaters and marine environments are presented in chapter "Viruses in polar lake and soil ecosystems" by Rastrojo and Alcamí and chapter "Viruses in marine ecosystems: From open waters to coral reefs" by Weynberg of this volume.

6.1 Introduction to Shotgun Metagenomics and High-Throughput Sequencing

At the start of the 21st century, the first metagenomics studies that addressed the composition of the marine virome were conducted using linker amplified shotgun libraries, cloning, and Sanger sequencing of a few hundreds to thousands of molecular clones. These studies revealed for the first time, the immense genetic diversity of viral communities in marine environments (Breitbart et al., 2002). A few years later, several NGS platforms emerged that enabled unbiased and high-throughput sequencing (i.e., of millions to hundreds of millions of sequences) of all of the nucleic acids in an investigated sample (Angly et al., 2006). Several different NGS platforms have now been developed; currently, Illumina is the dominant platform choice for research, due to its high throughput and low cost per sequenced nucleotide.

High-throughput sequencing methods soon became recognized as an invaluable tool in virology, especially due to their unbiased detection power and unprecedented resolution. Using NGS-based approaches, it is now possible to extract the total nucleic acids (either DNA or RNA, or both) from environmental samples, to prepare sequencing libraries (a collection of all the nucleic acids, fragmented and with the attached sequencing adapters), and finally to sequence these to identify all of the nucleic acids in the samples. Such an approach is also termed shotgun metagenomics, and it provides a relatively unbiased picture of the nucleic acids of the biological entities present in a sample. However, when focusing on viral communities specifically (i.e., on the virome), an unbiased approach will probably also result in an abundance of nonviral nucleic acids (e.g., from bacteria). Consequently, for virome studies of environmental water samples, viral enrichment methods are often used before NGS, to enhance the resolution of the virome part of the metagenome. These enrichment methods include different filtration, centrifugation, chromatography, and other methods, as reviewed in Section 5.1. In some cases, and especially for samples that are pure in terms of only containing viral nucleic acids, preamplification of the nucleic acids is also needed before NGS library preparation (e.g., Fernandez-Cassi et al., 2017; Kim et al., 2017). Since many plant, animal and human pathogenic viruses contain RNA as their genetic material (in contrary to predominantly studied DNA pages in water samples), it is important to sequence RNA (cDNA) or both, RNA and DNA pools, when focusing on such viruses.

6.2 Bioinformatics: Analysis of NGS Data for Virome Studies

Metagenomics studies of water samples necessarily generate large amounts of sequence data, and thus specific bioinformatics solutions are needed for their analysis (Fig. 2). The shotgun approaches, which are mostly used to study viromes, result in very diverse arrays of sequences. The efficient taxonomic classification of these often requires time-consuming comparisons of the data with the genome sequences of all known organisms. Nevertheless, when focusing only on viruses, the analysis time can be reduced by comparing the generated data only against all known viral species. Sequencing reads can be classified directly or they can first be de novo assembled into larger contigs, which can then be taxonomically classified in a second step. This latter reduces the number of comparisons, and thus also the computation time. A number of de novo assemblers of short reads have been developed (reviewed by Sohn and Nam, 2016), some of which specifically for metagenomics data (Deng et al., 2015; Haider et al., 2014; Nurk et al., 2017; Peng et al., 2012).

To determine the composition of the known part of the virome, the reads or the de novo assembled contigs can be mapped to all of the viral genomes using fast Burrows–Wheeler mapping approaches, such as bwa (Li and Durbin, 2009), bowtie2 (Langmead and Salzberg, 2012), and others. They can also be classified using the relatively slower BLAST-based algorithms (Altschul et al., 1990), or by the fast, but memory demanding, *k*-mer matching algorithms, such as Kraken (Wood and Salzberg, 2014). However, the biggest challenge in virome data analysis is the so-called viral

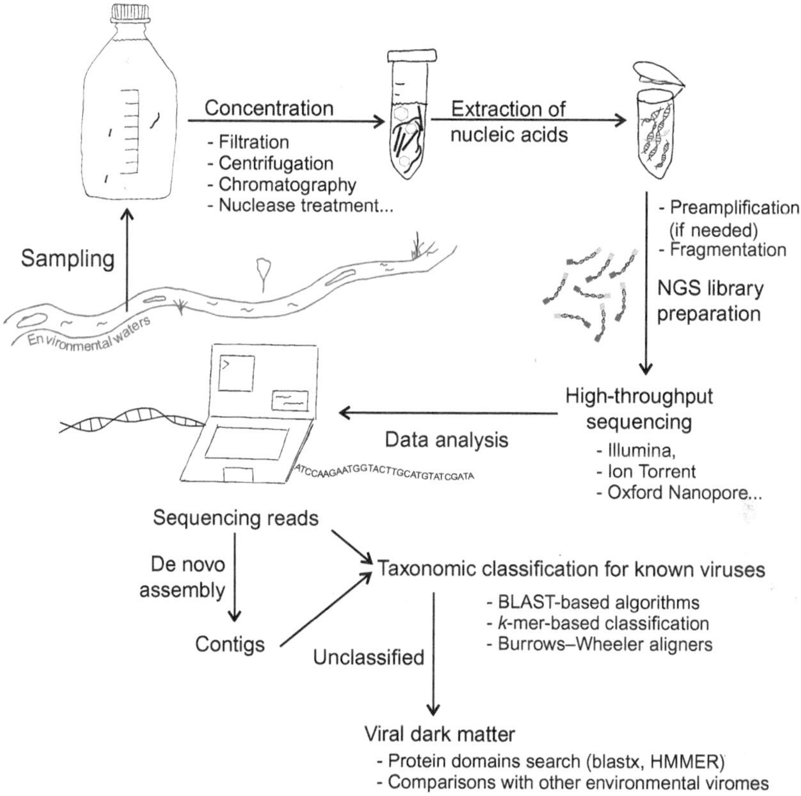

Fig. 2 Example of research pipeline for virome investigation of water samples representing different steps in the sample processing pipeline: concentration, isolation of nucleic acids, sequencing library preparation, high-throughput sequencing, and data analysis.

dark matter; i.e., the unknown part of the virome (previously undiscovered divergent viral sequences, which will sometimes represent the majority of the samples). At least part of this novel sequence pool can be deciphered through comparisons of the translated protein sequences (usually of the de novo assembled contigs) with the database of known proteins, using blastx and tblastx searches (Gish and States, 1993), or using Hidden Markov model-based protein domain searches (Eddy, 2009).

To foster more efficient analysis of shotgun metagenomics data, several computational tools and pipelines have been developed (Brady and Salzberg, 2009; Huson et al., 2016; Kim et al., 2016; Meyer et al., 2008; Rosen et al., 2008; Wood and Salzberg, 2014). Moreover, some pipelines have been developed specifically for virome analysis; e.g., VIROME (Wommack et al., 2012), Metavir 2 (Roux et al., 2014), ViromeScan (Rampelli et al., 2016), VIP (Li et al., 2016), Vipie (Lin et al., 2017), and VirusSeeker (Zhao et al., 2017).

Although over the last decade, the metagenomics studies of viral communities in environmental samples have generated a substantial amount of new data and partially uncovered the viral dark matter, recent studies have still reported substantial numbers of previously uncharacterized sequences for investigated water viromes (e.g., Coutinho et al., 2017). This illustrates the large amounts of undiscovered viral diversity that remains in such samples, and probably in part, this also reflects the fragmented nature of the metagenomics data repositories (Bolduc et al., 2016). The integration of data from metagenomics studies and their easier use will provide future contributions to more efficient virome research. Some of the analysis pipelines developed have already considered this (Roux et al., 2014; Wommack et al., 2012), and importantly, a recent iVirus resource infrastructure aims to integrate the available datasets and virome analysis tools (Bolduc et al., 2016). Moreover, the formal recognition of the viral taxa that have been discovered only through metagenomic approaches would aid in future research on viral diversity. A consensus statement that was endorsed by the International Committee on Taxonomy of Viruses has thus put forward a renewed virus classification scheme that includes the recognition of new viral taxa solely based on high-integrity metagenomics-based sequence data, without the need for additional biological characterization (Simmonds et al., 2017).

6.3 Lessons Learned by Studying the Viromes of Environmental Water Samples

The majority of water virome studies to date have studied marine viromes, primarily DNA viruses. The first study of this type pinpointed the high diversity of the viral communities (dominated by bacteriophages) in a marine environment (Breitbart et al., 2002). Further studies confirmed these observations on a global scale, and also revealed the global dispersion pattern of the viral species discovered (Angly et al., 2006). Moreover, shotgun sequencing of reverse-transcribed RNA from marine water samples showed that the oceans also contain diverse communities of RNA viruses (Culley et al., 2006). Marine viruses have thus became recognized as important components of ecological networks in the seas (Suttle, 2007). Consequently, several larger consortia are engaged in marine virome studies (reviewed by Brum and Sullivan, 2015) and have performed extensive sampling and metagenomics analyses of the oceans throughout the world, to better understand the diversity, distribution, and ecological roles of viruses in the marine environments (Suttle, 2016) (see chapter "Viruses in marine ecosystems: From open waters to coral reefs" by Weynberg, this volume). Most of these studies primarily focus on viruses that infect bacteria and less on plant, animal, and human-infecting viruses.

Characterization of freshwater viromes has also revealed the high diversities of previously unknown viruses in fresh waters. One of the first freshwater virome studies, characterized the virome of a freshwater lake in an anthropogenic area, with the focus on RNA viruses (Djikeng et al., 2009). The lake virome from this study contained known and novel viruses that can infect very different host species that were probably present in the lake, and also some viruses that are important for agriculture and public health. This indicated a potentially important role of freshwater bodies as repositories of a wide array of different viruses, and also the possible role of these repositories in viral spread in the environment. Another study revealed high diversity of eukaryotic DNA viruses in an Antarctic lake (López-Bueno et al., 2009), which was contrary to previously characterized water viromes that were dominated by bacteriophages (for more detailed discussion of viruses in polar freshwater lakes, see chapter "Viruses in polar lake and soil ecosystems" by Rostrajo and Alcamí, this volume). On the other hand, some later studies of freshwater lakes demonstrated high diversity and predominant abundance of bacteriophages also for these environments (Roux et al., 2012; Skvortsov et al., 2016). Interestingly, a metagenomics study that focused on DNA viruses in polar freshwater lakes reported a possible bipolar distribution of some related circular single-stranded DNA viruses, which indicated the possibility of the global spread of some viruses (Aguirre de Cárcer et al., 2015).

Metagenomic analyses of viruses in reclaimed water (Rosario et al., 2009a) have revealed large proportions of bacteriophages in analyzed samples. Moreover, the same study reported also the presence of some plant and human pathogenic viruses and proposed reclaimed water as a possible transmission source for stable viruses (especially plant viruses). A study investigating the DNA virome of a pond on a cattle farm assigned the majority of the viral reads to bacteriophages, but also found some human, animal, and plant pathogenic viruses (Rooks et al., 2010). A metagenomic investigation of raw sewage from United States, Spain, and Ethiopia (Cantalupo et al., 2011) resulted in characterization of the most diverse virome to date. It was also suggested that wastewater is a well-suited environment to study viral diversity on a broader scale; e.g., monitoring the viruses present in the water samples collected in the wastewater treatment plants can reflect the diversity and abundance of viruses in the waters flowing into it (Cantalupo et al., 2011). This idea was later reinforced by other studies that characterized the viromes of untreated sewage (Ng et al., 2012), wastewater treatment plants (Bibby and Peccia, 2013; Tamaki et al., 2012), and dairy lagoons wastewaters (Alhamlan et al., 2013). Finally, metagenomic analysis of sewage water was proposed as a tool for public health surveillance, indicating the presence of human, animal, and plant pathogenic viruses circulating in the environment (Fernandez-Cassi et al., 2018). Furthermore, also ballast water discharge was shown to be rich in diversity of viral species, and to also contain some potentially emerging pathogens of fish and shrimps. This thus indicated the possibility of the long-distance spread of pathogenic viruses in ship ballast waters (Kim et al., 2015).

Several different human enteric and pathogenic plant viruses have been detected in rivers using targeted detection methods (e.g., Hamza et al., 2011); however, only a few virome studies have investigated this environment. The virome of a subtropical river estuary in China contained high diversity of bacteriophage species and was more similar to marine viromes than freshwater ones (Cai et al., 2016). A recent virome study of river water in Spain indicated plant-infecting single-stranded RNA viruses as the most abundant viruses in the samples, followed by bacteriophages. Some human pathogenic viruses were also found in these samples, and further metagenomics analysis of the parsley plants irrigated with this water indicated the possible transfer of a wide diversity of virus families to the irrigated plants (Fernandez-Cassi et al., 2017).

Metagenomics sequencing approaches have provided better insights into the broad spectrum of viruses in water bodies and have revealed the hidden diversity of viral species in these environments. As NGS allows detection of all viruses in one test, it also appears to be a promising tool for surveillance and screening of water samples for pathogenic human, animal, and plant viruses (Aw et al., 2014; Fernandez-Cassi et al., 2018; Nieuwenhuijse and Koopmans, 2017; Nkili-Meyong et al., 2017). For efficient and reliable application of metagenomics approaches for such a purpose, several steps of the procedure need to be carefully monitored, including sampling, contamination prevention during sample preparation, efficiency of concentration methods, and bioinformatics analyses. Also, such NGS-based methods only allow detection of nucleic acids, and thus the presence of viable and infective viral particles in samples needs to be confirmed by other methods (as described in Section 7).

Research into water viromes is increasing and several exciting future research opportunities are on the horizon. On the one hand, the standardization of the procedures for reliable water quality surveillance for viruses using metagenomics will benefit the public health and agriculture sectors. On the other hand, more extensive virome research will enhance our understanding of viral ecology and the role of viruses in different ecosystems. This applies particularly to presently under sampled water environments, such as rivers and interfaces between river and marine environments, and the inclusion of longitudinal virome studies in time and space. From the technological perspective, novel approaches, such as single-virus genomics (Martinez-Hernandez et al., 2017), and application of new sequencing technologies (i.e., third-generation approaches, such as the MinIon system from Oxford Nanopore), are already providing new possibilities in this field. Indeed, the third-generation sequencing approaches can now provide sequencing of much longer molecules (Brown et al., 2017) and rapid in-the-field application of metagenomics studies (Quick et al., 2016).

7. DETERMINATION OF VIRUS VIABILITY

The information derived from electron microscopy, serological and molecular detection and identification assays, and metagenomic analysis can all indicate contamination of water with viruses. However, data on viral infectivity are needed to fully determine any risks to plants, animals, and humans. It has been shown that the infectivities of rotavirus and *Hepatitis A virus* in different type of waters (e.g., lake, creek, ground, tap, and mineral waters) decline at a faster rate than their antigenicities (Biziagos et al., 1988; Pancorbo et al., 1987). Nucleic acid amplification techniques can detect fragments of DNA or RNA, and therefore it can be expected that at least some viruses that can be detected in water environments using molecular techniques might be nonviable (Sobsey et al., 1998). The RNA of PepMV and PVY in water samples remained detectable using RT-qPCR many weeks after these viruses lost their infectivities (Mehle et al., 2014). Similarly, the persistence of the poliovirus genome in different water samples was shown to be longer than that of infectious polioviruses (Skraber et al., 2004; Wetz et al., 2004), and the degradation of the viral RNA correlated well with the loss of infectivity (Skraber et al., 2004). Staggemeier et al. (2015) showed that while human adenovirus was detected by qPCR in 87.3% of 55 water samples, infectious virus particles were found in only four of these samples (7.3%). Different strategies are being used to help the molecular methods to discriminate between infectious and noninfectious viruses, such as addition of different dyes prior to the PCR-based methods, to increase the specificity for viable viruses (Fuster et al., 2016; Leifels et al., 2015).

Ultimate confirmation of infectivity is by infection of the natural host and production of clinical disease symptoms. Mechanical inoculation of test plants with concentrated water samples is used for confirmation of the infectivity of plant viruses (Boben et al., 2007; Castello et al., 1995). In the case of animal and human viruses, it is not practical or ethical to inoculate hosts, although live human volunteers have been used in past studies of some viruses (Graham et al., 1994). Viruses that are confirmed as viable by infection of cell cultures are, therefore, considered as infectious, at least for practical purposes (Griffin et al., 2003).

The major drawbacks of these test-plant and cell-culture assays are that they are laborious and time consuming. Furthermore, it requires days to weeks to confirm both positive and negative results. In addition, universal test plants or cell lines that can be used for propagation and culturing of all plant and enteric viruses have not been established. There are many plant viruses that are not mechanically transmissible (Dijkstra and De Jager, 1998), and many human pathogenic viruses cannot be detected through cell-culture assays because they do not produce cytopathic effects, are extremely slow growing, or do not grow on the established cell lines (Haramoto et al., 2004; Irving and Smith, 1981; Tani et al., 1995; Xagoraraki et al., 2014).

Another important shortcoming here for the detection of infective waterborne viruses is the relatively low sensitivity. Although, in theory, one plant virus particle is sufficient to infect one plant cell, in practice, large numbers of particles (i.e., usually >500) are required to provide certainty of infection of one cell, with the formation of a visible lesion (Dijkstra and De Jager, 1998). This might be because only a small proportion of epidermis cells are wounded to such an extent that virus particles can enter the cytoplasm (Dijkstra and De Jager, 1998). Similarly, for cell culture, the viruses must come into contact with the cells for an infection to occur. Thus, the inclusion of additives that enhance the attachment of viruses to cells has been shown to increase the number of viruses detected (Benton and Hurst, 1986).

Several studies have combined cell-culture assays with subsequent PCR and have reported that this integrated cell-culture PCR (ICC-PCR) improves specific detection of infectious enteric viruses in water samples. After inoculation of a cell line, only the infectious viruses can propagate, if there are any present. The cells themselves can then be tested for viruses using PCR or qPCR before any cytopathic effects are noted (Ding et al., 2017; Fong and Lipp, 2005; Fongaro et al., 2013; Reynolds et al., 1996; Staggemeier et al., 2015). This is also appropriate for viruses that do not produce cytopathic effects, but do still infect and grow in cell lines (Lee and Kim, 2002). The sensitivity of ICC-PCR for detection of infectious enteroviruses in concentrated sewage, marine waters, and surface drinking waters has been shown to be greater compared to direct RT-PCR alone (Reynolds et al., 2001), which cannot discriminate between infectious and noninfectious. In the study of Reynolds et al. (2001), their data from ICC-PCR agreed with the cell cultures, but required only 48 h or less of incubation, compared to as long as 3 weeks for cytopathic effects following incubation with host cells. Chapron et al. (2000) used ICC-RT-nested PCR to show that 68.9% of their water samples were positive for an infectious virus (either astroviruses (family *Astroviridae*, genus *Mamastrovirus*), enteroviruses or adenoviruses), compared to 17.2% determined by conventional cell culture. Similarly, the environmental samples positive for adenoviruses (which included sewage and river waters) increased from 28.6% by conventional cell culture to 50.0% by ICC-PCR (Greening et al., 2002). However, carryover of nucleic acids of inactivated viruses inoculated onto cultured cells might result in false-positive data from samples containing no infectious viruses (Ko et al., 2003). In the case of adenoviruses, this can be solved by testing viral mRNA rather than DNA, because the viral mRNA is only transcribed by infectious adenoviruses during replication (Fongaro et al., 2013; Ko et al., 2003; Staggemeier et al., 2015). The drawbacks to ICC-PCR include the difficulty to adapt it to large-scale studies and automation, and as in standard cell culture, maintenance of the cell lines can be cost prohibitive (Griffin et al., 2003).

8. CONTROL MEASURES: ELIMINATION OF VIRUSES (CHLORINATION, CAVITATION, AND OTHERS)

The global need for water is increasing yearly, mostly due to climate changes. Similarly, and at the same time, the reuse of treated and untreated wastewaters for human-related practices is increasing, such as for crop irrigation and hydroponics (Thebo et al., 2017). As traditional water treatment facilities cannot clear infectious viruses (i.e., enteric viruses) from the water (Nwachcuku and Gerba, 2004; Simmons and Xagoraraki, 2011), it is crucial to implement additional disinfection methods if these waters are to be recycled. These kinds of methods are also required in households in certain countries, especially in the developing world, where there is no access to potable (drinking) water. They would also be helpful in the occurrence of natural disasters, such as earthquakes, floods, and tsunamis, when the potable and contaminated waters get mixed together.

A variety of methods have been used traditionally for water disinfection that are mainly based on two strategies: removal of the pathogens from the water and inactivation of the pathogens by either physical or chemical methods (Gibson, 2014; Ngwenya et al., 2013; Zhang et al., 2016). The first strategy here involves different levels of filtration, from domestic filtration devices to industrial scale membrane filtration and reverse osmosis (Ngwenya et al., 2013; Zhang et al., 2016). Examples of inactivation strategies are the widely used chlorination, and other treatments such as UV, ozone, or heat (Ding et al., 2017; Francy et al., 2012; Lim et al., 2010; http://www.who.int/water_sanitation_health/dwq/Boiling_water_01_15.pdf). Each water disinfection method affects one or several viral components. For example, while UV primarily inactivates viral nucleic acids, chlorine, and heat predominantly alter the native structure of the virus or of its recognition proteins (Wigginton and Kohn, 2012). All of these methods have their own advantages and disadvantages that are related to their efficiency, cost, scalability, and generation of disinfection by-products (Ngwenya et al., 2013). Thus while ozone is relatively efficient for virus inactivation, it is not easily applicable in the field at larger scales, and while application of chlorine is more straightforward, it results in disinfection by-products in the water that can be harmful to the health in the long term (Lyon et al., 2014; Richardson et al., 2007). In addition, viruses have their own strategies to avoid inactivation. Many of them can resist some of these different treatments, such as chlorination and UV radiation, often through their inherent tendency to attach organic particles in the water (Chahal et al., 2016), as described.

Until recently, the presence and concentration of coliforms in water was one of the only criteria to establish the level of contamination of a water body (Edberg et al., 2000; Rompré et al., 2002). Nowadays, it is widely accepted that the survival of waterborne viruses in an environment and their sensitivity to disinfection are not related to those of *Escherichia coli*, particularly as some of the traditionally used water cleaning methods that efficiently remove fecal coliforms do not also inactivate the viruses in the water (Blatchley et al., 2007). The chosen indicator organism is therefore an important feature in the efficient assessment of any virus inactivation method, with the recommended options at present being cultivable viruses, virus surrogates, or bacteriophages, such as MS2 (family *Leviviridae*; genus

Levivirus) (Emmoth et al., 2017; Havelaar, 1987; Huson et al., 2016). Moreover, it is essential that the method used to quantify the reduction of an indicator virus can discriminate between infectious and noninfectious viruses (see Section 7).

In this scenario, where different water-cleaning technologies with different characteristics coexist, any new method that can inactivate waterborne viruses in an efficient, clean, and easily scalable way is obviously welcome. Some of these new methods are improvements or innovations of existing ones, such as UV light-emitting diodes (Song et al., 2016), or they imply a new application of a previously existing technique, such as application of CIM monoliths not only to concentration of viruses but also to their removal (Rački et al., 2015). One promising alternative method is hydrodynamic cavitation. Cavitation occurs when the pressure of an initially homogeneous liquid medium is lowered below the vapor pressure, which generates the appearance of vapor cavities. These vapor structures often collapse violently when they reach a region of increased pressure, which can result in strong shear flows, jets, high local temperatures, shock waves, rapid depressurization, and supersonic flows (Kosel et al., 2017). Cavitation is being increasingly applied in biology, and it has the potential to be used for inactivation of bacteria and viruses in water (Dular et al., 2016). In a recent study, Kosel et al. (2017) demonstrated at low and medium scales that hydrodynamic cavitation can provide viability reductions in MS2 bacteriophages to the levels required by the US Environmental Protection Agency for water-cleaning devices (i.e., a 4-log reduction).

In conclusion here, the disinfection of water bodies intended for human use to remove both human and plant pathogenic viruses will be an increasingly necessary practice, at both the household and the water treatment plant levels. Combination of existing and new promising methods and strategies, together with the use of the correct viral indicators, will contribute to the establishment of clean and efficient procedures for water cleaning that can be adjusted to each need.

9. ECOLOGICAL ROLE OF WATER-MEDIATED TRANSMISSION OF VIRUSES

Large-scale metagenomics studies have highlighted the important ecological role of viruses in waters, which include their impact on the density of host populations (Coutinho et al., 2017; Rodriguez-Brito et al., 2010), and on nutrient cycling (Suttle, 2007), but they have been mostly focused on viruses of bacteria (see chapter "Viruses in marine ecosystems: From open waters to coral reefs" by Weynberg, this volume). Considering water as a transmission route for plant, animal, and human viruses, viruses can spread by water to distant environments that would not be possible for them to reach by other vectors. Consequently, pathogenic viruses can be rapidly transmitted to new areas, and the global incidence of diseases and their impacts on the ecosystem will increase more rapidly.

As well as the "classical" host–pathogen interactions, viruses have been shown to influence the living world also on other important levels, and especially by interspecies and intraspecies horizontal transfer of genetic material (Gilbert and Cordaux, 2017). As an example, a recent study of the virome of river water samples collected in Italy reported high abundance of antibiotic resistance genes in the virome datasets, which suggests a role for viromes as genetic reservoirs for microbial communities in the water (Colombo et al., 2017). Thus, viruses can also be seen as information vehicles, and their transmission by water can enhance the spread of genetic information in the ecosystem.

REFERENCES

Aguirre de Cárcer, D., López-Bueno, A., Pearce, D.A., Alcamí, A., 2015. Biodiversity and distribution of polar freshwater DNA viruses. Sci. Adv. 1, e1400127.

Alfaro-Fernandez, A., Cordoba-Selles, M.D.C., Herrera-Vasquez, J.A., Cebrian, M.D.C., Jorda, C., 2010. Transmission of pepino mosaic virus by the fungal vector olpidium virulentus. J. Phytopathol. 158, 217–226.

Alhamlan, F.S., Ederer, M.M., Brown, C.J., Coats, E.R., Crawford, R.L., 2013. Metagenomics-based analysis of viral communities in dairy lagoon wastewater. J. Microbiol. Methods 92, 183–188.

Altschul, S.F., Gish, W., Miller, W., Myers, E.W., Lipman, D.J., 1990. Basic local alignment search tool. J. Mol. Biol. 215, 403–410.

Angly, F.E., Felts, B., Breitbart, M., Salamon, P., Edwards, R.A., Carlson, C., Chan, A.M., Haynes, M., Kelley, S., Liu, H., Mahaffy, J.M., Mueller, J.E., Nulton, J., Olson, R., Parsons, R., Rayhawk, S., Suttle, C.A., Rohwer, F., 2006. The marine viromes of four oceanic regions. PLoS Biol. 4 (11), e368.

Aslan, A., Xagoraraki, I., Simmons, F.J., Rose, J.B., Dorevitch, S., 2011. Occurrence of adenovirus and other enteric viruses in limited-contact freshwater recreational areas and bathing waters. J. Appl. Microbiol. 111 (5), 1250–1261.

Aw, T.G., Howe, A., Rose, J.B., 2014. Metagenomic approaches for direct and cell culture evaluation of the virological quality of wastewater. J. Virol. Methods 210, 15–21.

Balasubramanian, M.N., Rački, N., Gonçalves, J., Kovač, K., Žnidarič, M.T., Turk, V., Ravnikar, M., Gutiérrez-Aguirre, I., 2016. Enhanced detection of pathogenic enteric viruses in coastal marine environment by concentration using methacrylate monolithic chromatographic supports paired with quantitative PCR. Water Res. 106, 405–414.

Beer, K.D., Gargano, J.W., Roberts, V.A., Hill, V.R., Garrison, L.E., Kutty, P.K., Hilborn, E.D., Wade, T.J., Fullerton, K.F., Yoder, J.S., 2015. Surveillance for

waterborne disease outbreaks associated with drinking water—United States, 2011–2012. MMWR 64 (31), 842–848.
Benton, W.H., Hurst, C.J., 1986. Evaluation of mixed cell types and 5-iodo-20-deoxyuridine treatment upon plaque assay titers of human enteric viruses. Appl. Environ. Microbiol. 51, 1036–1040.
Bibby, K., Peccia, J., 2013. Identification of viral pathogen diversity in sewage sludge by metagenome analysis. Environ. Sci. Technol. 47, 1945–1951.
Biziagos, E., Passagot, J., Crance, J.M., Deloince, R., 1988. Long-term survival of hepatitis-A virus and poliovirus type-1 in mineral water. Appl. Environ. Microbiol. 54, 2705–2710.
Blatchley 3rd, E.R., Gong, W.L., Alleman, J.E., Rose, J.B., Huffman, D.E., Otaki, M., Lisle, J.T., 2007. Effects of wastewater disinfection on waterborne bacteria and viruses. Water Environ. Res. 79, 81–92.
Boben, J., Kramberger, P., Petrovič, N., Cankar, K., Peterka, M., Štrancar, A., Ravnikar, M., 2007. Detection and quantification of tomato mosaic virus in irrigation waters. Eur. J. Plant Pathol. 118 (1), 59–71.
Bolduc, B., Youens-Clark, K., Roux, S., Hurwitz, B.L., Sullivan, M.B., 2016. iVirus: facilitating new insights in viral ecology with software and community data sets imbedded in a cyberinfrastructure. ISME J. 11, 7–14.
Borchardt, M.A., Bertz, P.D., Spencer, S.K., Battigelli, D.A., 2003. Incidence of enteric viruses in groundwater from household wells in Wisconsin. Appl. Environ. Microbiol. 69, 1172–1180.
Borchardt, M.A., Haas, N.L., Hunt, R.J., 2004. Vulnerability of drinking-water wells in La Crosse, Wisconsin, to enteric-virus contamination from surface water contributions. Appl. Environ. Microbiol. 70, 5937–5946.
Brady, A., Salzberg, S.L., 2009. Phymm and PhymmBL: metagenomic phylogenetic classification with interpolated Markov models. Nat. Methods 6, 673–676.
Breitbart, M., Salamon, P., Andresen, B., Mahaffy, J.M., Segall, A.M., Mead, D., Azam, F., Rohwer, F., 2002. Genomic analysis of uncultured marine viral communities. Proc. Natl. Acad. Sci. U. S. A. 99, 14250–14255.
Brinkman, N.E., Fout, G.S., 2009. Development and evaluation of a generic tag array to detect and genotype noroviruses in water. J. Virol. Methods 156 (1), 8–18.
Brown, B.L., Watson, M., Minot, S.S., Rivera, M.C., Franklin, R.B., 2017. MinION™ nanopore sequencing of environmental metagenomes: a synthetic approach. Gigascience 6, 1–10.
Brum, J.R., Sullivan, M.B., 2015. Rising to the challenge: accelerated pace of discovery transforms marine virology. Nat. Rev. Microbiol. 13, 147–159.
Büttner, C., Nienhaus, F., 1989. Virus contamination of waters in two forest districts of the Rhineland area (FRG). Eur. J. Forest Pathol. 19, 206–211.
Büttner, C., Jacobi, V., Koenig, R., 1987. Isolation of Carnation Italian ringspot virus from creek in a forested area south west of Bonn. J. Phytopathol. 118, 131–134.
Cai, L., Yang, Y., Jiao, N., Zhang, R., 2015. Evaluation of tangential flow filtration for the concentration and separation of bacteria and viruses in contrasting marine environments. PLoS One 10 (8), e0136741.
Cai, L., Zhang, R., He, Y., Feng, X., Jiao, N., 2016. Metagenomic analysis of Virioplankton of the subtropical Jiulong river estuary, China. Virus 8, 1–13.
Calder, L., Simmons, G., Thornley, C., Taylor, P., Pritchard, K., Greening, G., Bishop, J., 2003. An outbreak of hepatitis a associated with consumption of raw blueberries. Epidemiol. Infect. 131, 745–751.
Campbell, R.N., 1996. Fungal transmission of plant viruses. Annu. Rev. Phytopathol. 34, 87–108.

Cankar, K., Štebih, D., Dreo, T., Žel, J., Gruden, K., 2006. Critical points of DNA quantification by realtime PCR-effects of DNA extraction method and sample matrix on quantification of genetically modified organisms. BMC Biotechnol. 6, 37–55.

Cannon, J., Papafragkou, E., Park, G.W., Osborne, J., Jaykus, L.-A., Vinje, J., 2006. Surrogates for the study of norovirus stability and inactivation in the environment: a comparison of murine norovirus and feline calicivirus. J. Food Prot. 69 (11), 2761–2765.

Cantalupo, P.G., Calgua, B., Zhao, G., Hundesa, A., Wier, A.D., Katz, J.P., Grabe, M., Hendrix, R.W., Girones, R., Wang, D., Pipas, J.M., 2011. Raw sewage harbors diverse viral populations. mBio 2 (5), e00180-11.

Carducci, A., Morici, P., Pizzi, F., Battistini, R., Rovini, E., Verani, M., 2008. Study of viral removal efficiency in a urban wastewater treatment plant. Water Sci. Technol. 58, 893–897.

Carter, M.J., 2005. Enterically infecting viruses: pathogenicity, transmission and significance for food and waterborne infection. J. Appl. Microbiol. 98 (6), 1354–1380.

Casas, N., Suñén, E., 2002. Detection of enteroviruses, hepatitis A virus and rotaviruses in sewage by means of an immunomagnetic capture reverse transcription-PCR assay. Microbiol. Res. 157, 169–175.

Cashdollar, J.L., Wymer, L., 2013. Methods for primary concentration of viruses from water samples: a review and meta-analysis of recent studies. J. Appl. Microbiol. 115, 1–11.

Castello, J.D., Lakshman, D.K., Tavantzis, S.M., Rogers, S.O., Bachand, G.D., Jagels, R., Charlisle, J., Liu, Y., 1995. Detection of infectious tomato mosaic tobamovirus in fog and clouds. Phytopathology 85, 1409–1412.

Chahal, C., van den Akker, B., Young, F., Franco, C., Blackbeard, J., Monis, P., 2016. Pathogen and particle associations in wastewater: significance and implications for treatment and disinfection processes. Adv. Appl. Microbiol. 97, 63–119.

Chapron, C.D., Ballester, N.A., Fontaine, J.H., Frades, C.N., Margolin, A.B., 2000. Detection of astroviruses, enteroviruses, and adenovirus types 40 and 41 in surface waters collected and evaluated by the information collection rule and an integrated cell culture-nested PCR procedure. Appl. Environ. Microbiol. 66, 2520–2525.

Clark, B., McKendrick, M., 2004. A review of viral gastroenteritis. Curr. Opin. Infect. Dis. 17, 461–469.

Colombo, S., Arioli, S., Neri, E., Della Scala, G., Gargari, G., Mora, D., 2017. Viromes as genetic reservoir for the microbial communities in aquatic environments: a focus on antimicrobial-resistance genes. Front. Microbiol. 8, 1–13.

Costan-Longares, A., Moce-Llivian, L., Avellon, A., Jofre, J., Lucena, F., 2008. Occurrence and distribution of culture enteroviruses in wastewater and surface waters of northeastern Spain. J. Appl. Microbiol. 105 (6), 1945–1955.

Coutinho, F.H., Silveira, C.B., Gregoracci, G.B., Edwards, R.A., Brussaard, C.P.D., Dutilh, B.E., Thompson, F.L., 2017. Marine viruses discovered through metagenomics shed light on viral strategies throughout the oceans. Nat. Commun. 8, 1–12.

Crabtree, K.D., Gerba, C.P., Rose, J.B., Haas, C.N., 1997. Waterborne adenoviruses—a risk assessment. Water Sci. Technol. 35, 1–6.

Croci, L., Ciccozzi, M., De Medici, D., Di Pasquale, S., Fiore, A., Mele, A., Toti, L., 1999. Inactivation of hepatitis A virus in heat-treated mussels. J. Appl. Microbiol. 87, 884–888.

Culley, A.I., Lang, A.S., Suttle, C.A., 2006. Metagenomic analysis of coastal RNA virus communities. Science 312, 1795–1798.

De Keuckelaere, A., Baert, L., Duarte, A., Stals, A., Uyttendaele, M., 2013. Evaluation of viral concentration methods from irrigation and processing water. J. Virol. Methods 187, 294–303.

De Oliveira, L.K., Fleck, J.D., Comerlato, J., Kluge, M., Bergamaschi, B., Fabres, R.B., DaLuz, R.B., Da Silva, J.V.S., Rodrigues, M.T., Genro, J.L., Staggemeier, R.,

Baldasso, N., Spilki, F.R., 2012. Enteric viruses in water samples from Brazilian dairy farms. Agric. Water Manag. 111, 34–39.
Deetz, T.R., Smith, E.M., Goyal, S.M., Gerba, C.P., Vollet III, J.J., Tsai, L., Dupont, H.L., Keswick, B.H., 1984. Occurrence of rota- and enteroviruses in drinking and environmental water in a developing nation. Water Res. 18 (5), 567–571.
Deng, X., Naccache, S.N., Ng, T., Federman, S., Li, L., Chiu, Y., Delwart, E.L., 2015. An ensemble strategy that significantly improves de novo assembly of microbial genomes from metagenomic next-generation sequencing data. Nucleic Acids Res. 43, e46.
Dijkstra, J., De Jager, C.P., 1998. Practical Plant Virology—Protocols and Exercises. Springer-Verlag, Berlin, Heidelberg, p. 459.
Ding, N., Craik, S.A., Pang, X., Lee, B., Neumann, N.F., 2017. Assessing UV inactivation of adenovirus 41 using integrated cell culture real-time qPCR/RT-qPCR. Water Environ. Res. 89, 323–329.
Djikeng, A., Kuzmickas, R., Anderson, N.G., Spiro, D.J., 2009. Metagenomic analysis of RNA viruses in a fresh water lake. PLoS One 4 (9), e7264.
Donaldson, K.A., Griffin, D.W., Paul, J.H., 2002. Detection, quantitation and identification of enteroviruses from surface waters and sponge tissue from the Florida keys using real-time RT-PCR. Water Res. 36, 2505–2514.
Donovan, E., Unice, K., Roberts, J.D., Harris, M., Finley, B., 2008. Risk of gastrointestinal disease associated with exposure to pathogens in the water of the lower passaic river. Appl. Environ. Microbiol. 74 (4), 994–1003.
Dubois, E., Le Guyader, F., Haugarreau, L., Kopecka, H., Cormier, M., Pommepuy, M., 1997. Molecular epidemiological survey of rotaviruses in sewage by reverse transcriptase seminested PCR and restriction fragment length polymorphism assay. Appl. Environ. Microbiol. 63, 1794–1800.
Dular, M., Griessler-Bulc, T., Gutierrez-Aguirre, I., Heath, E., Kosjek, T., Krivograd Klemenčič, A., Oder, M., Petkovšek, M., Rački, N., Ravnikar, M., Šarc, A., Širok, B., Zupanc, M., Žitnik, M., Kompare, B., 2016. Use of hydrodynamic cavitation in (waste)water treatment. Ultrason. Sonochem. 29, 577–588.
Edberg, S.C., Rice, E.W., Karlin, R.J., Allen, M.J., 2000. Escherichia coli: the best biological drinking water indicator for public health protection. Symp. Ser. Soc. Appl. Microbiol. 29, 106–116.
Eddy, S.R., 2009. A new generation of homology search tools based on probabilistic inference. Genome Inform. 23, 205–211.
Edwards, R.A., Rohwer, F., 2005. Viral metagenomics. Nat. Rev. Microbiol. 3, 504–510.
Emmoth, E., Rovira, J., Rajkovic, A., Corcuera, E., Wilches Pérez, D., Dergel, I., Ottoson, J.R., Widén, F., 2017. Inactivation of viruses and bacteriophages as models for swine hepatitis E virus in food matrices. Food Environ. Virol. 9, 20–34.
Farrah, S.R., Goyal, S.M., Gerba, C.P., Wallis, C., Melnick, J.L., 1977. Concentration of enteroviruses from estuarine water. Appl. Environ. Microbiol. 33, 1192–1196.
Fernandez-Cassi, X., Timoneda, N., Gonzales-Gustavson, E., Abril, J.F., Bofill-Mas, S., Girones, R., 2017. A metagenomic assessment of viral contamination on fresh parsley plants irrigated with fecally tainted river water. Int. J. Food Microbiol. 257, 80–90.
Fernandez-Cassi, X., Timoneda, N., Martínez-Puchol, S., Rusiñol, M., Rodriguez-Manzano, J., Figuerola, N., Bofill-Mas, S., Abril, J.F., Girones, R., 2018. Metagenomics for the study of viruses in urban sewage as a tool for public health surveillance. Sci. Total Environ. 618, 870–880.
Fillhart, R.C., Bachand, G.D., Castello, J.D., 1998. Detection of infectious tobamovirus in forest soils. Appl. Environ. Microbiol. 64, 1430–1435.
Fong, T.-T., Lipp, E.K., 2005. Enteric viruses of humans and animals in aquatic environments: health risks, detection, and potential water quality assessment tools. Microbiol. Mol. Biol. Rev. 69 (2), 357–371.

Fongaro, G., Nascimento, M.A., Rigotto, C., Ritterbusch, G., Da Silva, A.D.A., Esteves, P.A., Barardi, C.R.M., 2013. Evaluation and molecular characterization of human adenovirus in drinking water supplies: viral integrity and viability assays. Virol. J. 10 (166), 1–9.

Francois, P., Tangomo, M., Hibbs, J., Bonetti, E.-J., Boehme, C.C., Notomi, T., Perkins, M.D., Schrenzel, J., 2011. Robustness of a loop-mediated isothermal amplification reaction for diagnostic applications. FEMS Immunol. Med. Microbiol. 62, 41–48.

Francy, D.S., Stelzer, E.A., Bushon, R.N., Brady, A.M., Williston, A.G., Riddell, K.R., Borchardt, M.A., Spencer, S.K., Gellner, T.M., 2012. Comparative effectiveness of membrane bioreactors, conventional secondary treatment, and chlorine and UV disinfection to remove microorganisms from municipal wastewaters. Water Res. 46, 4164–4178.

Fuchs, E., Schlufter, C., Kegler, H., 1996. Occurrence of a plant viruses in the Northern sea. Arch. Phytopathol. Plant Protect. 30, 365–366.

Fuhrman, J.A., 1999. Marine viruses and their biogeochemical and ecological effects. Nature 399, 541–548.

Fuhrman, J.A., Liang, X., Noble, R.T., 2005. Rapid detection of enteroviruses in small volumes of natural waters by real-time quantitative reverse transcriptase PCR. Appl. Environ. Microbiol. 71 (8), 4523–4530.

Fujioka, R.S., Yoneyama, B.S., 2002. Sunlight inactivation of human enteric viruses and fecal bacteria. Water Sci. Technol. 46, 291–295.

Fuster, N., Pintó, R.M., Fuentes, C., Beguiristain, N., Bosch, A., Guix, S., 2016. Propidium monoazide RTqPCR assays for the assessment of hepatitis a inactivation and for a better estimation of the health risk of contaminated waters. Water Res. 101, 226–232.

Gall, A.M., Mariñas, B.J., Lu, Y., Shisler, J.L., 2015. Waterborne viruses: a barrier to safe drinking water. PLoS Pathog. 11 (6), e1004867.

Gantzer, C., Dubois, E., Crance, J.-M., Billaudel, S., Kopecka, H., Schwartzbrod, L., Pommepuy, M., Le Guyader, F., 1998. Influence of environmental factors on the survival of enteric viruses in seawater. Oceanol. Acta 21, 983–992.

Gerba, C.P., Goyal, S.M., LaBelle, R.L., Cech, I., Bodgan, G.F., 1979. Failure of indicator bacteria to reflect the occurrence of enteroviruses in marine waters. Am. J. Public Health 69, 1116–1119.

Gerba, C.P., Rose, J.B., Hass, C.N., Crabtree, K.D., 1996. Waterborne rotavirus: a risk assessment. Water Res. 30, 2929–2940.

Gibson, K.E., 2014. Viral pathogens in water: occurrence, public health impact, and available control strategies. Curr. Opin. Virol. 4, 50–57.

Gilbert, C., Cordaux, R., 2017. Viruses as vectors of horizontal transfer of genetic material in eukaryotes. Curr. Opin. Virol. 25, 16–22.

Gish, W., States, D.J., 1993. Identification of protein coding regions by database similarity search. Nat. Genet. 3, 266–272.

Gosalvez, B., Navarro, J.A., Lorca, A., Botella, F., Sanchez-Pina, M.A., Pallas, V., 2003. Detection of Melon necrotic spot virus in water samples and melon plants by molecular methods. J. Virol. Methods 113, 87–93.

Goyal, S.M., Adams, W.N., O'Malley, M.L., Lear, D.W., 1984. Human pathogenic viruses at sewage sludge disposal sites in the Middle Atlantic region. Appl. Environ. Microbiol. 48, 758–763.

Graff, J., Ticehurst, J., Flehmig, B., 1993. Detection of hepatitis A virus in sewage sludge by antigen capture polymerase chain reaction. Appl. Environ. Microbiol. 59, 3165–3170.

Graham, D.Y., Jiang, X., Tanaka, T., Opekun, A.R., Madore, H.P., Estes, M.K., 1994. Norwalk virus infection of volunteers: new insights based on improved assays. J. Infect. Dis. 170 (1), 34–43.

Green, D.H., Lewis, G.D., 1999. Comparative detection of enteric viruses in wastewaters, sediments, and oysters by reverse transcription PCR and cell culture. Water Res. 33, 1195–1200.

Greening, G.E., Hewitt, J., Lewis, G.D., 2002. Evaluation of integrated cell culture-PCR (C-PCR) for virological analysis of environmental samples. J. Appl. Microbiol. 93, 745–750.

Griffin, D., Gibson, C.J., Lipp, E.K., Riley, K., Paul, J.H., Rose, J.B., 1999. Detection of viral pathogens by reverse transcriptase PCR and of microbial indicators by standard methods in the canals of the Florida keys. Appl. Environ. Microbiol. 65, 4118–4125.

Griffin, D.W., Donaldson, K.A., Paul, J.H., Rose, J.B., 2003. Pathogenic human viruses in coastal waters. Clin. Microbiol. Rev. 16, 129–143.

Gutiérrez-Aguirre, I., Kutnjak, D., Rački, N., Rupar, M., Ravnikar, M., 2018. Monolith chromatography as sample preparation step in virome studies of water samples. In: Pantaleo, V., Chiumenti, M. (Eds.), Viral Metagenomics: Methods and Protocols, Methods in Molecular Biology, vol. 1746. Humana Press, New York, pp. 63–75.

Gutiérrez-Aguirre, I., Banjac, M., Steyer, A., Poljšak-Prijatelj, M., Peterka, M., Štrancar, A., Ravnikar, M., 2009. Concentrating rotaviruses from water samples using monolithic chromatographic supports. J. Chromatogr. A 1216, 2700–2704.

Gutiérrez-Aguirre, I., Steyer, A., Banjac, M., Kramberger, P., Poljšak-Prijatelj, M., Ravnikar, M., 2011. On-site reverse transcription-quantitative polymerase chain reaction detection of rotaviruses concentrated from environmental water samples using methacrylate monolithic supports. J. Chromatogr. A 1218, 2368–2373.

Gutiérrez-Aguirre, I., Rački, N., Dreo, T., Ravnikar, M., 2015. Droplet digital PCR for absolute quantification of pathogens. In: Lacomme, C. (Ed.), Plant Pathology: Techniques and Protocols, Methods in Molecular Biology, vol. 1302, Springer Science+Business Media, New York, pp. 331–347.

Haider, B., Ahn, T.H., Bushnell, B., Chai, J., Copeland, A., Pan, C., 2014. Omega: an overlap-graph de novo assembler for metagenomics. Bioinformatics 30, 2717–2722.

Hamza, I.A., Jurzik, L., Überla, K., Wilhelm, M., 2011. Evaluation of pepper mild mottle virus, human picobirnavirus, and Torque teno virus as indicators of fecal contamination in river water. Water Res. 45 (3), 1358–1368.

Haramoto, E., Katayama, H., Ohgaki, S., 2004. Detection of noroviruses in tap water in Japan by means of a new method for concentrating enteric viruses in large volumes of freshwater. Appl. Environ. Microbiol. 70, 2154–2160.

Haramoto, E., Katayama, H., Oguma, K., Yamashita, H., Tajima, A., Nakajima, H., Ohgaki, S., 2006. Seasonal profiles of human noroviruses and indicator bacteria in a wastewater treatment plant in Tokyo, Japan. Water Sci. Technol. 54, 301–308.

Havelaar, A.H., 1987. Bacteriophages as model organisms in water treatment. Microbiol. Sci. 4, 362–364.

Helmi, K., Skraber, S., Gantzer, C., Willame, R., Hoffmann, L., Cauchie, H.M., 2008. Interactions of *Cryptosporidium parvum, Giardia lamblia*, vaccinal poliovirus type 1, and bacteriophages phi X174 and MS2 with a drinking water biofilm and a wastewater biofilm. Appl. Environ. Microbiol. 74, 2079–2088.

Hundesa, A., Maluquer De Motes, C., Bofill-Mas, S., Albinama-Gimenez, N., Girones, R., 2006. Identification of human and animal adenoviruses and polyoma viruses for determination of sources of fecal contamination in the environment. Appl. Environ. Microbiol. 72, 7886–7893.

Huson, D.H., Beier, S., Flade, I., Górska, A., El-Hadidi, M., Mitra, S., Ruscheweyh, H.J., Tappu, R., 2016. MEGAN Community Edition—Interactive exploration and analysis of large-scale microbiome sequencing data. PLoS Comput. Biol. 12, 1–12.

Ikner, L.A., Gerba, C.P., Bright, K.R., 2012. Concentration and recovery of viruses from water: a comprehensive review. Food Environ. Virol. 4, 41–67.

Irving, L.G., Smith, F.A., 1981. One-year survey of enteroviruses, adenoviruses, and reoviruses isolated from effluent at an activated-sludge purification plant. Appl. Environ. Microbiol. 41, 51–59.

Iwai, M., Hasegawa, S., Obara, M., Nakamura, K., Horimoto, E., Takizawa, T., Kurata, T., Sogen, S., Shiraki, K., 2009. Continuous presence of noroviruses and sapoviruses in raw sewage reflects infections among inhabitants of Toyama, Japan (2006 to 2008). Appl. Environ. Microbiol. 75, 1264–1270.

Jacobi, V., Castello, J.D., 1991. Isolation of tomato mosaic virus from waters draining forest strands in New York State. Phytopathology 81, 1112–1117.

Jiang, S.C., Noble, R., Chu, W., 2001. Human adenoviruses and coliphages in urban runoff-impacted coastal waters of Southern California. Appl. Environ. Microbiol. 67, 179–184.

Johnson, D.C., Enriquez, C.E., Pepper, I.L., Davis, T.L., Gerba, C.P., Rose, J.B., 1997. Survival of giardia, cryptosporidium, poliovirus, and salmonella in marine waters. Water Sci. Technol. 35, 261–268.

Jones, R.A.C., 2004. Using epidemiological information to develop effective integrated virus disease management strategies. Virus Res. 100, 5–30.

Jones, T.H., Muehlhauser, V., Thériault, G., 2014. Comparison of ZetaPlus 60S and nitrocellulose membrane filters for the simultaneous concentration of F-RNA coliphages, porcine teschovirus and porcine adenovirus from river water. J. Virol. Methods 206, 5–11.

Katayama, H., Shimasaki, A., Ohgaki, S., 2002. Development of a virus concentration method and its application to detection of enterovirus and Norwalk virus from coastal seawater. Appl. Environ. Microbiol. 68, 1033–1039.

Keswick, B.H., Gerba, C.P., 1980. Viruses in groundwater. Environ. Sci. Technol. 14 (11), 1290–1297.

Kim, Y., Aw, T.G., Teal, T.K., Rose, J.B., 2015. Metagenomic investigation of viral communities in Ballast Water. Environ. Sci. Technol. 49, 8396–8407.

Kim, D., Song, L., Breitwieser, F.P., Salzberg, S.L., 2016. Centrifuge: rapid and sensitive classification of metagenomic sequences. Genome Res. 26, 1721–1729.

Kim, Y., Van Bonn, W., Aw, T.G., Rose, J.B., 2017. Aquarium viromes: viromes of human-managed aquatic systems. Front. Microbiol. 8, 1–14.

Kitajima, M., Haramoto, E., Phanuwan, C., Katayama, H., Furumai, H., 2012. Molecular detection and genotyping of human noroviruses in influent and effluent water at a wastewater treatment plant in Japan. J. Appl. Microbiol. 112, 605–613.

Ko, G., Cromeans, T.L., Sobsey, M.D., 2003. Detection of infectious adenovirus in cell culture by mRNA reverse transcription-PCR. Appl. Environ. Microbiol. 69, 7377–7384.

Koenig, R., 1986. Plant viruses in rivers and lakes. Adv. Virus Res. 31, 321–333.

Koenig, R., 1988. Detection in surface waters of plant viruses with known and unknown natural hosts. In: Cooper, J.I., Asher, M.J.C. (Eds.), Developments in Applied Biology II. Association of Applied Biologists, Wellesbourne, UK, pp. 305–313.

Koenig, R., Lesemann, D.E., 1985. Plant viruses in German rivers and lakes. Phytopathol. Z. 112, 105–116.

Koenig, R., Lesemann, D.E., Burgermeister, W., 1988. Isolation of Carnation ringspot virus from a canal near a sewage plant: cDNA hybridization analysis, serology and cytopathology. J. Phytopathol. 121, 346–356.

Koenig, R., Rüdel, M., Lesemann, D.E., 1989. Detection of Petunia asteroid mosaic, Carnation ring spot and tobacco necrosis viruses in ditches and drainage canals in a grapevine-growing area in West Germany. J. Phytopathol. 127, 169–172.

Koenig, R., Pfeilstetter, E., Kegler, H., Lesemann, D.E., 2004. Isolation of two strains of a new Tombusvirus (Havel river virus, HaRV) from surface waters in Germany. Eur. J. Plant Pathol. 110, 429–433.

Kogovšek, P., Hodgetts, J., Hall, J., Prezelj, N., Nikolić, P., Mehle, N., Lenarčič, R., Rotter, A., Dickinson, M., Boonham, N., Dermastia, M., Ravnikar, M., 2015. LAMP assay and rapid sample preparation method for on-site detection of flavescence doree phytoplasma in grapevine. Plant Pathol. 64, 286–296.

Kosel, J., Gutierrez Aguirre, I., Dreo, T., Rački, N., Ravnikar, M., Dular, M., 2017. Efficient inactivation of MS-2 virus in water by hydrodynamic cavitation. Water Res. 124, 465–471.

Kovač, K., Gutiérrez-Aguirre, I., Banjac, M., Peterka, M., Poljšak-Prijatelj, M., Ravnikar, M., Mijovski, J.Z., Schultz, A.C., Raspor, P., 2009. A novel method for concentrating hepatitis a virus and caliciviruses from bottled water. J. Virol. Methods 162, 272–275.

Kramberger, P., Petrovič, N., Štrancar, A., Ravnikar, M., 2004. Concentration of plant viruses using monolithic chromatographic supports. J. Virol. Methods 120, 51–57.

Kramberger, P., Urbas, L., Štrancar, A., 2015. Downstream processing and chromatography based analytical methods for production of vaccines, gene therapy vectors, and bacteriophages. Hum. Vaccin. Immunother. 11, 1010–1021.

Krczal, G., Albouy, J., Damy, I., Kusiak, C., Deogratias, J.M., Moreau, J.P., Berkelmann, B., Wohanka, W., 1995. Transmission of pelargonium flower break virus (PFBV) in irrigation systems and by Thrips. Plant Dis. 79, 163–166.

Krikelis, V., Markoulatos, P., Spyrou, N., Serie, C., 1985. Detection of indigenous enteric viruses in raw sewage effluents of the city of Athens, Greece, during a two year survey. Water Sci. Technol. 17, 159–164.

Kukkula, M., Arstila, P., Klossner, M.L., Maunula, L., Bonsdorff, C.H.V., Jaatinen, P., 1997. Waterborne outbreak of viral gastroenteritis. Scand. J. Infect. Dis. 29 (4), 415–418.

Labelle, R.L., Gerba, C.P., 1980. Influence of estuarine sediment on virus survival under field conditions. Appl. Environ. Microbiol. 39 (4), 749–755.

Langmead, B., Salzberg, S.L., 2012. Fast gapped-read alignment with Bowtie 2. Nat. Methods 9, 357–359.

Lee, S.H., Kim, S.J., 2002. Detection of infectious enteroviruses and adenoviruses in tap water in urban areas in Korea. Water Res. 36, 248–256.

Lee, S.H., Lee, C., Lee, K.W., Cho, H.B., Kim, S.J., 2005. The simultaneous detection of both enteroviruses and adenoviruses in environmental water samples including tap water with an integrated cell culture-multiplex-nested PCR procedure. J. Appl. Microbiol. 98, 1020–1029.

Lehtola, M.J., Torvinen, E., Kusnetsov, J., Pitk, A., Nen, T., Maunula, L., Bonsdorffch, V., Martikainen, P.J., Wilks, S.A., Keevil, C.W., Miettinen, I.T., 2007. Survival of *Mycobacterium avium, Legionella pneumophila, Escherichia coli* and caliciviruses in drinking water-associated biofilms grown under high-shear turbulent flow. Appl. Environ. Microbiol. 73, 2854–2859.

Leifels, M., Jurzik, L., Wilhelm, M., Hamza, I.A., 2015. Use of ethidium monoazide and propidium monoazide to determine viral infectivity upon inactivation by heat, UV-exposure and chlorine. Int. J. Hyg. Environ. Health 218, 686–693.

Lemarchand, K., Masson, L., Brousseau, R., 2004. Molecular biology and DNA microarray technology for microbial quality monitoring of water. Crit. Rev. Microbiol. 30, 145–172.

Li, H., Durbin, R., 2009. Fast and accurate short read alignment with Burrows–Wheeler transform. Bioinformatics 25, 1754–1760.

Li, D., Gu, A.Z., Yang, W., He, M., Hu, X.-h., Shi, H.-C., 2010. An integrated cell culture and reverse transcription quantitative PCR assay for detection of infectious rotaviruses in environmental waters. J. Microbiol. Methods 82 (1), 59–63.

Li, D., Gu, A.Z., Zeng, S.-Y., Yang, W., He, M., Shi, H.-C., 2011. Monitoring and evaluation of infectious rotaviruses in various wastewater effluents and receiving waters revealed correlation and seasonal pattern of occurrences. J. Appl. Microbiol. 110 (5), 1129–1137.

Li, Y., Wang, H., Nie, K., Zhang, C., Zhang, Y., Wang, J., Niu, P., Ma, X., 2016. VIP: an integrated pipeline for metagenomics of virus identification and discovery. Sci. Rep. 6, 23774.

Lim, M.Y., Kim, J.M., Lee, J.E., Ko, G., 2010. Characterization of ozone disinfection of murine norovirus. Appl. Environ. Microbiol. 76, 1120–1124.

Lin, J., Kramna, L., Autio, R., Hyöty, H., Nykter, M., Cinek, O., 2017. Vipie: web pipeline for parallel characterization of viral populations from multiple NGS samples. BMC Genomics 18, 378.

Lipp, E.K., Farrah, S.R., Rose, J.B., 2001a. Assessment and impact of microbial fecal contamination and human enteric pathogens in a coastal community. Mar. Pollut. Bull. 42, 286–293.

Lipp, E.K., Kurz, R., Vincent, R., Rodriguez-Palacios, C., Farrah, S.R., Rose, J.B., 2001b. The effects of seasonal variability and weather on microbial fecal pollution and enteric pathogens in a subtropical estuary. Estuaries 24, 238–258.

Lodder, W.J., de Roda Husman, A.M., 2005. Presence of noroviruses and other enteric viruses in sewage and surface waters in the Netherlands. Appl. Environ. Microbiol. 71, 1453–1461.

López-Bueno, A., Tamames, J., Velázquez, D., Moya, A., Quesada, A., Alcamí, A., 2009. High diversity of the viral community from an Antarctic Lake. Science 326 (5954), 858–861.

Lyon, B.A., Milsk, R.Y., DeAngelo, A.B., Simmons, J.E., Moyer, M.P., Weinberg, H.S., 2014. Integrated chemical and toxicological investigation of UV-chlorine/chloramine drinking water treatment. Environ. Sci. Technol. 48, 6743–6753.

Mandahar, C.L., 1990. Virus Transmission. In: Mandahar, C.L. (Ed.), Plant Viruses. Volume II Pathology. CRC Press, Inc., FL, pp. 208–210.

Martinez-Hernandez, F., Fornas, O., Lluesma Gomez, M., Bolduc, B., de la Cruz Peña, M.J., Martínez, J.M., Anton, J., Gasol, J.M., Rosselli, R., Rodriguez-Valera, F., Sullivan, M.B., Acinas, S.G., Martinez-Garcia, M., 2017. Single-virus genomics reveals hidden cosmopolitan and abundant viruses. Nat. Commun. 8, 15892.

Matthews, R.E.F., 1992. Fundamentals of Plant Virology. Academic Press, USA, p. 403.

McMinn, B.R., Korajkic, A., 2015. A small volume procedure for viral concentration from water. J. Vis. Exp. 3 (96), e51744.

Mehle, N., Ravnikar, M., 2012. Plant viruses in aqueous environment: survival, water mediated transmission, and detection. Water Res. 46, 4902–4917.

Mehle, N., Gutiérrez-Aguirre, I., Prezelj, N., Delić, D., Vidic, U., Ravnikar, M., 2014. Survival and transmission of potato virus Y, pepino mosaic virus, and potato spindle tuber viroid in water. Appl. Environ. Microbiol. 80 (4), 1455–1462.

Mehle, N., Kogovšek, P., Rački, N., Jakomin, T., Gutiérrez-Aguirre, I., Kramberger, P., Ravnikar, M., 2017. Filling the gaps in diagnostics of pepino mosaic virus and potato spindle tuber viroid in water and tomato seeds and leaves. Plant Pathol. 66, 1191–1201.

Melnick, J.L., 1984. Etiological Agents and Their Potential for Causing Water-Borne Virus Diseases. In: Melnick, J.L. (Ed.), Enteric Viruses in Water, vol. 15, Kargel, Basel, Switzerland, pp. 1–16.

Meschke, J.S., Sobsey, M.D., 1998. Comparative adsorption of norwalk virus, poliovirus 1, and F+ RNA coliphage MS2 to soils suspended in treated wastewater. Water Sci. Technol. 38, 187–189.

Meyer, F., Paarmann, D., D'Souza, M., Olson, R., Glass, E., Kubal, M., Paczian, T., Rodriguez, A., Stevens, R., Wilke, A., Wilkening, J., Edwards, R., Venter, J., Remington, K., Heidelberg, J., Halpern, A., Rusch, D., Eisen, J., Wu, D., Paulsen, I., Nelson, K., Nelson, W., Tyson, G., Chapman, J., Hugenholtz, P.,

Allen, E., Ram, R., Richardson, P., Solovyev, V., Rubin, E., Rokhsar, D., Banfield, J., Huse, S., Huber, J., Morrison, H., Sogin, M., Welch, D., Overbeek, R., Begley, T., Butler, R., Choudhuri, J., Diaz, N., Chuang, H.-Y., Cohoon, M., de Crécy-Lagard, V., Disz, T., Edwards, R., McNeil, L., Reich, C., Aziz, R., Bartels, D., Cohoon, M., Disz, T., Edwards, R., Gerdes, S., Hwang, K., Kubal, M., Margulies, M., Egholm, M., Altman, W., Attiya, S., Bader, J., Bemben, L., Berka, J., Braverman, M., Chen, Y., Chen, Z., Aziz, R., Bartels, D., Best, A., DeJongh, M., Disz, T., Edwards, R., Formsma, K., Gerdes, S., Glass, E., Kubal, M., Field, D., Morrison, N., Selengut, J., Sterk, P., Altschul, S., Madden, T., Schaffer, A., Zhang, J., Zhang, Z., Miller, W., Lipman, D., Jarvie, T., DeSantis, T., Hugenholtz, P., Larsen, N., Rojas, M., Brodie, E., Keller, K., Huber, T., Dalevil, D., Hu, P., Andersen, G., Cole, J., Chai, B., Farris, R., Wang, Q., Wuyts, J., de Peer, Y., Winkelmans, T., De Wachter, R., Leplae, R., Hebrant, A., Wodak, S., Toussaint, A., Meyer, F., Overbeek, R., Rodriquez, A., Tringe, S., von Mering, C., Kobayashi, A., Salamov, A., Chen, K., Chang, H., Podar, M., Short, J., Mathur, E., Detter, J., Rodriguez-Brito, B., Rohwer, F., Edwards, R., McHardy, A., Martin, H., Tsirigos, A., Hugenholtz, P., Rigoutsos, I., Edwards, R., Rodriguez-Brito, B., Wegley, L., Haynes, M., Breitbart, M., Peterson, D., Saar, M., Alexander, S., Alexander, E., Rohwer, F., Fierer, N., Breitbart, M., Nulton, J., Salamon, P., Lozupone, C., Jones, R., Robeson, M., Edwards, R., Felts, B., Rayhawk, S., Wegley, L., Edwards, R., Rodriguez-Brito, B., Liu, H., Rohwer, F., Mou, X., Edwards, R., Hodson, R., Moran, M., Dinsdale, E., Edwards, R., Hall, D., Angly, F., Breitbart, M., Brulc, J., Furlan, M., Desnues, C., Haynes, M., Li, L., Krause, L., Diaz, N., Bartels, D., Edwards, R., Puhler, A., Rohwer, F., Meyer, F., Stoye, J., Liang, F., Holt, I., Pertea, G., Karamycheva, S., Salzberg, S., Quackenbush, J., Rohwer, F., 2008. The metagenomics RAST server—a public resource for the automatic phylogenetic and functional analysis of metagenomes. BMC Bioinformatics 9, 386.

Middleton, P.J., 1996. Viruses that multiply in the gut and cause endemic and epidemic gastroenteritis. Clin. Diagn. Virol. 6, 93–101.

Monpoeho, S., Dehee, A., Mignotte, B., Schwartzbrod, L., Marechal, V., Nicholas, J.C., Billaudel, S., Ferre, V., 2000. Quantification of enterovirus RNA in sludge samples using single tube real-time RT PCR. Biotechniques 29, 88–93.

Navarro, J.A., Torok, V.A., Vetten, H.J., Pallas, V., 2005. Genetic variability in the coat protein genes of Lettuce big-vein associated virus and Mirafiori lettuce big-vein virus. Arch. Virol. 150, 681–694.

Ng, T.F.F., Marine, R., Wang, C., Simmonds, P., Kapusinszky, B., Bodhidatta, L., Oderinde, B.S., Wommack, K.E., Delwart, E., 2012. High variety of known and new RNA and DNA viruses of diverse origins in untreated sewage. J. Virol. 86, 12161–12175.

Ngwenya, N., Ncube, E.J., Parsons, J., 2013. Recent advances in drinking water disinfection: successes and challenges. Rev. Environ. Contam. Toxicol. 222, 111–170.

Nieuwenhuijse, D.F., Koopmans, M.P.G., 2017. Metagenomic sequencing for surveillance of food- and waterborne viral diseases. Front. Microbiol. 8, 1–11.

Nkili-Meyong, A.A., Bigarré, L., Labouba, I., Vallaeys, T., Avarre, J.-C., Berthet, N., 2017. Contribution of next-generation sequencing to aquatic and fish virology. Intervirology 59, 285–300.

Nordgren, J., Matussek, A., Mattsson, A., Svensson, A.P., Lindgren, E., 2009. Prevalence of norovirus and factors influencing virus concentrations during one year in a full-scale wastewater treatment plant. Water Res. 43, 1117–1125.

Nurk, S., Meleshko, D., Korobeynikov, A., Pevzner, P.A., 2017. MetaSPAdes: a new versatile metagenomic assembler. Genome Res. 27, 824–834.

Nwachcuku, N., Gerba, C.P., 2004. Emerging waterborne pathogens: can we kill them all? Curr. Opin. Biotechnol. 15, 175–180.

Pancorbo, O.C., Evanshen, B.G., Campbell, W.F., Lambert, S., Curtis, S.K., Woolley, T.W., 1987. Infectivity and antigenicity reduction rates of human rotavirus strain Wa in fresh waters. Appl. Environ. Microbiol. 53, 1803–1811.

Papapetropoulou, M., Vantarakis, A.C., 1998. Detection of adenovirus outbreak at a municipal swimming pool by nested PCR amplification. J. Infect. 36 (1), 101–103.

Pares, R.D., Gunn, L.V., Cresswell, G.C., 1992. Tomato mosaic virus infection in a recirculating nutrient solution. J. Phytopathol. 135, 192–198.

Peng, Y., Leung, H.C.M., Yiu, S.M., Chin, F.Y.L., 2012. IDBA-UD: a de novo assembler for single-cell and metagenomic sequencing data with highly uneven depth. Bioinformatics 28, 1420–1428.

Piazzolla, P., Castellano, M.A., De Stradis, A., 1986. Presence of plant viruses in some rivers of southern Italy. J. Phytopathol. 116, 244–246.

Pina, S., Puig, M., Lucena, F., Jofre, J., Girones, R., 1998. Viral pollution in the environment and in shellfish: human adenovirus detection by PCR as an index of human viruses. Appl. Environ. Microbiol. 64, 3376–3382.

Pleše, N., Juretić, N., Mamula, D., Polak, Z., Krajačić, M., 1996. Plant viruses in soil and water of forest ecosystems in Croatia. Phyton (Horn, Austria) 36 (1), 135–143.

Powell, K.L., Taylor, R.G., Cronin, A.A., Barrett, M.H., Pedley, S., Sellwood, J., Trowsdale, S.A., Lerner, D.N., 2003. Microbial contamination of two urban sandstone aquifers in the UK. Water Res. 37, 339–352.

Quick, J., Loman, N.J., Duraffour, S., Simpson, J.T., Severi, E., Cowley, L., Bore, J.A., Koundouno, R., Dudas, G., Mikhail, A., Ouédraogo, N., Afrough, B., Bah, A., Baum, J.H.J., Becker-Ziaja, B., Boettcher, J.P., Cabeza-Cabrerizo, M., Camino-Sánchez, Á., Carter, L.L., Doerrbecker, J., Enkirch, T., Dorival, I.G., Hetzelt, N., Hinzmann, J., Holm, T., Kafetzopoulou, L.E., Koropogui, M., Kosgey, A., Kuisma, E., Logue, C.H., Mazzarelli, A., Meisel, S., Mertens, M., Michel, J., Ngabo, D., Nitzsche, K., Pallasch, E., Patrono, L.V., Portmann, J., Repits, J.G., Rickett, N.Y., Sachse, A., Singethan, K., Vitoriano, I., Yemanaberhan, R.L., Zekeng, E.G., Racine, T., Bello, A., Sall, A.A., Faye, O., Faye, O., Magassouba, N., Williams, C.V., Amburgey, V., Winona, L., Davis, E., Gerlach, J., Washington, F., Monteil, V., Jourdain, M., Bererd, M., Camara, A., Somlare, H., Camara, A., Gerard, M., Bado, G., Baillet, B., Delaune, D., Nebie, K.Y., Diarra, A., Savane, Y., Pallawo, R.B., Gutierrez, G.J., Milhano, N., Roger, I., Williams, C.J., Yattara, F., Lewandowski, K., Taylor, J., Rachwal, P., Turner, J., Pollakis, G.D., Hiscox, J.A., Matthews, D.A., Shea, M.K.O., Johnston, A.M., Wilson, D., Hutley, E., Smit, E., Di Caro, A., Wölfel, R., Stoecker, K., Fleischmann, E., Gabriel, M., Weller, S.A., Koivogui, L., Diallo, B., Keïta, S., Rambaut, A., Formenty, P., Günther, S., Carroll, M.W., 2016. Real-time, portable genome sequencing for Ebola surveillance. Nature 530, 228–232.

Rački, N., Dreo, T., Gutierrez-Aguirre, I., Blejec, A., Ravnikar, M., 2014a. Reverse transcriptase droplet digital PCR shows high resilience to PCR inhibitors from plant, soil, and water samples. Plant Methods 10 (42), 1–10.

Rački, N., Morisset, D., Gutierrez-Aguirre, I., Ravnikar, M., 2014b. One-step RT-droplet digital PCR: a breakthrough in the quantification of waterborne RNA viruses. Anal. Bioanal. Chem. 406, 661–667.

Rački, N., Kramberger, P., Steyer, A., Gašperšič, J., Štrancar, A., Ravnikar, M., Gutierrez-Aguirre, I., 2015. Methacrylate monolith chromatography as a tool for waterborne virus removal. J. Chromatogr. A 1381, 118–124.

Rampelli, S., Soverini, M., Turroni, S., Quercia, S., Biagi, E., Brigidi, P., Candela, M., 2016. ViromeScan: a new tool for metagenomic viral community profiling. BMC Genomics 17, 165.

Rao, V.C., Seidel, K.M., Goyal, S.M., Metcalf, T.G., Melnick, J.L., 1984. Isolation of enteroviruses from water, suspended solids, and sediments from Galveston Bay: survival of poliovirus and rotavirus adsorbed to sediments. Appl. Environ. Microbiol. 48, 404–409.

Raphael, R.A., Sattar, S.A., Springthorpe, V.S., 1985. Long-term survival of human rotavirus in raw and treated river water. Can. J. Microbiol. 31, 124–128.

Reynolds, K.A., Gerba, C.P., Pepper, I.L., 1995. Detection of enteroviruses in marine waters by direct RT-PCR and cell culture. Water Sci. Technol. 31, 322–328.

Reynolds, K.A., Gerba, C.P., Pepper, I.L., 1996. Detection of infectious enteroviruses by an integrated cell culture-PCR procedure. Appl. Environ. Microbiol. 62, 1424–1427.

Reynolds, K.A., Gerba, C.P., Abbaszadegan, M., Pepper, L.L., 2001. ICC/PCR detection of enteroviruses and hepatitis a virus in environmental samples. Can. J. Microbiol. 47, 153–157.

Reynolds, K.A., Mena, K.D., Gerba, C.P., 2008. Risk of waterborne illness via drinking water in the United States. Rev. Environ. Contam. Toxicol. 192, 117–158.

Richardson, S.D., Plewa, M.J., Wagner, E.D., Schoeny, R., Demarini, D.M., 2007. Occurrence, genotoxicity, and carcinogenicity of regulated and emerging disinfection by-products in drinking water: a review and roadmap for research. Mutat. Res. 636, 178–242.

Rigotto, C., Victoria, M., Moresco, V., Kolesnikovas, C.K., Corrêa, A.A., Souza, D.S.M., Miagostovich, M.P., Simões, C.M.O., Barardi, C.R.M., 2010. Assessment of adenovirus, hepatitis a virus and rotavirus presence in environmental samples in Florianopolis, South Brazil. J. Appl. Microbiol. 109, 1979–1987.

Robertson, J.B., Edberg, S.C., 1997. Natural protection of spring and well drinking water against surface microbial contamination. I. Hydrogeological parameters. Crit. Rev. Microbiol. 23 (2), 143–178.

Rodriguez-Brito, B., Li, L., Wegley, L., Furlan, M., Angly, F., Breitbart, M., Buchanan, J., Desnues, C., Dinsdale, E., Edwards, R., Felts, B., Haynes, M., Liu, H., Lipson, D., Mahaffy, J., Martin-Cuadrado, A.B., Mira, A., Nulton, J., Pasić, L., Rayhawk, S., Rodriguez-Mueller, J., Rodriguez-Valera, F., Salamon, P., Srinagesh, S., Thingstad, T.F., Tran, T., Thurber, R.V., Willner, D., Youle, M., Rohwer, F., 2010. Viral and microbial community dynamics in four aquatic environments. ISME J. 4, 739–751.

Rompré, A., Servais, P., Baudart, J., de-Roubin, M.R., Laurent, P., 2002. Detection and enumeration of coliforms in drinking water: current methods and emerging approaches. J. Microbiol. Methods 49, 31–54.

Rooks, D.J., Smith, D.L., Mcdonald, J.E., Woodward, M.J., Mccarthy, A.J., Allison, H.E., 2010. 454-pyrosequencing: a molecular battiscope for freshwater viral ecology. Genes (Basel) 1, 210–226.

Rosario, K., Nilsson, C., Lim, Y.W., Ruan, Y., Breitbart, M., 2009a. Metagenomic analysis of viruses in reclaimed water. Environ. Microbiol. 11, 2806–2820.

Rosario, K., Symonds, E.M., Sinigalliano, C., Stewart, J., Breitbart, M., 2009b. Pepper mild mottle virus as an indicator of fecal pollution. Appl. Environ. Microbiol. 75 (22), 7261–7267.

Rose, J.B., Mullinax, R.L., Singh, S.N., Yates, M.V., Gerba, C.P., 1987. Occurrence of rotaviruses and enteroviruses in recreational waters of Oak Creek, Arizona. Water Res. 21 (11), 1375–1381.

Rosen, G., Garbarine, E., Caseiro, D., Polikar, R., Sokhansanj, B., 2008. Metagenome fragment classification using N-mer frequency profiles. Adv. Bioinforma. 2008, 205969.

Roux, S., Enault, F., Robin, A., Ravet, V., Personnic, S., Theil, S., Colombet, J., Sime-Ngando, T., Debroas, D., 2012. Assessing the diversity and specificity of two freshwater viral communities through metagenomics. PLoS One 7, e33641.

Roux, S., Tournayre, J., Mahul, A., Debroas, D., Enault, F., 2014. Metavir 2: new tools for viral metagenome comparison and assembled virome analysis. BMC Bioinformatics 15, 76.

Rutjes, S.A., Italiaander, R., van den Berg, H.H.J.L., Lodder, W.J., de Roda Husman, A.M., 2005. Isolation and detection of enterovirus RNA from large-volume water samples by using the NucliSens miniMAG system and real-time nucleic acid sequence-based amplification. Appl. Environ. Microbiol. 71 (7), 3734–3740.

Rutjes, S.A., Lodder, W.J., Docters van Leeuwen, A., de Roda Husman, A.M., 2009. Detection of infectious rotavirus in naturally contaminated source waters for drinking water production. J. Appl. Microbiol. 107, 97–105.

Rzeżutka, A., Cook, N., 2004. Survival of human enteric viruses in the environment and food. FEMS Microbiol. Rev. 28 (4), 441–453.

Schwarz, D., Beuch, U., Bandte, M., Fakhro, A., Büttner, C., Obermeier, C., 2010. Spread and interaction of Pepino mosaic virus (PepMV) and *Pythium aphanidermatum* in a closed nutrient solution recirculation system: effects on tomato growth and yield. Plant Pathol. 59 (3), 443–452.

Sedmak, G., Bina, D., Macdonald, J., Couillard, L., 2005. Nine-year study of the occurrence of culturable viruses in source water for two drinking water treatment plants and the influent and effluent of a wastewater treatment plant in Milwaukee, Wisconsin (August 1994 through July 2003). Appl. Environ. Microbiol. 71, 1042–1050.

Shieh, Y.-S.C., Baric, R.S., Sobsey, M.D., 1997. Detection of low levels of enteric viruses in metropolitan and airplane sewage. Appl. Environ. Microbiol. 63, 4401–4407.

Simmonds, P., Adams, M.J., Benko, M., Breitbart, M., Brister, J.R., Carstens, E.B., Davison, A.J., Delwart, E., Gorbalenya, A.E., Harrach, B., Hull, R., King, A.M.Q., Koonin, E.V., Krupovic, M., Kuhn, J.H., Lefkowitz, E.J., Nibert, M.L., Orton, R., Roossinck, M.J., Sabanadzovic, S., Sullivan, M.B., Suttle, C.A., Tesh, R.B., van der Vlugt, R.A., Varsani, A., Zerbini, F.M., 2017. Consensus statement: virus taxonomy in the age of metagenomics. Nat. Rev. Microbiol. 15 (3), 161–168.

Simmons, F.J., Xagoraraki, I., 2011. Release of infectious human enteric viruses by full-scale wastewater utilities. Water Res. 45, 3590–3598.

Skraber, S., Gassilloud, B., Schwartzbrod, L., Gantzer, C., 2004. Survival of infectious poliovirus-1 in river water compared to the persistence of somatic coliphages, thermotolerant coliforms and poliovirus-1 genome. Water Res. 38, 2927–2933.

Skvortsov, T., De Leeuwe, C., Quinn, J.P., McGrath, J.W., Allen, C.C.R., McElarney, Y., Watson, C., Arkhipova, K., Lavigne, R., Kulakov, L.A., 2016. Metagenomic characterisation of the viral community of lough neagh, the largest freshwater lake in Ireland. PLoS One 11, e0150361.

Sobsey, M.D., Battigelli, D.A., Shin, G.-A., Newland, S., 1998. RT-PCR amplification detects inactivated viruses in water and wastewater. Water Sci. Technol. 38 (12), 91–94.

Sohn, J., Nam, J.-W., 2016. The present and future of de novo whole-genome assembly. Brief. Bioinform. bbw096

Song, K., Mohseni, M., Taghipour, F., 2016. Application of ultraviolet light-emitting diodes (UV-LEDs) for water disinfection: a review. Water Res. 94, 341–349.

Soule, H., Genoulaz, O., Gratacap-Cavallier, B., Chevallier, P., Liu, J.X., Seigneurin, J.M., 2000. Ultrafiltration and reverse transcription-polymerase chain reaction: an efficient process for poliovirus, rotavirus and hepatitis a virus detection in water. Water Res. 34, 1063–1067.

Staggemeier, R., Bortoluzzi, M., da Silva Heck, T.M., da Luz, R.B., Fabres, R.B., Soliman, C.M., Rigotto, C., Baldasso, N.A., Spilki, F.R., de Matos Almeida, S.E., 2015. Animal and human enteric viruses in water and sediment samples from dairy farms. Agric. Water Manag. 152, 135–141.

Stanghellini, M.E., Rasmussen, S.L., 1994. Hydroponics: a solution for zoosporic pathogens. Plant Dis. 78, 1129–1137.
Steinmann, J., 1981. Detection of rotavirus in sewage. Appl. Environ. Microbiol. 41, 1043–1045.
Stewart-Wade, S.M., 2011. Plant pathogens in recycled irrigation water in commercial plant nurseries and greenhouses: their detection and management. Irrig. Sci. 29 (4), 267–297.
Steyer, A., Gutiérrez-Aguirre, I., Rački, N., Beigot Glaser, S., Brajer Humar, B., Stražar, M., Škrjanc, I., Poljšak-Prijatelj, M., Ravnikar, M., Rupnik, M., 2015. The Detection Rate of Enteric Viruses and Clostridium difficile in a Waste Water Treatment Plant Effluent. Food Environ. Virol. 7, 164–172.
Suttle, C.A., 2007. Marine viruses-major players in the global ecosystem. Nat. Rev. Microbiol. 5, 801–812.
Suttle, C.A., 2016. Environmental microbiology: viral diversity on the global stage. Nat. Microbiol. 1, 16205.
Tamaki, H., Zhang, R., Angly, F.E., Nakamura, S., Hong, P.Y., Yasunaga, T., Kamagata, Y., Liu, W.T., 2012. Metagenomic analysis of DNA viruses in a wastewater treatment plant in tropical climate. Environ. Microbiol. 14, 441–452.
Tani, N., Shimamoto, K., Ichimura, K., Nishi, Y., Tomita, S., Oda, Y., 1992. Enteric virus levels in river water. Water Res. 26 (1), 45–48.
Tani, N., Dohi, Y., Jurumatani, N., Yonemasu, K., 1995. Seasonal distribution of adenoviruses, enteroviruses and reoviruses in urban river water. Microbiol. Immunol. 39, 577–580.
Teakle, D.S., Morris, T.J., 1981. Transmission of Southern bean mosaic virus from soil to bean seeds. Plant Dis. 65, 599–600.
Thebo, A.L., Drechsel, P., Lambin, E.F., Nelson, K., 2017. A global, spatially-explicit assessment of irrigated croplands influenced by urban wastewater flows. Environ. Res. Lett. 12, 074008.
Thurston-Enriquez, J.A., Haas, C.N., Jacangelo, J., Riley, K., Gerba, C.P., 2003. Inactivation of feline calicivirus and adenovirus type 40 by UV radiation. Appl. Environ. Microbiol. 69, 577–582.
Tomlinson, J.A., Faithfull, E.M., 1980. Studies on the control of lettuce big-vein disease in recirculated nutrient solutions. Acta Hortic. 98, 325–332.
Tomlinson, J.A., Faithfull, E.M., 1984. Studies on the occurrence of tomato bushy stunt virus in english rivers. Ann. Appl. Biol. 104, 485–495.
Tomlinson, J.A., Thomas, B.J., 1986. Studies on melon necrotic spot virus disease of cucumber and on the control of the fungus vector (*Olpidium radicale*). Ann. Appl. Biol. 108 (1), 71–80.
Tomlinson, J.A., Faithfull, E.M., Flewett, T.H., Beards, G., 1982. Isolation of infective tomato bushy stunt virus after passage through the human alimentary tract. Nature 300, 637–638.
Tomlinson, J.A., Faithfull, E.M., Webb, M.J.W., Fraser, R.S.S., Seeley, N.D., 1983. Chenopodium necrosis: a distinctive strain of *Tobacco necrosis virus* isolated from river water. Ann. Appl. Biol. 102, 135–147.
Tošić, M., Tošić, D., 1984. Occurrence of tobacco mosaic virus in water of the danube and sava rivers. J. Phytopathol. 110, 200–202.
Toze, S., 2006. Reuse of effluent water—benefits and risks. Agric. Water Manag. 80, 147–159.
Tsai, Y.-L., Tran, B., Sangermano, L.R., Palmer, C.J., 1994. Detection of poliovirus, hepatitis a virus, and rotavirus from sewage and ocean water by triplex reverse transcriptase PCR. Appl. Environ. Microbiol. 60, 2400–2407.
Tsai, Y.L., Tran, B., Palmer, C.J., 1995. Analysis of viral RNA persistence in seawater by reverse transcriptase-PCR. Appl. Environ. Microbiol. 61, 363–366.

Van der Vlugt, R.A.A., 2009. Pepino mosaic virus. Hellenic Plant Prot. J. 2, 47–56.
Vasickova, P., Pavlik, P., Verani, M., Carducci, A., 2010. Issues concerning survival of viruses on surfaces. Food Environ. Virol. 2, 24–34.
Wetz, J.J., Lipp, E.K., Griffin, D.W., Lukasik, J., Wait, D., Sobsey, M.D., Scott, T.M., Rose, J.B., 2004. Presence, infectivity, and stability of enteric viruses in seawater: relationship to marine water quality in the Florida keys. Mar. Pollut. Bull. 48, 698–704.
Wigginton, K.R., Kohn, T., 2012. Virus disinfection mechanisms: the role of virus composition, structure, and function. Curr. Opin. Virol. 2, 84–89.
Wommack, K.E., Bhavsar, J., Polson, S.W., Chen, J., Dumas, M., Srinivasiah, S., Furman, M., Jamindar, S., Nasko, D.J., 2012. VIROME: a standard operating procedure for analysis of viral metagenome sequences. Stand. Genomic Sci. 6, 427–439.
Wood, D.E., Salzberg, S.L., 2014. Kraken: ultrafast metagenomic sequence classification using exact alignments. Genome Biol. 15, R46.
Xagoraraki, I., Kuo, D.H., Wong, K., Wong, M., Rose, J.B., 2007. Occurrence of human adenoviruses at two recreational beaches of the great lakes. Appl. Environ. Microbiol. 73 (24), 7874–7881.
Xagoraraki, I., Yin, Z., Svambayev, Z., 2014. Fate of viruses in water systems. J. Environ. Eng. 04014020, 1–18.
Yarwood, C.E., 1960. Release and preservation of virus by roots. Phytopathology 50, 111–114.
Yates, M.V., Gerba, C.P., Kelley, L.M., 1985. Virus persistence in groundwater. Appl. Environ. Microbiol. 49 (4), 778–781.
Yi, L., Lesemann, D.E., Koenig, R., Rüdel, M., Pfeilstetter, E., 1992. Isometric plant viruses in ditches and streams in agricultural areas: recovery of previously found viruses and identification of hitherto unrecorded carmo- and tombusviruses including grapevine Algerian latent virus. J. Phytopathol. 134, 121–132.
Zhang, T., Breitbart, M., Lee, W.H., Run, J.Q., Wei, C.L., Soh, S.W.L., Hibberd, M.L., Liu, E.T., Rohwer, F., Ruan, Y., 2006. RNA viral community in human feces: prevalence of plant pathogenic viruses. PLoS Biol. 4 (1), 108–118. e3.
Zhang, C.M., Xu, L.M., Xu, P.C., Wang, X.C., 2016. Elimination of viruses from domestic wastewater: requirements and technologies. World J. Microbiol. Biotechnol. 32, 69.
Zhao, G., Wu, G., Lim, E.S., Droit, L., Krishnamurthy, S., Barouch, D.H., Virgin, H.W., Wang, D., 2017. VirusSeeker, a computational pipeline for virus discovery and virome composition analysis. Virology 503, 21–30.

FURTHER READING

Hodgson, K.R., Torok, V.A., Turnbull, A.R., 2017. Bacteriophages as enteric viral indicators in bivalve mollusc management. Food Microbiol. 65, 284–293.

CHAPTER FIVE

Evaluating the Importance of Environmental Persistence for *Ranavirus* Transmission and Epidemiology

Jesse L. Brunner[1], Christian M. Yarber
School of Biological Sciences, Washington State University, Pullman, WA, United States
[1]Corresponding author: e-mail address: jesse.brunner@wsu.edu

Contents

Abstract

Viruses persist outside their hosts in a variety of forms, from naked virions to virus protected in sloughed tissues or carcasses, and for a range of times, all of which affect the likelihood and importance of transmission from the environment. This review synthesizes the literature on environmental persistence of viruses in the genus *Ranavirus* (family *Iridoviridae*), which are large double-stranded DNA viruses of ectothermic, often aquatic or semiaquatic vertebrates. Ranaviruses have been associated with mass mortality events in natural and captive settings around the world, and with population and community-wide declines in Europe. Early work suggested ranaviruses are environmentally robust and transmission from the environment should be common. More recent work has shown a large effect of temperature and microbial action on persistence times, although other aspects of the environment (e.g., water chemistry) and aquatic communities (e.g., zooplankton) may also be important. Ranaviruses may persist in the carcasses of animals that have died of infection, and so decomposing organisms and invertebrate scavengers may reduce these persistence times. The question is, do persistence times vary enough to promote or preclude substantial transmission from the environment. We built an epidemiological model with transmission from contacts,

Advances in Virus Research, Volume 101
ISSN 0065-3527
https://doi.org/10.1016/bs.aivir.2018.02.005

free virus in water, and carcasses, to explore the conditions in which environmental persistence could be important for ranavirus epidemiology. Based on prior work, we expected a substantial amount of transmission from the water and that longer persistence times would make this route of transmission dominant. However, neither water-borne nor transmission from carcasses played an important role in the simulated epidemics except under fairly restrictive conditions, such as when there were high rates of virus shedding or high rates of scavenging on highly infectious carcasses. While many aspects of environmental persistence of ranaviruses are being resolved by experiments, key parameters such as viral shedding rates are virtually unknown and will need to be empirically constrained if we are to determine whether environmental persistence and transmission from the environment are essential or insignificant features of *Ranavirus* epidemiology. We conclude by emphasizing the need to place environmental persistence research in an epidemiological framework.

1. INTRODUCTION

Viruses of aquatic vertebrates are known primarily from their impacts on host populations. The earliest known animal virus, lymphocystis disease virus, for instance, was initially studied for the wart-like disease it causes in fish (Weissenberg, 1965). The epizootic hematopoietic necrosis virus (EHNV), one of the first viruses of aquatic vertebrates listed by the World Organization for Animal Health (OIE), was of particular concern because of its capacity to cause massive fish die-offs (Langdon and Humphrey, 1987; Langdon et al., 1986). Both of these viruses are members of the *Alphairidovirinae* subfamily of the *Iridoviridae* family—large, icosahedral viruses with double-stranded DNA genomes—which includes three genera, *Lymphocystivirus* and *Megalocytivirus*, which infect fish, and *Ranavirus*, which infects fish, amphibians, and reptiles (Chinchar et al., 2017). Ranaviruses in particular have received a great deal of attention for causing die-offs of tens to tens of thousands of fish and aquatic or semiaquatic amphibians and turtles around the world (Duffus et al., 2015). The recent emergence of a ranavirus in Spain has led to declines in amphibian communities (Price et al., 2014). Ranaviruses also continue to threaten aquaculture and fish hatcheries (e.g., Huang et al., 2011; Kayansamruaj et al., 2017; Waltzek et al., 2014).

Beyond an impressively broad host range, experiments have shown that ranaviruses can be transmitted through several routes, including via direct contact, consumption of infectious tissues (cannibalism and necrophagy), and adsorption of virions from the water (reviewed in Brunner et al., 2015). The host's gastrointestinal tract is probably the most common site of entry (Robert et al., 2011). Dose–response studies in fish, salamanders, and frogs (e.g., Brunner et al., 2005; Forzan et al., 2015; Plumb and Zilberg, 1999b) are reasonably fit by a simple model assuming independent action of each virion (J.L. Brunner, unpublished analysis), which suggests a simple per virion hazard of infection (Ben-Ami et al., 2010; Regoes et al., 2002; van der Werf et al., 2011). Different routes of transmission are therefore more or less effective because they deliver more or fewer virions. What is interesting about this perspective is that environmental factors can influence transmission, perhaps strongly, by promoting or limiting the accumulation of infectious viral particles in the environment. If this is the case, it might help explain the sporadic occurrence of *Ranavirus* epidemics and die-offs in endemic regions (e.g., O'Connor et al., 2016; Whittington et al., 2010, #42952). It might also afford some control over the epidemiology of these emerging viruses; by preventing ranaviruses from persisting in the environment, one could perhaps prevent epidemics from occurring.

2. PERSISTENCE ON A CONTINUUM: A FRAMEWORK

We find it useful to think of environmental persistence of viruses along a continuum. Most viruses persist outside a living host for a time, but that time may be measured in hours (e.g., Ebola virus in sterilized waste water; Bibby et al., 2015) or days (Human Adenovirus 2 in groundwater; Ogorzaly et al., 2010) or months (avian influenza in distilled water; Brown et al., 2007). Longer persistence times facilitate transmission from the environment (Kotwal and Cannon, 2014), whereas shorter times restrict it, often to a small enough role that it can be effectively ignored. And so, we posit, for viruses such as ranaviruses that infect aquatic animals that the relative importance of environmental transmission is in large part a function of environmental persistence times. Persistence times, in turn, are a function of the environment in which the virus finds itself and the adaptations it may have to persist in that environment. The form of free virus also spans a continuum from naked virus particles to large quantities of virus within newly deceased hosts (Fig. 1). Naked particles are clearly more vulnerable to the chemical and physical environment, extracellular enzymes secreted by microbes, and filter feeding organisms (e.g., zooplankton) while viruses contained in carcasses are somewhat protected from these elements but may be more vulnerable to scavenging or simply being lost from the water column. Propagules embedded in or on sloughed tissues (e.g., skin, stomach lining) or feces from an infected host likely experience intermediate protection.

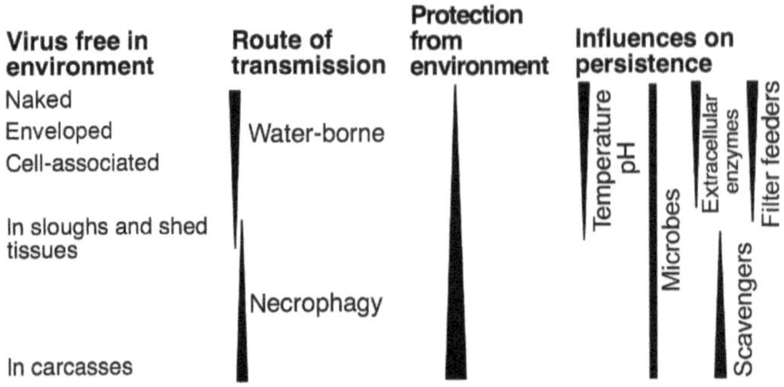

Fig. 1 Viruses such as *Ranavirus* spp. can be free in the environment in a variety of forms, from naked virions to virions held within carcasses of infected hosts. The form has consequences for the route by which this free virus can infect susceptible hosts as well as how it persists in the environment. Virions in carcasses, for instance, are largely protected from the physical environment (e.g., water pH, extracellular enzymes), but may be consumed by invertebrate scavengers or necrophagous hosts.

We can also map onto this continuum the route by which propagules are transmitted from the environment to naïve hosts. Because they are small and easily dispersed by Brownian motion or currents, viruses shed directly into water, or associated with small bits of sloughed tissues and fomites (e.g., virus adhered to floating particulate matter), contribute to water-borne transmission. Greater rates of shedding and longer environmental persistence times allow large amounts of virus to accumulate in the water and thus likely promote water-borne transmission. As the sloughs get larger, up to and including infectious carcasses, we can no longer assume that these sources of infection are well mixed in the water column, but rather that susceptible hosts encounter virus in these forms by consumption. Many animals, even putative herbivores like anuran tadpoles, engage in a degree of scavenging, which might lead to consuming infectious materials in the environment as they feed (Altig et al., 2007; Getz, 2011; Wilson and Wolkovich, 2011). At the extreme, some animals, including many fish and amphibians, are necrophagous, seeking out and consuming conspecific carcasses, which can be a very efficient means of transferring large doses of virus (e.g., Diamant, 1997). High viral titers in carcasses, longer persistence times of the carcasses, and high rates of scavenging or necrophagy, and especially sharing of carcasses (Rudolf and Antonovics, 2007), all promote transmission by necrophagy.

While these general patterns seem intuitive, the actual importance of environmental persistence for virus transmission in nonhuman systems is rarely considered quantitatively, especially across the spectrum of forms of free virus and routes of transmission. Moreover, outside of sporadic studies in groundwater or wastewater (e.g., Bibby et al., 2015; Brown et al., 2007; Ogorzaly et al., 2010), most research on persistence of animal viruses focuses on persistence in hospital settings (e.g., Cook et al., 2015) or on hands or food surfaces (e.g., Kotwal and Cannon, 2014) or other human dominated settings. As such, there is a need to better understand the persistence of nonhuman animal viruses in nature. Here, we explore the factors that affect the persistence of ranaviruses and try to relate persistence times directly to ranavirus transmission. Note that we largely ellipse over environmental persistence between epidemics, although though this is an important issue deserving future attention.

3. ENVIRONMENTAL INFLUENCES ON PERSISTENCE TIMES

Until recently, most of what we knew about ranaviral persistence came from anecdotes or small studies with ranavirus grown in culture held under various laboratory conditions (e.g., Jancovich et al., 1997; Langdon, 1989; Speare and Smith, 1992). The Bohle Iridovirus, for instance, can survive 6 h at 25°C and 37°C in culture medium, but is rapidly inactivated at 56°C and titers are dramatically reduced at pH 2, 3, 10, and 11 (Speare and Smith, 1992). EHNV maintained its infectiousness for at least 97 days in sterile water at 15°C and at least 300 days when held at 4°C in tissue culture (Langdon, 1989). While these results suggest that ranaviruses are inherently robust, it is difficult to extrapolate these results from essentially aseptic conditions to natural environments where microbes and other organisms actively or inadvertently inactivate virus particles (Ward et al., 1986).

A series of studies beginning with Reinauer et al. (2005) and Nazir et al. (2012) brought much needed ecological realism to our understanding of persistence of ranaviruses in the environment. These first two studies used "germ carriers," comprised of virus in tissue culture adsorbed to a filter and sealed within a polycarbonate membrane "sandwich" with a 10-nm pore size, much smaller than most bacteria (Stevik et al., 2004; Young, 2006). This apparatus allows extracellular enzymes to pass through and act on the virus, but prevents contamination of the culture by other microbes. These germ carriers were then placed in soil or in drinking water or pond water that had been filter sterilized or left unmanipulated. If we assume that drinking water lacks appreciable microbes and that filter-sterilizing leaves the extracellular enzymes initially present, these studies collectively provide estimates of ranavirus persistence under three conditions: (1) without any microbial action; (2) with the extracellular enzymes initially present, but no further action; and (3) with extracellular enzymes continually produced by microbes. However, to understand the role of direct microbial action (e.g., adsorption or endocytosis) on *Ranavirus* persistence, we must look to studies that spiked ranavirus in culture directly into water and then attempted to reisolate and titer the virus (Brunner *unpublished data*, referenced in Brunner et al., 2007; Reinauer, 2004; Johnson and Brunner, 2014; Munro et al., 2016). [Note that Johnson and Brunner's (2014) results are primarily based on quantities of viral DNA rather than live virus titers as viral titers rapidly fell below the limits of detection.] Importantly, there is little if any evidence that ranavirus isolates, even those from amphibians and reptiles, vary in their ability to persist in the environment (Nazir et al., 2012). It is possible that certain strains have been subject to selection for greater environmental resistance, but measurable differences have not been found. We therefore include studies with both amphibian and reptile ranavirus isolates in our synthesis. (Unfortunately, persistence of fish ranaviruses has not been studied quantitatively.) These studies have found a great deal of variation in persistence times, measured as the time it takes for viral titers to be reduced by 90% (T90), but several patterns emerge when the evidence is considered collectively (Fig. 2).

First, soil and pond substrate may reduce persistence. Ranaviruses appear to degrade more rapidly in soil or pond substrate than in water (Fig. 2), especially at warmer temperatures (discussed later). While ranavirus may persist in capsules in the soil or in sterilized pond substrate, as in the experiment by Brunner et al. (2007), retaining its ability to infect 86% of naïve salamander larvae placed in it, when *Ranavirus* is dried in the substrate and then rehydrated, it loses its activity entirely (0% infection; Brunner et al., 2007). This stands in sharp contrast to the long persistence times found when virus was dried on plastic or metal (Langdon, 1989; Nazir et al., 2012) and suggests that soil or sediments may act on the virus directly. Indeed, Munro et al. (2016) found that adding sediments, even autoclave sterilized sediment, to water increased rates of degradation (Fig. 2), or at least reduced viral recovery—perhaps irreversibly binding to it or somehow chemically inactivating it. Viruses can rapidly adsorb to soil particles, at a rate dependent on soil properties, pH, water content, etc. (Sobsey et al., 1980). In any case, it appears that ranavirus is unlikely to persist long in the soil or sediment under warm conditions or when dried but may be able to persist throughout a winter if kept moist and cool. Whether persistence in or on soil plays much if any role in ranavirus transmission within epidemics remains unknown.

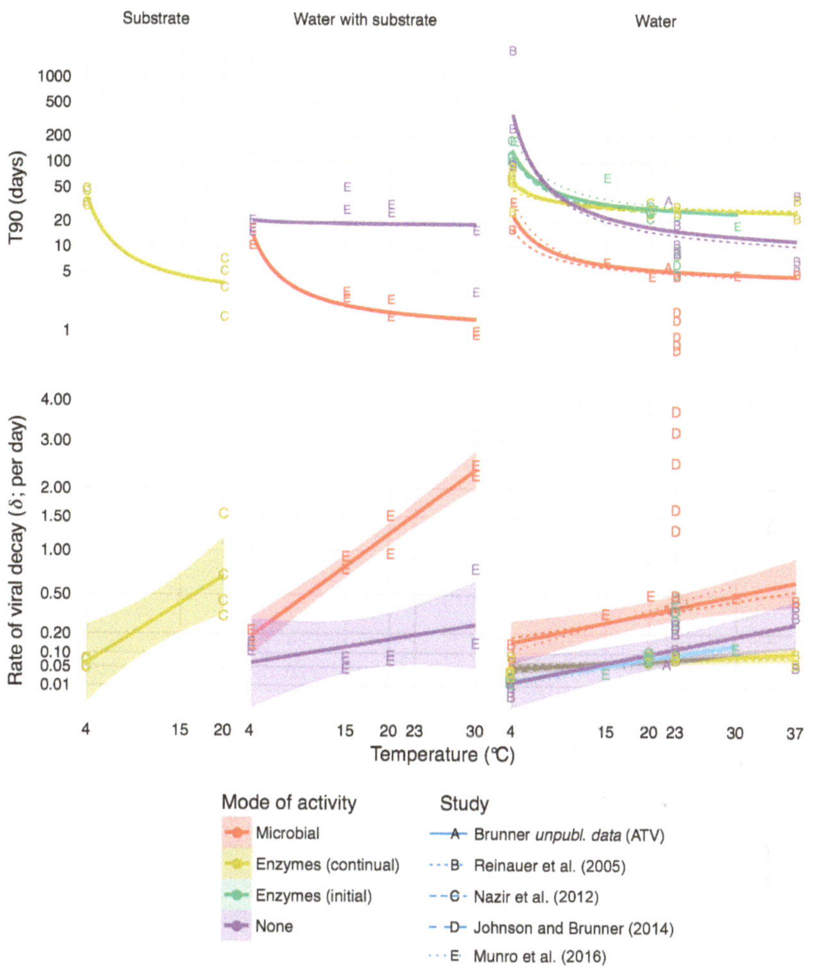

Fig. 2 Estimates of the time it takes for viral titers to be reduced by 90% (T90; *top*) or, conversely, the rate of viral decay [$\delta = \text{T90}^{-1}/\log_{10}e$] as a function of temperature in soil, water with substrate, or water. We classify treatments by the mode of biotic action they allow, from no action (e.g., autoclaved pond water or tap water) to extracellular enzymes initially in the water (when water is filter-sterilized) or continually excreted (when water is left unmanipulated) to allowing direct interaction with microbes. *Lines* are hyperbolas ($y = b_0 + b_1/°\text{C}$) fit to the T90 times and linear fit to the decay rates by study (*thin dashed* or *dotted lines*) or to all studies together (*thick solid lines*), excluding Johnson and Brunner (2014), which focused on viral DNA rather than infectious virus.

Second, other microbes, presumably bacteria and fungi, appear to play a critical role in the inactivation of ranavirus, at least in water (Fig. 2). In particular, when additional microbes are allowed to directly interact with ranavirus, persistence times are four to five times shorter than when the only mode of action is extracellular enzymes (Fig. 2). Surprisingly, while bacterial proteolytic enzymes may be the primary means of viral inactivation in water (Ward et al., 1986), there were only minor differences between treatments that exposed ranavirus to the action of extracellular enzymes and those that used tap water (Reinauer, 2004) or autoclaved water (Brunner, unpublished data; Fig. 2). Johnson and Brunner (2014) demonstrated a similar role for zooplankton. Concentrations of viral DNA and titers of active ranavirus declined rapidly in spring when *Daphnia pulex* were present (T90 < 1 day) relative to virtually no degradation in spring water alone. Moreover, viral titers on the *Daphnia* themselves declined during the experiment, suggesting these filter-feeding cladocerans do not simply concentrate active virus, which might infect their predators, but rather inactivate the ranavirus they filter. Collectively, these results suggest that aquatic microbes, which interact directly with virions, and the zooplankton that filter them, can play an important role in the inactivation of ranaviruses in freshwater. The degree to which aquatic community composition and abundance influence ranavirus persistence remains an open question.

Lastly, similar to other animal viruses in water (e.g., avian influenza virus; Brown et al., 2007, 2009), there is a clear pattern of increasing rates of degradation of *Ranavirus*, and thus shorter persistence times, with increasing temperatures across substrates and experimental manipulations (Fig. 2). T90 times in soil were on the order of several days under summer conditions (≥20°C) relative to T90 times >30 days in the cool spring (Nazir et al., 2012; Reinauer et al., 2005). Persistence times in water vary by an order of magnitude across an ecologically relevant range, from several weeks or more at 4°, which is a common temperature in small bodies of water in the winter or early spring, to less than a week at temperatures ≥20°C (Fig. 2), which are commonly achieved in the summer. Larger water bodies can also have dramatic temperature clines from deep to shallow waters and so we would expect persistence times to vary a great deal within a single habitat.

The effects of other aspects of the physical aquatic environments (e.g., pH, turbidity) on ranavirus persistence have not been well explored, but are likely to be important (Brown et al., 2009). Indeed, viral persistence times varied dramatically among ponds in Johnson and Brunner's (2014) study, even in filtered water under common garden conditions. A key question is whether and how much this variation in viral persistence times in the environment matter to ranavirus transmission and dynamics. We return to this question later, but first we consider rates of degradation in host tissues.

4. PERSISTENCE IN TISSUES AND CARCASSES

One critique of the studies presented earlier is that they all used virus grown in cell culture, which includes nonenveloped, enveloped, and cell-associated ranavirus particles, but not virus shed from host animals or shed tissues (e.g., sloughed skin or intestinal lining). Virus particles inside of host tissues or sloughs are likely protected to a degree from the action of microbes, plankton, and water chemistry and so the estimates of persistence times in prior experiments (Fig. 2) may underestimate the persistence of naturally shed ranavirus. The production of sloughs can be considerable during late stages of ranaviral disease, and their consumption, intentional or not, can be a very effective route of transmission (Brunner et al., 2007; Jancovich et al., 1997). Extending this reasoning to carcasses of hosts killed by ranavirus infections, we see that the route of transmission and the mode of virus inactivation are likely to change when virus occurs in host tissues rather than free floating in the water.

Infectious carcasses can contain very high viral titers (e.g., Brunner and Collins, 2009), which makes them very capable of infecting scavenging or necrophagous hosts, such as conspecifics or other ectothermic vertebrates in the habitat (Ariel and Owens, 1997; Brunner et al., 2007; Harp and Petranka, 2006; Pearman et al., 2004). Scavenging or inadvertent necrophagy is a common feeding pattern, even among putative herbivores like tadpoles (Altig et al., 2007; Getz, 2011; Wilson and Wolkovich, 2011) and so this route of transmission may be more important than typically considered, at least for pathogens like ranaviruses that remain infectious in dead hosts.

Carcasses are removed from ecosystems by the detrital or decomposer community, through the combined actions of obligate and facultative vertebrate and invertebrate scavengers and microbes (Parmenter and Lamarra, 1991; Regester and Whiles, 2006; Stevenson and Childers, 2004). Rates at which hosts decompose vary substantially depending on the species considered and the environment in which rates are tested. Decomposition rates of freshwater fish biomass vary from T90 times of less than 2 days (mosquito fish in a Florida marsh; Stevenson and Childers, 2004) to over 2 months (bluegill in a Wisconsin lake; Kitchell et al., 1975; reviewed in Regester and Whiles, 2006). Decomposition rates of amphibians are on the shorter end of this range. Regester and Whiles (2006), for instance, estimated that spotted salamander (*Ambystoma maculatum*) larvae have a T90 time of 3 days. Their and other studies enclosed the carcasses in water in wire mesh of varying sizes, which likely excluded the action of larger scavengers and so may underestimate rates of decay (Regester and Whiles, 2006). Le Sage et al. (in review), for instance, found that a single diving beetle (Dytiscidae) larva or dragonfly (Aeschnidae) naiad in aquaria consumed 84% or 62%, respectively, of the mass of a larval *Ambystoma macrodactylum* carcass within 2 days, and larval Notonectidae and Gammaridae also consumed some carcass mass. The entire community of scavenging invertebrate animals in addition to microbes plays an important role in removing carcasses (Le Sage et al., in review; Parmenter and Lamarra, 1991; Regester and Whiles, 2006; Stevenson and Childers, 2004). Whether carcass removal plays an important role in ranavirus transmission, however, has been unexplored until very recently.

To test whether scavengers might play an important role in *Ranavirus* transmission, Le Sage et al. (in review) conducted an experiment in the laboratory in which larval salamanders, *A. macrodactylum*, either had access to the carcass of a conspecific that had died of ranavirus infection or was prevented from contacting (and consuming) the carcass by a mesh barrier. As expected, most (68%) of larvae that could contact and feed on the carcass for 24 h became infected while only 16% of larvae were infected when this contact was prevented. When the carcass was first scavenged by a dytiscid (diving beetle) larva for 24 h before the salamander larvae were introduced, only 23% of salamander larvae became infected. In other words, the scavenging beetle effectively reduced transmission to levels equivalent to those observed when transmission could only occur indirectly, by virions shed into the water. Moreover, scavenging did not appear to spread infectious particles into the water as had been initially hypothesized (a scavenged carcass behind a screen infected 20% of the larvae). Le Sage et al. (in review) also estimated the relative densities of salamander larvae and dytiscid larvae, and the functional response of dytiscid larvae to carcasses, and found that over most of the larval period, the scavengers had the capacity to consume carcasses from the entire salamander population, were all individuals to die at once. While this is only one study, it does suggest that invertebrate scavengers have the potential to minimize or even prevent transmission from carcasses and so dramatically reduce ranavirus transmission within amphibian breeding ponds.

5. AN EPIDEMIC MODEL TO EVALUATE THE IMPORTANCE OF PERSISTENCE TIMES

While there is still uncertainty about the physical, chemical, and biotic factors that control ranaviruses persistence in various environments, it is clear that persistence times vary considerably, often by orders of magnitude. Perhaps, the most pressing question is thus whether this variation in persistence times is important for ranavirus transmission and epidemiology. In other words, do long persistence times increase ranavirus transmission through water or from carcasses, relative to short persistence times? We next develop an epidemiological model to help address this question.

First, let us assume for simplicity a single large, well-mixed population of a single host species in a pond or lake with no birth and no natural deaths (i.e., deaths due to factors besides disease). With these assumptions, susceptible hosts are lost only to infections, and so we focus our attention on the change in density of infected hosts:

$$\frac{dI}{dt} = Sb\left(\kappa\eta\frac{I}{N} + \mu\lambda_{\text{h}}C + \omega V\right) - \alpha I.$$

The variables *S*, *I*, and *V* are the densities of susceptible and infected hosts, and ranavirus in the water, respectively. Susceptible hosts become infected by exposure to virions via (in order of the terms in parentheses) (i) direct casual contact with infected hosts; (ii) consumption of infectious carcasses (at density *C* in the pond); or (iii) absorption or ingestion of virus in the water. Susceptible hosts (*S*) have a per virion hazard of infection, *b*, and so this formulation makes clear that these three routes of transmission differ only in the concentration of virions they provide to susceptible hosts per unit time. Direct casual contact with an infected individual delivers κ virions per contact, and we assume this route of transmission is frequency dependent, i.e., contacts occur at a constant rate, η, only I/N of which are with infected hosts (Brunner et al., 2017). Infectious carcasses house μ virions, which is presumably much higher than κ, and carcasses are consumed at some rate, λ_{h}. Exposure to virus particles in the water occurs as hosts move through ω volume of water, which has virus concentration, *V*. Lastly, infectious hosts (*I*) die from infection at rate α and enter the "carcass" class, *C*.

The density of infectious carcasses can be modeled as,

$$\frac{dC}{dt} = \alpha I - [\lambda_{\text{h}}(S + I) + \lambda_{\text{d}}D]C,$$

which includes the gain from disease-induced mortality (αI) and the losses through consumption by hosts (the first term in the brackets) and by microbial decomposition and scavenging by the decomposer community (the second term in the brackets), which we assume exists at a constant density, *D*. While scavenging rates by hosts (λ_{h}) and by invertebrates (λ_{d}) are likely saturating functions of carcass density (a type II functional response; Le Sage et al., in review), for our current purposes we assume that these are simply constant, linear functions. In any case, this equation emphasizes a sort of competition for carcasses between necrophagous hosts (both infected and uninfected) and nonhost decomposers or scavengers. If carcasses decompose or are scavenged inefficiently, this allows carcasses to be eaten by and infect naïve hosts.

Lastly, we model the change in the density of ranavirus in the water as

$$\frac{dV}{dt} = \rho(I + C) - \delta V,$$

where both infected hosts and carcasses shed ranavirus at some rate, ρ, which we assume is the same for simplicity—there are few data on shedding rates of live hosts, let alone carcasses, though they appear similarly infectious (Brunner et al., 2007). These virions in the water are then lost at a constant rate, δ (Fig. 2). Again, if the virus is degraded slowly relative to the rate of shedding, virus will accumulate in the water and lead to more infections.

The dynamics of this model (with no recovery of susceptible hosts or introduction of new susceptibles, and no resistant hosts), are such that if the infection spreads, it infects the entire population. Therefore, while we numerically solved the equations of this model using the simecol package in R (Petzoldt and Rinke, 2007) to determine the density of hosts in each class through time, we present these data as the *proportion* of the population infected at the end of the epidemic by each route of transmission (code available at https://github.com/JesseBrunner/RV_Persistence). It is important to note that this is a simple model that assumes that all individuals within a class are identical (e.g., no stage-specific susceptibility or tolerance; Warne et al., 2011); that infections become immediately transmissible (no latent period or changes in shedding through time; Brunner et al., 2007); that infected hosts do not recover to an immune or still susceptible class (Brunner et al., 2004); and that the exposure dose does not affect the timing of mortality (Brunner et al., 2005). All these assumptions are empirically suspect and relaxing them may change some of the resulting dynamics. However, even this relatively simple model can shed light on the importance of environmental persistence for ranavirus transmission.

The principle result of this modeling exercise is that the two routes of environmental transmission (water-borne and carcass consumption) of ranaviruses become important only under a narrow set of conditions. Water-borne transmission does not play a substantial role (i.e., comprise >0.2 of overall transmission) until shedding rates are high (>10^5 virions per day) and rates of viral decay are low (Fig. 3). There is currently a great

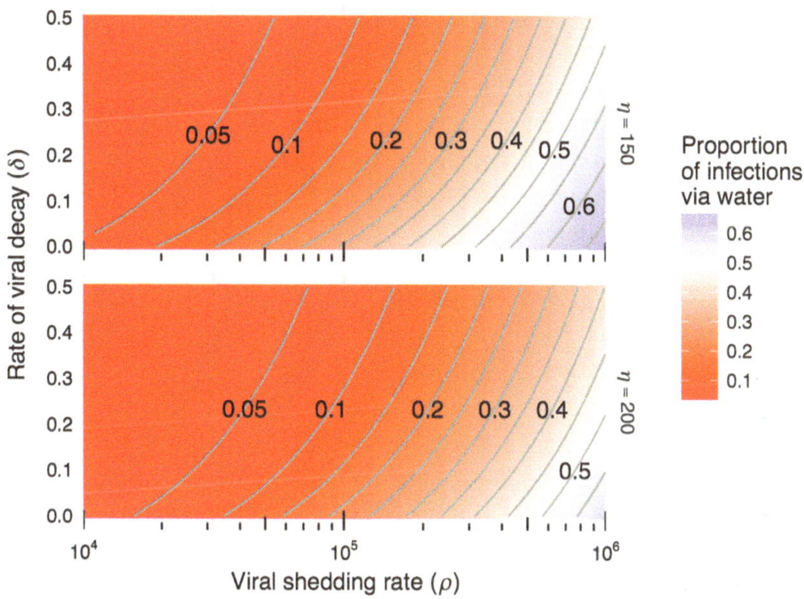

Fig. 3 The proportion of infections caused by water-borne transmission as a function of viral shedding rate (ρ; virions per day per infected host or carcass) and the rate of viral decay (δ; per day, as in Fig. 2). The *top panel* is for $\eta = 150$ contacts per host per day, which produces slow epidemics by direct contact, and the *bottom* is for $\eta = 200$, which produces epidemics that peak at about 50 days. The baseline conditions for the other parameters are $b = 2.5 \times 10^{-5}$, which corresponds to an ID_{50} of $\sim 10^{4.44}$ pfu in a 1-day exposure; $\kappa = 50$ virions are transferred per casual contact; infected hosts die at a rate of $\alpha = 1/18$ per day; carcasses are removed at a rate of $\lambda_d D = 0.7$ per day by invertebrate scavengers and microbes, and a rate of $\lambda_h = 0.25$ per day per host, but hosts do not become infected as carcasses had $\mu = 0$ virions. Hosts sweep through $\omega = 1$L day ($= 10^4$ cm long path $\times$ 0.1 cm^2 cross section); this parameter was not manipulated as it enters the model as a product with ρ.

deal of uncertainty in shedding rates and there may well be order-of-magnitude variation among individuals of a species (Brunner et al., in revision). Thus, understanding whether water-borne transmission is important or inconsequential requires additional investigation focused on understanding shedding rates, but it may be highly context dependent. In any case, it appears that at a given shedding rate (ρ), the rate of viral decay (δ) can strongly effect the relative proportion of transmission from the water (e.g., from >0.3 with $\delta \approx 0$ to <0.1 when $\delta \approx 0.5$ with $\rho = 10^5$ and $\eta = 150$; Fig. 3). To achieve such a range of outcomes in nature, however, would require large changes in δ, on the order of seasonal changes in temperature in addition to changes in microbial activity, pH, or similar environmental impacts (e.g., lake turnover). More moderate differences in viral decay should lead to relatively smaller impacts on the rate of water-borne transmission.

This simple model also suggests that transmission via necrophagy by hosts is largely unimportant over a wide range of parameters, and hence rates of invertebrate scavenging and decomposition of carcasses, which limit opportunities for host necrophagy, also may play only minor roles in transmission dynamics. Indeed, only with high virulence ($\alpha = 1/9$), large amounts of virus transferred by host necrophagy ($\lambda_h \mu > 5 \times 10^6$), and low removal by other microbes or invertebrate scavengers ($\lambda_d D < 0.5$) does transmission by host necrophagy become important (Fig. 4). This is largely because transmission by necrophagy occurs after hosts have been infected by direct transmission and then die. Since the simulated epidemics usually proceed quickly relative to the average time to death (dynamics not shown), there is not sufficient time or susceptible hosts (S) remaining for necrophagous feedbacks to contribute much additional transmission. All that said,

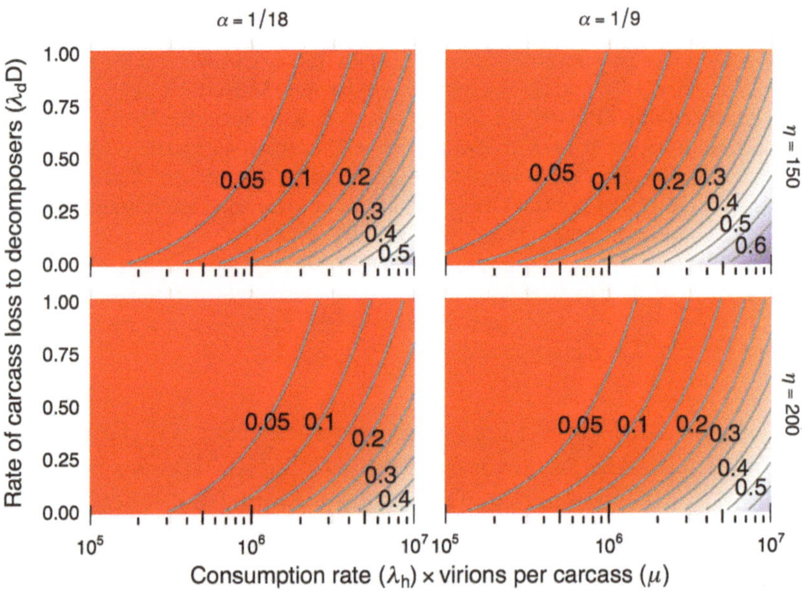

Fig. 4 The proportion of infections caused by necrophagy as a function of the product of the rate of carcass consumption by hosts (λ_h; per host per day) and virions per carcass (μ), and the rate of loss to decomposition and invertebrate scavenging ($\lambda_d D$; per day). The *top row* is $\eta = 150$ and the *bottom* is $\eta = 200$ contacts per host per day, and the *left column* for the baseline virulence of $\alpha = 1/18$ per day and the *right column* for the much faster $\alpha = 1/9$ per day. The other parameters are as in the caption of Fig. 3, except that $\rho = 0$, so that there is no water-borne transmission. The *color scale* is the same as Fig. 3.

until more empirical data are collected, there remains great uncertainty in these parameters and these restrictive conditions may well be met in some settings or species. Viral titers in carcasses, for instance, are often $>10^2$ virions and sometimes $>10^7$ in just a small sample used for DNA extractions (Brunner and Collins, 2009; Hoverman et al., 2010), suggesting that the whole carcass may have very high titers. At any given (high) level of virus transfer, the rate of decomposition and scavenging by invertebrates can strongly influence transmission rates (e.g., from nearly 0.6 to with $\lambda_d D \approx 0$ to 0.2 when $\lambda_d D \approx 1$ with $\lambda_h \mu = 5 \times 10^6$ and $\eta = 150$; Fig. 4). These are very large changes in rates of decomposition and scavenging, but this may well occur seasonally or between ponds, especially if they have very different aquatic invertebrate scavenger communities. It is also worth noting that because of the handling time associated with scavenging a carcass and resultant saturating functional response (Le Sage et al., in review), as well as the fact that scavengers may become sated during die-off events, which can occur very suddenly (Hall et al., 2016; Wheelwright et al., 2014), scavenging may become inefficient in epidemics. Thus, this rate, $\lambda_d D$, may be rather low and ineffectual when there is a great deal of host mortality. This suggests that transmission by necrophagy is more important during die-off events, when carcass removal by scavengers may be saturated, than in the earlier phases of ranavirus epidemics. However, while transmission by necrophagy may play a minor role during ranavirus epidemics of a single species, this may be an important route of transmission between host taxa (Belzer and Seibert, 2011). Future work should specifically consider the routes of spillover between species that may not come into contact with each other (hence, $\eta_{AB} \approx 0$ between species A and B). It may be that that environmental persistence and scavenging become especially important in determining whether other host species become involved in *Ranavirus* epidemics.

6. CONCLUSIONS

Ranaviruses have long had a reputation for environmental resistance (Langdon, 1989; Whittington et al., 2010). They survive desiccation, freezing, and, under the right conditions, high temperatures in the laboratory. This view has led to the expectation that ranaviruses could generally persist in the environment when hosts were not present and perhaps be translocated with fomites, and that transmission from the environment is important (reviewed in Gray et al., 2009). Simply recognizing that viruses occur free in the environment in a range of forms from naked virions to virus protected from the physical environment within shed tissues or carcasses suggests that the reality is more complex (Fig. 1). Indeed, a growing number of studies under increasingly realistic conditions have demonstrated that ranavirus persistence times vary by orders of magnitude with environmental conditions (e.g., temperature) and with the activity of the aquatic microbial and planktonic community. Persistence times on the order of a few days would seem to preclude survival in the environment between epidemics, which occur at most annually in highly seasonal environments (Brunner et al., 2015), and likely support only small amounts of environmental transmission. Persistence over several weeks or even months, on the other hand, makes the environment an important source of inoculum for transmission within an epidemic as well as for sparking new epidemics in subsequent seasons or years. As such, we should expect the relative importance of environmental persistence to vary among and even within sites and across seasons. But just how important is environmental persistence for transmission?

Incorporating free virus in the water and in infectious carcasses—two ends of the continuum of persistence—into a simple epidemiological model provided initial assessments of the conditions under which environmental persistence could affect ranavirus transmission and epidemiology. The model indicates that water-borne transmission only becomes a dominant route of infection with high rates of viral shedding and long persistence times. The conditions in which transmission by host necrophagy of infectious carcasses becomes important are even more restrictive, requiring very high rates of necrophagy by susceptible hosts consuming carcasses with very high titers, and low rates of carcass removal by decomposing microbes and invertebrate scavengers. These conditions may well be met, but, we cannot say with certainty. Ironically, this exercise in understanding the importance of environmental persistence in ranavirus epidemiology has shown that we do not know enough about the key quantities related to virus transmission, including the rates of ranavirus shedding, titers in whole carcasses, or rates of host necrophagy. Future studies that better quantify these parameters will help us understand the role that environmental persistence plays in ranavirus epidemiology.

Lastly, it is worth recognizing that even if environmental persistence proves to be unimportant to ranavirus transmission in most situations, it is likely that ranaviruses can persist over the winter, between epidemics, in the environment, at least in some settings, which is critical for the longer-term and larger scale epidemiology of ranaviruses. Persistence times of free virions in water extend to weeks and months at 4°C, and may be much longer if the virus is protected in host tissues (e.g., frozen in carcasses; Langdon, 1989; Plumb and Zilberg, 1999a) or otherwise protected from microbial action. Ranaviruses may also persist for long periods in soil if kept cool and moist (Brunner et al., 2007; Nazir et al., 2012; Reinauer et al., 2005), although whether and how ranavirus in soil can infect susceptible hosts is still unclear. Indeed, as the previous modeling exercise makes clear, simply demonstrating that long persistence times in the environment are possible is a necessary first step, but is not sufficient to demonstrate that this is an important mechanism of long-term survival for the virus. We hope this review encourages a greater focus on considering the influence of environmental persistence within an epidemiological framework. It is critical to determine not only what is possible, but what is important.

ACKNOWLEDGMENTS

Many thanks to Rachel Marschang for providing the original data from Reinauer (2004) and Reinauer et al. (2005) and helpful discussions over the years on ranaviral persistence, to Matt Gray for reminding J.L.B. that the relative importance of different routes of transmission is far from settled, and to Carolyn Malmstrom for the invitation to think more deeply about environmental persistence as well as many helpful suggestions for the manuscript.

REFERENCES

Altig, R., Whiles, M.R., Taylor, C.L., 2007. What do tadpoles really eat? Assessing the trophic status of an understudied and imperiled group of consumers in freshwater habitats. Freshw. Biol. 52, 386–395.

Ariel, E., Owens, L., 1997. Epizootic mortalities in tilapia *Oreochromis mossambicus*. Dis. Aquat. Organ. 29, 1–6.

Belzer, W., Seibert, S., 2011. A natural history of *Ranavirus* in an eastern box turtle population. Turtle Tortoise Newsl. 15, 18–25.

Ben-Ami, F., Ebert, D., Regoes, R.R., 2010. Pathogen dose infectivity curves as a method to analyze the distribution of host susceptibility: a quantitative assessment of maternal effects after food stress and pathogen exposure. Am. Nat. 175, 106–115.

Bibby, K., et al., 2015. Persistence of Ebola virus in sterilized wastewater. Environ. Sci. Technol. Lett. 2, 245–249.

Brown, J.D., et al., 2007. Persistence of H5 and H7 avian influenza viruses in water. Avian Dis. 51, 285–289.

Brown, J.D., et al., 2009. Avian influenza virus in water: infectivity is dependent on pH, salinity and temperature. Vet. Microbiol. 136, 20–26.

Brunner, J.L., Collins, J.P., 2009. Testing assumptions of the trade-off theory of the evolution of parasite virulence. Evol. Ecol. Res. 11, 1169–1188.

Brunner, J.L., et al., 2004. The role of an intraspecific reservoir in the persistence of a lethal ranavirus. Ecology 85, 560–566.

Brunner, J.L., Richards, K., Collins, J.P., 2005. Dose and host characteristics influence virulence of ranavirus infections. Oecologia 144, 399–406.

Brunner, J.L., Schock, D.M., Collins, J.P., 2007. Transmission dynamics of the amphibian ranavirus *Ambystoma tigrinum* virus. Dis. Aquat. Organ. 77, 87–95.

Brunner, J.L., et al., 2015. Ranavirus ecology and evolution: from epidemiology to extinction. In: Gray, M.J., Chinchar, V.G. (Eds.), Ranaviruses: Lethal Pathogens of Ecothermic Vertebrates. Springer International Publishing, Cham, Switzerland, pp. 71–104.

Brunner, J.L., et al., 2017. Heterogeneities in the infection process drive ranavirus transmission. Ecology 98, 576–582.

Brunner, J.L. et al. (in revision) Performance of nonlethal methods of detecting *Ranavirus* infections in captivity and trade. *Dis. Aquat. Organ.*

Chinchar, V.G., Waltzek, T.B., Subramaniam, K., 2017. Ranaviruses and other members of the family Iridoviridae: their place in the virosphere. Virology 511, 259–271.

Cook, B.W., et al., 2015. Evaluating environmental persistence and disinfection of the Ebola virus Makona variant. Virus 7, 1975–1986.

Diamant, A., 1997. Fish-to-fish transmission of a marine myxosporean. Dis. Aquat. Organ. 30, 99–105.

Duffus, A.L.J., et al., 2015. Distribution and host range of ranaviruses. In: Gray, M.J., Chinchar, V.G. (Eds.), Ranaviruses: Lethal Pathogens of Ecothermic Vertebrates. Springer International Publishing, Cham, Switzerland, pp. 9–57.

Forzan, M.J., et al., 2015. Clinical signs, pathology and dose-dependent survival of adult wood frogs, *Rana sylvatica*, inoculated orally with frog virus 3 (*Ranavirus* sp., Iridoviridae). J. Gen. Virol. 96, 1138–1149.

Getz, W.M., 2011. Biomass transformation webs provide a unified approach to consumer-resource modelling. Ecol. Lett. 14, 113–124.

Gray, M.J., Miller, D.L., Hoverman, J.T., 2009. Ecology and pathology of amphibian ranaviruses. Dis. Aquat. Organ. 87, 243–266.

Hall, E.M., et al., 2016. Evaluating environmental DNA-based quantification of ranavirus infection in wood frog populations. Mol. Ecol. Resour. 16, 423–433.

Harp, E.M., Petranka, J.W., 2006. Ranavirus in wood frogs (*Rana sylvatica*): potential sources of transmission within and between ponds. J. Wildl. Dis. 42, 307–318.

Hoverman, J.T., Gray, M.J., Miller, D.L., 2010. Anuran susceptibilities to ranaviruses: role of species identity, exposure route, and a novel virus isolate. Dis. Aquat. Organ. 89, 97–107.

Huang, X.H., et al., 2011. Singapore grouper iridovirus, a large DNA virus, induces nonapoptotic cell death by a cell type dependent fashion and evokes ERK signaling. Apoptosis 16, 831–845.

Jancovich, J.K., et al., 1997. Isolation of a lethal virus from the endangered tiger salamander *Ambystoma tigrinum stebbinsi*. Dis. Aquat. Organ. 31, 161–167.

Johnson, A.F., Brunner, J.L., 2014. Persistence of an amphibian ranavirus in aquatic communities. Dis. Aquat. Organ. 111, 129–138.

Kayansamruaj, P., et al., 2017. Outbreaks of ulcerative disease associated with ranavirus infection in barcoo grunter, *Scortum barcoo* (McCulloch & Waite). J. Fish Dis. 40, 1341–1350.

Kitchell, J.F., Koonce, J.F., Tennis, P.S., 1975. Phosphorus flux through fishes. Verh. Int. Ver. Theor. Angew. Limnol. 19, 2478–2484.

Kotwal, G., Cannon, J.L., 2014. Environmental persistence and transfer of enteric viruses. Curr. Opin. Virol. 4, 37–43.

Langdon, J.S., 1989. Experimental transmission and pathogenicity of epizootic haematopoietic necrosis virus (EHNV) in redfin perch, *Perca fluviatilis* L., and 11 other teleosts. J. Fish Dis. 12, 295–310.

Langdon, J.S., Humphrey, J.D., 1987. Epizootic haematopoietic necrosis, a new viral disease in redfin perch, *Perca fluviatilis* L., in Australia. J. Fish Dis. 10, 289–297.

Langdon, J.S., et al., 1986. First virus isolation from Australian fish: an iridovirus-like pathogen from redfin perch, *Perca fluviatilis* L. J. Fish Dis. 9, 263–268.
Le Sage, M.J., Towey, B.D., Brunner, J.L., in review Are scavengers good for your health? An experimental test of the effect of invertebrate scavengers on ranavirus transmission. Funct. Ecol.
Munro, J., et al., 2016. Survival of frog virus 3 in freshwater and sediment from an English lake. J. Wildl. Dis. 52, 138–142.
Nazir, J., Spengler, M., Marschang, R.E., 2012. Environmental persistence of amphibian and reptilian ranaviruses. Dis. Aquat. Organ. 98, 177–184.
O'Connor, K.M., Rittenhouse, T.A.G., Brunner, J.L., 2016. Ranavirus is common in wood frog (*Lithobates sylvaticus*) tadpoles throughout Connecticut. Herpetol. Rev. 47, 394–397.
Ogorzaly, L., et al., 2010. Occurrence, survival, and persistence of human adenoviruses and F-specific RNA phages in raw groundwater. Appl. Environ. Microbiol. 76, 8019–8025.
Parmenter, R.R., Lamarra, V.A., 1991. Nutrient cycling in a freshwater marsh: the decomposition of fish and waterfowl carrion. Limnol. Oceanogr. 36, 976–987.
Pearman, P.B., et al., 2004. Response of the Italian agile frog (Rana latastei) to a Ranavirus, frog virus 3: a model for viral emergence in naive populations. J. Wildl. Dis. 40, 660–669.
Petzoldt, T., Rinke, K., 2007. Simecol: an object-oriented framework for ecological modeling in R. J. Stat. Softw. 22, 1–31.
Plumb, J.A., Zilberg, D., 1999a. Survival of largemouth bass iridovirus in frozen fish. J. Aquat. Anim. Health 11, 94–96.
Plumb, J.A., Zilberg, D., 1999b. The lethal dose of largemouth bass virus in juvenile largemouth bass and the comparative susceptibility of striped bass. J. Aquat. Anim. Health 11, 246–252.
Price, S., et al., 2014. Collapse of amphibian communities due to an introduced *Ranavirus*. Curr. Biol. 24, 2586–2591.
Regester, K.J., Whiles, M.R., 2006. Decomposition rates of salamander (*Ambystoma maculatum*) life stages and associated energy and nutrient fluxes in ponds and adjacent forest in southern Illinois. Copeia 2006, 640–649.
Regoes, R.R., Ebert, D., Bonhoeffer, S., 2002. Dose-dependent infection rates of parasites produce the allee effect in epidemiology. Proc. R. Soc. Lond. Ser. B Biol. Sci. 269, 271–279.
Reinauer, S., 2004. Tenazitaet Dreier Schildkroetenviren Unter Naturnahen Bedingungen (German) Tenacities of three Turtle Viruses under Close to Nature Conditions (German). Diploma thesis thesis, University of Hohenheim, Stuttgart, Germany.
Reinauer, S., Bohm, R., Marschang, R.E., 2005. Inactivation of tortoise viruses in the environment. J. Herpetol. Med. Surg. 15, 10–15.
Robert, J., et al., 2011. Waterborne infectivity of the Ranavirus frog virus 3 in *Xenopus laevis*. Virology 417, 410–417.
Rudolf, V.H., Antonovics, J., 2007. Disease transmission by cannibalism: rare event or common occurrence. Proc. Biol. Sci. 274, 1205–1210.
Sobsey, M.D., et al., 1980. Interactions and survival of enteric viruses in soil materials. Appl. Environ. Microbiol. 40, 92–101.
Speare, R., Smith, J.R., 1992. An iridovirus-like agent isolated from the ornate burrowing frog *Limnodynastes ornatus* in northern Australia. Dis. Aquat. Organ. 14, 51–57.
Stevenson, C., Childers, D.L., 2004. Hydroperiod and seasonal effects on fish decomposition in an oligotrophic Everglades marsh. Wetlands 24, 529–537.
Stevik, T.K., et al., 2004. Retention and removal of pathogenic bacteria in wastewater percolating through porous media: a review. Water Res. 38, 1355–1367.
van der Werf, W., et al., 2011. Heterogeneous host susceptibility enhances prevalence of mixed-genotype micro-parasite infections. PLoS Comput. Biol. 7, e1002097.

Waltzek, T.B., et al., 2014. New disease records for hatchery-reared sturgeon. I. Expansion of frog virus 3 host range into *Scaphirhynchus albus*. Dis. Aquat. Organ. 111, 219–227.
Ward, R.L., Knowlton, D.R., Winston, P.E., 1986. Mechanism of inactivation of enteric viruses in fresh water. Appl. Environ. Microbiol. 52, 450–459.
Warne, R.W., Crespi, E.J., Brunner, J.L., 2011. Escape from the pond: stress and developmental responses to ranavirus infection in wood frog tadpoles. Funct. Ecol. 25, 139–146.
Weissenberg, R., 1965. Fifty years of research on the lymphocystis virus disease of fishes (1914–1964). Ann. N. Y. Acad. Sci. 126, 362–374.
Wheelwright, N.T., et al., 2014. Sudden mass die-off of a large population of wood frog (*Lithobates sylvaticus*) tadpoles in Maine, USA, likely due to ranavirus. Herpetol. Rev. 45, 240–242.
Whittington, R.J., Becker, J.A., Dennis, M.M., 2010. Iridovirus infections in finfish—critical review with emphasis on ranaviruses. J. Fish Dis. 33, 95–122.
Wilson, E.E., Wolkovich, E.M., 2011. Scavenging: how carnivores and carrion structure communities. Trends Ecol. Evol. 26, 129–135.
Young, K.D., 2006. The selective value of bacterial shape. Microbiol. Mol. Biol. Rev. 70, 660–703.

FURTHER READING

Kimble, S.J., et al., 2015. Mosquitoes as a potential vector of ranavirus transmission in terrestrial turtles. Ecohealth 12, 334–338.

CHAPTER SIX

Plant and Insect Viruses in Managed and Natural Environments: Novel and Neglected Transmission Pathways

Roger A.C. Jones[*,†,1]
[*]Institute of Agriculture, Faculty of Science, University of Western Australia, Crawley, WA, Australia
[†]Department of Primary Industries and Regional Development, South Perth, WA, Australia
[1]Corresponding author: e-mail address: roger.jones@uwa.edu.au

Contents

Abstract

The capacity to spread by diverse transmission pathways enhances a virus' ability to spread effectively and survive when circumstances change. This review aims to improve understanding of how plant and insect viruses spread through natural and managed environments by drawing attention to 12 novel or neglected virus transmission pathways whose contribution is underestimated. For plant viruses, the pathways reviewed are vertical and horizontal transmission via pollen, and horizontal transmission by

Advances in Virus Research, Volume 101
ISSN 0065-3527
https://doi.org/10.1016/bs.aivir.2018.02.006

parasitic plants, natural root grafts, wind-mediated contact, chewing insects, and contaminated water or soil. For insect viruses, they are transmission by plants serving as passive "vectors," arthropod vectors, and contamination of pollen and nectar. Based on current understanding of the spatiotemporal dynamics of virus spread, the likely roles of each pathway in creating new primary infection foci, enlarging previously existing infection foci, and promoting generalized virus spread are estimated. All pathways except transmission via parasitic plants, root grafts, and wind-mediated contact transmission are likely to produce new primary infection foci. All 12 pathways have the capability to enlarge existing infection foci, but only to a limited extent when spread occurs via virus-contaminated soil or vertical pollen transmission. All pathways except those via parasitic plant, root graft, contaminated soil, and vertical pollen transmission likely contribute to generalized virus spread, but to different extents. For worst-case scenarios, where mixed populations of host species occur under optimal virus spread conditions, the risk that host species jumps or virus emergence events will arise is estimated to be "high" for all four insect virus pathways considered, and, "very high" or "moderate" for plant viruses transmitted by parasitic plant and root graft pathways, respectively. To establish full understanding of virus spread and thereby optimize effective virus disease management, it is important to examine all transmission pathways potentially involved, regardless of whether the virus' ecology is already presumed to be well understood or otherwise.

1. INTRODUCTION

Transmission to new hosts is one of the most important processes in plant and insect virus ecology as it ensures the long-term survival of a viral pathogen (Carter, 1961; Cooper and Jones, 2006; Cory, 2010, 2015; Harrison, 1981; Jones, 2009, 2014a; Malmstrom et al., 2011; Power, 2008; Rothman and Myers, 2000; Thresh, 1980, 1981, 1991). Moreover, transmission between hosts and multiplication within hosts constitute the two key components of virus fitness, and knowledge of how these two components are linked is critical to understanding virus evolution (Cooper and Jones, 2006; Sacristán et al., 2011). However, there are significant gaps in our understanding of the virus transmission pathways employed by plant and insect viruses, which is largely based on knowledge of well-known viruses in managed systems. Nevertheless, we know plant and insect viruses adopt diverse transmission pathways to infect their hosts successfully, and these pathways influence their disease cycles and patterns of spread in managed and natural environments. Such pathways belong to one or other of two principal transmission types, vertical and horizontal. Vertical transmission involves virus transmission between parents and their progeny, and horizontal transmission involves transmission to new individuals that are neither progeny nor parents (e.g., Chen et al., 2006a; Cory, 2015; Hull, 2014). These terms have been used widely in insect virus transmission studies, but less so with plant viruses, although they apply equally well with both.

Horizontal transmission of plant viruses commonly occurs through activities of specific flying insects or wind-transported mites, ectoparasitic nematodes or parasitic fungus-like organisms (chytrids and protists) that parasitize roots, or contact transmission where mechanical damage occurs. In horizontal transmission, viruses enter healthy plants through tiny wounds in their cell walls created by vector activity or other damage. Wounding is also required for vertical transmission that involves viral contamination of the seed coat, but not for vertical transmission via seeds with infected embryos or via vegetative propagation from virus-infected planting material (such as cuttings, rootstocks, tubers, bulbs, and corms). Vector, vegetative propagation, and seed transmission pathways have been explained in depth in many reviews and textbooks (e.g., Bawden, 1964; Grogan and Campbell, 1966; Hull, 2014; Johansen et al., 1994; Jones, 2009, 2014a, 2016; Jones and Barbetti, 2012; Matthews, 1970; Mink, 1993; Wilson, 2014), and so are not considered further here.

Contact transmission constitutes an important horizontal transmission pathway for plant viruses that requires more attention, particularly in natural environments. Most of the novel or neglected plant virus transmission pathways described in this review include some form of contact transmission. Contact-transmitted viruses generally produce particles that are stable outside the infected cell, remain infectious on contaminated surfaces, and reach high concentrations (titer) in infected plants. Viroids also demonstrate these characteristics despite lacking encapsidated particles (e.g., Hull, 2014; Matthews, 1970). In a study of plant virus transmission by leaf contact, Sacristán et al. (2011) described how microwounds on leaf surfaces both liberate virus particles from the infected source leaf and produce susceptible sites in the healthy recipient leaf through which it becomes infected. In managed systems, such wounds develop when (i) leaves of infected plants rub against leaves of healthy ones, or against virus- or viroid-contaminated plant debris, or a wide range of contaminated surfaces (e.g., contaminated machinery, equipment, tools, and clothing); (ii) large animals move through plant stands; or (iii) grazing and trampling by stock or mowing takes place (e.g., Dombrovsky et al., 2017; Hull, 2014; Jones and Barbetti, 2012; Matthews, 1970; McKirdy et al., 1998). However, few contact-transmitted virus epidemics have been studied quantitatively (Sacristán et al., 2011), and none of these studies involved natural environments. In wild herbaceous plant communities, contact transmission can occur when virus infected and healthy leaves rub against each other or against contaminated plant debris, or when animals trample, or brush past plants. Accumulating evidence indicates that contact transmission also occurs among virus genera that produce less stable particles and reach lower within-plant concentrations, such as *Potyviruses* (Congdon et al., 2016; Coutts and Jones, 2015; Coutts et al., 2013; Draper and Gudmestad, 1992; Wintermantel, 2011) and *Ilarviruses* (Pethybridge and Madden, 2003; Pethybridge et al., 2000, 2002).

Among the important but often overlooked mechanisms for the spread of plant viruses is pollen transmission, which is well studied in few pathosystems and mostly disregarded otherwise. Pollen transmission constitutes vertical transmission when it occurs from parental plants to their progeny but horizontal transmission when pollen recipients are not progeny plants (Card et al., 2007; Hull, 2014; Mink, 1993). Plant viruses are spread by several additional novel (known only since year 2000) or neglected (mostly disregarded) transmission pathways that are rarely considered (e.g., Hamelin et al., 2016; Sacristán et al., 2011). These generally overlooked strategies include horizontal transmission via parasitic plants, natural root-to-root grafts, wind-mediated contact, chewing insects, and contaminated soil and water (e.g., Epstein, 1978; Gooding and Todd, 1976; Hosford, 1967; Kleinhempel and Kegler, 1982; Matthews, 1970; Mehle and Ravnikar, 2012; Sarra et al., 2004; Shukla et al., 1979; Teakle, 1986; Walters, 1951, 1952).

For insect viruses, much current knowledge is based on study of (i) viruses as biological control agents (e.g., *Baculoviruses*) or (ii) virus diseases of silkworms (*Bombyx mori*), honey bee (*Apis mellifera*), bumble bee (*Bombus* spp.) pollinators, and common fruit flies (*Drosophila melanogaster*) (e.g., Cory, 2010, 2015; Cory and Myers, 2003; Miller and Ball, 1998). Knowledge is lacking about the ecologies of most other viruses from insect hosts, especially insect RNA viruses that cause chronic infections. With insect viruses, vertical transmission can be transovaum (on the egg surface) or transovarial (inside the egg) and sometimes involves male as well as female parents. Horizontal transmission often occurs when insects eat contaminated food, but also involves other pathways, such as cannibalism and sexual or air-borne infection. Insect feces constitute an important virus contamination source for insect food. When viruses infect insects, infected individuals may die, break open, and spill virus particles into the environment. New infections then occur when healthy insects make contact with contaminated surfaces of plants or soil. Surfaces may also be contaminated with insect viruses by the feces of insect predators (e.g., birds, other insects), or by water run-off, rain-splash, and air-borne soil particles (Bonning and Miller, 2010; Burand et al., 2012; Chen and Siede, 2007; Chen et al., 2006a,b; Cory, 2010, 2015; Fuller et al., 2012; Longdon and Jiggins, 2012; McNeil, 2010; Strand and Burke, 2013; Vilaplana et al., 2010; Virto et al., 2013, 2014). Additional neglected or novel transmission pathways are sometimes involved in the transmission of insect viruses. Such pathways include horizontal transmission via virus-contaminated pollen and nectar, arthropod vectors, and plants acting as passive virus "vectors" (Ban et al., 2007; Bonning and Miller, 2010; Cory, 2015; Ribiere et al., 2007; Singh et al., 2010; Wamonje et al., 2017).

This review considers transmission of plant and insect viruses by the novel and neglected pathways highlighted above and estimates the likely significance of each using a broad, holistic scale. The approach adopted draws on long-established understanding of the principles underlying the spatiotemporal dynamics of virus spread (e.g., Bwye et al., 1994; Coutts et al., 2004; Jones, 2005; Matthews, 1970; Moreno et al., 2007; Nutter, 1997; Thresh, 1974, 1976, 1980, 1981, 1983). This approach was used previously to dissect the various ways diverse types of virus control measures operate in managed situations (Jones, 2006). Here, it was employed to estimate the likely roles of each individual transmission pathway in contributing toward (i) creation of new primary infection foci, (ii) enlargement of primary and secondary virus infection foci, and (iii) the subsequent phase of generalized virus spread. In addition, for each pathway, the risk that host species jumps or virus emergence events will occur in mixed species populations was evaluated by considering worst-case scenarios where multiple host species and virus infection sources are abundant and virus transmission circumstances are optimal. With plant viruses, such conditions are most likely to occur at the interface between natural and managed vegetation from which viruses of native plants emerge to infect cultivated plants, and cultivated plant viruses spread to new hosts in native plant communities (e.g., Alexander et al., 2014; Cooper and Jones, 2006; Jones, 2006, 2009, 2014a; Jones and Coutts, 2015; Malmstrom and Alexander, 2016; Malmstrom et al., 2011; Power, 2008; Roossinck and Garcia-Arenal, 2015; Thresh, 1980, 1981, 1991). With insect viruses, they are most likely to apply where host populations include a mixture of managed and natural insect species that use the same food source and viruses spread between the two groups (e.g., Faillace et al., 2017; Furst et al., 2014; Longdon et al., 2014; Manley et al., 2015).

2. PLANT VIRUSES

2.1 Vertical and Horizontal Transmission by Pollen

The extent to which virus-contaminated pollen constitutes a significant factor in virus spread in the field has not been assessed thoroughly (Hull, 2014). Indeed, so far pollen transmission has been studied only in a few commercially important perennial or annual crops, and rarely, if ever, in wild plants growing in natural environments. It is known that pollen transmission requires more than mere association between viruses and pollen grains (Mink, 1993). Pollen transmission takes place where fertilization with infected plant pollen leads to infection of the seed that develops on the recipient plant (Card et al., 2007). This in turn, results in embryo infection within the developing seed and infection of seedlings that germinate from them, such transmission being vertical when it occurs between parental plants and their own progeny, but horizontal when this is not the case. Hamelin et al. (2016) considered transmission between parental plants and their own progeny to be direct vertical pollen transmission, but transmission to the progeny of any other plant to be indirect vertical pollen transmission. As the latter does not fit the classical definition of vertical transmission, the term horizontal pollen transmission will be used for it here. Less frequently, horizontal transmission occurs when the plant pollinated itself becomes infected via the fertilized flower (e.g., George and Davidson, 1963; Gilmer, 1965). This requires virus movement to the maternal tissues of the pollinated plant, either assisted or unassisted by pollinators (Card et al., 2007). Unassisted horizontal pollen transmission requires virus passage through the callose layer separating the embryo from the mother plant without direct cellular contact or vascular connection (Card et al., 2007; Hull, 2014; Mink, 1993). The connection between the seed coat (testa) and suspensor of the developing seed is the postulated interface that enables such outward movement of the virus (or viroid) into the mother plant (Roberts et al., 2003; Wang and Maule, 1992). Unassisted horizontal pollen transmission is reported for the *Nepovirus Blueberry leaf mottle virus* (BLMV) (Boylan-Pett et al., 1991) and the *Idaeovirus Raspberry bushy dwarf virus* (RBDV) (Murant et al., 1974).

Assisted horizontal pollen transmission requires presence of mechanical damage caused by the feeding pollinator when it creates wounds and so constitutes a form of contact transmission (Greber et al., 1991, 1992; Sdoodee and Teakle, 1993). Evidence for assisted horizontal pollen transmission of several viruses by *Thrips tabaci* and other thrips species comes from several studies (e.g., Greber et al., 1991, 1992; Hardy and Teakle, 1992; Milne and Walter, 2003; Mink, 1993; Sdoodee and Teakle, 1987, 1993). For example, pollen contaminated with the *Ilarvirus Tobacco streak virus* is carried both internally and externally by the thrips, and infection occurs via wounds caused by thrip feeding (Sdoodee and Teakle, 1993). Likewise, the *Ilarvirus Prunus necrotic ringspot virus* (PNRSV) is transmitted by a combination of infected pollen and injury due to thrip feeding (Greber et al., 1991, 1992).

Bees play a major role in pollen transmission (for more on bee transmission, see chapter "Emerging viruses in bees: From molecules to ecology" by McMahon et al., this volume). For example, by transferring virus-contaminated pollen from flowers produced by infected plants to flowers on healthy plants, honey bees transmitted the *Ilarviruses* PNRSV, *Prune dwarf virus* and *Blueberry shock virus* (BlShV), and the *Nepovirus* BLMV (Bristow and Martin, 1999; Childress and Ramsdell, 1987; George and Davidson, 1964; Mink, 1993). In the field, BlShV spread from diseased to healthy blueberries only during bloom and then only when honey bees transferred virus-contaminated pollen from infected to healthy plants (Bristow and Martin, 1999). Long-distance spread occurred when colonies containing PNRSV- and BLMV-contaminated pollen were transported between distant locations (Boylan-Pett et al., 1991; Mink, 1993). Under protected cropping conditions, the *Tobamoviruses Cucumber green mottle mosaic virus* (CGMMV) and *Tobacco mosaic virus* (TMV), and *Potexvirus Pepino mosaic virus* (PepMV), were spread by pollinating bumble bees or honey bees (Darzi et al., 2018; Lacasa et al. 2003; Okada et al., 2000; Shipp et al., 2008). Two viroids, *Tomato chlorotic dwarf viroid* (TCDVd) and *Tomato apical stunt viroid* (TASVd), were also spread by pollinating bumble bees (Antignus et al., 2007; Matsuura et al., 2010). When foraging bees visit flowers, such horizontal virus or viroid transmission occurs via wounds in flower tissues or when mother plant infection develops via the fertilized ovule (Antignus et al., 2007; Boylan-Pett et al., 1991). As TMV, PepMV, TCDVd, and TASVd are all highly stable, reach high concentrations in the plant, and spread readily by contact, transmission via wounds seems the most likely explanation for their infections. This is probably also so with CGMMV, but Liu et al. (2014a) reported that its transmission from infected pollen to flowers and fruits occurred following artificial pollination. More research is required to establish definitively (i) which transmission pathway occurs most frequently when bee visits result in infection of the mother plants of the flowers they fertilize and (ii) the roles pollinating native bees play in transmitting viruses to wild plants in natural vegetation.

Pollen transmission occurs in multiple plant virus genera. Card et al. (2007) listed 39 viruses belonging to at least 15 plant virus genera and 5 viroids, as pollen transmitted. Vertical transmission occurred with all 39 viruses, horizontal transmission also occurring with 17 of them. Six other viruses were tentatively reported to be pollen borne. Most pollen-borne viruses were in the genera *Alphacryptovirus*, *Ilarvirus*, *Nepovirus*, and *Potyvirus*. All seven *Ilarviruses* listed were transmitted both vertically and horizontally, and this also occurred with 5/9 and 1/5 *Nepoviruses* and *Potyviruses*, respectively. By contrast, the seven *Alphacryptoviruses* listed were only transmitted vertically. Subsequently, Hull (2014) listed 17 viral and 3 viroid genera in which pollen transmission occurs. As metagenomics analyses discover new viruses, the list of pollen-transmitted viruses will likely expand.

No studies have investigated virus transmission by pollen to seeds or mother plants by noninsect pollinators, such as birds or marsupials. Indeed, the lack of information about the importance of pollen transmission of plant viruses in natural ecosystems, and the paucity of such information in most managed systems is not only surprising given how long pollen transmission has been known but also urgently in need of rectification.

It has been suggested that pollen transmission is more likely to be important for woody perennials than for annual plants (Hull, 2014). The importance for woody perennials is clear when the pollen transmission results in infection of the entire plant pollinated rather than solely its seeds. However, pollen transmission solely to seeds may be equally important for many annual or short-lived perennial plants, because it results in infected seedlings which then serve as infection foci for further virus spread and help initiate annual epidemics in managed or unmanaged systems. Table 1 provides estimates of the likely roles of vertical and horizontal pollen transmission in creation of new primary infection foci, enlargement of primary and secondary virus infection foci, and generalized virus spread. Both pollen transmission pathways are likely to contribute to new primary infection foci arising from seed-infected annual plants. When infected plants that provide the pollen are surrounded by their progeny and vertical pollen transmission via insect pollinator activity causes these progeny to become systemically infected, the primary infection foci will enlarge. Owing to their longer lifespans, such a scenario seems most likely to be important in perennial vegetation. In contrast, when pollinators facilitate horizontal pollen transmission and pollinated plants become systemically infected, both primary and secondary

Table 1 Novel and Neglected Plant and Insect Virus Transmission Pathways: Insect Pollinator Involvement, Patterns of Spread, Spread Between Host Species and Likely Evolutionary Significance[a]

				Likely Role in Virus Spread				
Pathway	**Novel or Neglected**	**Transmission Type**	**Insect Pollinator Involvement (Flowering Plants)**	**Creates New Primary Infection Foci**	**Enlarges Infection Foci Locally**	**Spreads Virus More Generally**	**Spread Between Different Host Species**	**Potential for Host Species Jumps/New Virus Emergence**
Plant viruses								
Pollen	Neglected	Vertical	+[b]	+	(+)	–	–	No
Pollen	Neglected	Horizontal	+	+	+	(+)	(+)	Low
Parasitic plant	Neglected	Horizontal via haustoria	–	–	+	–	+	Very high
Root grafts	Neglected	Horizontal via natural grafts	–	–	+	–	(+)	Moderate
Wind-mediated	Novel	Horizontal via contact	–	–	+	(+)	+	Low
Chewing insect[c]	Neglected	Horizontal via contact	–	+	+	(+)	+	Low
Water	Neglected	Horizontal via contact	–	+	+	(+)	+	Low
Soil	Neglected	Horizontal via contact	–	+	(+)	–	+	Low

Continued

Table 1 Novel and Neglected Plant and Insect Virus Transmission Pathways: Insect Pollinator Involvement, Patterns of Spread, Spread Between Host Species and Likely Evolutionary Significance—cont'd

				Likely Role in Virus Spread				
Pathway	**Novel or Neglected**	**Transmission Type**	**Insect Pollinator Involvement (Flowering Plants)**	**Creates New Primary Infection Foci**	**Enlarges Infection Foci Locally**	**Spreads Virus More Generally**	**Spread Between Different Host Species**	**Potential for Host Species Jumps/New Virus Emergence**
Insect viruses								
Plant vector[d]	Novel	Horizontal plant mediated	–	+	+	+	+	High
Arthropod vector	Novel	Horizontal parasite mediated	–	+	+	+	+	High
Pollen	Novel	Horizontal plant mediated	+	+	+	+	+	High
Nectar[e]	Novel	Horizontal plant mediated	+	+	+	+	+	High

[a]"Novel" refers to a transmission pathways only known since the year 2000 or afterward. "Neglected" means mostly disregarded. ( ) = occurs infrequently.
[b]Both plant self-pollination and pollination by an insect pollinator can lead to mother plant fertilization that gives rise to virus-infected progeny plants.
[c]Refers to nonspecific virus transmission by insects with chewing mouthparts. Wounding arising from insect pollinator activity excluded.
[d]Refers to virus transmission to herbivorous insects feeding on plants in which insect viruses are circulating via the phloem.
[e]Nectar transmission pathway based on circumstantial evidence provided by insect virus contamination of pollen within bee colonies.

infection foci infection could enlarge. Depending on the lifespan of the plant population, more extensive virus spread by pollen transmission would then take place. For example, Murant et al. (1974) attributed widespread RBDV infection in field plots of healthy raspberry plants entirely to horizontal pollen transmission because within 2 years most plants that were allowed to flower became RBDV infected. When insect pollinators carry virus-infected pollen between two plant species, infection of the recipient can occur through wounds created by pollinator activity. In worst-case scenarios, where insect pollinators and virus-infected plants are abundant, and many plant species present, such virus spread could contribute toward a host species jump or even the emergence of a new virus. The likelihood of this occurring with horizontal pollen transmission is estimated to be "low" (Table 1) as a continuous connection between the virus donor and the recipient plant is absent, unlike the situation with parasitic plants or natural root grafts (see Sections 2.2 and 2.3). It would not occur with vertical pollen transmission as, by definition, this cannot involve plants belonging to other host species.

2.2 Transmission by Parasitic Plants

Bennett (1940, 1944) showed the parasitic plant dodder (*Cuscuta* spp., family *Cuscutaceae*) could transmit viruses from plant-to-plant via the haustoria that connect its vascular tissue with that of its host. The stem of the dodder plant forms a bridge that allows virus to pass from the infected donor plant to the healthy recipient plant. *Cuscuta* spp. possesses no roots or fully expanded leaves, and their vegetative portion appears to be a stem only; they live as stem holoparasites on other plants (Kaiser et al., 2015). Between 1940 and 1960, many viruses were shown to be transmissible by dodder (Hosford, 1967), but its use in virus transmission studies declined subsequently such that it is now rarely utilized. Dodder transmission resembles graft transmission, but with the important difference that some dodder species transmit viruses between plants belonging to taxonomically disparate plant families (Desjardins et al., 1969). By contrast, graft inoculation can only transmit viruses between plant species in the same family. Moreover, although some viruses transmitted by dodder can multiply within its tissues, dodder sometimes serves simply as a passive pipeline connecting donor and recipient plants. Thus, when dodder coinfected with the *Cucumovirus* *Cucumber mosaic virus* (CMV) and the *Tobamovirus* (TMV) was grown on plant species immune to both viruses, CMV multiplied within it, and therefore persisted, but TMV did not (Bennett, 1940). There are about 200 species of dodder, some of which have extensive host ranges; for example, *Cuscuta campestris* parasitizes over 100 plant species, including members of both monocotyledonous and dicotyledonous plant families (Hosford, 1967; Kaiser et al., 2015). Many of these dodder species transmit viruses between host plants (e.g., Bennett, 1940, 1944; Cochran, 1947; Hosford, 1967; Johnson, 1941; Mikona and Jelkmann, 2010). Since dodder hosts include crop, pasture, and wild plant species, understanding dodder's importance in spreading viruses in managed and natural environments represents a subject of considerable interest for future investigation.

Besides dodder, many other parasitic angiosperms are generalists with broad host ranges. It is therefore probable that, like dodder, many of these parasitic plants can transmit viruses between taxonomically disparate plant families. For example, Gal-On et al. (2009) found that several viruses were translocated from infected hosts to the parasitic plant broomrape (*Phelipanche (Orobanche) aegyptica*), although whether the broomrape then passed the virus on to other hosts was apparently not investigated. Thus, the importance of parasitic plants in spreading viruses in managed and natural environments, deserves further study, particularly in areas where they are abundant and in biodiversity hot spots, such as the southwest Australian floristic region (Hopper and Gioia, 2004; Webster et al., 2007) and the Western Cape region of southern Africa (Linder, 2003). Parasitic plant-mediated virus transmission between neighboring host plants seems unlikely to create new primary virus infection foci as evidence of virus seed transmission is lacking, but it can enlarge existing primary or secondary virus infection foci (Table 1). Such an enlargement of the virus source for subsequent spread by vectors would accelerate the exponential phase of epidemic development resulting in widespread infection. Moreover, because many parasitic plants are generalists with wide natural host ranges and can provide continuous connections between virus donors and recipients, the risk of new host species jumps and emergence of new viruses seems exceptionally high, particularly in diverse plant communities where parasitic and virus-infected plants are abundant. This situation thus warrants a "very high" risk estimate (Table 1). The potential of parasitic plants to contribute toward interfamily host species jumps and the emergence of new viruses merits substantial attention not only in diverse natural plant communities and managed mixed species situations (pastures, gardens, and subsistence farming) but also at the interface between natural and managed vegetation.

2.3 Transmission by Natural Root Grafts

Underground natural root grafts have received little attention as a horizontal virus transmission pathway, even though they occur frequently in natural forest and plantation tree communities, as well as among shrubs and herbaceous plants in managed and natural environments (e.g., Cooper, 1993; Epstein, 1978; Goldschmidt, 2014; Graham and Bormann, 1966; Lev-Yadun, 2011). Intraspecific root grafts occur over a wide range of environmental conditions but with greater frequency in shallower soils (Graham and Bormann, 1966). Interspecific root grafts seem less common (Epstein, 1978). Early research found that in apple trees the *Ilarvirus Apple mosaic virus* (ApMV) spread slowly from plant-to-plant by natural root–root grafts (Hunter et al., 1958), and the *Robigovirus Cherry twisted leaf virus* spread in this way in cherry (Keane and May, 1963). Later, Pethybridge et al. (2002) found that ApMV was spread inefficiently in hop plantations by natural root grafts. When healthy and virus-infected roses were planted in close proximity to each other, Golino et al. (2005) reported "significant" spread of the *Ilarviruses* PNRSV and ApMV by natural root grafting. Park et al. (1999) reported that, when root tip grafts were present, the *Tobamovirus* TMV was transmitted readily between infected and healthy tobacco, tomato, and pepper plants growing in hydroponic conditions. Working with the *Tobamovirus Yellow tailflower mild mottle virus* (YTMMV), which infects the southwest Australian native plant yellow tailflower (*Anthocercis littorea*), Koh et al. (2018) found root-to-root transmission of this virus occurred occasionally, but no systemic movement was detected and whether root grafts were present was not stated. With the vast majority of viruses, information on their possible plant–plant transmission by root grafts in the absence of vectors is unavailable for both managed and natural systems. This situation needs to be rectified.

Although it cannot create new primary virus infection foci, natural root graft-mediated virus transmission between neighboring host plants could contribute toward the enlargement of existing primary and secondary infection foci (Table 1). As with parasitic plant-mediated virus transmission, such an enlargement of the virus source for subsequent spread by vectors would likely accelerate the exponential phase of epidemic development resulting in widespread infection. However, despite the continuous connection they create between virus donor and recipient plants and their more frequent occurrence, the likely risk of new host species jumps and emergence of new viruses seems lower with root grafts than with parasitic plants, because roots grafts appear to provide connections between hosts of different species less frequently. A "moderate" risk estimate for root graft transmission thus seems appropriate for diverse communities where virus-infected plants are abundant (Table 1). Therefore, the natural root–root graft transmission pathway warrants attention in future virus host shift and virus emergence research.

2.4 Wind-Mediated Contact Transmission

Another much neglected aspect of horizontal virus transmission is wind-mediated contact transmission. Despite its likely wide-scale importance in facilitating plant-to-plant virus spread in the field, only two studies have focused on it, one with the *Sobemovirus Rice yellow mottle virus* (RYMV) and the other with the *Potyvirus Pea seed-borne mosaic virus* (PSbMV). Sarra et al. (2004) demonstrated RYMV transmission by simulated wind using oscillating fans to blow neighboring virus-infected and healthy rice plants against each other. They also demonstrated it in the field by showing that RYMV spread was much greater in wind-exposed plots than in plots protected from wind by 1.5 m high plastic sheets. Congdon et al. (2016) found that simulated wind transmitted PSbMV from plant-to-plant when oscillating fans blew against intertwining healthy and infected pea plants, but no passive contact transmission occurred without wind. They also used abrasive to demonstrate a possible role for wind-borne dust and sand in enhancing wind-mediated spread. Wind-mediated contact transmission was less when higher temperatures reduced virus concentration in plants growing under warm conditions.

The lack of any other studies on wind-mediated contact transmission of plant viruses is surprising, especially because there are so many economically important viruses with stable particles that reach high concentrations in infected plants and rely on spread by contact without vector involvement. The same applies to viroids with their stable-naked RNAs. Viruses that behave like this occur in the *Potexvirus*, *Sobeomovirus*, *Tobamovirus*, and *Tombusvirus* genera. Where wind strength is considerable and plants lack shelter, wind-mediated horizontal contact transmission seems likely to play a major role in their spread in the field. In such circumstances, wind would not only help expand their primary and secondary infection virus foci, thereby magnifying the virus source for further contact transmission and enabling subsequent spread of infection to occur generally (by activities like movement of farm machinery, hand-weeding and grazing, and trampling by large herbivores), but also it could itself spread virus infection more generally (Table 1), as demonstrated by Sarra et al. (2004) with RYMV spread in rice.

The same is also likely to be so with wind-mediated spread of viruses of native vegetation, such as the *Tobamovirus* YTMMV (Koh et al., 2018). Moreover, even in pathosystems that rely principally on insect vector transmission for spread but are also contact transmitted (e.g., the *Potyviruses* PSbMV in pea (Congdon et al., 2016); *Zucchini yellow mosaic virus* in cucurbits (Coutts et al., 2013); and *Potato virus Y* (PVY) in potato (Coutts and Jones, 2015)), or in pathosystems that rely on both contact and vector transmission (e.g., *the Carlavirus Potato virus S* (Santillan et al., 2018)), wind-mediated contact transmission is likely to significantly influence virus spread early in the life of susceptible crops or susceptible wild plants. Thus, by expanding initial primary virus infection foci such that more virus source plants are available when vectors arrive, wind-mediated contact transmission is likely to accelerate the exponential phase of virus spread. In annual crops or in natural annual plant populations, the result would be greater virus incidence, disease-induced damage, and losses in crop seed production or wild plant fitness.

Stable, high-titer viruses that spread solely by contact are highly infectious and likely to spread readily with strong winds. Therefore, they pose a higher risk for new host species jumps and emergence of new viruses than do other solely contact-transmitted viruses that are less stable and lower in titer. However, without living vectors, stable high-titer viruses seem less likely to spread well over distance than less stable low-titer viruses that are contact transmitted but also vector borne. Viruses known to be contact transmitted are often generalists, but some like PSbMV are specialists with narrow host ranges (Congdon et al., 2016; Vincent et al., 2014). Fewer host species jumps arising from encounters with other potential host species would occur with specialist viruses than with generalists possessing broader host ranges. Assuming mixed plant species populations are involved, virus-infected plants are abundant and wind strengths considerable, the risk that wind-mediated horizontal contact transmission would contribute toward new host species jumps and new virus emergence is estimated to be "low" with both types of virus since a continuous connection between virus donor and recipient plants is absent (Table 1).

2.5 Nonspecific Transmission by Chewing Insects

Nonspecific horizontal virus transmission by chewing insects constitutes another much neglected virus transmission pathway likely to be important in some managed and natural environments. Virus transmission by insects with chewing mouthparts (e.g., grasshoppers, earwigs, beetles, and the larvae of moths/butterflies and flies) was reported in early research on contact-transmitted viruses with stable particles and high titers, such as *Tobamoviruses* and *Potexviruses* (e.g., Hoggan, 1931, 1934; Matthews, 1970; Orlob, 1963; Smith, 1965; Walters, 1951, 1952). Similar transmission was also demonstrated then with viroids (Goss, 1931; Schultz and Folsom, 1925, both cited by Matthews, 1970). With chewing insects, such nonspecific transmission mostly occurred when insect mouthparts were contaminated with infective sap and ceased as soon as this sap was cleaned away during feeding (Orlob, 1963; Walters, 1952). Transmission was also reported to occur when virus particles were carried on the feet of grasshoppers walking from virus infected to healthy plants (Smith, 1965). In the same era, aphids (sucking insects) were reported to occasionally transmit two *Sobemoviruses* (*Squash mosaic virus* and *Turnip crinkle virus*) thought not to be aphid transmitted (Freitag, 1941; Martini, 1958). Likewise, Teakle and Sylvester (1962) showed that aphids feeding on leaves previously sprayed with another nonaphid-transmitted virus (the *Tobamovirus* TMV) were in fact able to cause infection at low levels, and Lojek and Orlob (1969) found that aphids could acquire TMV from virus-coated leaves and inoculate it inefficiently to healthy leaves. In both instances, these occasional transmissions occurred not via typical pathways of aphid transmission but rather via wounds created by their stylets while feeding. In addition, Bradley and Harris (1972) suggested that aphids occasionally transmit virus nonspecifically by making tiny wounds when they claw leaf surfaces. The forgotten nonspecific virus transmission by chewing insect pathway deserves a new research effort to reveal the true extent to which it occurs.

What does the existence of nonspecific virus transmission by chewing insects imply regarding patterns of virus spread? Since contact-transmitted viruses with stable particles and high titer are involved and assuming sufficient numbers of chewing insects are present, nonspecific horizontal virus transmission by such insects likely will both expand primary and secondary virus infection foci (thereby magnifying the virus source for other types of contact transmission), and also help spread virus infection more generally (Table 1). In addition, unlike spread by wind-mediated transmission, chewing insects transmitting infection nonspecifically can create new primary virus infection foci by dispersing to and feeding on the other plant populations. Because the viruses involved are highly infectious and often generalists, but a continuous connection between virus donor and recipient plants is absent, they pose similar risk for new host species jumps and emergence of new viruses as that posed by wind-mediated contact transmission. Therefore, in worst-case scenarios, in diverse plant communities where large numbers of chewing insects are present and virus infection is prevalent, a "low" level risk estimate still seems appropriate (Table 1).

2.6 Transmission via Contaminated Water

Despite knowledge that water can act as both a viral inoculum source and a means of virus dispersal, few studies have investigated the importance of virus-contaminated water as a means of virus spread to plants when vectors are absent (Mehle and Ravnikar, 2012) (for more discussion, see chapter "Water-mediated transmission of plant, animal, and human viruses" by Mehle et al., this volume). For more than 30 years, plant viruses have been known to occur widely in natural water sources, such as lakes and rivers (e.g., Büttner and Nienhaus, 1989b; Koenig, 1986; Koenig and Leseman, 1985; Piazzolla et al., 1986; Tomlinson and Faithfull, 1984; Tomlinson et al., 1983; Vani and Varma, 1993). More recently, studies found plant viruses in irrigation systems (Hong and Moorman, 2005) and recirculated nutrient solutions, such as hydroponic systems (Stewart-Wade, 2011). In addition, virome studies on freshwater lakes and reclaimed water found pathogenic plant viruses (e.g., Djikeng et al., 2009; Rosario et al., 2009; Roux et al., 2012). In their review of water-borne virus research, Mehle and Ravnikar (2012) reported the presence of 16 terrestrial plant viruses representing 7 genera (*Tobamoviruses*, *Potexviruses*, *Tombusviruses*, *Carmoviruses*, *Cucumoviruses*, *Necroviruses*, and *Danthoviruses*), as well as viroids, in freshwater systems.

Most terrestrial plant viruses that occur frequently in water have stable particles and occur at high concentrations in infected plants. They are released from infected plant roots and infect plants without utilizing vectors; many of these viruses have wide host ranges (Mehle and Ravnikar, 2012). Adsorption of virus particles to organic or inorganic colloidal particles, especially clays, makes the particles more resistant to inactivation than free virus (Koenig, 1986; Piazzolla et al., 1986). Horizontal transmission to healthy roots occurs via tiny wounds and so represents another form of contact transmission. Pares et al. (1992) found that the *Tobamovirus Tomato mosaic virus* (ToMV) remained infectious in nutrient solutions for at least 6 months. An increase in flooding from storms would accelerate virus dispersion via irrigation and drainage channels, streams, and rivers (Jones and Barbetti, 2012). Some irrigation methods, such as the hydroponic systems used in protected cropping and flood irrigation of open fields, would favor virus dispersion more than others.

In more recent research on water-borne virus transmission, tomato plants growing under glasshouse conditions became infected with ToMV after receiving irrigation water contaminated with this virus (Boben et al., 2007). Li et al. (2015) demonstrated water-borne transmission of the *Tobamovirus* CGMMV to watermelon plants by drip and flow irrigation. Mehle et al. (2014) studied water-borne transmission of the *Potyvirus* PVY, *Potexvirus* PepMV, and viroid *Potato spindle tuber viroid* (PSTVd). When PepMV and PVY were released from infected roots into aqueous nutrient solutions, they infected tomato plants via their roots. When water containing PSTVd was added to developing potato tubers, they became infected with PSTVd. Contaminated water remained infectious for 7, 3, and 1 weeks for PSTVd, PepMV, and PVY, respectively. For managed systems, the paucity of information on water-borne virus or viroid survival or their transmission to plants, and the lack of it for natural systems, suggests further research on water-mediated virus transmission is urgently required.

Water-mediated horizontal contact transmission seems likely to resemble wind-mediated virus transmission and nonspecific contact transmission by chewing insects as these also mostly involve viruses with stable particles (see earlier), so their patterns of spread through managed and natural environments are likely to be similar. However, presence of the virus inoculum in a water medium also means there are differences. For example, virus-contaminated irrigation or flood water from an external source might create new primary infection foci within otherwise healthy plant populations (Table 1). When new foci arise in this way, in young plants in particular, major virus epidemics could be initiated as secondary spread could occur via water, and multiple other pathways. In contrast, wind-mediated contact transmission cannot initiate new primary infection foci. However, water-mediated virus transmission resembles these two other pathways in terms of the possibilities of enlargement of primary or secondary virus infection foci. Enlargement of infection foci would occur locally when virus particles from an initially infected plant leak into the soil and infect neighboring healthy plants via their roots. Moreover, irrigation water or natural flooding could spread virus infection further initiating new secondary infection foci distant from initial infection sites. How frequently such water-mediated secondary spread is likely to occur would be expected to vary considerably depending on the extent, type, and timing of irrigation or natural flood events, and thus would most likely occur infrequently. Because of this likely infrequency and the lack of a continuous direct connection between virus donor and recipient plants, the risk of contributing toward new host species jumps and new virus emergence scenarios still seems "low," even under the worst-case scenario (Table 1).

2.7 Transmission via Direct Soil Contamination

In some cases, plants can acquire virus infection directly from virus-contaminated soils. Using bait plants, several plant viruses with unknown vectors, such as *Tobamoviruses* and *Potexviruses*, were shown to occur in soils in forest ecosystems in Europe. *Potexviruses* were the most common and included well-known viruses like *Potato virus X* and *Narcissus mosaic virus* (Büttner and Nienhaus, 1989a). The *Tobamoviruses* TMV and ToMV, and the *Potyvirus* PVY were also found (Büttner and Nienhaus, 1989a). Baiting involves transplanting healthy seedlings into potentially virus-contaminated soils, where they can acquire infection via tiny root wounds created during planting. This constitutes a form of assisted horizontal contact transmission associated with mechanical damage caused by human activity. In natural environments, unwittingly transplanting seedlings or larger plants into virus-contaminated soil is likely to occur during revegetation programs. Research is needed to establish whether unassisted horizontal contact transmission of viruses, such as the *Tobamovirus* TYFMV, is possible from contaminated soil in undisturbed natural environments. It seems likely that friction created by abrasion against gravel or other soil particles during normal root growth through soil sometimes results in sufficient wounding of delicate surface cells, e.g., root hairs, to allow such transmission to occur.

Owing to their virion stability and high titer, *Tobamoviruses*, *Potexviruses*, *Tombusviruses*, and *Sobemoviruses* are obvious candidates for abiotic transmission via virus-contaminated soil. Teakle (1986) concluded that this type of abiotic transmission of stable plant viruses in soil is probably much more common than previously realized. As evidence of frequent transmission in soil in the apparent absence of virus vectors, he cited (i) his own study with soil transmission of the *Sobemovirus Southern bean mosaic virus* to common bean (Teakle, 1986) and (ii) similar studies with the *Tombusviruses Cymbidium ringspot virus* to *Nicotiana clevelandii* (Hollings et al., 1977) and *Tomato bushy stunt virus* to *Celosia argentea* (Kleinhempel and Kegler, 1982), and the *Gallantivirus Galinsoga mosaic virus* (family *Tombusviridae*) to *Galinsoga parviflora* (Shukla et al., 1979). Subsequently, Traore et al. (2008) demonstrated abiotic transmission of the *Sobemovirus* RYMV from contaminated soil to rice plants. For *Tobamoviruses*, such transmission was shown in early studies with TMV, ToMV, and CGMMV in which root infection from contaminated soil resulted in infected tobacco, tomato, and cucurbit crops, respectively (Broadbent, 1976; Broadbent and Fletcher, 1963; Gooding and Todd, 1976; Hollings et al., 1975). Infectivity of *Tobamoviruses* can persist for long periods in fallow soil. For example, ToMV infectivity persisted for 2 years in dry soil and in root debris in soil for up to 22 months (Broadbent, 1976). With CGMMV and *Cucumber fruit mottle mosaic virus* (another *Tobamovirus*), more recent studies confirmed that transplanting cucurbit plants into contaminated soil resulted in infection via their roots (Antignus et al., 2005; Li et al., 2015). TMV was also transmitted from contaminated soil directly to leaves (Allen, 1981). Bond and Pirone (1970) provided evidence that such transmission can likewise occur with *Potyviruses* as *Sugarcane mosaic virus* was occasionally transmitted from contaminated soil to sorghum plants. Also, as mentioned earlier, the *Potyvirus* PVY was found in forest soils (Büttner and Nienhaus, 1989a), and, although less stable than *Potexviruses* and *Tobamoviruses*, resembles them in being contact transmissible (Coutts and Jones, 2015). With the vast majority of viruses, the availability of information on soil-borne virus infection of plants in the absence of vectors is minimal for both managed and natural systems, a deficiency that needs to be addressed.

Demonstrated horizontal contact transmission via contaminated soil involves the same groups of viruses (*Tobamoviruses, Potexviruses, Tombusviruses, Sobemoviruses, Potyviruses*) as those involved in spread by wind-mediated contact transmission and nonspecific contact transmission by chewing insects. However, the resulting patterns of spread seen in the field are likely to differ because of the much greater restrictions imposed by the soil medium. The principal role a soil inoculum source is likely to play is in creation of new primary infection foci (Table 1). Where the rooting zones of neighboring plants intermingle, leakage of virus particles from the primarily infected plant into the soil in this zone might enlarge such primary infection foci somewhat. However, no other virus spread role for inoculum in the soil seems possible unless localized movement of freshly contaminated soil occurs through natural flooding, human intervention (e.g., within crop cultivation or irrigation), or the activity of soil animals such as moles and worms. Obviously, with managed systems in which seedlings are transplanted into uniformly virus-infested soil, numerous primary infection foci can still result in high virus incidences within a crop. In worst-case scenarios, where mixed plant species communities are involved, virus-contaminated soil is abundant, and seedlings are transplanted, the likelihood of this transmission pathway contributing toward new host species jumps or emergence of new viruses still seems "low" as a continuous connection between virus donor and recipient plants is absent.

3. INSECT VIRUSES

3.1 Transmission by Plants Serving as Passive "Vectors"

Horizontal virus transmission via passive plant "vectors" provides a novel and intriguing way for some viruses of herbivorous insects to enhance their survival and dispersal. Although these insect viruses cannot multiply within plants, they can circulate within the plants on which their insect hosts feed. The presence of these viruses distributed among the organs of such plants greatly increases the available extent of the "vector" from which healthy insects can become infected. The aphid-infecting *Cripaviruses Aphid lethal paralysis virus* (ALPV) and *Rhopalosiphum padi virus* (RhPV) both employ this transmission strategy (Bonning and Miller, 2010; D'Arcy et al., 1981; Gildow and D'Arcy, 1990; Liu et al., 2014b; Maina et al., 2017). For example, RhPV circulates throughout the barley plants its host aphid *R. padi* feeds on, reaching their roots in little as 7 days (Ban et al., 2007). Wamonje et al. (2017) suggested that the permissiveness of plants to similarly behaving *Cripaviruses* may have evolved to allow them to benefit from their insect-lethal properties which help control the aphid populations feeding on them. In return, the plants provide the *Cripaviruses* with greater access to their aphid hosts. *Cripaviruses* could therefore be used as natural biopesticides (Bonning and Miller, 2010), which would be deployed in both natural and managed environments to decrease feeding damage caused by insect pests and dampen the spread of insect-vectored plant viruses. Clearly, horizontal virus transmission via passive "vector" plants constitutes an important and fascinating arena for future insect virus research.

What are the implications of horizontal virus transmission via passive plant "vectors" on the patterns of *Cripavirus* spread within herbivorous insect populations? If a *Cripavirus*-infected herbivorous insect, such as a susceptible aphid species, flies or is blown to a new location in which the *Cripavirus* is absent, and feeds on a plant species capable of allowing the virus to circulate within it, soon afterward this plant will become the first passive "vector" for subsequent virus acquisition by susceptible aphids. When the first susceptible aphid from the new location alights and feeds on this newly viruliferous "vector" plant, it becomes infected, creating the first primary infection focus in the new location (Table 1). Every time the newly infected aphid alights and feeds on another plant, this plant soon becomes an additional passive "vector." Thus, every time susceptible aphids become infected from such "vector" plants, new secondary infection foci aphids will be created. As this process continues, the *Cripavirus* epidemic will enter its exponential phase within all susceptible aphid species populations present at the location, benefitting the plant "vector" species. This scenario could unfold equally well in both managed and natural environments. In situations where aphid populations include a mixture of different susceptible aphid species each occurring in high numbers and plant species in which *Cripaviruses* are circulating are abundant, the likelihood of this transmission pathway contributing toward new host species jumps or emergence of new viruses is estimated to be "high" because a continuous supply of donor "infector" plants would be present from which numerous recipient aphids could acquire the virus from (Table 1).

3.2 Transmission by Arthropod Vectors

Horizontal insect virus transmission by arthropod vectors is considered by insect virologists to be an indirect horizontal transmission pathway as it involves another biological agent (Chen et al., 2006a,b). In contrast to the situation with plant viruses, with one exception arthropod vector pathways for insect virus transmission have received little attention despite the abundance of potential ectoparasitic arthropod vectors (mites and insects) found afflicting insects. The sole exception to this is an obligate parasite of honey bee, the varroa mite, *Varroa destructor*. Varroa mites use their piercing mouthparts to penetrate the bodies of bees and reproduce on honey bee brood. They then emerge and attach to adult bees, or infest larvae or pupae. Uninfested colonies become infested naturally if bees from within them rob-infested hives, or bees from infested colonies "drift" over to (i.e., join) uninfested colonies, which tends to happen especially with drones but sometimes also with female worker bees (Anonymous, 2011). Working with the *Cripavirus Kasmir bee virus* (KBV) and varroa mite, Chen et al. (2004) provided the first unequivocal evidence of vector transmission of an insect virus. The mite transmits viruses, such as KBV, the *Aparavirus Israeli acute paralysis virus* (IAPV), and *Iflaviruses Deformed wing virus* (DWV) and *Sacbrood virus* (SBV) from virus infected to healthy bees, and these viruses possibly multiply within the mite vector itself as well as in their insect hosts (Chen et al., 2006a,b; Di Prisco et al., 2011; Martin et al., 2013; Shen et al., 2005a,b). Outbreaks of several bee viruses, such as DWV, KBV, SBV, and the *Triatovirus Black queen cell virus* (BQCV), have been associated with this mite (Chen et al., 2006a,b) and a key role in bee viral disease emergence was attributed to it (Martin et al., 2013). The mites' rapid global spread is hypothesized to have selected more virulent DWV variants; it has become one of the world most widely distributed and contagious insect viruses (Martin et al., 2013) (see also chapter "Emerging viruses in bees: From molecules to ecology" by McMahon et al., this volume).

Bumble bees and other social bees, and also native solitary bees host a wide diversity of mites (Eickwort, 1994; Klimov et al., 2007; Park et al., 2009). It is likely that these mites play viral transmission roles similar to the varroa mite in honey bee, and such transmission is important in spreading insect viruses through managed and natural environments. Whether insect parasites of other insects also help transmit the viruses of their hosts is unknown, and so would constitute an important subject for future investigation.

How might horizontal transmission of insect viruses via arthropod vectors influence patterns of virus spread within populations of honey bee pollinators? A colony of honey bee pollinators free of vectors or virus infection could become a new primary virus infection focus if an individual honey bee robs a vector-infested, virus-infected colony in another location, becomes infested with viruliferous vectors, and then returns and introduces vector infestation and virus infection to its own colony (Table 1). The same thing applies when a forager or drone from a vector-infested and virus-infected colony joins a colony in a new location lacking such infestation or infection. When viruliferous vector-infested honey bees from the new primary infection focus infest healthy colonies in the same location, secondary infection foci are created. As this process of colony infection continues, the virus epidemic will enter its exponential phase with further honey bee colony infection occurring at increasing speed. This scenario could unfold equally well in managed and natural environments. In situations where pollinator insect populations include a mixture of species co-occurring with many vector-infested, virus-infected honey bee colonies, the likelihood of this transmission pathway contributing toward new host species jumps or emergence of new insect pollinator viruses is estimated to be "high" (Table 1). This is because under this scenario a continuous supply of many vector-infested, virus-infected colonies would be present from which colonies of diverse pollinator species could acquire virus. However, in the specific case of varroa mites, which infect only honey bees, the initial route of arthropod-mediated horizontal transmission from honey bees to other bee species, or to other pollinators more generally, must include passive or active transfer via other ectoparasites, not many of which are known to be common to honey bees and other pollinators. Horizontal transmission of virus infection from honey bees to other species is much more likely to occur via shared flower use, as expanded upon in the next section.

3.3 Transmission via Contaminated Pollen

Insect pollinators are critically important for the wellbeing of managed and natural systems as many flowering and wild plants depend on them (Klein et al., 2007; Ollerton et al., 2011). Because many pollinators are generalist flower visitors (Waser et al., 1996), flower sharing provides a major route for both intraspecies and interspecies virus transmission by the oral-fecal route (Manley et al., 2015). In the early 2000s, six honeybee viruses were found in pollen (Chen et al., 2006a,b; Shen et al., 2005a). Soon afterward, Singh et al. (2010) studied RNA virus transmission by hymenopteran pollinators. When virus-contaminated pollen and honey were given to healthy honey bee colonies, the queen became infected and laid-infected eggs. The *Iflaviruses* DWV and SBV were detected in the pollen loads of uninfected foraging honeybees showing that they were carrying virus-contaminated pollen they had picked elsewhere back to their hives. In addition, when virus sequences from virus-infected foraging bees and their pollen pellets were examined separately, in only a small number did the virus sequences of the forager and her pollen pellet match. This finding implied that pollen was a virus source for infection of healthy colonies, and that, mediated by pollen from shared flowers, one pollinator species could become infected by a virus introduced to the flower by another pollinator species. Interestingly, virus-infected pollen pellets contained pollen from diverse plant species suggesting no restriction on the virus-pollen association arising from different flowering plant species. Singh et al. (2010) also demonstrated that, when they shared the same flowers as food sources, the *Aparavirus* IAPV was passed from infected bumble bees to uninfected honeybees and vice versa. Random deposition of virus-contaminated feces onto flowers was the most likely mechanism underlying intraspecies and interspecies virus transmission via pollen. They concluded that pollen serves as one of the major routes of inter-taxa virus transmission in the pollinator community.

Furst et al. (2014) emphasized that pathogen spillover from managed pollinators, such as commercial honey bee colonies, poses a potentially serious threat to wild pollinator species worldwide, citing presence of the same DWV strains among them. Sharing of diverse foraging sites facilitated interspecific virus transmission. Manley et al. (2015) highlighted that, in addition to DWV, another 23 honey bee viruses have broad host ranges infecting bumble bee, solitary bee, wasp, ant, and hoverfly species, and accidental release of virus-infected commercial bumble bees poses a serious risk of transmission to wild pollinators. Both studies advocated improved virus and other pathogen control in managed bee species and emphasized the need for more research to fill critical knowledge gaps over the global extent, influence and other aspects of emerging infectious virus, and other pathogens in managed and wild insect pollinators (see also chapter "Emerging viruses in bees: From molecules to ecology" by McMahon et al., this volume).

How might horizontal virus transmission via contaminated pollen influence patterns of virus spread occurring within insect pollinator populations? If a virus-infected insect pollinator flies to a new virus-free location and defecates within the flowers of plants it collects pollen from, the remaining pollen, or new pollen produced later, will become virus contaminated. The first local pollinator that subsequently arrives would then collect and return with contaminated pollen to its previously healthy nest. Any offspring in the local nest that consumes this pollen could then become infected and the offspring would serve as the first new primary infection focus in the new location (Table 1). Insects from this primary infection focus could then spread virus through local flowers to other insects, and their nests would then become secondary virus infection foci. As this process continues, the virus epidemic would enter its exponential phase with pollinator infection occurring with increasingly speed. In situations where pollinator insect populations include a mixture of different species each occurring in high numbers and virus infection is widespread, the likelihood of this transmission pathway contributing toward new host species jumps or emergence of new insect pollinator viruses is estimated to be "high" (Table 1). This is because a continuous supply of virus-contaminated flowers belonging to different species of flowering plants would be present from which diverse pollinator species could acquire the virus from on multiple occasions. Support for this "high" estimate was provided in a study by McMahon et al. (2015) who applied network analysis to examine the risk of emerging virus transmission occurring between managed and wild pollinator bee species. They found widespread transmission of RNA viruses between the two groups. Common RNA viruses previously associated with honey bees were widespread in bumble bee populations, and the different viruses varied greatly in terms of individual pathogen load and population-level prevalence. They suggested that virus disease spillover may be an important general aspect of RNA virus epidemiology in bee pollinators.

3.4 Transmission via Contaminated Nectar

Nectar serves as another vehicle for transmission of insect viruses. The *Iflaviruses* DWV and SBV, *Cripavirus* KBV, and *Triatovirus* BQCV, have all been found contaminating honey (Chen et al., 2006a,b; Shen et al., 2005a). Honey commences as nectar collected from flowers by honey bees, which store it inside honeycombs where it gets broken down into simple sugars. Constant fanning by bee wings causes evaporation, creating the sweet honey liquid. Insect viruses are spread by feeding contaminated honey to young bee larvae. Nurse bees also become virus infected while cleaning out cells containing infected larvae (e.g., Capinera, 2008). Detection of insect viruses in honey (Chen et al., 2006a,b; Shen et al., 2005a) suggests that the virus contamination might originate in the nectar collected from flowers by bees before using it to make the honey. However, this is likely a less frequent route for honey contamination. Honey bees process nectar for honey by mixing it with secretions from the hypopharyngeal glands in their heads, which are full of enzymes (e.g., invertase) used to invert sucrose to glucose and fructose. In infected bees, these glands can be filled with virus (Lanzi et al., 2006 and references therein) and thus glandular secretions are the most likely source of virus contamination of honey. It is also possible that honey sometimes becomes contaminated with virus within the colony by small amounts of virus-contaminated pollen. In cases where virus transfer to honey via nectar occurs, it would result from random deposition of contaminated pollinator feces within flowers. In pollinator species other than honey bees, it is possible that nectar serves as a more important transmission pathway and merits further investigation.

The implications of horizontal virus transmission occurring via virus-contaminated nectar upon the patterns of virus spread occurring within insect populations likely resemble those for pollen transmission (see above and Table 1). In situations where insect pollinator populations include a mixture of different species each occurring in high numbers and virus infection is widespread, the likelihood of this transmission pathway contributing toward new host species jumps or emergence of new insect viruses is estimated to be "high" (Table 1). This is because of the presence of a continuous supply of contaminated flowers belonging to different species of flowering plants from which numerous colonies and diverse insect species could repeatedly acquire virus.

4. ECOLOGICAL SIGNIFICANCE OF TRANSMISSION PATHWAYS

The capacity for transmission by different pathways enhances the ability of a virus to spread and survive when circumstances change. Multiple transmission pathways are useful for a virus when, for example, a formerly relied upon specific vector becomes rare (Hamelin et al., 2016). The need for diverse transmission pathways varies with the relative complexity and stability of the pathosystem. Selection for multiple transmission pathways as a way to survive changing circumstances is likely to be higher in natural communities containing diverse plant and insect species than in managed monocultures with steady access to water and nutrition, and where arthropod vectors are uncontrolled. Generalist viruses are likely to adapt more readily to fluctuating-natural communities, whereas specialist viruses may have a selective advantage in cultivated vegetation managed under standard conditions. At least with plant viruses, managed communities of mixed species, such as pastures, market or back gardens, and subsistence farms, would likely apply intermediate selection pressures. By contrast, disturbed natural vegetation at the managed and natural vegetation interface would likely exert the greatest selection pressure for plant virus transmission pathway diversity, as circumstances where the greatest diversity of managed and native insect pollinators rely on the same food source would do for insect pollinator viruses. As a result, host species jumps and new virus emergence events for both plant and insect viruses seem most likely to occur in these same situations where selection pressure is high. Unstable weather scenarios associated with climate change and human activities will accentuate such selection pressures (Jones, 2009, 2016; Jones and Barbetti, 2012; Potts et al., 2010; Vanbergen, 2013).

This review estimated the likelihood that each of the plant and insect virus transmission pathways considered would create new primary infection foci, enlarge previously existing infection foci, and/or contribute to more generalized virus spread (Table 1). Such information about virus spread patterns is important if we are to understand the circumstances where each of the different transmission pathways would promote the fitness of individual viruses. Brunner and Yarber (see chapter "Evaluating the importance of environmental persistence for *Ranavirus* transmission and epidemiology" by Brunner and Yarber, this volume) used a simple modeling approach with ranaviruses that could be employed in the future to assess the relative importance of different transmission pathways when sufficient data are available to parameterize a model.

This review further estimated the likelihood that each individual virus transmission pathway might contribute toward development of host species jumps and virus emergence events (Table 1). The likelihood of jumps and emergence were estimated for worst-case scenario situations in which multiple host species were present and virus transmission circumstances optimal. Such information is significant if we are to understand the circumstances that favor host species jumps or virus emergence events where mixtures of host species intermingle, especially in disturbed vegetation at the managed and natural vegetation interface, but also in undisturbed natural vegetation, managed pasture communities, gardens, and subsistence farms.

Experimental data to explore these ideas are beginning to be developed. For example, Bernardo et al. (2018) used a geometagenomics approach to examine the spatial and temporal distributions of plant virus populations within managed and natural vegetation components of agroecological landscapes in the Western Cape region of southern Africa and southwest France. They found that the viral diversity and incidence of novel viruses identified were both greater in natural vegetation, but overall virus prevalence was greater, and some virus families were more common, in managed vegetation. Virus patterns in natural vegetation apparently reflected those communities' greater host diversity, whereas patterns in managed vegetation apparently reflected greater plant host abundance, adaptation of some virus families to the cultivated plant species present, and the reduction in host species diversity characteristic of crop monocultures. Emerging viruses were mostly generalists that "spill-over" into new host populations from ecologically connected host reservoir populations in which the virus was maintained permanently. Identification of reservoir potential is central to understanding virus emergence, which is greater in short-lived hosts than long-lived hosts as the latter invest more in "costly" pathogen defenses (e.g., Hily et al., 2014).

In another relevant example, Li et al. (2014) found that exposure of insect pollinators to pollen contaminated with plant virus increased the risk of viral host shifts from plants to insect pollinators. They reported that under some conditions, the plant-infecting *Nepovirus Tobacco ringspot virus* (TRSV) was surprisingly able to replicate and produce virions within different honey bee body parts. TRSV-infected bees were continually present within some monitored honey bee colonies, which were negatively impacted (but were also infected with DWV), raising the prospect of prolonged virus transmission by individuals from contaminated colonies. TRSV was found inside varroa mites suggesting that such mites facilitate TRSV spread within the bee colonies. Flenniken (2014) provided a commentary on this discovery that a plant virus can replicate in honey bees, varroa mite may play a role in its horizontal transmission, and across kingdom infection was involved. This fascinating scenario provides plant and insect virologists a new arena for collaborative research on virus transmission pathways and host shifts involving insect pollinators.

The fact that many well-known viruses often remain difficult to control suggests that the common assumption that their "disease cycles" are sufficiently studied may be unwarranted. Revisiting previous ecological findings, and combining that information with knowledge concerning the likely roles of each of the virus transmission pathways considered in this review, will help to provide greater understanding of how plant and insect viruses survive and spread. Furthermore, this type of holistic approach to virus ecology can help fill critical information gaps that enable the effective targeting of previously unknown vulnerabilities in virus "disease cycles." This in turn, helps make possible the development of improved integrated virus disease management procedures, forecasting models, and decision support systems that permit interventions to manage virus spread more effectively (e.g., Jones, 2004, 2006, 2009; Jones et al., 2010; Thresh, 2006). Deployment of effective integrated virus disease management, forecasting models, and decision support systems in this way is becoming increasingly necessary as virus epidemics become more difficult to control due to climate change and anthropogenic activities (Furst et al., 2014; Jones, 2009, 2016; Jones and Barbetti, 2012; Manley et al., 2015; Potts et al., 2010; Vanbergen, 2013). Fortunately, technological innovation is providing opportunities to enhance research in virus ecology and epidemiology, and improving abilities to collect and process information relevant to interventions that target virus diseases across different scales (e.g., Jones, 2014a,b; Roossinck et al., 2015; see also chapter "Emerging viruses in bees: From molecules to ecology" by McMahon et al., this volume).

5. CONCLUSION

In the current era of burgeoning human population growth and rapid climate change, it is becoming increasingly important to fully understand the ecologies of plant and insect viruses, and the threats they pose to global food security and natural communities. Such knowledge is essential for successful management of virus disease outbreaks across landscapes. For plant viruses, the current understanding of the ecologies of well-known viruses in managed systems, often established many years ago, is not necessarily the whole story, as some transmission pathways have received little attention. Even less is known about how most insect viruses survive and spread. Therefore, regardless of whether virus threats are already presumed to be well understood, it is important to examine any additional transmission pathway parameters likely to be involved. This is because the types of viruses likely to depend on such alternative pathways include pathogens that currently cause economic losses and others likely to threaten plant and beneficial insect populations in the future. Despite this, little attention is currently being paid to the roles of the 12 virus transmission pathways covered in this review, all of which, in specific situations, are likely to play significant roles in virus disease spread across managed and natural landscapes. The neglected pathways all deserve renewed attention and research, and the novel pathways much more investigation.

ACKNOWLEDGMENTS

I thank Brenda Coutts, Benjamin Congdon, and Craig Webster for helpful discussions about virus transmission pathways.

REFERENCES

Alexander, H.M., Mauck, K.E., Whitfield, A.E., Garrett, K.A., Malmstrom, C.M., 2014. Plant–virus interactions and the agro-ecological interface. Eur. J. Plant Pathol. 138, 529–547.

Allen, W.R., 1981. Dissemination of tobacco mosaic virus from soil to plant leaves under glasshouse conditions. Can. J. Plant Pathol. 3, 163–168.

Anonymous, 2011. Varroa Mites. BeeAware, Plant Health Australia, Canberra, Australia. http://beeaware.org.au/archive-pest/varroa-mites.

Antignus, Y., Lachman, O., Pearlsman, A., Koren, A., 2005. Containment of *Cucumber fruit mottle mosaic virus* (CFMMV) infection through roots by planting into a virus-free intermediating medium. Phytoparasitica 33, 85–87.

Antignus, Y., Lachman, O., Pearlsman, M., 2007. Spread of *Tomato apical stunt viroid* (TASVd) in greenhouse tomato crops is associated with seed transmission and bumble bee activity. Plant Dis. 91, 47–50.

Ban, L., Didon, A., Jonsson, L.M., Glinwood, R., Delp, G., 2007. An improved detection method for the *Rhopalosiphum padi virus* (RhPV) allows monitoring of its presence in aphids and movement within plants. J. Virol. Methods 142, 136–142.

Bawden, F.C., 1964. Plant Viruses and Virus Diseases, fourth ed. Ronald Press, New York.

Bennett, C.W., 1940. Acquisition and transmission of viruses by dodder (*Cuscuta subinclusa*). Phytopathology 30, 649–656.

Bennett, C.W., 1944. Studies of dodder transmission of plant viruses. Phytopathology 34, 905–932.

Bernardo, P., Charles-Dominique, T., Barakat, M., Ortet, P., Fernandez, E., et al., 2018. Geometagenomics illuminates the impact of agriculture on the distribution and prevalence of plant viruses at the ecosystem scale. ISME J. 12, 173–184.
Boben, J., Kramberger, P., Petrovič, N., Cankar, K., Peterka, M., Štrancar, A., Ravnikar, M., 2007. Detection and quantification of *Tomato mosaic virus* in irrigation waters. Eur. J. Plant Pathol. 118, 59–71.
Bond, W.P., Pirone, T.P., 1970. Evidence for soil transmission of sugarcane mosaic virus. Phytopathology 60, 437–440.
Bonning, B.C., Miller, W.A., 2010. Dicistroviruses. Annu. Rev. Entomol. 55, 129–150.
Boylan-Pett, W., Ramsdell, D.C., Hoopingarner, R.A., Hancock, J.F., 1991. Honeybee foraging behavior, in hive survival of infectious, pollen-borne blueberry leaf mottle virus and transmission of the virus in highbush blueberry. Phytopathology 81, 1407–1412.
Bradley, R.H.E., Harris, K.F., 1972. Aphids can inoculate plants with tobacco mosaic virus by clawing. Virology 50, 615–618.
Bristow, P.R., Martin, R.R., 1999. Transmission and the role of honeybees in field spread of blueberry shock ilarvirus, a pollen-borne virus of highbush blueberry. Phytopathology 89, 124–130.
Broadbent, L., 1976. Epidemiology and control of tomato mosaic virus. Annu. Rev. Phytopathol. 14, 75–96.
Broadbent, L., Fletcher, J.T., 1963. The epidemiology of tomato mosaic. Ann. Appl. Biol. 52, 233–241.
Burand, J.P., Kim, W., Afonso, C.L., Tulman, E.R., Kutish, T.F., Lu, Z., Rock, D.L., 2012. Analysis of the genome of a sexually transmitted insect virus *Helicoverpa zea nudivirus 2*. Viruses 4, 28–61.
Büttner, C., Nienhaus, E., 1989a. Virus contamination of soils in forest ecosystems of the Federal Republic of Germany. For. Pathol. 19, 47–53.
Büttner, C., Nienhaus, F., 1989b. Virus contamination of waters in two forest districts of the Rhineland area (FRG). For. Pathol. 19, 206–211.
Bwye, A.M., Jones, R.A.C., Proudlove, W., 1994. Effects of sowing seed with different levels of infection, plant density and the growth stage at which plants first develop symptoms on cucumber mosaic-virus infection of narrow-leafed lupins (*Lupinus angustifolius*). Aust. J. Agric. Res. 45, 1395–1412.
Capinera, J.L. (Ed.), 2008. Encyclopedia of Entomology. Springer, Dordrecht.
Card, S.D., Pearson, M.N., Clover, G.R.G., 2007. Plant pathogens transmitted by pollen. Australas. Plant Pathol. 36, 455–461.
Carter, W., 1961. Ecological aspects of plant virus transmissions. Annu. Rev. Entomol. 6, 347–370.
Chen, Y.P., Siede, R., 2007. Honey bee viruses. Adv. Virus Res. 70, 33–80.
Chen, Y.P., Pettis, J.S., Evans, J.D., Kramer, M., Feldlaufer, M.F., 2004. Transmission of *Kashmir bee virus* by the ectoparasitic mite *Varroa destructor*. Apidologie 35, 441–448.
Chen, Y.P., Evans, J., Feldlaufer, M., 2006a. Horizontal and vertical transmission of viruses in the honey bee, Apis mellifera. J. Invertebr. Pathol. 92, 152–159.
Chen, Y., Pettis, J., Collins, A., Feldlaufer, M.F., 2006b. Prevalence and transmission of honeybee viruses. Appl. Environ. Microbiol. 72, 606–611.
Childress, A.M., Ramsdell, D.C., 1987. Bee-mediated transmission of blueberry leaf mottle virus via infected pollen in highbush blueberry. Phytopathology 77, 167–172.
Cochran, G.W., 1947. The dodder graft a new method of using dodder to transmit plant viruses. Phytopathology 37, 5.
Congdon, B.S., Coutts, B.A., Renton, M., Jones, R.A.C., 2016. *Pea seed-borne mosaic virus:* stability and wind-mediated contact transmission in field pea. Plant Dis. 100, 953–958.
Cooper, J.I., 1993. Virus Diseases of Trees and Shrubs. Institute of Terrestrial, Natural Environment Research Institute, Oxford.

Cooper, J.I., Jones, R.A.C., 2006. Wild plants and viruses: under-investigated ecosystems. Adv. Virus Res. 67, 1–47.

Cory, J.S., 2010. The ecology of baculoviruses. In: Asgari, S., Johnson, K.N. (Eds.), Insect Virology. Academic Press, Cambridge, MA; New York, pp. 405–421.

Cory, J.S., 2015. Insect virus transmission: different routes to persistence. Curr. Opin. Insect Sci. 8, 130–135.

Cory, J.S., Myers, J.H., 2003. The ecology and evolution of insect baculoviruses. Annu. Rev. Ecol. Evol. Syst. 34 (1), 239–272.

Coutts, B.A., Jones, R.A.C., 2015. *Potato virus Y*: contact transmission, stability on surfaces, and inactivation with disinfectants. Plant Dis. 99, 387–394.

Coutts, B.A., Thomas-Carroll, M.L., Jones, R.A.C., 2004. Patterns of spread of *Tomato spotted wilt virus* in field crops of lettuce and pepper: spatial dynamics and validation of control measures. Ann. Appl. Biol. 145, 231–245.

Coutts, B.A., Kehoe, M.A., Jones, R.A.C., 2013. *Zucchini yellow mosaic virus*: contact transmission, stability on surfaces, and inactivation with disinfectants. Plant Dis. 97, 765–771.

D'Arcy, C.J., Burnett, P.A., Hewings, A.D., 1981. Detection, biological effects and transmission of a virus of the aphid *Rhopalosiphum padi*. Virology 114, 268–272.

Darzi, E., Smith, E., Shargil, D., Lachman, O., Ganot, L., Dombrovsky, A., 2018. The honeybee *Apis mellifera* contributes to *Cucumber green mottle mosaic virus* spread via pollination. Plant Pathol. 67, 244–251.

Desjardins, P.R., Drake, R.J., French, J.V., 1969. Transmission of citrus ringspot virus to citrus and non-citrus hosts by dodder (*Campestris subinclusa*). Plant Dis. Rep. 53, 969.

Di Prisco, G., Pennacchio, F., Caprio, E., Boncristiani Jr., H.F., Evans, J.D., Chen, Y., 2011. *Varroa destructor* is an effective vector of *Israeli acute paralysis virus* in the honeybee, *Apis mellifera*. J. Gen. Virol. 92, 151–155.

Djikeng, A., Kuzmickas, R., Anderson, N.G., Spiro, D.J., 2009. Metagenomic analysis of RNA viruses in a fresh water lake. PLoS ONE 4, e7264.

Dombrovsky, A., Tran-Nguyen, L.T., Jones, R.A.C., 2017. Cucumber green mottle mosaic virus: rapidly increasing global distribution, etiology, epidemiology, and management. Annu. Rev. Phytopathol. 55, 231–256.

Draper, M.A., Gudmestad, N.C., 1992. Mechanical transmission of PVY by seed cutting. Am. Potato J. 69, 577–578.

Eickwort, G.C., 1994. Evolution and life-history patterns of mites associated with bees. In: Houck, M.A. (Ed.), Mites: Ecological and Evolutionary Analyses of Life-History Patterns. Chapman and Hall, New York, pp. 218–251.

Epstein, A.H., 1978. Root graft transmission of tree pathogens. Annu. Rev. Phytopathol. 16, 181–192.

Faillace, C.A., Lorusso, N.S., Duffy, S., 2017. Overlooking the smallest matter: viruses impact biological invasions. Ecol. Lett. 20, 524–538.

Flenniken, M.L., 2014. Honey bee-infecting plant virus with implications on honey bee colony health. mBio 5 (2), e00877–14.

Freitag, J.H., 1941. Insect transmission, host range and properties of squash-mosaic virus. Insect transmission, host range and properties of squash-mosaic virus. Phytopathology 31, 8. (Abstr.).

Fuller, E., Elderd, B.D., Dwyer, G., 2012. Pathogen persistence in the environment and insect–baculovirus interactions: disease-density thresholds, epidemic burn out and insect outbreaks. Am. Nat. 179, E70–E96.

Furst, M.A., McMahon, D.P., Osborne, J.L., Paxton, R.J., Brown, M.J.F., 2014. Disease associations between honeybees and bumblebees as a threat to wild pollinators. Nature 506, 364–366.

Gal-On, A., Naglis, A., Leibman, D., Ziadna, H., Kathiravan, K., Papayiannis, L., Holdengraber, V., Guenoune-Gelbert, D., Lapidot, M., Aly, R., 2009. Broomrape can acquire viruses from its hosts. Phytopathology 99, 1321–1329.

George, J.A., Davidson, T.R., 1963. Pollen transmission of sour cherry yellows viruses from tree to tree. Can. J. Plant Sci. 43, 276–288.

George, J.A., Davidson, T.R., 1964. Further evidence of pollen transmission of necrotic ringspot and sour cherry yellows viruses in sour cherry. Can. J. Plant Sci. 44, 383–384.

Gildow, F.E., D'Arcy, C.J., 1990. Cytopathology and experimental host range of *Rhopalosiphum padi virus*, a small isometric RNA virus infecting cereal grain aphids. J. Invertebr. Pathol. 55, 245–257.

Gilmer, R.M., 1965. Additional evidence of tree-to-tree transmission of sour cherry yellows virus by pollen. Phytopathology 55, 482(Abstr.).

Goldschmidt, E.E., 2014. Plant grafting: new mechanisms, evolutionary implications. Front. Plant Sci. 5, 727. https://doi.org/10.3389/fpls.2014.00727.

Golino, D.A., Cunningham, M., Rowhani, A., Sim, S.T., 2005. Transmission of rose mosaic viruses. Acta Hortic. 751, 217–224.

Gooding Jr., G.V., Todd, F.A., 1976. Soil-borne tobacco mosaic virus as an inoculum source for flue-cured tobacco. Tob. Sci. 20, 140–142.

Goss, R.W., 1931. Infection Experiments With Spindle Tuber and Unmottled Curly Dwarf of the Potato. Research Bulletin of the Agricultural Experiment Station of Nebraska. No. 53.

Graham, B.F., Bormann, F.H., 1966. Natural root grafts. Bot. Rev. 32, 255–292.

Greber, R.S., Klose, M.J., Milne, R.S., Teakle, D.S., 1991. Transmission of prunus necrotic ringspot virus using plum pollen and thrips. Ann. Appl. Biol. 118, 589–593.

Greber, R.S., Teakle, D.S., Milne, R.S., 1992. Thrips-facilitated transmission of prune dwarf and prunus necrotic ringspot viruses from cherry pollen to cucumber. Plant Dis. 76, 1039–1041.

Grogan, R.G., Campbell, R.N., 1966. Fungi as vectors and hosts of viruses. Annu. Rev. Phytopathol. 4, 29–52.

Hamelin, F.M., Allen, L.J., Prendeville, H.R., Hajimorad, M.R., Jeger, M.J., 2016. The evolution of plant virus transmission pathways. J. Theor. Biol. 396, 75–89.

Hardy, V.G., Teakle, D.S., 1992. Transmission of sowbane mosaic virus by *Thrips tabaci* in the presence and absence of virus-carrying pollen. Ann. Appl. Biol. 121, 315–320.

Harrison, B.D., 1981. Plant virus ecology: ingredients, interactions and environmental influences. Ann. Appl. Biol. 99, 195–209.

Hily, J.M., García, A., Moreno, A., Plaza, M., Wilkinson, M.D., Fereres, A., Fraile, A., García-Arenal, F., 2014. The relationship between host lifespan and pathogen reservoir potential: an analysis in the system *Arabidopsis thaliana-Cucumber mosaic virus*. PLoS Pathog. 10 (11), e1004492.

Hoggan, I.A., 1931. Further studies on aphid transmission of plant viruses. Phytopathology 21, 199–212.

Hoggan, I.A., 1934. Transmissibility by aphids of the tobacco mosaic virus from different hosts. J. Agric. Res. 49, 1135–1142.

Hollings, M., Komuro, Y., Tochihara, H., 1975. Cucumber green mottle mosaic virus. In: Descriptions of Plant Viruses, vol. 154. CMI/AAB, Wellesbourne.

Hollings, M., Stone, O.M., Barton, R.J., 1977. Pathology, soil transmission and characterization of cymbidium ringspot, a virus from cymbidium orchids and white clover (*Trifolium repens*). Ann. Appl. Biol. 85, 233–248.

Hong, C.X., Moorman, G.W., 2005. Plant pathogens in irrigation water: challenges and opportunities. Crit. Rev. Plant Sci. 24, 189–208.

Hopper, S.D., Gioia, P., 2004. The southwest Australian floristic region: evolution and conservation of a global hot spot of biodiversity. Annu. Rev. Ecol. Evol. Syst. 35, 623–650.

Hosford, R.M., 1967. Transmission of plant viruses by dodder. Bot. Rev. 33, 387–406.

Hull, R., 2014. Mathews' Plant Virology, fifth ed. Academic Press, London.

Hunter, J.A., Chamberlain, E.E., Atkinson, J.D., 1958. Note on transmission of apple mosaic by natural root grafting. N. Z. J. Agric. Res. 1, 80–82.

Johansen, E., Edwards, M.C., Hampton, R.O., 1994. Seed transmission of viruses: current perspectives. Annu. Rev. Phytopathol. 32, 363–386.

Johnson, F., 1941. Transmission of plant viruses by dodder. Phytopathology 31, 649–656.

Jones, R.A.C., 2004. Using epidemiological information to develop effective integrated virus disease management strategies. Virus Res. 100, 5–30.

Jones, R.A.C., 2005. Patterns of spread of two non-persistently aphid-borne viruses in lupin stands under four different infection scenarios. Ann. Appl. Biol. 146, 337–350.

Jones, R.A.C., 2006. Control of plant virus diseases. Adv. Virus Res. 67, 205–244.

Jones, R.A.C., 2009. Plant virus emergence and evolution: origins, new encounter scenarios, factors driving emergence, effects of changing world conditions, and prospects for control. Virus Res. 141, 113–130.

Jones, R.A.C., 2014a. Plant virus ecology and epidemiology: historical perspectives, recent progress and future prospects. Ann. Appl. Biol. 164, 320–347.

Jones, R.A.C., 2014b. Trends in plant virus epidemiology: opportunities from new or improved technologies. Virus Res. 186, 3–19.

Jones, R.A.C., 2016. Future scenarios for plant virus pathogens as climate change progresses. Adv. Virus Res. 95, 87–147.

Jones, R.A.C., Barbetti, M.J., 2012. Influence of climate change on plant disease infections and epidemics caused by viruses and bacteria. CAB Rev. 7 (22), 1–32. http://www.cabi.org/cabreviews.

Jones, R.A.C., Coutts, B.A., 2015. Spread of introduced viruses to new plants in natural ecosystems and the threat this poses to plant biodiversity. Mol. Plant Pathol. 16, 541–545.

Jones, R.A.C., Salam, M.U., Maling, T., Diggle, A.J., Thackray, D.J., 2010. Principles of predicting epidemics of plant virus disease. Annu. Rev. Phytopathol. 48, 179–203.

Kaiser, B., Vogg, G., Fürst, U.B., Albert, M., 2015. Parasitic plants of the genus *Cuscuta* and their interaction with susceptible and resistant host plants. Front. Plant Sci. 6, 45.

Keane, F.W.L., May, J., 1963. Natural root grafting in cherry, and spread of cherry twisted leaf virus. Can. Plant Dis. Surv. 43, 54–61.

Klein, A.M., Vaissiere, B.E., Cane, J.H., Steffan-Dewenter, I., Cunningham, S.A., Kremen, C., Tscharntke, T., 2007. Importance of pollinators in changing landscapes for world crops. Proc. Roy. Soc. Ser. B 274, 303–313.

Kleinhempel, H., Kegler, G., 1982. Transmission of tomato bushy stunt virus without vectors. Acta Phytopathol. Acad. Sci. Hung. 17, 17–21.

Klimov, P.B., O'Connor, B., Knowles, L.L., 2007. Museum specimens and phylogenies elucidate ecology's role in coevolutionary associations between mites and their bee hosts. Evolution 61, 1368–1379.

Koenig, R., 1986. Plant viruses in rivers and lakes. Adv. Virus Res. 31, 321–333.

Koenig, R., Leseman, D.E., 1985. Plant viruses in German rivers and lakes. I. Tombusviruses, a potexvirus and carnation mottle virus. Phytopathology 112, 105–116.

Koh, S.H., Li, H., Sivasithamparam, K., Admiraal, R., Jones, M.G.K., Wylie, S.J., 2018. Low root-to-root transmission of a tobamovirus, yellow tailflower mild mottle virus, and resilience of its virions. Plant Pathol. 67, 651–659.

Lacasa, A., Guerrero, M.M., Hita, I., Martınez, M.A., Jorda, C., Bielza, P., Contreras, J., Alcazar, A., Cano, A., 2003. Implication of bumble bees (Bombus spp.) on *Pepino mosaic virus* (PepMV) spread on tomato crops. Plagas 29, 393–403.

Lanzi, G., de Miranda, J.R., Boniotti, M.B., Cameron, C.E., Lavazza, A., Capucci, L., Camazine, S.M., Rossi, C., 2006. Molecular and biological characterization of deformed wing virus of honeybees (*Apis mellifera* L.). J. Virol. 80, 4998–5009.
Lev-Yadun, S., 2011. Why should trees have natural root grafts? Tree Physiol. 31, 575–578.
Li, J.L., Cornman, R.S., Evans, J.D., Pettis, J.S., Zhao, Y., Murphy, C., Peng, W.J., Wu, J., Hamilton, M., Boncristiani, H.F., Zhou, L., 2014. Systemic spread and propagation of a plant-pathogenic virus in European honeybees. Apis mellifera. MBio 5 (1), e00898-13.
Li, J.X., Liu, S.S., Gu, Q.S., 2015. Transmission efficiency of *Cucumber green mottle mosaic virus* via seeds, soil, pruning and irrigation water. J. Phytopathol. (5), 300–309.
Linder, H.P., 2003. The radiation of the cape flora, southern Africa. Biol. Rev. 78, 597–638.
Liu, H.W., Luo, L.X., Li, J.Q., Liu, P.F., Chen, X.Y., Hao, J.J., 2014a. Pollen and seed transmission of *Cucumber green mottle mosaic* virus in cucumber. Plant Pathol. 63, 72–77.
Liu, S., Vijayendran, D., Carrillo-Tripp, J., Miller, W.A., Bonning, B.C., 2014b. Analysis of new *Aphid lethal paralysis virus* (ALPV) isolates suggests evolution of two ALPV species. J. Gen. Virol. 95, 2809–2819.
Lojek, J.S., Orlob, G.B., 1969. Aphid transmission of tobacco mosaic virus. Science 164, 1407–1408.
Longdon, B., Jiggins, F.M., 2012. Vertically viral endosymbionts of insects: do sigma viruses walk alone? Proc. Roy. Soc. Ser. B 279, 3889–3898.
Longdon, B., Brockhurst, M.A., Russell, C.A., Welch, J.J., Jiggins, F.M., 2014. The evolution and genetics of virus host shifts. PLoS Pathog. 10 (11), e1004395.
Maina, S., Edwards, O.R., de Almeida, L., Ximenes, A., Jones, R.A.C., 2017. Metagenomic analysis of cucumber RNA from East Timor reveals an aphid lethal paralysis virus genome. Genome Announc. 5 (2), e01445-16.
Malmstrom, C.M., Alexander, H.M., 2016. Effects of crop viruses on wild plants. Curr. Opin. Virol. 19, 30–36.
Malmstrom, C.M., Melcher, U., Bosque-Pérez, N.A., 2011. The expanding field of plant virus ecology: historical foundations, knowledge gaps and research directions. Virus Res. 159, 84–94.
Manley, R., Boots, M., Wilfert, L., 2015. Emerging viral disease risk to pollinating insects: ecological, evolutionary and anthropogenic factors. J. Appl. Ecol. 52, 331–340.
Martin, S.J., Highfield, A.C., Brettell, L., Villalobos, E.M., Budge, G.E., Powell, M., Nikaido, S., Schroeder, D.C., 2013. Global honey bee viral landscape altered by a parasitic mite. Science 336, 1304–1306.
Martini, C., 1958. Proceedings of the Third Conference on Potato Virus Diseases, Lisse-Wageningen. p. 106.
Matsuura, S., Matsushita, Y., Kozuka, R., Shimizu, S., Tsuda, S., 2010. Transmission of *Tomato chlorotic dwarf viroid* by bumblebees (*Bombus ignitus*) in tomato plants. Eur. J. Plant Pathol. 126, 111–115.
Matthews, R.E.F., 1970. Plant Virology. Academic Press, New York.
McKirdy, S.J., Jones, R.A.C., Sivasithamparam, K., 1998. Determining the effectiveness of grazing and trampling by livestock in transmitting white clover mosaic and subterranean clover mottle viruses. Ann. Appl. Biol. 132, 91–105.
McMahon, D.P., Fürst, M.A., Caspar, J., Theodorou, P., Brown, M.J., Paxton, R.J., 2015. A sting in the spit: widespread cross-infection of multiple RNA viruses across wild and managed bees. J. Anim. Ecol. 84, 615–624.
McNeil, J., 2010. Viruses as biological control agents of insect pests. Org. Agric. eOrganic 2525. http://articles.extension.org/pages/18927/viruses-as-biological-control-agents-of-insect-pests.
Mehle, N., Ravnikar, M., 2012. Plant viruses in aqueous environment–survival, water mediated transmission and detection. Water Res. 46, 4902–4917.

Mehle, N., Gutierrez-Aguirre, L., Prezelj, N., Delic, D., Vidic, U., Ravnikar, M., 2014. Survival and transmission of *Potato virus Y*, *Pepino mosaic virus*, and *Potato spindle tuber viroid* in water. Appl. Environ. Microbiol. 80, 1455–1462.

Mikona, C., Jelkmann, W., 2010. Replication of grapevine leafroll-associated virus-7 (GLRaV-7) by *Cuscuta* species and its transmission to herbaceous plants. Plant Dis. 94, 471–476.

Miller, L.K., Ball, L.A. (Eds.), 1998. The Insect Viruses. Plenum Press, New York.

Milne, J.R., Walter, G.H., 2003. The coincidence of thrips and dispersed pollen in PNRSV-infected stonefruit orchards—a precondition for thrips-mediated transmission via infected pollen. Ann. Appl. Biol. 142, 291–298.

Mink, G.I., 1993. Pollen and seed-transmitted viruses and viroids. Annu. Rev. Phytopathol. 31, 375–402.

Moreno, A., Nebreda, M., Diaz, B.M., García, M., Salas, F., Fereres, A., 2007. Temporal and spatial spread of *Lettuce mosaic virus* in lettuce crops in central Spain: factors involved in *Lettuce mosaic virus* epidemics. Ann. Appl. Biol. 150, 351–360.

Murant, A.F., Chambers, J., Jones, A.T., 1974. Spread of raspberry bushy dwarf virus by pollination, its association with crumbly fruit, and problems of control. Ann. Appl. Biol. 77, 271–281.

Nutter Jr., F.W., 1997. Quantifying the temporal dynamics of plant virus epidemics: a review. Crop Prot. 16, 603–618.

Okada, K., Kusakari, S., Kawaratani, M., Negoro, J., Ohki, S.T., Osaki, T., 2000. Tobacco mosaic virus is transmissible from tomato to tomato by pollinating bumblebees. J. Gen. Plant Pathol. 66, 71–74.

Ollerton, J., Winfree, R., Tarrant, S., 2011. How many flowering plants are pollinated by animals? Oikos 120, 321–326.

Orlob, G.B., 1963. Reappraisal of transmission of tobacco mosaic virus by insects. Phytopathology 53, 822–830.

Pares, R.D., Gunn, L.V., Cresswell, G.C., 1992. Tomato mosaic virus infection in a recirculating nutrient solution. J. Phytopathol. 135, 192–198.

Park, W.M., Lee, G.P., Ryu, K.H., Park, K.W., 1999. Transmission of tobacco mosaic virus in recirculating hydroponic system. Sci. Hortic. 79, 217–226.

Park, Y.-L., Kondo, V., White, J., West, T., McConnell, B., McCutcheon, T., 2009. Nest-to-nest dispersal of *Chaetodactylus krombeini* (Acari, Chaetodactylidae) associated with *Osmia cornifrons* (Hym., Megachilidae). J. Appl. Entomol. 133, 174–180.

Pethybridge, S.J., Madden, L.V., 2003. Analysis of spatiotemporal dynamics of virus spread in an Australian hop garden by stochastic modeling. Plant Dis. 87, 56–62.

Pethybridge, S.J., Wilson, C.R., Ferrandino, F.J., Leggett, G.W., 2000. Spatial analyses of viral epidemics in Australian hop gardens: implications for mechanisms of spread. Plant Dis. 84, 513–515.

Pethybridge, S.J., Wilson, C.R., Hay, F.S., Leggett, G.W., Sherriff, L.J., 2002. Mechanical transmission of *Apple mosaic virus* in Australian hop (*Humulus lupulus*) gardens. Ann. Appl. Biol. 141, 77–85.

Piazzolla, P., Castellano, M.A., Stradis, A.D., 1986. Presence of plant viruses in some rivers of southern Italy. J. Phytopathol. 116, 244–246.

Potts, S.G., Biesmeijer, J.C., Kremen, C., Neumann, P., Schweiger, O., Kunin, W.E., 2010. Global pollinator declines: trends, impacts and drivers. Trends Ecol. Evol. 25, 345–353.

Power, A.G., 2008. Community ecology of plant viruses. In: Roossinck, M.J. (Ed.), Plant Virus Evolution. Springer, Berlin, Germany, pp. 15–26.

Ribiere, M., Lallemand, P., Iscache, A.L., Schurr, F., Celle, O., Blanchard, P., Olivier, V., Faucon, J.P., 2007. Spread of infectious chronic bee paralysis virus by honeybee (*Apis mellifera* L.) feces. Appl. Environ. Microbiol. 73, 7711–7716.

Roberts, I.M., Wang, D., Thomas, C.L., Maule, A.J., 2003. *Pea seed-borne mosaic virus* seed transmission exploits novel symplastic pathways to infect the pea embryo and is, in part, dependent upon chance. Protoplasma, 31–43.

Roossinck, M.J., Garcia-Arenal, F., 2015. Ecosystem simplification, biodiversity loss and plant virus emergence. Curr. Opin. Virol. 10C, 56–62.

Roossinck, M.J., Martin, D.P., Roumagnac, P., 2015. Plant virus metagenomics: advances in virus discovery. Phytopathology 105, 716–727.

Rosario, K., Nilsson, C., Wei Lim, Y., Ruan, Y., Breitbart, M., 2009. Metagenomic analysis of viruses in reclaimed water. Environ. Microbiol. 11, 2806–2820.

Rothman, L.D., Myers, J.H., 2000. Ecology of Insect Viruses. Academic Press, San Diego, CA.

Roux, S., Enault, F., Robin, A., Ravet, V., Personnic, S., Theil, S., Colombet, J., Sime-Ngando, T., Debroas, D., 2012. Assessing the diversity and specificity of two freshwater viral communities through metagenomics. PLoS One 7, e33641.

Sacristán, S., Díaz, M., Fraile, A., García-Arenal, F., 2011. Contact transmission of *Tobacco mosaic virus*: a quantitative analysis of parameters relevant for virus evolution. J. Virol. 85, 4974–4981.

Santillan, F., Fribourg, C., Adams, I., Gibbs, A.J., Boonham, N., Kehoe, M., Maina, S., Jones, R.A.C., 2018. The biology and phylogenetics of potato virus S isolates from the Andean region of South America. Plant Dis. https://doi.org/10.1094/PDIS-09-17-1414-RE (in press).

Sarra, S., Oevering, P., Guindo, S., Peters, D., 2004. Wind-mediated spread of *Rice yellow mottle virus* (RYMV) in irrigated rice crops. Plant Pathol. 53, 148–153.

Schultz, E.S., Folsom, D., 1925. Infection and dissemination experiments with degeneration diseases of potatoes. Observations in 1923. J. Agric. Res. 30, 493–528.

Sdoodee, R., Teakle, D.S., 1987. Transmission of tobacco streak virus by *Thrips tabaci* a new method of plant virus transmission. Plant Pathol. 36, 377–380.

Sdoodee, R., Teakle, D.S., 1993. Studies on the mechanism of transmission of pollen-associated tobacco streak ilarvirus virus by *Thrips tabaci*. Plant Pathol. 42, 88–92.

Shen, M., Cui, L., Ostiguy, N., Cox-Foster, D., 2005a. Intricate transmission routes and interactions between picorna-like viruses (Kashmir bee virus and sacbrood virus) with the honeybee host and the parasitic varroa mite. J. Gen. Virol. 86, 2281–2289.

Shen, M., Yang, X., Cox-Foster, D., Cui, L., 2005b. The role of varroa mites in infections of *Kashmir bee virus* (KBV) and *Deformed wing virus* (DWV) in honey bees. Virology 342, 141–149.

Shipp, J.L., Buitenhuis, R., Stobbs, L., Wang, K., Kim, W.S., Ferguson, G., 2008. Vectoring of *Pepino mosaic virus* by bumble-bees in tomato greenhouses. Ann. Appl. Biol. 153, 149–155.

Shukla, D.D., Shanks, G.J., Teakle, D.S., Behncken, G.M., 1979. Mechanical transmission of galinsoga mosaic virus in soil. Aust. J. Biol. Sci. 32, 267–276.

Singh, R., Levitt, A.L., Rajotte, E.G., Holmes, E.C., Ostiguy, N., Lipkin, W.I., Toth, A.L., Cox-Foster, D.L., 2010. RNA viruses in hymenopteran pollinators: evidence of inter-taxa virus transmission via pollen and potential impact on non-*Apis* hymenopteran species. PLoS One 5 (12), e14357.

Smith, K.M., 1965. Plant virus vector relationship. Adv. Virus Res. 11, 61–93.

Stewart-Wade, S.M., 2011. Plant pathogens in recycled irrigation water in commercial plant nurseries and greenhouses: their detection and management. Irrig. Sci. 29, 267–297.

Strand, M.R., Burke, G.R., 2013. Polydnavirus-wasp associations: evolution, genome organization, and function. Curr. Opin. Virol. 3, 587–594.

Teakle, D.S., 1986. Abiotic transmission of southern bean mosaic virus in soil. Aust. J. Biol. Sci. 39, 353–360.

Teakle, D.S., Sylvester, E.S., 1962. Infection of plants by the feeding of aphids on leaves sprayed with tobacco mosaic virus. Virology 16 (4), 363–365.
Thresh, J.M., 1974. Temporal patterns of virus spread. Annu. Rev. Phytopathol. 12, 111–128.
Thresh, J.M., 1976. Gradients of plant virus diseases. Ann. Appl. Biol. 82, 381–406.
Thresh, J.M., 1980. The origins and epidemiology of some important plant virus diseases. Appl. Biol. 5, 1–65.
Thresh, J.M. (Ed.), 1981. Pests, Pathogens and Vegetation. Pitman, London.
Thresh, J.M., 1983. Progress curves of plant virus disease. Adv. Appl. Biol. 8, 1–85.
Thresh, J.M., 1991. The ecology of tropical plant virus diseases. Plant Pathol. 40, 324–339.
Thresh, J.M., 2006. Control of tropical plant virus diseases. Adv. Virus Res. 67, 245–295.
Tomlinson, J.A., Faithfull, E.M., 1984. Studies on the occurrence of tomato bushy stunt virus in English rivers. Ann. Appl. Biol. 104, 485–495.
Tomlinson, J.A., Faithfull, E.M., Webb, M.J.W., Fraser, R.S.S., Seeley, N.D., 1983. Chenopodium necrosis: a distinctive strain of tobacco necrosis virus isolated from river water. Ann. Appl. Biol. 102, 135–147.
Traore, M.D., Traore, V.S.E., Galzi-Pinel, A., Fargette, D., Konate, G., Traore, A.S., Traore, O., 2008. Abiotic transmission of *Rice yellow mottle virus* through soil and contact between plants. Pak. J. Biol. Sci. 11, 900–904.
Vanbergen, A.J., 2013. Threats to an ecosystem service: pressures on pollinators. Front. Ecol. Environ. 11, 251–259.
Vani, S., Varma, A., 1993. Properties of cucumber green mottle mosaic virus isolated from water of river Jamuna. Indian Phytopathol. 46, 18–22.
Vilaplana, L., Wilson, K., Redman, E.M., Cory, J.S., 2010. Pathogen persistence in migratory insects: high levels of vertically-transmitted virus infection in field populations of the African armyworm. Evol. Ecol. 24, 147–160.
Vincent, S.J., Coutts, B.A., Jones, R.A.C., 2014. Effect of introduced and indigenous viruses on native plants: exploring their disease causing potential at the agro-ecological interface. PLoS ONE 9 (3), e91224.
Virto, C., Zárate, C.A., López-Ferber, M., Murillo, R., Caballero, P., Williams, T., 2013. Gender-mediated differences in vertical transmission of a nucleopolyhedrovirus. PLoS ONE 8 (8), e70932.
Virto, C., Navarra, M., Mar Tellez, M., Herrero, T., Williams, R., Murillo, R., Caballero, P., 2014. Natural populations of Spodoptera exigua are infected by multiple viruses that are transmitted to their offspring. J. Invertebr. Pathol. 122, 22–27.
Walters, H.J., 1951. Grasshopper transmission of three plant viruses. Science 113, 36–37.
Walters, H.J., 1952. Some relationships of three plant viruses to the differential grasshopper, *Melanoplus differentialis*, Thos. Phytopathology 42, 355–362.
Wamonje, F.O., Michuki, G.N., Braidwood, L.A., Njuguna, J.N., Mutuku, J.M., Djikeng, A., Harvey, J.J., Carr, J.P., 2017. Viral metagenomics of aphids present in bean and maize plots on mixed-use farms in Kenya reveals the presence of three dicistroviruses including a novel *Big Sioux River virus*-like dicistrovirus. Virol. J. 14, 188.
Wang, D., Maule, A.J., 1992. Early embryo invasion as a determinant in pea of the seed transmission of pea seed-borne mosaic virus. J. Gen. Virol. 73, 1615–1620.
Waser, N.M., Chittka, L., Price, M.V., Williams, N.M., Ollerton, J., 1996. Generalization in pollination systems, and why it matters. Ecology 77, 1043–1060.
Wilson, C.R., 2014. Applied Plant Virology. CABI Press, Wallingford.
Webster, C.G., Coutts, B.A., Jones, R.A.C., Jones, M.G.K., Wylie, S.J., 2007. Virus impact at the interface of an ancient ecosystem and a recent agroecosystem: studies on three legume-infecting potyviruses in the South West Australian Floristic Region. Plant Pathol. 56, 729–742.

Wintermantel, W.M., 2011. A comparison of disinfectants to prevent spread of potyviruses in greenhouse tomato production. Plant Health Prog. Online publication. https://doi.org/10.1094/PHP-2011-0221-01-RS.

FURTHER READING

Amari, K., Burgos, L., Pallas, V., Sanchez-Pina, M.A., 2007. *Prunus necrotic ringspot virus* early invasion and its effects on apricot pollen grain performance. Phytopathology 97, 892–899.

Bowen-Walker, P.L., Martin, S.J., Gunn, A., 1999. The transmission of deformed wing virus between honeybees (*Apis mellifera* L.) by the ectoparasitic mite *Varroa jacobsoni* Oud. J. Invertebr. Pathol. 73, 101–106.

Broadbent, L., 1965. The epidemiology of tomato mosaic. 8. Virus infection through tomato roots. Ann. Appl. Biol. 55, 57–66.

Cameron, R.R., Milbraith, J.A., Tate, L.A., 1973. Pollen transmission of prunus necrotic ringspot in prune and sour cherry orchards. Plant Dis. Rep. 57, 241–243.

Dombrovsky, A., Luria, N., 2013. The Nerium oleander aphid *Aphis nerii* is tolerant to a local isolate of *Aphid lethal paralysis virus* (ALPV). Virus Genes 46, 354–361.

Erdiller, G., Akbas, B., 1994. Plant viruses in Ankara rivers and lakes. J. Turk. Phytopathol. 23, 119–126.

Fouks, B., Lattorff, H.M.G., 2011. Recognition and avoidance of contaminated flowers by foraging bumblebees (*Bombus terrestris*). PLoS ONE 6, e26328.

Holdengreber, V., Guenoune-Gelbert, D., Lapidot, M., Aly, R., 2009. Broomrape can acquire viruses from its hosts. Phytopathology 99, 1321–1329.

Kelley, R.D., Cameron, H.R., 1986. Location of prune dwarf and Prunus necrotic ringspot viruses associated with sweet cherry pollen and seed. Phytopathology 76, 317–322.

Klose, M.J., Sdoodee, R., Teakle, D.S., Milne, J.R., Greber, R.S., Walter, G.H., 1996. Transmission of three strains of tobacco streak ilarvirus by different thrips species using virus-infected pollen. J. Phytopathol. 144, 281–284.

Levitt, A.L., Singh, R., Cox-Foster, D.L., Rajotte, E., Hoover, K., Ostiguy, N., Holmes, E.C., 2013. Cross-species transmission of honey bee viruses in associated arthropods. Virus Res. 176, 232–240.

Ryabov, E.V., Keane, G., Naish, N., Evered, C., Winstanley, D., 2009. Densovirus induces winged morphs in asexual clones of the rosy apple aphid, *Dysaphis plantaginea*. Proc. Natl. Acad. Sci. U. S. A. 106, 8465–8470.

CHAPTER SEVEN

Evolutionary Determinants of Host and Vector Manipulation by Plant Viruses

Kerry E. Mauck*[,1], Quentin Chesnais*, Lori R. Shapiro†
*Department of Entomology, University of California, Riverside, Riverside, CA, United States
†Department of Applied Ecology, North Carolina State University, Raleigh, NC, United States
[1]Corresponding author: e-mail address: kerry.mauck@ucr.edu

Contents

Abstract

Plant viruses possess adaptations for facilitating acquisition, retention, and inoculation by vectors. Until recently, it was hypothesized that these adaptations are limited to virus proteins that enable virions to bind to vector mouthparts or invade their internal tissues. However, increasing evidence suggests that viruses can also manipulate host plant

Advances in Virus Research, Volume 101
ISSN 0065-3527
https://doi.org/10.1016/bs.aivir.2018.02.007

phenotypes and vector behaviors in ways that enhance their own transmission. Manipulation of vector–host interactions occurs through virus effects on host cues that mediate vector orientation, feeding, and dispersal behaviors, and thereby, the probability of virus transmission. Effects on host phenotypes vary by pathosystem but show a remarkable degree of convergence among unrelated viruses whose transmission is favored by the same vector behaviors. Convergence based on transmission mechanism, rather than phylogeny, supports the hypothesis that virus effects are adaptive and not just by-products of infection. Based on this, it has been proposed that viruses manipulate hosts through multifunctional proteins that facilitate exploitation of host resources and elicitation of specific changes in host phenotypes. But this proposition is rarely discussed in the context of the numerous constraints on virus evolution imposed by molecular and environmental factors, which figure prominently in research on virus–host interactions not dealing with host manipulation. To explore the implications of this oversight, we synthesized available literature to identify patterns in virus effects among pathogens with shared transmission mechanisms and discussed the results of this synthesis in the context of molecular and environmental constraints on virus evolution, limitations of existing studies, and prospects for future research.

1. INTRODUCTION

Vector-borne plant viruses are obligate intracellular parasites that can fundamentally change the physiology of their host plants. The outcomes of these changes, such as reductions in crop yield or quality, have driven much of the research on virus–host interactions. But virus effects on host plants extend well beyond agronomically relevant metrics. There is now increasing evidence that viruses can alter aspects of the host plant phenotype (cues) that mediate interactions with other organisms, including the mobile insect vectors responsible for much of virus transmission (Casteel and Falk, 2016; Eigenbrode and Bosque-Perez, 2016; Mauck, 2016; Mauck et al., 2012, 2016). These cues include visual and tactile characteristics, odors, induced defenses, secondary metabolites, sugars, free amino acids, and likely other undescribed factors (Bosque-Pérez and Eigenbrode, 2011; Casteel et al., 2014; Mauck et al., 2014a,b). Insect vectors make their initial foraging decisions by integrating visual and odor cues, which convey information about plant presence, identity, and quality. After contacting a host plant, vectors assess additional cues from leaf or stem surfaces, parenchyma, and vascular tissues through olfactory and gustatory sensory systems. The insects' probing, feeding, and dispersal behaviors in response to plant cues directly determine the probability that virions will be acquired, retained, and transported (Fereres, 2016; Fereres and Collar, 2001; Hogenhout et al., 2008;

Madden et al., 2000; Moreno et al., 2012; Ng and Falk, 2006). Thus, if a plant virus alters aspects of the host plant that provide cues for its herbivorous vectors, these changes have potential to influence rates of host–vector contact and vector-feeding behaviors that determine virus transmission. Given that transmission is critical to the fitness of vector-borne plant viruses, it has been proposed that viruses evolve traits that induce (or at least maintain) host phenotypes and effects on vectors that encourage virus spread.

Consistent with this hypothesis, there are now more than 100 published reports of plant viruses purportedly "manipulating" host plant phenotypes to increase vector attraction to infected plants, or elicit transmission-conducive feeding behaviors (reviewed in Casteel and Falk, 2016; Eigenbrode and Bosque-Perez, 2016; Fereres and Moreno, 2009; Heil, 2016; Mauck, 2016; Mauck et al., 2012, 2016). More recently, there is evidence that some plant viruses can also manipulate vector behaviors to favor virus transmission to new hosts by interacting with the vector's tissues following acquisition from infected hosts or artificial substrates (Ingwell et al., 2012; Moreno-Delafuente et al., 2013; Rajabaskar et al., 2014; Stafford et al., 2011). The idea that plant viruses can manipulate hosts and vectors to enhance transmission is not unique. Manipulation of host phenotypes by parasites is well documented across a wide range of taxonomic groups (Heil, 2016). The fitness advantages of manipulation as a strategy are captured by the "adaptive host manipulation hypothesis," which proposes that parasites can evolve to control elements of their host's phenotype that help maintain or enhance rates of transmission (Poulin, 2010). Thus, "manipulated" hosts exhibit additional hallmarks of infection beyond those associated with the basic need for a parasite to attenuate host immunity and use host resources for reproduction (Lefèvre et al., 2009). Although there are hundreds of reports of putative host manipulation by parasites (Lafferty and Shaw, 2013; Lefèvre et al., 2009; Mauck et al., 2012, 2016; Poulin, 2010), only a handful of these studies have made progress in pinpointing the *parasite* as the "manipulator"—that is, the organism having genetic control over the altered host phenotype. An equally likely explanation is that the phenotype of the infected host is due to an immune response under genetic control of the host. Alternatively, observed phenotypes could represent by-products of pathology, or even the residual influence of inherited ancestral traits that were adaptive in one host–parasite context but have become maladaptive in another context (Heil, 2016). Parsing these explanations has proven difficult for eukaryotic parasites that provide the most charismatic examples of host manipulation, but prove to be intractable laboratory models. The growing evidence of host and vector manipulation by plant viruses provides new opportunities to explore the adaptive significance of parasite manipulation in the context of environmental variation using pathosystems that are more amenable to experimental methods involving functional genomics.

Although we do not yet have a thorough mechanistic understanding of plant virus genes that confer manipulative traits, there is still evidence to support the hypothesis that host and vector manipulation is adaptive for plant viruses. Theoretical studies demonstrate that "manipulative" viruses inducing transmission-enhancing effects in hosts (or vectors) will spread more rapidly, and from lower starting frequencies, relative to viruses that have neutral effects, or viruses that elicit changes that deter virus acquisition by vectors (Jeger et al., 2004; McElhany et al., 1995; Roosien et al., 2013; Shaw et al., 2017; Sisterson, 2008). Thus, under ideal conditions, manipulative virus genotypes are expected to enjoy higher fitness than nonmanipulative virus genotypes. Additionally, the nature of virus influence on vectors generally corresponds with the virus transmission mechanism regardless of virus phylogeny (Mauck et al., 2012). In other words, virus-induced changes in host cues and quality for vectors are not uniform, but differ depending on the requirements for virion uptake and transmission that are inherent to a given virus (Mauck et al., 2012, 2016). Each insect-borne plant virus is classified into one of the four transmission mechanism groups depending on requirements for acquisition, retention, and inoculation (Table 1). These requirements are based on virus localization within hosts and the nature of associations with vectors, which range from transient binding to cuticular surfaces of vector mouthparts (noncirculative, nonpersistent viruses), to invasion of the hemocoel (circulation) and retention (persistence) of ingested virions in salivary glands (circulative-persistent viruses). Within the circulative-persistent category, some viruses undergo active replication in salivary glands and other internal vector tissues (propagative), while others localize to salivary glands but do not replicate (nonpropagative) (Table 1). These associations determine what vector behavioral sequences most favor efficient virus acquisition, retention, and eventual inoculation. Thus, transmission mechanism groups can serve as a basis for generating predictions about how any given virus might be expected to alter host phenotypes (or vector physiology) to influence probing, feeding, and dispersal behaviors in ways that are conducive to its own transmission. The adaptive significance of virus effects on hosts and vectors can then be explored by evaluating evidence for convergence in virus effects across phylogenetically diverse viruses that share a transmission mechanism group.

Table 1 Characteristics of Plant Virus Transmission Mechanism Groups

	Circulative		**Noncirculative**	
Transmission Modes	**Persistent Propagative**	**Persistent Nonpropagative**	**Semipersistent**	**Nonpersistent**
Acquisition time[a]	Minutes to hours		Seconds to hours	Seconds to minutes
Retention time[b]	Days to months		Minutes to hours	Seconds to minutes
Inoculation time[c]	Minutes to hours		Seconds to hours	Seconds to minutes
Association with vectors[d]	Internal		External	
Replication in vectors	Yes	No	No	
Association with plants	Restricted to phloem		Some restricted to phloem, others not	Not restricted to phloem
Abbreviation	C-P-Prop	C-P-NProp	NC-SPer	NC-NPer

[a]Time required for a vector to efficiently acquire virus particles following initiation of probing or feeding.
[b]Time during which the virus remains infectious within its vector following acquisition.
[c]Time required for a vector to efficiently inoculate infectious virus particles to a new healthy plant.
[d]Internal means that the virus enters the inner body of its vector, passing through cellular barriers. External means that the virus binds to cuticular surfaces (stylet or foregut) and never passes through cellular barriers.
Adapted from Brault, V., Uzest, M., Monsion, B., Jacquot, E., Blanc, S., 2010. Aphids as transport devices for plant viruses. C. R. Biol. 333 (6–7), 524–538. https://doi.org/10.1016/j.crvi.2010.04.001.

We took this approach for our earlier quantitative synthesis of 55 papers reporting putative instances of virus manipulation (Mauck et al., 2012), which was the first to show that phylogenetically divergent plant viruses transmitted via the same sequences of vector behavior induce similar phenotypes in their host plants. By demonstrating convergent effects based on transmission mechanism group, this synthesis provided support for the hypothesis that virus effects are the result of adaptations and not just by-products of infection. Since the publication of this synthesis, the number of empirical reports of putative plant virus manipulation has more than doubled. Despite the popularity of this topic and its clear relevance for understanding virus epidemiology, there has not been any subsequent attempt to comprehensively reevaluate virus effects with regard to transmission mechanism groups, or to place this body of work within the context of constraints on virus evolution imposed by molecular and environmental factors, which should not be lightly dismissed. Plant viruses have small genomes that often encode less than 10 functional proteins, sometimes through overlapping open reading frames. Virus proteins perform multiple functions in the host plant, interact with each other extensively, and may play a dual role in facilitating interactions with both plant and vector tissues. These features enable rapid replication and maintain vector transmissibility but impose major limitations on virus evolution because most mutations are likely to be deleterious and will be rapidly purged. Molecular constraints will further interact with environmental factors to shape virus evolution. In a field context, plant viruses are subject to heterogeneous host environments at intraspecific and interspecific levels, as well as variation in the frequency of transmission-conducive contacts with vectors (Elena et al., 2014; Gutiérrez et al., 2013; Pagán et al., 2012; Rodelo-Urrego et al., 2013; Roossinck and García-Arenal, 2015). These constraints will influence the evolution and maintenance of manipulative traits in plant viruses, and thus all reports of putative manipulation of hosts and vectors must be considered within the context of these constraints.

To explore the extent to which the existing literature considers molecular and environmental axes of virus evolution, and to revisit the question of whether viruses exhibit convergence in effects within each transmission mechanism group, we performed a comprehensive review and quantitative synthesis of all studies reporting putative instances of virus manipulation of hosts and vectors following the guidelines used in Mauck et al. (2012). Here, we discuss this synthesis in the context of the methodologies employed and molecular and environmental factors that may facilitate, or hinder, the evolution of manipulative functions. Our results provide evidence of convergence in virus effects within transmission mechanism groups while revealing a number of inadequacies in the current literature that provide a roadmap for future research directions.

2. VIRUS EFFECTS ON HOST PHENOTYPES AND VECTOR BEHAVIOR

Virus manipulation of vector behavior can occur via two mechanisms that are not mutually exclusive. The most reported mechanism, and the first discussed in our synthesis, involves changes in aspects of the host phenotype that influence vector orientation to, and feeding behaviors on, virus-infected plants (Eigenbrode and Bosque-Perez, 2016; Fereres and Moreno, 2009;

Mauck et al., 2012, 2016). This pathway is *indirect* because it is mediated by the host resource being shared by both the virus and its vector(s). A second, more recently described behavioral modification occurs when the virus is acquired by the vector and *directly* interacts with vector tissues (Ingwell et al., 2012; Stafford et al., 2011). These more intimate associations create opportunities for viruses to alter vector physiology in ways that affect behaviors related to virus transmission (e.g., relative preferences for host cues mediating orientation to, and feeding on, infected or healthy hosts). The question of whether viruses can evolve these functions in natural or agricultural settings is central to our understanding of the ecological and epidemiological importance of host and vector manipulation. To begin to address this question, we provide a conceptual outline of predictions for adaptive virus effects that is based on the transmission requirements outlined in Table 1 and Fig. 1, and then present a quantitative synthesis of all experiments on plant virus manipulation of vector behavior performed to date. Our goal is to create a framework for discussing the mechanisms underlying the evolution of manipulative functions, molecular and environmental constraints on these mechanisms, limitations of the existing studies, and possible avenues for future research.

2.1 Transmission Mechanisms and Predicted Virus Effects

Much emphasis is placed on the host plant as a selective agent driving virus evolution. This supposition is based on the idea that the host is the fundamental environment in which the virus resides (Elena et al., 2014). While this is true, plant virus fitness ultimately depends on the virus capacity to infect more than one host, which is often achieved via the feeding activities of mobile insect vectors (Whitfield et al., 2015). As discussed briefly in the Introduction, plant viruses are classified into different transmission mechanism groups according to their associations with hosts and vectors (Table 1) and the corresponding vector–host interactions required for transmission. Arthropod-borne circulative-persistent viruses (designated here as C-P) can be either propagative (C-P-Prop viruses) or nonpropagative (C-P-NProp viruses) and are transmitted by insects and arachnids with variations on the piercing–sucking mode of feeding (aphids, whiteflies, leafhoppers, planthoppers, and mites) (Table 1; Fig. 1). Propagative viruses replicate within their vectors, while nonpropagative viruses circulate and reside in specific tissues but do not engage in replication (Fig. 1). Both are generally phloem restricted and acquired during long-term phloem sap ingestion (Table 1).

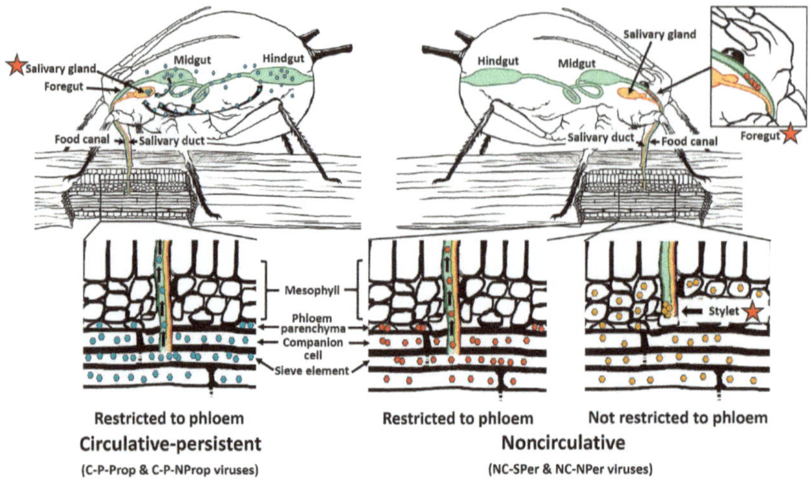

Fig. 1 Schematic representations of plant virus transmission mechanisms. Retention sites for each virus are indicated by a *red star* adjacent to the labeled area. *Left*: Schematic representation of associations between hosts, vectors, and circulative-persistent propagative (C-P-Prop) viruses or circulative-persistent nonpropagative (C-P-NProp) viruses. Nearly all circulative-persistent viruses are restricted to the phloem, and so acquisition requires phloem sap ingestion by the vector. Virions are retained in salivary glands after acquisition and, for propagative viruses, within several other vector tissues. *Right*: Schematic representation of associations between hosts, vectors, and noncirculative viruses. Depending on where the virus is localized within the plant, virions of noncirculative viruses will be acquired during long-term phloem sap ingestion (like phloem-restricted circulative-persistent viruses) or during brief probes of epidermal or mesophyll tissue prior to phloem contact. Noncirculative semipersistent (NC-SPer) viruses exhibit both types of association with plants, whereas noncirculative nonpersistent (NC-NPer) viruses are only acquired from epidermal and mesophyll cells and are lost if the vector feeds in the phloem.

Noncirculative viruses (designated as NC) can be either semipersistent (NC-SPer viruses) or nonpersistent (NC-NPer viruses). NC-SPer viruses are transmitted by aphids, whiteflies, mealybugs, beetles, and leafhoppers (Fereres and Raccah, 2009) and are usually retained in the foregut—a chitinous anterior region of the alimentary canal that immediately precedes the gut (Fig. 1). Virion retention is typically measured in hours (Table 1) and virions can be inoculated to multiple hosts following a single acquisition event. Some NC-SPer viruses are phloem-limited (Ng and Zhou, 2015), so the vector behaviors that favor efficient acquisition (long-term phloem ingestion) are similar to those required for circulative-persistent viruses (Fig. 1). Other NC-SPer viruses are not restricted to the phloem and can also occupy mesophyll tissue (Fig. 1). These viruses are acquired during brief ingestion events from nonvascular tissues (similar to NC-NPer viruses) (Fig. 1). NC-NPer viruses are exclusively transmitted by aphids and are acquired rapidly during brief probes in the nonvascular tissues of the plant *before* phloem ingestion occurs. The virus binds to target sites on the aphid stylet (Hogenhout et al., 2008; Martin et al., 1997; Ng and Falk, 2006) and virions can be lost if the vectors proceed to initiate sustained phloem sap ingestion on the infected host plant (Ng and Falk, 2006) (Fig. 1). Thus, in contrast to phloem-restricted C-P viruses, transmission of NC viruses is favored by dispersal of a vector from an infected plant shortly after initial probing to assess host suitability. Following NC-NPer virion acquisition, vectors typically remain viruliferous (i.e., carry virus) for a brief period (minutes) and are capable of inoculating only a few plants before losing viruliferous status.

Each virus transmission mechanism is characterized by different host and vector associations, and requirements for acquisition and inoculation. But among taxa that share a transmission mechanism group, the same patterns of vector behavior are required for virus spread. We used these shared requirements for transmission as a basis for generating predictions about how phylogenetically diverse viruses within a group might alter host phenotypes and/or vector physiology to enhance specific probing, feeding, and dispersal behaviors in ways that are generally conducive to transmission. These predictions are described in Fig. 2 and tested using our quantitative synthesis in the ensuing sections.

2.2 Host-Mediated Effects of Plant Viruses on Vector Behavior

Viruses can alter multiple biochemical pathways in plants, each of which might disrupt cues that are important for vector foraging. Identifying each of the cues mediating a putative manipulation, and the virus genes responsible, would be extremely challenging. Instead of taking this approach, we used previously published methods (Mauck et al., 2012) to find evidence of convergence in virus effects on host phenotypes and vector behavior across distantly related taxa that share a transmission mechanism group. Here, we present the results of this expanded quantitative synthesis, which includes all literature from 1960 to 2017. In brief, we searched Google Scholar and the ISI Web of Science for publications related to virus–host–vector interactions following the criteria for manuscript identification and selection described by Mauck et al. (2012). Our first review summarized 55 papers published between 1960 and 2012. Here, we consider an additional 67 newer papers (for a total of 122). We parsed each study into individual experiments, each

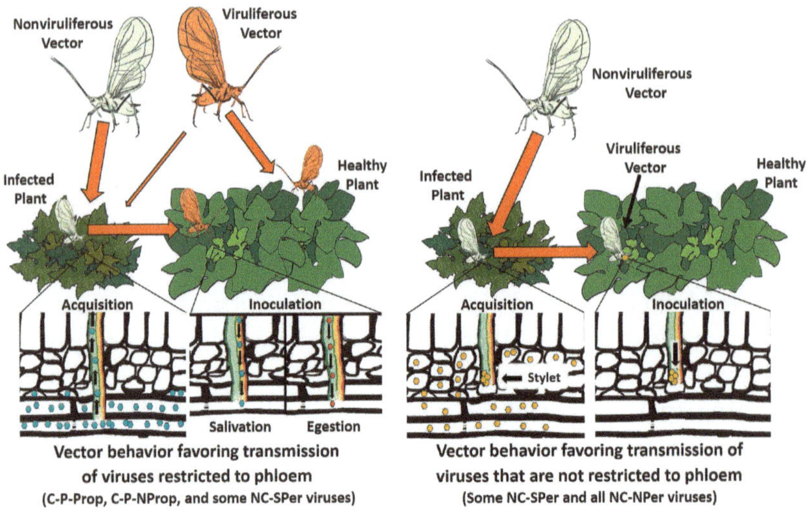

Fig. 2 Graphical representations of predictions for transmission-conducive effects of viruses on the behavior of their insect vectors. We expect to see convergence of virus effects across phylogenetically distant virus groups that share the same transmission mechanism, and, therefore, are transmitted via similar sequences of vector probing, feeding, and dispersal behavior (Fig. 1; Table 1). *Prediction 1*: Consistent with the need to maintain or increase vector contact rates with infected hosts, we expect both viruses that are restricted to phloem (*left*) and viruses that are not phloem restricted (*right*) will have neutral to positive effects on plant cues mediating vector orientation preference. *Prediction 2*: Viruses that are restricted to phloem diverge from viruses that are not restricted to phloem in requirements for vector probing/feeding frequencies and durations (Fig. 1; Table 1). We expect phloem-restricted viruses (C-P-Prop, C-P-NProp, and some NC-SPer viruses) to have neutral to positive effects on plant palatability cues that encourage vector settling and feeding prior to dispersal (to facilitate virion uptake from the phloem) (*left*). In contrast, we expect viruses that are not restricted to phloem (*right*) to reduce plant palatability in ways that discourage sustained feeding to ensure vectors disperse after probing and acquiring virions, and before virions are lost during subsequent salivation events (Ng and Falk, 2006). This category includes all NC-NPer viruses and some NC-SPer viruses. *Prediction 3*: We expect to see plant quality effects that are consistent with virus-induced changes in host palatability. For phloem-restricted viruses (C-P-Prop, C-P-NProp, and some NC-SPer viruses), enhanced host quality may lead to higher reproductive rates, crowding, accelerated use of host resources, and eventual dispersal of viruliferous vectors (*left*). For viruses that are not restricted to the phloem (all NC-NPer viruses and some NC-SPer viruses), we expect to see reductions in plant quality, which will increase vector restlessness and opportunities for virus acquisition (*right*). *Predictions regarding direct effects*: For C-P-Prop and C-P-NProp viruses that are phloem-restricted (*left*), dispersal is beneficial for the virus only if it occurs after sufficient virions have been acquired to establish a viruliferous state (either by retention or by replication). Therefore, we expect that these viruses may evolve mechanisms to encourage dispersal following several hours to days of feeding on infected hosts, and to discourage viruliferous vectors from visiting virus-infected hosts (indicated by a *smaller*

addressing a single virus strain or isolate by host species (or cultivar) by vector interaction (440 experiments total). Each experiment was categorized according to the virus transmission mechanism group (Table 1; Fig. 1) and the vector response component (orientation preference, settling/feeding preference, or performance) (Fig. 2). For experiments measuring preferences (orientation or settling/feeding), results were classified as preference for virus-infected plants, preference for healthy plants or no preference. For measurements of vector performance, results were classified as a positive, neutral, or negative effect of virus infection in a host plant on the vector(s). Compiled results were evaluated for departure from an expected even distribution of effects (1:1:1 for *infected/no preference/healthy* within behavioral effect categories, and *positive/neutral/negative* within the performance effect categories) across the three interaction types using chi-square tests followed by pairwise comparisons (*P*-value adjustment method: "fdr," R package "RVAideMemoire"). To test our hypothesis that neutral effects should also be considered adaptive for the virus, we then evaluated results using two-way chi-square tests by grouping the "no preference" or "neutral" categories with the appropriate "transmission-enhancing" category for each transmission mechanism group. Individual groupings for each transmission mechanism group are described within the predictions below.

Prediction 1: Both circulative-persistent viruses and noncirculative viruses will have neutral to positive effects on plant cues mediating vector orientation.

Our analysis of the 52 available experiments related to this first prediction indicates that C-P-NProp viruses often induce changes in host phenotype that result in enhanced attraction of vectors to virus-infected plants when presented in a choice scenario vs healthy plants (Tables 2 and 3). An important caveat of this result is that almost all the experiments studying this effect for C-P-NProp viruses focused on Luteoviridae, with the exception of one study on a Begomovirus showing that the whitefly vector, *Bemisia tabaci*, exhibited no orientation preference between *Tomato severe rugose virus*-infected (*ToSRV*, Geminiviridae) and healthy tomatoes (Fereres et al., 2016). For the NC-NPer viruses, while the number of available studies is more limited, our

orientation arrow toward infected hosts, and a *larger orientation arrow* to healthy hosts). These mechanisms could include direct effects of retained/replicating viruses on vector behavior (e.g., perception or processing of host cues), indirect effects on host plants (e.g., changes in palatability or quality that only occur following vector feeding), or a combination of both of these mechanisms.

Table 2 Quantitative Synthesis of Overall Host-mediated Effects of Plant Viruses on Vector Behavior and Performance

			Effect on Orientation Preference			Effect on Settling/Feeding Preference			Effect on Performance		
Mode of Transmission	**Virus taxon**	**Genome**	**Infected**	**No Pref.**	**Healthy**	**Infected**	**No Pref.**	**Healthy**	**Positive**	**Neutral**	**Negative**
C-P-Prop viruses	Tospoviridae	(−) seg. RNA	0	0	0	21	3	1	8	8	3
	Reoviridae	ds RNA	4	1	0	1	1	3	3	2	2
		Sum	4	1	0	22(a)	4(b)	4(b)	11	10	5
			Chi-square = not tested			*Chi-square* $= 21.6$, $P < 0.001$			*Chi-square = NS*		
C-P-NProp viruses	Luteoviridae	(+) ss RNA	20	8	2	24	13	3	37	10	6
	Geminiviridae	ss circ. DNA	0	1	0	10	2	2	37	20	14
	Nanoviridae	ss circ. DNA	0	0	0	1	0	0	1	0	0
		Sum	20(a)	9(b)	2(b)	35 (a)	15(b)	5(c)	75(a)	30 (b)	20(b)
			Chi-square $= 15.94$, $P < 0.001$			*Chi-square* $= 25.45$, $P < 0.001$			*Chi-square* $= 41.20$, $P < 0.001$		

Mode of Transmission	Virus taxon	Genome	Effect on Orientation Preference: Infected	Effect on Orientation Preference: No Pref.	Effect on Orientation Preference: Healthy	Effect on Settling/Feeding Preference: Infected	Effect on Settling/Feeding Preference: No Pref.	Effect on Settling/Feeding Preference: Healthy	Effect on Performance: Positive	Effect on Performance: Neutral	Effect on Performance: Negative
NC-SPer viruses	Closteroviridae	(+) ss RNA	0	1	2	8	2	1	6	3	1
	Sobemovirus	(+) ss RNA	0	0	0	1	0	0	1	0	0
	Caulimoviridae	ds DNA	0	0	0	0	1	1	0	0	2
	Secoviridae	ss RNA	1	0	0	4	0	0	1	1	1
		Sum	1	1	2	13(a)	3(b)	2(b)	8	4	4
			Chi-square = not tested			*Chi-square = 12.33, P = 0.002*			*Chi-square = NS*		
NC-NPer viruses	Bromoviridae	(+) ss RNA	4	2	0	3	5	7	3	5	15
	Potyviridae	(+) ss RNA	3	3	0	14	12	12	17	12	13
		Sum	7	5	0	17	17	19	20	17	28
			Chi-square = 6.5, P = 0.039			*Chi-square = NS*			*Chi-square = NS*		

Transmission group abbreviations: C-P-NProp, circulative-persistent nonpropagative viruses; *C-P-Prop*, circulative-persistent propagative viruses; *NC-NPer*, noncirculative nonpersistent viruses; *NC-SPer*, noncirculative semipersistent viruses. Genome abbreviations: *dsRNA*, double-stranded RNA viruses; *(−) seg. RNA*, negative sense, segmented, single-stranded RNA viruses; *ss circ. DNA* = single-stranded circular DNA viruses; *(+) ssRNA* = positive sense, single-stranded RNA viruses; *ssRNA*, single-stranded RNA viruses, sense unknown. Different lower case letters indicate statistically significant differences between effects (pairwise comparisons using chi-squared tests, $P < 0.05$).

Table 3 Quantitative Synthesis of Adaptive vs Nonadaptive Host-mediated Effects of Plant Viruses on Vector Behavior and Performance

			Effect on Orientation Preference		Effect on Settling/ Feeding Preference		Effect on Performance	
Mode of transmission	**Virus Taxon**	**Genome**	**Adaptive**	**Maladaptive**	**Adaptive**	**Maladaptive**	**Adaptive**	**Maladaptive**
C-P-Prop viruses	Tospoviridae	(−) seg. RNA	0	0	24	1	16	3
	Reoviridae	ds RNA	5	0	2	3	5	2
		Sum	5	0	26	4	21	5
			Chi-square = not tested		*Chi-square = 16.13, $P < 0.001$*		*Chi-square = 9.85, $P = 0.002$*	
C-P-NProp viruses	Luteoviridae	(+) ss RNA	28	2	37	3	47	6
	Geminiviridae	ss circ. DNA	1	0	12	2	57	14
	Nanoviridae	ss circ. DNA	0	0	1	0	1	0
		Sum	29	2	50	5	105	20
			Chi-square = 23.52, $P < 0.001$		*Chi-square = 36.81, $P < 0.001$*		*Chi-square = 57.79, $P < 0.001$*	

			Effect on Orientation Preference		Effect on Settling/ Feeding Preference		Effect on Performance	
Mode of transmission	**Virus Taxon**	**Genome**	**Adaptive**	**Maladaptive**	**Adaptive**	**Maladaptive**	**Adaptive**	**Maladaptive**
NC-SPer viruses	Closteroviridae	(+) ss RNA	1	2	10	1	9	1
	Sobemovirus	(+) ss RNA	0	0	1	0	1	0
	Caulimoviridae	ds DNA	0	0	1	1	0	2
	Secoviridae	ss RNA	1	0	4	0	2	1
		Sum	2	2	16	2	12	4
			Chi-square = not tested		*Chi-square = 10.89,* $P = 0.001$		*Chi-square = 4.00,* $P = 0.045$	
NC-NPer viruses	Bromoviridae	(+) ss RNA	6	0	12	3	20	3
	Potyviridae	(+) ss RNA	6	0	24	14	25	17
		Sum	12	0	36	17	45	20
			Chi-square = 12.00, $P < 0.001$		*Chi-square = 6.81,* $P = 0.009$		*Chi-square = 9.61,* $P = 0.019$	

The "adaptive" category for each transmission mechanism group includes experiments demonstrating neutral effects plus the effect predicted to enhance transmission for viruses within that group (according to group characteristics in Table 1 and predictions in Fig. 2). Transmission group abbreviations: *C-P-NProp*, circulative-persistent nonpropagative viruses; *C-P-Prop*, circulative-persistent propagative viruses; *NC-NPer*, noncirculative nonpersistent viruses; *NC-SPer*, noncirculative semipersistent viruses. Genome abbreviations: *dsRNA*, double-stranded RNA viruses; *(−) seg. RNA*, negative sense, segmented, single-stranded RNA viruses; *ss circ. DNA*, single-stranded circular DNA viruses; *(+) ssRNA*, positive sense, single-stranded RNA viruses; *ssRNA*, single-stranded RNA viruses, sense unknown.

synthesis suggests that virus infections in host plants either enhanced or had no effect on orientation preference by vectors (Table 2). If we combine the two categories that would be considered "adaptive" for the virus (orientation preference toward infected plants + no preference) (Table 3), there is evidence of selection *against* virus genotypes that reduce host attractiveness to vectors. This pattern is unlikely to be the result of publication bias, since an adverse effect of a virus on its own transmission is still of ecological interest (Chen et al., 2015a,b). Concerning the two other transmission mechanism groups (C-P-Prop viruses and NC-SPer viruses), too few studies have been conducted to perform a statistical analysis; however for C-P-Prop viruses, it is again notable that no experiments reported orientation preferences for healthy plants (Tables 2 and 3). Thus, the pattern for C-P-Prop viruses is consistent with the other transmission classes, and the prediction described in Fig. 2; both circulative-persistent and noncirculative viruses have neutral to positive effects on plant cues mediating vector orientation.

Prediction 2: Nonviruliferous vectors prefer to settle and feed on plants infected with viruses acquired during phloem sap ingestion. We identified 156 experiments exploring vector settling preferences for virus-infected or healthy host plants. Consistent with our prediction (Fig. 2), our analysis suggests that nonviruliferous vectors preferred to settle and feed on plants infected with C-P-Prop viruses, C-P-NProp viruses, and NC-SPer viruses more than on healthy plants (Tables 2 and 3). Within the NC-SPer viruses, the pattern is driven largely by phloem-restricted Closteroviridae, which require sustained feeding for acquisition just like C-P-Prop and C-P-NProp viruses (Table 2). Experiments with Closteroviridae largely report settling preferences for infected plants (eight experiments) or no preference (two experiments), with just one experiment showing a preference for healthy plants. However, the intriguing patterns observed within the NC-SPer viruses must be considered with caution as there are only a limited number of studies (18 experiments) for this transmission mechanism group. For NC-NPer viruses, effects were evenly distributed among all three choices (Table 2). This pattern suggests that viruses within the NC-NPer transmission mechanism group may be less likely to manipulate hosts in ways that *enhance* vector dispersal following virion acquisition (Fig. 2). However, in the two-way comparison of adaptive vs maladaptive outcomes, there is evidence of selection *against* NC-NPer viruses that increase host palatability and thus inhibit vector dispersal and virus transmission (Table 3). Furthermore, the two virus families that make up the NC-NPer group differ in their effects. The Bromoviridae tend to induce adaptive changes in host palatability that enhance or at least do not reduce the probability of vectors dispersing after acquiring virions (*chi-square* = 4.26, $P = 0.039$), while effects of Potyviridae are more evenly distributed (*chi-square* = 2.14, $P = 0.144$). This result is notable because we did not observe the same degree of divergence in effects according to virus family within the other transmission mechanism groups.

Prediction 3: Vectors perform better on plants infected by viruses restricted to phloem, and poorly on plants infected by viruses that are not restricted to phloem.

Among the 232 experiments addressing this prediction, our analyses suggest that C-P-Prop viruses and C-P-NProp viruses have mostly neutral to positive effects on vector performance (i.e., survival, fecundity, or both). This pattern is congruent with effects of these transmission classes on palatability (Tables 2 and 3) and supports our prediction (Fig. 2). Positive effects of virus infection are most evident for C-P-NProp viruses (Tables 2 and 3). For C-P-Prop viruses and NC-SPer viruses the analysis did not show significant differences from the expected even distribution (Table 2). But if neutral to positive effects are considered together (both of these being adaptive for the virus) (Table 3), there is evidence of selection against viruses that reduce vector performance (Table 3). It is particularly notable that C-P-Prop viruses have 21/26 experiments in the neutral or positive categories considering that vectors feeding on C-P-Prop virus-infected plants will also serve as hosts for virus replication (Barandoc-Alviar et al., 2016; Hogenhout et al., 2008). This is evidence of selection for genotypes that maintain or enhance host quality, and against genotypes that are overly pathogenic within vectors feeding on these infected hosts. For NC-SPer viruses, effects on performance seem to track with effects on host palatability, but this pattern must be considered with caution as the overall sample size is low (16 experiments). Phloem-restricted Closteroviridae represent more than half of the available experiments (10/16) and show strong evidence of adaptive effects on host plant quality for vectors (9/10 studies). This is consistent with the expectation that phloem-limited NC-SPer viruses acquired during long-term feeding should induce host phenotypes that are similar to those generated by circulative-persistent virus infection (Fig. 2). While the number of available studies is limited for nonphloem-restricted NC-SPer viruses (such as *Cauliflower mosaic virus* [CaMV, Caulimoviridae]), it is notable that the trend for this virus, which is acquired much more rapidly, seems to be the opposite.

For NC-NPer viruses, effects on host plant quality for vectors are evenly distributed across the three performance categories (Table 2) but fall largely in the "adaptive" category in the two-way comparison of neutral + negative effects vs positive effects (Table 3). Notably, as was evident for settling preferences, the two virus families in the NC-NPer group diverge with respect to the distribution of experiments into adaptive vs maladaptive categories (Table 3). Viruses in the family Bromoviridae have mostly neutral + negative effects on plant quality (*chi-square* $=12.56$, $P<0.001$), while effects were evenly distributed between neutral + negative effects and positive effects for Potyviridae (*chi-square* $=1.52$, $P=0.217$) (Table 3). The divergence of effects on both plant palatability and plant quality between these two families within the NC-NPer group may reflect other biological characteristics of viruses within each taxon. For example, the Bromoviridae studied thus far are multipartite viruses that depend on vectors acquiring and retaining each encapsidated portion of the genome. Requirements for dispersal following probing may be more stringent for these multipartite viruses relative to monopartite Potyviridae, and this may favor more extreme reductions in plant quality to encourage vector restlessness and probing behavior. In fact, there is evidence from one Potyviridae system (*Turnip mosaic virus*, TuMV) that vectors are still capable of transmitting even after successfully colonizing the infected host plant (Casteel et al., 2014). This is unusual for an NC-NPer virus because virions are acquired during brief probes in nonvascular tissue and are generally lost from the stylet if the vector proceeds to colonize the plant and initiate phloem ingestion (Fig. 1; Table 1). Thus, while transmission mechanisms provide a good metric for exploring potentially adaptive virus effects on host phenotypes, our synthesis reveals interesting patterns *within* transmission mechanisms that warrant further exploration.

2.3 Mechanisms Underlying Host-Mediated Effects

If transmission-conducive effects on host phenotype are the product of virus adaptations, then the targets of manipulation should frequently include plant cues known to mediate interactions with vectors. However, unlike clear cases of parasite manipulation involving protozoan or metazoan parasites of animals, it may be difficult in individual plant virus pathosystems to distinguish adaptive indirect (host-mediated) effects from by-products of pathology because plant viruses can induce subtle changes in suites of cues that are already produced by healthy hosts. A fraction of the studies in our analysis profiled plant cues and/or transcriptional responses of hosts to viruses, and an even smaller subset used functional genomics approaches to identify virus proteins involved in the induction of specific host phenotypes. Here, we review these studies to identify points of convergence in the mechanisms underlying virus effects on host phenotypes and track overall progress toward the goal of demonstrating that such effects are the product of virus adaptations rather than generalized host responses to pathogen infection.

2.3.1 Virus Effects on Long-Range Cues

Many sucking insects that transmit plant viruses (i.e., aphids, whiteflies, thrips, planthoppers, leafhoppers), as well as chewing vectors (i.e., beetles), use visual and olfactory cues to orient toward potential host plants. Aphids are the most well-represented vectors in the studies included in our synthesis (55.7% of virus–host–vector combinations), and both winged and wingless aphid morphs make use of volatile cues while walking and during initial host plant assessments (Eigenbrode et al., 2002; Mauck et al., 2010; Webster, 2012). Thus, we expect that these cues will frequently be targets of manipulation by plant viruses of all transmission classes (Figs. 1 and 2). Out of nine studies in our analysis that profiled volatile metabolites of infected and healthy plants and odor-based vector preferences, seven found that vectors preferred the odor source that emitted a greater quantity of volatiles per unit of leaf tissue—usually the infected host (Eigenbrode et al., 2002; Jiménez-Martínez et al., 2004b; Lu et al., 2016; Mauck et al., 2010, 2014a; Rajabaskar et al., 2014; Werner et al., 2009). Volatile blends of infected hosts were characterized by enhanced emissions of volatiles produced constitutively as a result of normal physiological activity, rather than induction of novel compounds or major changes in blend component ratios. In one case, virus infection suppressed volatile emissions (BPMV infecting soybean), which corresponded with reduced attraction of vectors (Peñaflor et al., 2016). While not adaptive for the virus, this result is still consistent with the pattern of higher emitting plants being more attractive. In two cases out of the nine total studies, virus infection caused variable changes in the quantities of each volatile emitted rather than overall enhanced emissions of all compounds, with divergent outcomes for vector attraction. Wu et al. (2014) reported that infection of peas by either of two different viruses in the Luteoviridae (*Bean leafroll virus* and *Pea enation mosaic virus*) increased the ratio of green leaf volatiles (*trans*-3 hexen-1-ol, *cis*-3-hexen-1-ol, 1-hexanol, and *cis*-3-hexenyl acetate) to monoterpenes (β-pinene and β-ocimene). This change was associated with increased attractiveness of infected plants to the vector (*Acyrthosiphon pisum*) and is consistent with the possibility that the viruses manipulate the lipoxygenase pathway, in which multiple green leaf volatiles are synthesized from hydroperoxy intermediates derived from linolenic acid (Dudareva et al., 2013). In Mauck et al. (2014a), a squash-adapted isolate of CMV infecting a novel host (pepper) induced changes in the percent composition of seven compounds within the volatile blend and had no overall effect on total volatile emissions. Aphids were not more attracted to this volatile blend but were attracted to the overall enhanced volatile blend emitted by squash (the primary host) infected with this same CMV genotype. Several studies also report virus effects on visual cues (spectral reflectance or light polarization) that may enhance vector attraction to infected hosts (Ajayi and Dewar, 1983; Maxwell et al., 2017), and synergism between enhanced volatile cues and altered visual cues (Fereres et al., 2016).

Although not numerous, these studies suggest that induction of a "supernormal stimulus" (i.e., exaggeration of existing cues) might be a common mechanism by which viruses enhance attractiveness of hosts to vectors (Dawkins and Krebs, 1979). This makes sense given that viruses are often transmitted by more than one vector species, and certainly by different genotypes within a species, making manipulation of the "host present" signal a better strategy than changes in blend composition that might compromise attractiveness to a subset of vectors. The range of potential manipulations could also be limited by molecular constraints on multifunctional proteins (discussed in Section 3). Unfortunately, there is little information about how the viruses studied thus far might induce overall volatile increases. We can speculate that manipulation of green leaf volatiles might arise as a derivation of virus effects on plastid membranes that create protected sites of replication. Membranes contain the precursors for the lipoxygenase pathway (linolenic and linoleic acids) and virus proteins are already adapted for interacting with these molecules. Manipulation of terpenoid compounds might involve virus effects on stomatal apertures, which are known release sites for sesquiterpenes (Seidl-Adams et al., 2015), and possibly other volatiles. For example, the CMV 2b protein, a virus suppressor of RNA silencing, has been shown to interfere with abscisic acid signaling to increase stomatal permeability in *Arabidopsis thaliana* (Westwood et al., 2013b), and it also induces an attractive odor phenotype in this same host (Wu et al., 2017). Uncovering the mechanisms by which viruses alter host volatile emissions will require a functional genomics approach that explores the role of individual proteins alone, and in combination, across different host environments.

2.3.2 Virus Effects on Contact and Palatability Cues

Following contact with a potential host, insect vectors detect leaf surface cues and ingest small quantities of plant material to rapidly discriminate between suitable and unsuitable host plants (Margaritopoulos et al., 2005; Powell et al., 2006). The outcome of these assessments is initiation of dispersal behavior (from unsuitable hosts) or initiation of subsequent feeding activities (on suitable hosts). A variety of plant characteristics and cues are thought to mediate host plant acceptance, although a comprehensive understanding of host assessment is lacking for most piercing–sucking vector insects (Powell et al., 2006). Leaf toughness, thickness, trichome density, and phloem accessibility will influence the number and duration of probing events. Specialists may respond to just a few plant cues (e.g., a particular class of secondary metabolites) that enable identification of a limited range of suitable hosts (Gabrys and Tjallingii, 2002), while generalists use a variety of hosts that presumably vary widely in secondary chemistry. Nutritional cues, such as free amino acids and sugars, are expected to play a larger role in host discrimination by generalists (Douglas, 2003; Tosh et al., 2003). These cues may interact with plant defense status. Stylets distort the cell walls during penetration, which leads to induction of rapid changes in ion permeability of the plasma membrane, ion exchange (Ca^{2+} and H^+ in, K^+ and Cl^- out), and production of reactive oxygen species as a first line of defense (Mai et al., 2013; Powell et al., 2006). Vectors, and especially those that sample intracellular contents of parenchyma, might be sensitive to these rapidly induced defenses and could be using them as additional measures of host quality.

Experiments on vector settling/feeding preferences are numerous in our quantitative synthesis, but only a handful of these studies quantified plant cues. Most do not separate the relative influence of nutritional vs defense-related cues because it is logistically difficult to do so, and pathway mutants are not always available for susceptible hosts. In general, vector preferences for infected plants are correlated with increased quantities of free amino acids in leaf tissue (Casteel et al., 2014; Cui et al., 2016; Fiebig et al., 2004; McMenemy et al., 2012), but strong relationships are not always apparent. Costa et al. (1991) reported that whiteflies preferred several species of host plants infected with Geminiviridae or Closteroviridae, but found no relationship between whitefly oviposition preferences and total quantities of free amino acids in parenchymal tissues. Blua et al. (1994) showed that squash plants infected with the NC-NPer viruses, *Zucchini yellow mosaic virus* (ZYMV, Potyviridae), had higher levels of free amino acids in leaf tissue at all stages of disease progression, but aphids had difficulty feeding on this tissue only when plants had been infected for 4 weeks. Using microscopy, metabolomics, and techniques for monitoring stylet activities, it was subsequently shown that infected plants in this late stage of disease progression had higher trichome densities and a reduction in the ratio of sucrose to total free amino acids (Blua and Perring, 1992b; Blua et al., 1994). This nutrient ratio is known to be a key mediator of hemipteran host preferences (Abisgold et al., 1994), and it may be the reason why some studies focusing only on absolute amino acid concentrations did not find a clear relationship with vector preference.

Nutritional changes can also interact with virus-induced alterations of plant defenses. *Tomato yellow leaf curl virus* (TYLCV, Geminiviridae) increases both free amino acids and sugars in phloem sap of infected tomatoes and attenuates induction of defenses against the vector, *B. tabaci* (Su et al., 2015). Whiteflies prefer to settle on TYLCV-infected tomatoes (Legarrea et al., 2015) and acquisition-associated feeding behaviors are enhanced by TYLCV infection (Liu et al., 2013). Attenuation of defenses also occurs in tobacco infected by the related *Tomato yellow leaf curl china virus* (TYLCCNV, Geminiviridae) (Luan and Yao, 2013; Zhang et al., 2012), which improves whitefly performance, but the same enhancement of amino acids in the phloem sap of TYLCCNV-infected tobacco was not detected (Wang et al., 2012). Studies across both geminivirus pathosystems illustrate the complexity of virus-induced changes in host phenotypes, as well as the difficulties associated with identifying whether changes in suites of metabolites are under the genetic control of the pathogen or represent immune responses on the part of the host plant.

Among the NC-NPer viruses *Cucumber mosaic virus* (CMV, Bromoviridae) is one of the best-studied systems with regard to mechanisms underlying virus effects on host phenotypes because it is tractable in the laboratory and amenable to functional genomics studies. As with the ZYMV pathosystem (Blua et al., 1994), Mauck et al. (2014b) found that the ratio of sugars to amino acids in leaf tissue of squash plants infected with the Fny strain of CMV played a role in mediating the rapid dispersal of aphid vectors from infected hosts following probing (Mauck et al., 2010). Using this same strain of CMV in *A. thaliana*, Westwood et al. (2013a) showed that, in this host, rapid dispersal behavior is instead mediated by a secondary metabolite (the glucosinolate compound 4-methoxy-indol-3-yl-methylglucosinolate), which is a mild aphid feeding deterrent. A reverse genetics approach identified the CMV 2a protein (RNA-dependent-RNA-polymerase) as the virus factor responsible for production of the glucosinolate compound by activating defensive signaling in the host (Westwood et al., 2013a). The 2a protein works in concert with the 2b protein, which enhances attractiveness of *Arabidopsis* plants to aphid vectors via effects on volatile cues (Westwood et al., 2013a; Wu et al., 2017), to produce a host phenotype that deceives vectors into visiting plants that they will ultimately find unsuitable after acquiring virions (Mauck et al., 2010; Westwood et al., 2013a). However, on tobacco, the Fny strain of CMV neither renders plants more attractive to aphid vectors (Tungadi et al., 2017) nor induces an unpalatable phenotype (Ziebell et al., 2011).

A reverse genetics approach was also employed to characterize virus proteins mediating effects of TuMV (Potyviridae) on host phenotypes and aphid vector behavior (Bak et al., 2017; Casteel et al., 2014, 2015). TuMV is an NC-NPer virus that apparently has relaxed restrictions on vector behaviors favoring transmission relative to other NC-NPer viruses. While CMV transmission is clearly reduced when aphid vectors engage in phloem feeding (Fereres and Collar, 2001; Martin et al., 1997), transmission of TuMV can occur following aphid colonization of the host plant (Casteel et al., 2014). Correspondingly, Casteel et al. (2014) reported that expression of the NIa-Pro virus protease protein in *Nicotiana benthamiana* enhanced palatability of host plants for *Myzus persicae* by increasing free amino acid content in leaf tissue and decreasing deposition of feeding-deterrent callose tissue in response to aphid feeding. In an elegant study, Bak et al. (2017) further showed that long-term aphid vector colonization and performance are due to rapid relocalization of NIa-Pro to the vacuole after perception of local signals associated with aphid feeding. Relocalization was not as strongly induced when the attacking insect was a nonvector, and it was also host specific, occurring in *N. benthamiana* but not *Nicotiana tabacum*. Together with work on CMV, these studies highlight the complexity of virus effects on hosts, and the difficulty in demonstrating that such effects are the product of virus adaptations rather than host responses to pathogen infection. Changes in host phenotype can result from the effects of single or multiple virus proteins and can vary depending on the host plant. These findings suggest that molecular and environmental constraints on virus evolution may limit the emergence of manipulative functions (discussed in Sections 3 and 4). Future mechanistic studies should combine functional genomics approaches with artificial selection experiments to elucidate factors favoring the evolution of host plant manipulation.

2.4 Changes in Vector Behavior Following Virus Acquisition

The best-documented cases of putative host manipulation involve protozoan and metazoan parasites. Some of these parasites directly influence the behavior of intermediate animal hosts in ways that enhance the probability that the parasite will subsequently be ingested by primary hosts, where it can complete its life cycle and, if possible, reproduce sexually (Moore, 2013; Poulin, 2010; Thomas et al., 2005). Although the mechanisms underlying these effects are still poorly characterized, the prevailing hypothesis is that parasites change the behaviors of animal hosts by *directly* targeting the host's neural, endocrine, neuromodulatory, and immunomodulatory systems during infections (Lafferty and Shaw, 2013). Plant viruses with circulative-persistent transmission mechanisms (C-P-Prop viruses and C-P-NProp viruses) may also be capable of targeting these systems in their insect vectors during interactions with vector tissues following acquisition (Fig. 1; Hogenhout et al., 2008). C-P-NProp viruses encounter and traverse a diversity of membrane barriers in different tissue systems during the journey from the midgut to the salivary glands, while C-P-Prop viruses use vector resources directly by replicating in gut epithelium, muscle tissue, salivary glands, fat bodies, and the nervous system. Consistent with these intimate associations, a growing number of studies provide evidence of *direct* manipulation of insect vector behavior by C-P-Prop and C-P-NProp viruses.

As demonstrated in Section 2, viruses frequently induce changes in host plants that render them more attractive to insect vectors (Tables 2 and 3), which is hypothesized to increase virus spread (Roosien et al., 2013; Sisterson, 2008). However, once vectors acquire virions and become viruliferous, there are no additional benefits (for the virus) if vectors continue visiting infected hosts (Table 1; Fig. 1) (Hogenhout et al., 2008; McElhany et al., 1995). Changes in vector preferences following virion acquisition have been named *conditional vector preferences* because the relative attractiveness of plant cues associated with infected and healthy plants is conditional on the viruliferous status of the vector. There are several published reports of conditional vector preferences for C-P-Prop and C-P-NProp pathosystems. For example, the white-backed planthopper (*Sogatella furcifera*), which transmits the C-P-Prop *Southern rice black-streaked dwarf virus* (SRBSDV, Reoviridae), and the brown planthopper (*Nilaparvata lugens*), vector of the C-P-Prop, *Rice ragged stunt virus* (RRSV, Reoviridae), are both differentially attracted to healthy or infected plants depending on their viruliferous status, the infection status of the host, and the stage of disease progression relative to a same-age healthy host (Lu et al., 2016). Preference changes are also evident in C-P-NProp virus pathosystems. Aphids carrying *Potato leafroll virus* (PLRV, Luteoviridae) preferred volatiles from healthy plants, whereas nonviruliferous aphids preferred odor cues from infected plants (Rajabaskar et al., 2014). The same effect was observed for a whitefly-transmitted C-P-NProp virus

(*Tomato severe rugose virus* [ToSRV, Geminiviridae]), where viruliferous whiteflies preferred the odors of mock-inoculated healthy plants over those of ToSRV-infected plants (Fereres et al., 2016). Theoretical explorations of conditional vector preferences demonstrated a substantial positive influence on virus spread (Roosien et al., 2013; Shaw et al., 2017). Indeed, while a constant preference for infected hosts is expected to increase virus spread only when infected plants are rare, a conditional preference eliminates this trade-off and increases virus spread at all stages of epidemic progression (Roosien et al., 2013; Shaw et al., 2017).

Conditional vector preferences based on plant palatability cues are also evident for several C—P virus pathosystems, with the most well-studied group being aphid-transmitted viruses in the Luteoviridae. Carmo-Sousa et al. (2016) demonstrated that *Aphis gossypii* carrying *Cucurbit aphid-borne yellows virus* (CABYV, Luteoviridae) preferred to settle on healthy cucumber plants over CABYV-infected plants, while no settling preference was observed for nonviruliferous aphids. In the PLRV system described earlier, Rajabaskar et al. (2014) found that viruliferous aphids have a clear settling preference for healthy potato plants, whereas nonviruliferous aphids preferred settling on infected plants. And using the well-studied *Barley yellow dwarf virus* (BYDV)-wheat pathosystem, Ingwell et al. (2012) observed the same preference reversal for viruliferous aphids as seen in the PLRV system. As a further step, this study demonstrated that the reversal occurred even when aphids acquired BYDV virions by feeding on an artificial medium containing virions without contact with the infected plants. This provides additional evidence that the observed shifts in behavior are the result of a direct effect of the plant virus on the aphid vector and not mediated by proteins or chemical compounds associated with infected plants. A whitefly-transmitted virus (TYLCV (Geminiviridae) also induces palatability-based conditional vector preferences following acquisition (Legarrea et al., 2015; Moreno-Delafuente et al., 2013), although these effects appear to be specific to the virus genotype by vector combination under study (Fang et al., 2013). Beyond plant viruses, similar conditional vector preferences have been documented for several psyllid-transmitted phytoplasmas in the *Candidatus liberibacter* group (Mann et al., 2012; Martini et al., 2015; Mas et al., 2014). Evidence of convergence among viruses and phytoplasmas that both reside and replicate within their vectors lends further support to the hypothesis that these effects may be the product of pathogen adaptations.

The studies above propose that transmission-conducive direct effects are under genetic control of the virus because some of the outcomes appear to be detrimental for the vector (e.g., discouraging visitation to, and feeding on, nutritionally superior virus-infected hosts). While there has been no attempt to dissect the roles of individual virus proteins in induction of direct effects, other mechanistic studies provide some insight into specific behavioral processes that change in response to virus acquisition and retention. The electrical penetration graphing (EPG) technique is a powerful tool for monitoring the probing and feeding behaviors of piercing–sucking vectors during interactions with plant hosts (Tjallingii, 1988). EPG systems translate stylet position in the plant (mesophyll, xylem, or phloem tissues) and associated activities (intercellular stylet progression, intracellular penetration, salivation, or ingestion) into distinct waveforms that can be analyzed quantitatively, making this tool particularly valuable for dissecting transmission-relevant behavioral differences between viruliferous and nonviruliferous vectors. For example, in the SRBSDV system, where Lu et al. (2016) observed a preference of viruliferous white-backed planthoppers for healthy plants, Lei et al. (2016) showed that viruliferous white-backed planthoppers spent more time in salivation and phloem sap ingestion on healthy plants than nonviruliferous insects. These behaviors are expected to increase SRBSDV inoculation. In another C-P-Prop virus pathosystem (*Tomato spotted wilt virus* [TSWV, Tospoviridae]), viruliferous males of *Frankliniella occidentalis* performed a greater number of probes into plant tissue, and specifically more noningestion probes that consist of salivation without extensive cellular damage, which is expected to increase the probability of virus establishment in living cells (Stafford et al., 2011). Among C-P-NProp virus pathosystems, Moreno-Delafuente et al. (2013) showed that *B. tabaci* carrying TYLCV had a larger number of phloem contacts and longer salivation phases in phloem sieve elements than nonviruliferous whiteflies. Viruliferous whiteflies also moved half as quickly as nonviruliferous whiteflies and tended to settle and feed more readily. This arrestment behavior is likely related to the enhanced probing and feeding of viruliferous whiteflies on healthy plants. Minor effects on prephloem activities (mean durations of nonprobing, intercellular pathway, and intracellular punctures) were also observed in the CABYV-cucumber pathosystem discussed earlier (Carmo-Sousa et al., 2016). There is even evidence of one foregut-borne NC-SPer virus *Cucurbit chlorotic yellows virus* (CCYV, Closteroviridae) inducing earlier and longer salivation phases in viruliferous male whiteflies relative to nonviruliferous whiteflies (Lu et al., 2017).

While examples are not numerous, the studies discussed earlier suggest that salivation and feeding behaviors are targets for direct manipulations by C-P plant viruses. Virus proteins from both C-P-Prop viruses and C-P-NProp viruses interact with a variety of host proteins while residing or replicating in the salivary glands. Mar et al. (2014) demonstrated that SRBSDV coimmunoprecipitates with at least 18 proteins from its insect vector, *S. furcifera*, including proteins in the nervous system and the ubiquitin proteasome system, which regulates numerous cellular processes. A differential proteomics analysis of *F. occidentalis* infected with TSWV showed abundant expression of proteins involved in signaling, stress and defense responses, translation, and lipid metabolism (Ogada et al., 2017). Similarly, Luteovirids and Geminivirids (C-P-NProp viruses) both perturb hundreds of biochemical pathways during retention in their respective vectors (Gray et al., 2014; Hasegawa et al., 2018; Luan et al., 2011). Many of these perturbations involve suppression of vector immune responses that might limit virus invasion of vector tissues, providing evidence that C-P-NProp viruses can evolve to manipulate core pathways in their insect vectors. Therefore, it may not be such an evolutionary leap for C-P viruses to evolve adaptations for manipulating vector-feeding behavior. One potential mechanism is via virus effects on the expression levels of salivary effector proteins that modify the defense status and suitability of plants for vector feeding (Elzinga and Jander, 2013). Mutti et al. (2008) demonstrated that this route is feasible by showing that reduced expression of a putative salivary effector protein in pea aphids resulted in a lowered frequency and duration of phloem feeding on a suitable host, as measured by EPG recordings. There are dozens of other known salivary protein effectors in aphids (Elzinga and Jander, 2013) and hundreds of putative secretory proteins have been identified in whiteflies (Su et al., 2012) and planthoppers (Ji et al., 2013). As these proteins are characterized and more functional genomics tools become available for important herbivorous vectors, it will be interesting to explore how C—P virus retention influences salivary effector delivery and efficacy.

3. MOLECULAR CONSTRAINTS ON THE EVOLUTION OF MANIPULATIVE FUNCTIONS

Section 2 provides evidence that viruses can evolve to manipulate vector behavior indirectly, via effects on host plant cues (Tables 2 and 3) (Sections 2.1–2.3), and directly, via effects on vectors (Section 2.4). The handful of studies identifying putative virus effectors of host phenotypes provide additional evidence that virus proteins can evolve manipulative functions. But these same studies also suggest that there are constraints on the evolution of these functions. For example, manipulation of host phenotype is species specific (Bak et al., 2017; Mauck et al., 2014a; Westwood et al., 2013a,b; Ziebell et al., 2011) and can even be limited to certain *genotypes* within a species (Rajabaskar et al., 2013a,b). This suggests that plant viruses cannot evolve manipulative functions that induce a transmission-conducive phenotype in all possible hosts. Limitations on the evolution of manipulative functions are likely governed by the same molecular constraints that limit virus evolution and adaptation generally. There are numerous reviews on the mechanisms underlying the generation and fixation of mutations in virus genomes, particularly for RNA viruses, which are thought to have relatively high mutation rates and extreme limitations on genome size (Belshaw et al., 2007; Duffy et al., 2008; Gilbertson et al., 2003). Here, we provide a short overview of several key constraints on virus evolution (genome size, pleiotropy and epistasis, and maintenance of vector transmission) as a framework for understanding barriers to the evolution of secondary functions in virus proteins that may confer manipulative traits.

3.1 Genome Size and Secondary Structure

Plant viruses have some of the smallest genomes of any organism (4–20 kb). Restrictions on plant virus genome size are imposed by several factors. Rigid plant cell walls force most plant viruses to use plasmodesmata as a means of establishing systemic infections (Lucas, 2006). Both DNA and RNA viruses accomplish this by actively increasing the size-exclusion limits of plasmodesmata and exploiting the endogenous RNA trafficking mechanisms of the host using movement proteins (Lucas, 2006). But the size-exclusion limit typically cannot be increased beyond a certain point (large enough for passage of a 10 kDa protein) (Hanley-Bowdoin et al., 2000; Lucas, 2006). This constraint limits the size of the virus genome, which must pass through the modified plasmodesmata with the help of viral proteins (Lucas, 2006). A second factor that limits the size of plant virus genomes is mutation. If the mutation rate is too high, increasingly large genomes will experience progressive decay due to rapid accumulation of lethal mutations. This is one mechanism to explain the finding that RNA plant viruses tend to have smaller genomes than DNA plant viruses (Belshaw et al., 2008). RNA-dependent RNA polymerases encoded by RNA viruses lack 3′ to 5′ exonuclease activity and therefore do not proofread during synthesis of new RNA molecules (Drake et al., 1998). The existence of secondary RNA structure also poses challenges for polymerase fidelity and may lead to deletions during template slippage (Duffy et al., 2008). The realities of RNA genome structure and replication have led to the hypothesis that RNA plant viruses experience higher overall mutation rates relative to DNA plant viruses (Drake et al., 1998). However, recent studies with plant viruses having single-stranded DNA (ssDNA) genomes suggest that these pathogens experience similar mutation rates to those observed in RNA plant viruses even though they are using host plant DNA polymerases for replication. For example, the monopartite whitefly-transmitted begomovirus, TYLCV, and the bipartite begomovirus, *East African cassava mosaic virus*, both exhibit nucleotide substitution rates within the range of $\sim 1.56 \times 10^{-3}$ to 4.63×10^{-4} substitutions per site, per year, which are in line with estimates of nucleotide substitution rates for RNA viruses (Duffy and Holmes, 2008, 2009).

Regardless of the mechanisms underlying restrictions on genome size in plant viruses, the outcome of such restrictions is the tight packing of coding sequences within the available space for information storage. Many plant viruses with RNA genomes exhibit some level of gene overlap, where there are simultaneously two open reading frames (ORFs) coded by the same set of nucleotides. In an analysis of 701 reference RNA virus genomes, Belshaw et al. (2007) found that 56% of the sequences exhibited some gene overlap. Of this 56%, approximately 29% (116 of 392 reference genomes) belonged to plant viruses. A subsequent analysis of the Potyviridae found that this group (represented by 54 sequences in Belshaw et al., 2007) also has gene overlap (Chung et al., 2008). Across all virus families, viruses with smaller genomes tended to have proportionally more gene overlap than viruses with genomes larger than the median length (Belshaw et al., 2007). It is hypothesized that gene overlap benefits the virus because it allows more information to be encoded in a smaller size molecule (e.g., as terminal overlaps of the 3′ end of one gene with the 5′ end of a second gene). It also accommodates the creation of new genes (as internal overlaps) without a concurrent increase in genome size. Maintenance of a smaller genome via gene overlap may facilitate higher replication rates and lower mutation rates, but is also hypothesized to impose additional constraints on virus evolution (Sanjuán and Elena, 2006; Sanjuán et al., 2004, 2005). Even though mutation rates of plant viruses may be high enough to generate the necessary diversity for the evolution of alternative protein functions (including manipulative functions), mutations that occur in overlapping regions are more likely to be detrimental and may subsequently be rapidly purged from the population. Plant viruses also experience extreme reductions in population sizes (bottlenecks) during the initiation of systemic infections (Gutiérrez et al., 2010; Li and Roossinck, 2004; Monsion et al., 2008) and during inoculation by vectors (Ali et al., 2006). Bottlenecks introduce an additional level of stochasticity in the selection process by reducing the chances that any given mutation will proliferate within the population (Ali et al., 2006; Power, 2000). The frequency with which plant viruses experience population bottlenecks suggests that even beneficial mutations, including those that confer manipulative functions, may be regularly purged by chance events.

3.2 Pleiotropy and Epistasis

Most plant viruses are capable of infecting multiple host species and certainly multiple host genotypes within a species. Therefore, mutations that do go to fixation during infection of one host (because they are either beneficial or, more rarely, neutral) might be detrimental during infection of a subsequent host if the mutation abolish infectivity (host-selective lethality) or limit within-host fitness. This phenomenon is categorized as a type of antagonistic pleiotropy because the same gene controls virus fitness outcomes in two different hosts. Thus, pleiotropic effects impose constraints on virus evolution by limiting the number of host environments available for infection (potentially selecting for specialist viruses), or by imposing costs on viruses that have evolved a generalist lifestyle (Agudelo-Romero et al., 2008; García-Arenal and Fraile, 2013; Malpica et al., 2006). Plant viruses serve as good models for basic research on host trade-offs because many of them can be mechanically passaged through homogeneous or heterogeneous host environments in a logistically simple, replicated design. Studies of this nature repeatedly show the evolution of species-specific mutations, many of which confer reductions in fitness in novel hosts relative to the primary host (Bedhomme et al., 2012; Elena, 2016; Miyashita et al., 2016). In an alternative approach, the artificial introduction of point mutations into infectious clones of plant viruses demonstrates how mutational effects interact with host identity (Elena and Lalić, 2013; Lalić et al., 2011). This approach was used to explore the effects of 20 single point mutations on the fitness of the Solanaceae-specialist RNA virus, *Tobacco etch potyvirus* (TEV), in its primary host environment (*N. tabacum*), other related Solanaceae, and phylogenetically distant hosts in the Asteraceae and Amaranthaceae (Lalić et al., 2011). Effects of mutations on virus fitness varied depending on the host environment, providing clear evidence of pleiotropic effects. Looking across the axis of host relatedness (*N. tabacum*), the fraction of beneficial mutations was slightly higher in non-Solanaceae hosts. Thus, the negative outcomes of antagonistic pleiotropy depend on the phylogenetic distance between two possible host species (Lalić et al., 2011). In the context of the evolution of secondary, manipulative functions in virus proteins, the widespread occurrence of antagonistic pleiotropy suggests that any mutations that confer manipulative ability to a virus in one host have the potential to confer maladaptive effects in a second host, with such effects depending on the phylogenetic distance between host species.

Plant viruses with tightly packed genomes with overlapping ORFs encoding multifunctional proteins will also experience significant evolutionary constraints due to nonmultiplicative interactions among mutations, a phenomenon known as epistasis (Lalić and Elena, 2013; Sanjuán et al., 2004; Torres-Barceló et al., 2010). The importance of epistasis as a factor limiting the evolutionary pathways available to viruses is evident from numerous experimental evolution and double-mutant interaction studies, most of which employ RNA viruses (Bedhomme et al., 2015). In a follow-up study to the one described earlier, Lalić and Elena (2013) explored epistasis effects using 10 novel TEV genotypes each having a different combination of 2 mutations whose effects in single occurrence were known (Lalić et al., 2011). They found that the magnitude (fitness value) and sign (positive or negative) of epistatic interactions among mutations varied depending on the host background. In hosts of the same family as the origin host (Solanaceae), TEV mutants experienced fewer independent fitness effects of epistasis relative to TEV infections in distantly related hosts (Asteraceae or Amaranthaceae) (Lalić and Elena, 2013). Epistatic effects included both within-host fitness reductions and host-dependent lethality, and mutant pairs in different proteins led to similar epistatic effects in several cases. These examples suggest that epistatic interactions among virus genes are expected to further limit pathways available for the evolution of manipulative functions, particularly for multihost viruses that infect phylogenetically divergent host species.

3.3 Maintenance of Vector Transmission

Most studies exploring constraints on plant virus evolution do not include insect vectors. This is primarily due to logistical constraints (mechanical inoculations are more consistent), not lack of interest in the interactive effects of selection pressures imposed by host plants and vectors. Nonetheless, we can hypothesize that the need to maintain vector transmissibility will further constrain plant virus evolution. Consistent with the multifunctional nature of most virus proteins, there is evidence that mutations in protein domains mediating virion interactions with vectors can have effects on within-host replication and systemic spread. In the well-studied *Barley yellow dwarf virus* pathogen (BYDV, Luteoviridae), a nonstructural protein encoded by ORF 4 is required for both aphid transmission and systemic movement in plants. Mutation of the read-through domain of this protein abolished the capacity for virions to invade aphid salivary glands and also reduced the accumulation of virus in infected plants (Chay et al., 1996). In a similar scenario, a single amino acid change in the coat protein of CMV reduced aphid transmissibility and induced necrosis in tobacco plants (Ng et al., 2005). The consequences of this mutation will have a twofold effect on virus fitness: the probability of transmission will be reduced *directly* as a result of inefficiencies in virion binding to aphid mouthparts, and *indirectly* as a result of lethal necrotic symptoms that limit the duration of time that the host can serve as a source of inoculum. In both examples, effects are pleiotropic because each respective mutation influences multiple seemingly unrelated traits (host infection and vector transmissibility). But unlike the examples earlier (Bedhomme et al., 2012; Lalić et al., 2011), the signs of pleiotropic effects do not change when the consequences of the mutations are explored in host and vector environments. The cumulative negative effects of mutations that simultaneously disrupt both host–virus and vector–virus interactions will impose additional limits on pathways available for evolution of novel protein functions, including functions that result in the induction of specific host phenotypes.

Host phenotype and vector transmissibility can also be modified by the presence of subviral RNAs—molecular parasites of viruses that reproduce and proliferate using virus-encoded proteins (Simon et al., 2004). Plant viruses are associated with a particular type of subviral RNA (satellites), which are unique in that their sequences are mostly or entirely unrelated to those of the helper virus (Simon et al., 2004). For example, various satellite RNAs of CMV (Bromoviridae) interact with the primary CMV genotype to modify virus accumulation and symptom expression in the host. Satellite RNAs that induce a necrotic phenotype in the host plant will only persist at high vector densities (when there are more opportunities for transmission) because aphid vectors are less likely to perform transmission-conducive behaviors on plants infected with CMV + a necrogenic satellite RNA (Escriu et al., 2000, 2003). Association with a satellite RNA can therefore mask any manipulative effects of the "host" virus genotype on plant phenotype, effectively eliminating selection for, or against, mutations conferring manipulative functions.

4. ENVIRONMENTAL CONSTRAINTS ON THE EVOLUTION OF MANIPULATIVE FUNCTIONS

Molecular constraints on virus evolution operate in the context of environmental factors that have both spatial and temporal dimensions, and that are more or less stochastic depending on anthropogenic influences and abiotic variability. Virus variants that manipulate host phenotype and vector behavior may be selected against if the mutations responsible reduce virus fitness in some environments. Environmental features such as host diversity and vector population size will impose genetic structure on plant virus populations by increasing or decreasing the stringency of molecular constraints, and by augmenting the number of bottlenecks (reductions in genetic diversity) and founder events (chance survival of one particular genotype) (Ali et al., 2006; Betancourt et al., 2008; Geoghegan et al., 2016; Gutiérrez et al., 2012; Roossinck and García-Arenal, 2015). Furthermore, abiotic stressors, nonvector herbivores, and competing pathogens will impact host (and by extension, virus) survival. Thus, transmission-conducive effects of plant viruses will be favored only if they do not increase plant vulnerability to abiotic and biotic threats because plants that succumb to these stressors will no longer serve as sources of inoculum. Here, we review key environmental factors shaping virus evolution in a community context in order to generate predictions about how these factors might interact with molecular constraints to favor, or hinder, the evolution of secondary manipulative functions.

4.1 Host Community Composition

Fitness trade-offs among hosts have been repeatedly documented using laboratory experiments that explore virus evolution during serial transmission events, or the distribution of mutational fitness effects across host environments following artificial introductions of point mutations (discussed in Section 3). These studies demonstrate that adaptation to a primary host can result in reduced fitness in novel hosts due to antagonistic pleiotropy, although there are also documented cases of no apparent costs to virus fitness in novel hosts (Bedhomme et al., 2012). When trade-offs are evident, reductions in virus fitness tend to be more severe with increased phylogenetic distance between primary and novel hosts (Lalić and Elena, 2013; Lalić et al., 2011). Physiological aspects of host life history that determine reservoir potential (life span and investment in immune defenses) will also determine whether a host is "permissive" or "restrictive" for virus infection and replication (Cronin et al., 2010, 2014; Hily et al., 2014). Based on this experimental evidence, it is logical to expect that the diversity of hosts in the landscape functions as one major axis along which the evolution of manipulative functions will be directed and possibly constrained depending on pathogen and host characteristics.

This expectation has not been explored empirically, either through artificial evolution experiments in the laboratory or through characterization of the phenotypic effects of virus isolates from different host communities. But it is still possible to generate predictions regarding the mechanisms by which host composition might influence evolution of manipulative functions. One scenario under which manipulation is expected to enhance parasite fitness is when opportunities for transmission are constrained to a small-time window (Heil, 2016). In regard to plant viruses, this scenario is characteristic of agricultural monocultures that have an abundance of genetically uniform hosts, some of which may only be susceptible for a brief period during development due to age-related resistance (Panter and Jones, 2002; Sigvald, 1985). Use of chemical insecticides that target vectors, or targeted removal of infected hosts, could impose additional limitations on opportunities for virus acquisition. Additionally, annual crops with rapid life cycles and poor immune defenses are considered "permissive" host environments for virus replication (Cronin et al., 2010, 2014; Hily et al., 2014), which will lead to higher virus replication rates and more opportunities for mutations that could confer manipulative functions. Annual crops are also dead-end hosts for the virus due to harvesting or natural senescence. Based on these features, it is expected that annual monocultures will favor the evolution of manipulative functions that increase the probability of vectors acquiring, retaining, and transporting viruses from crops to alternative reservoirs before crop destruction.

Within these uniform host environments, we might expect to see more frequent evolution of manipulative functions for viruses with circulative-persistent transmission mechanisms (Table 1; Figs. 1 and 2). This is because most circulative-persistent viruses are exclusively transmitted by a limited suite of colonizing vectors that must engage in long-term feeding to acquire and retain the pathogen. The window of opportunity for circulative-persistent virus acquisition and inoculation is therefore constrained not only by the availability of susceptible hosts but also by the presence of colonizing vectors (Fereres and Moreno, 2009; Hogenhout et al., 2008). Selection will tend to favor virus genotypes that enhance the number of contacts with these vectors and encourage feeding behaviors conducive to transmission (Fig. 2).

In contrast, noncirculative viruses, particularly NC-NPer viruses, are often transmitted by a large number of species, many of which may be transient visitors but nonetheless efficient vectors (Perring et al., 1999; Pirone and Perry, 2002; Radcliffe and Ragsdale, 2002). For example, there are more than 50 known aphid vectors of the NC-NPer viruses, *Potato virus Y* (PVY), but only a few of these species actually colonize the solanaceous crop hosts that are susceptible to PVY (Radcliffe and Ragsdale, 2002). Spread of PVY in field environments is not driven by the abundance of colonizing vectors, but rather by the number of those that are transient, noncolonizing visitors engaging in rapid probing and dispersal behaviors conducive to NC-NPer virus transmission (Mondal et al., 2016; Radcliffe and Ragsdale, 2002) (Fig. 2). Spread of two other NC-NPer viruses, *Watermelon mosaic virus* and *Papaya ringspot virus*, in cucurbit crops was also linked to abundance of several noncolonizers, but not of colonizing aphids (Angelella et al., 2015). These less-specific vector–host relationships create more opportunities for successful transmission events and might "dilute" the features of annual monoculture environments that favor the evolution of manipulative functions.

Mixed host plant communities are expected to be less conducive to the evolution of manipulative functions relative to communities with low host diversity because the potential for antagonistic pleiotropy is greater (Bedhomme et al., 2015; Elena, 2016). Multihost viruses that induce transmission-conducive effects in one host might induce transmission-limiting effects in a second host, which will favor the spread of nonmanipulative genotypes that have neutral effects on host–vector interactions. Furthermore, direct effects of viruses on vector behavioral responses are unlikely to be robust across host plants with vastly different chemistries (Carrasco et al., 2015). Even among hosts in the same species, phenotypic effects may vary with genotype, competitive status, or other heterogeneous factors in diverse, unmanaged plant communities. These constraints suggest that the evolution of manipulative functions in mixed-host environments should be limited to specialist viruses that have narrow host ranges and tightly coevolved vector relationships. If susceptible hosts are few or patchily distributed, opportunities for transmission will be rare—a condition that is hypothesized to favor the evolution of host and vector manipulation (Heil, 2016). Specialist viruses may evolve manipulative functions in such environments if the cues responsible for mediating host–vector contacts are very specific and consistent (e.g., secondary metabolites associated with a specific plant family) and if the host range of the vectors is also similarly specialized. But the majority of characterized plant viruses naturally infect hosts across two or more plant families. Thus, heterogeneous host environments are expected to constrain, rather than favor, the evolution of manipulative functions.

4.2 Prevalence of Competent Vectors

If competent vectors are prevalent, these vectors are expected to contact and colonize host plants frequently regardless of plant infection status or phenotype. Agricultural monocultures are particularly likely to have large vector populations. Vectors with high reproductive rates, such as aphids and whiteflies, will go through successive generations on one crop host and emigrate in large numbers to other, more appealing crops following declines in initial host quality or harvesting (Carrière et al., 2017; Mondal et al., 2016; Thomas et al., 1993). Vector amplification and emigration are such a reliable phenomenon in agriculture that continent-wide, suction trap networks are in place in many locations to monitor vector emigrants and predict optimal planting dates (Harrington et al., 2004). This can be an effective strategy for mitigating virus impacts because there is ample evidence that rates of pathogen spread are strongly tied to vector abundance (Jeger et al., 2011; Madden et al., 2000; McElhany et al., 1995; Shaw et al., 2017; Sisterson, 2008). If planting dates align with high vector numbers, there are more vectors per host niche and almost every plant is contacted or colonized, creating opportunities for transmission. This will favor the spread of both manipulative and nonmanipulative virus genotypes and reduce the relative fitness benefits of host and vector manipulation. If planting dates do not align with high vector numbers, there are fewer vectors per host niche, and manipulated hosts should serve as sources of inoculum more often than nonmanipulated hosts.

Frequency of virus transmission is not only influenced by vector prevalence. Potential vectors must also be *competent* (capable of transmitting the virus) and *efficient* (transmit at high enough rates to contribute significantly to virus spread). Vector competence refers to the ability of an organism to acquire, retain, and inoculate a virus (or other microbial entity). Transmission efficiency refers to the probability of a competent vector transmitting a virus from one host to the next and is determined by characteristics of the virus, host, and vector as well as external factors, such as temperature, that modify relationships among these players (Anhalt and Almeida, 2008; Bosquee et al., 2016; Chatzivassiliou et al., 2002). Vector behavior is one major determinant of transmission efficiency (Fereres, 2016; Fereres and

Collar, 2001; Madden et al., 2000; Martin et al., 1997; Roosien et al., 2013; Shaw et al., 2017; Sisterson, 2008; Wang and Ghabrial, 2002). If the most abundant vectors have poor efficiency because they do not perform behaviors conducive to acquisition or inoculation, then transmission events will be infrequent (Madden et al., 2000; Shaw et al., 2017). Evolution of manipulative functions may be favorable under these conditions if such functions selectively increase transmission-conducive interactions with efficient vectors. Alternatively, if efficient vectors are abundant and most plants experience vector contacts independent of infection status, selection for manipulative functions will be weaker because there are more opportunities for transmission of all virus genotypes.

4.3 Off-Target Effects of Virus-Induced Host Phenotypes

As discussed in Section 2, plant viruses can alter plant nutritional quality, defense responses, and plant-derived sensory cues, either as part of an adaptive strategy of indirect manipulation of vector behavior or as a by-product of pathology. These often-significant impacts on host plant phenotypes are expected to strongly influence broader community interactions, with implications for the longevity, survival, and fitness of the infected host and the virus that depends on its resources (Alexander et al., 2013; Mauck et al., 2015). Nonvector arthropods are sensitive to many of the same cues and plant quality changes that mediate host plant interactions with insect vectors. And virus effects on conserved phytohormone signaling pathways will alter how hosts respond to both biotic and abiotic stressors (Aguilar et al., 2017). Therefore, evolution of manipulative functions in plant viruses will potentially be limited to effects on plant phenotype that maintain or enhance transmission-conducive vector behaviors without significantly increasing host vulnerability to nonvector organisms (off-target effects).

Limited explorations of the off-target effects of putative manipulations have been reported for several of the pathosystems covered in our quantitative synthesis (Tables 2 and 3). In the well-studied *Tomato spotted wilt virus*-pepper-thrips pathosystem (Tospoviridae), Belliure et al. (2010) reported that virus infection increased host susceptibility to a nonvector herbivore (*Tetranychus urticae*) that feeds in a similar manner as the thrips vector. Kersch-Becker and Thaler (2013) also demonstrated increased susceptibility of tomato plants infected with *Potato virus Y* (Potyviridae) to two chewing herbivores, including a Solanaceae specialist (*Leptinotarsa decemlineata*). More recently, Ángeles-López et al. (2016, 2017) showed that infection of chili plants by *Pepper golden mosaic virus* (Geminiviridae) induces a phenotype that is expected to increase transmission by its whitefly vector (*B. tabaci*), but which also increases host susceptibility to, and quality for, a nonvector whitefly (*Trialeurodes vaporariorum*). The authors additionally report that *T. vaporariorum* feeding significantly reduces virus titers in infected host plants and attenuates aspects of the (putatively) transmission-conducive phenotype induced by the virus (Ángeles-López et al., 2017). Coinfections with other pathogens may similarly disrupt transmission-conducive phenotypes (Peñaflor et al., 2016), although one report suggests that when coinfecting viruses are transmitted by the same vector, selection may favor genotypes of the "nonmanipulating" virus species that have minimal effects on the transmission-enhancing functions of the manipulative virus species (Salvaudon et al., 2013). Together, this small set of studies reveals at least three scenarios under which selection will disfavor manipulative virus genotypes in a community context: when virus-induced susceptibility to nonvectors (i) reduces host tissue volume or survival, (ii) abolishes the transmission-conducive phenotype, or (iii) directly reduces within-host virus fitness.

Manipulative genotypes are more likely to emerge and spread if the virus-induced host phenotype confers additional benefits to the host plant or at least has neutral effects on host resistance to other stressors. Recent studies highlight a few pathosystems where this appears to be the case. CMV (Bromoviridae) induces a transmission-enhancing phenotype in several host plants (*Cucurbita pepo*, *Cucumis sativus*, *A. thaliana*, *N. tabacum*) (Carmo-Sousa et al., 2014; Mauck et al., 2010, 2014a,b; Shi et al., 2016; Westwood et al., 2013a; Wu et al., 2017). In at least some of these hosts, the transmission-enhancing phenotype also confers drought tolerance (Westwood et al., 2013a; Xu et al., 2008) and reduces susceptibility to generalist and specialist nonvector herbivores (Mauck et al., 2015; Saad et al., 2017), both of which will increase the length of time infected hosts can persist as sources of inoculum. The putative host manipulation and the beneficial effects of virus infection on host resistance to drought and nonvector herbivores have been partially attributed to the functions of the 2b silencing suppressor protein (Westwood et al., 2013a,b). Several other reports demonstrate selectivity in the effects of virus-induced host phenotypes, with benefits only being realized for the vector, and not for common, co-occurring nonvectors (Cassone et al., 2014; Peñaflor et al., 2016; Sadeghi et al., 2016). But for the majority of pathosystems summarized in Tables 2 and 3, we have no knowledge of how putative virus manipulations of host phenotypes are likely to influence resistance to biotic and abiotic stressors.

5. LIMITATIONS OF EXISTING STUDIES IN THE CONTEXT OF MOLECULAR AND ENVIRONMENTAL CONSTRAINTS

Our quantitative synthesis of the literature (Tables 2 and 3) supports the predictions about convergent effects of plant viruses on host phenotypes and vector behavior outlined in Fig. 2, and first proposed in Mauck et al. (2012). Regardless of transmission mechanism, vectors are generally attracted to virus-infected plants over noninfected plants (Tables 2 and 3). Viruses that require sustained feeding for acquisition (C-P-Prop, C-P-NProp, and most NC-SPer viruses) generally enhance palatability and quality, whereas viruses that require rapid dispersal following acquisition (NC-NPer viruses) have more mixed effects, with a greater proportion of experiments demonstrating no change or a reduction in host palatability and quality (Tables 2 and 3). Convergence in the effects of viruses from multiple lineages that share a common transmission mechanism suggests that such effects are adaptive, or at least that viruses are subject to selection *against* the evolution of traits that have transmission-limiting effects (Poulin, 2010; Thomas et al., 2005). The few studies that have explored putative virus effectors of host phenotypes indicate that just one or two virus proteins can influence whole suites of host cues in ways that influence vector behavior. Despite this evidence, the broader literature on the molecular and environmental factors shaping virus evolution (Sections 3 and 4) suggests that there are only a few scenarios that favor the emergence and maintenance of manipulative traits, and many scenarios that do not. In light of these constraints, we performed a second quantitative synthesis of the same set of literature (Fig. 3) to determine how many studies considered the molecular and environmental factors discussed in Sections 3 and 4, and to identify target areas for future research to resolve multiple unknowns regarding virus manipulation of hosts and vectors in environments outside of the laboratory.

5.1 Pathogen Provenance

Nearly every study in our quantitative synthesis proposes that transmission-conducive effects are evidence of adaptive manipulation on the part of the virus. Presumably, this means that manipulative functions evolved in, and were selected for, an environment where they conferred fitness benefits to the pathogen. But more than 60% of the studies used virus strains that have been in laboratory culture for many years, or even decades (Fig. 3A). A significant proportion of studies with laboratory strains report culture

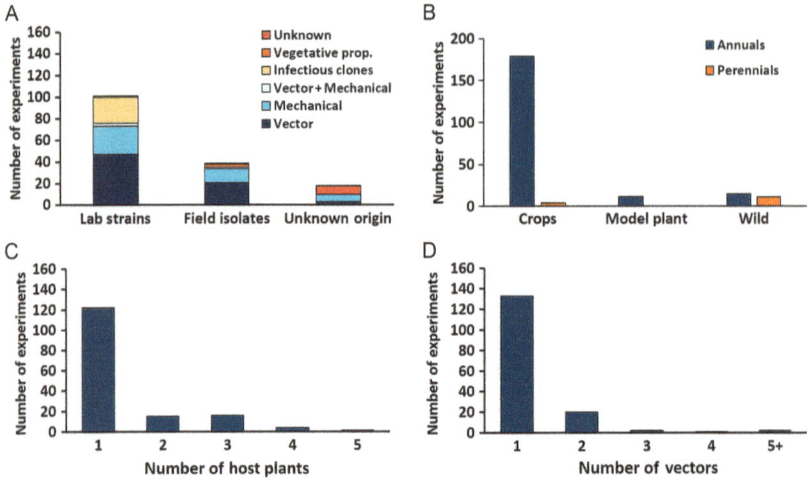

Fig. 3 Pathosystem features of the studies included in the quantitative synthesis (Tables 2 and 3). (A) Virus provenance; (B) host physiological phenotype and domestication status; (C) number of plant hosts examined for a single focal virus strain or isolate; (D) number of vector species or biotypes studied for each virus-host combination examined. For (A), (C), and (D), we parsed each article (122) into individual experiments, each addressing a single virus strain or isolate (158 experiments total). For (B), individual experiments represent a single virus (strain or isolate) studied on a single host (species or cultivar) (221 experiments total).

maintenance via serial transfer using vectors (44.9% of studies with lab strains) or mechanical transmission (29.1% of studies with lab strains). There is little to no methodological detail provided regarding these culture techniques (e.g., host age at inoculation, age of the inoculum source, number of vectors used to transmit, etc.). But we know from the experimental evolution studies discussed in Section 3 that repeated passage in a single host can drive virus evolution, including the fixation of mutations that influence host phenotype and constrain pathways for evolving novel functions (Bedhomme et al., 2012, 2013; Elena, 2016; Miyashita et al., 2016). Drastic changes in host phenotype can be the result of a single amino acid change in just one virus protein (Lewsey et al., 2009). Mutations that occur in overlapping ORFs may have proportionally larger effects on multiple aspects of the virus–host interaction (Belshaw et al., 2007), while epistatic interactions among virus proteins will further shape the impacts of any one mutation (Lalić and Elena, 2013; Sanjuán et al., 2004; Sanjuán and Elena, 2006). In the absence of selection pressures imposed by environmental factors, bottlenecks during vector transmission and systemic invasion of the host plant are expected to lead to genetic drift and, potentially, to the random loss of manipulative functions that might have evolved in the original environment from which a virus was isolated (Ali et al., 2006; Li and Roossinck, 2004). By the same mechanism, maladaptive mutations might emerge and become fixed because among-host virus fitness is under the watchful eye of the researcher, and not dependent on vector behavior.

Under these experimental conditions, it is difficult to assert that transmission-conducive effects are the result of selection for manipulative functions. Serial passage in a single culture host is far more likely to select for mutations that enhance exploitation of that host environment regardless of effects on host phenotype (e.g., higher replication rates, higher virus titer, more rapid systemic colonization). Indeed, higher virus titers have been positively associated with transmission-enhancing virus effects in several studies with *Cucumber mosaic virus* (Bromoviridae) (Mauck et al., 2014a; Shi et al., 2016), *Barley yellow dwarf virus* (Luteoviridae) (Fereres et al., 1989; Jiménez-Martínez et al., 2004a,b; Medina-Ortega et al., 2009), and *Tomato yellow leaf curl virus* (Geminiviridae) (Legarrea et al., 2015). But the majority of studies in our synthesis did not consider virus titer or other metrics of infection (virulence, symptom expression), and very few have taken a comparative approach among virus genotypes that differ in levels of host exploitation and degree of phenotypic manipulation. To fully explore alternative explanations for putative manipulations, future studies should compare closely related viruses that induce transmission-enhancing and transmission-limiting phenotypes and collect data on changes in host phenotype that are apparently unrelated to vector manipulation, but important for virus fitness.

5.2 Host Physiological Phenotype

The most permissive hosts for plant viruses are short-lived species that have poor immune defenses, high nutrient levels (especially phosphorus), low leaf mass per area, and fast metabolisms—all of which are features that make them susceptible to both pathogen exploitation and vector feeding (Cronin et al., 2010; Elser et al., 2010; Hily et al., 2014; Lind et al., 2013; Reich, 2014). Domestication has exacerbated these features for many annual crops by breeding out functional pathways for production of secondary metabolites and other defenses as a means of increasing productivity, edibility or palatability for human consumption (Chen et al., 2015a; Gaillard et al., 2018). Almost all studies exploring putative instances of plant virus manipulation used domesticated annual plants or laboratory models (*N. benthamiana* and *A. thaliana*) (Fig. 3B). This is likely because fast-growing annual crops and model hosts are logistically feasible for laboratory work, and funding is available to support research on mitigating virus impacts on annual crops. As discussed in Section 4, annual monocultures are expected to favor the evolution of manipulative functions in plant viruses, so use of annuals as model hosts is logical given this expectation. However, it also means that we cannot rule out the possibility that the host physiological phenotype is playing a role in determining some of the patterns observed in our quantitative synthesis.

With the current data set, there are not enough studies using wild or perennial hosts to test predictions regarding variation in virus effects along a continuum of host physiological phenotypes (e.g., fast vs slow lifestyles). But within studies using annual hosts, the few that have explored changes in virus effects over the course of disease progression provide evidence that host physiological phenotype does play a role. Plants have age-related resistance to virus infection (Panter and Jones, 2002). Young plants are typically more susceptible to viruses and other pathogens, and this susceptibility decreases as the plant progresses through different phenological stages. We identified 12 studies that explicitly examined virus effects at different stages of disease progression and host phenology (Blua and Perring, 1992a,b; Blua et al., 1994; Higashi and Bressan, 2013; Legarrea et al., 2015; Lu et al., 2016; Mann et al., 2008; Rajabaskar et al., 2013b; Shi et al., 2016; Wang et al., 2012; Werner et al., 2009; Williams, 1995). Several studies with NC-NPer viruses showed that palatability and quality for vectors were enhanced early in disease progression but reduced at later stages (Blua and Perring, 1992a,b; Blua et al., 1994; Shi et al., 2016). Surprisingly, this same pattern was observed in studies with C-P-NProp viruses. Werner et al. (2009) demonstrated that volatiles from potato plants infected with PLRV (Luteoviridae) were more attractive to vectors at 4 weeks postinoculation, but not at 8 or 10 weeks postinoculation. Rajabaskar et al. (2013a,b) further demonstrated that attraction to PLRV-infected plants at 4 weeks postinoculation depends on infection occurring between 1 and 3 weeks posttransplanting and is not induced when older plants are inoculated. Legarrea et al. (2015) found that whiteflies preferred TYLCV-infected tomato plants at 6 weeks postinfection, but not at 9 weeks post-infection. According to this limited suite of reports, virus effects on host attractiveness or palatability may be transient. Early stages of infection in younger plants are associated with greater palatability or attractiveness, while later stages are associated with reduced palatability or attractiveness, regardless of the virus transmission mechanism. This pattern clearly requires further study, but it does suggest that host physiological phenotype may determine *whether* a virus can manipulate a host, as well as how long transmission-enhancing phenotypic changes are expressed by the host. It will be interesting to explore how other determinants of plant traits and host physiological phenotypes, such as nutrient supplies and ratios, influence virus-induced changes in host plant cues and quality (Borer et al., 2010; Cebrian et al., 2009). This approach would be particularly useful if integrated with comparisons of wild and cultivated congeners, or perennial and annual hosts, which also differ widely in plant traits that influence interactions with vectors.

5.3 Pathosystem Complexity

As discussed in Section 3, antagonistic pleiotropy limits the extent to which plant viruses can equally exploit multiple host plants (Agudelo-Romero et al., 2008; García-Arenal and Fraile, 2013; Malpica et al., 2006). The widespread occurrence of antagonistic pleiotropy suggests that any mutations that confer manipulative ability to a virus in one host have the potential to confer maladaptive effects in a second host, with such effects potentially depending on the phylogenetic distance between host species (Lalić et al., 2011). Epistatic interactions among virus proteins may also vary from one host environment to the next, resulting in different outcomes for virus effects on host phenotype and vector behavior (Lalić and Elena, 2013). Many of the pathogens included in our quantitative synthesis can infect multiple hosts, sometimes across several families. Yet only about 23% of the studies employed more than one plant species or genotype (Fig. 3C). Of those that did explore virus infections across multiple hosts, over 57% report that virus-induced changes in host phenotype are species specific. Some studies even report variation in effects among different genotypes within the same host species (Araya and Foster, 1987; Fereres et al., 1990; Hily et al., 2014; Liu et al., 2014; Rajabaskar et al., 2013a,b). In general, when effects diverge depending on host environment, the shift is from a transmission-enhancing phenotype to a phenotype that has neutral effects on transmission probability, but shifts to maladaptive phenotypes have also been reported (Mauck et al., 2014a). It is also possible that some of the potentially maladaptive phenotypes revealed in our quantitative synthesis represent instances of a lack of virus adaptation to a particular host (Tables 2 and 3). But the insufficient information on virus provenance and use of laboratory strains in most studies limit our ability to interpret these maladaptive effects within an evolutionary context. Although few, these studies support our expectation that heterogeneous host environments are likely to disfavor the evolution of manipulative functions because mutations conferring these functions may reduce transmission from a subset of susceptible hosts (Section 4). Moving forward, it will be particularly important to incorporate landscape heterogeneity and temporal limitations on virus effects (discussed in Section 4) into models that describe the epidemiological outcomes of virus manipulations (Roosien et al., 2013; Shaw et al., 2017).

The number of vectors studied per host combination is also quite limited (Fig. 3D) even though many plant viruses are transmitted by multiple vector species, especially NC-NPer viruses (e.g., CMV and PVY are each transmitted by more than 50 species of aphid), but also a number of the most well-studied C-P-NProp viruses (e.g., B/CYDVs, PLRV, TSWV each have more than five vector species). Only about 15.8% of studies tested the behavioral or performance-related responses of multiple vector species, biotypes, or genotypes. Of these 15.8%, the majority (68%) show at least one instance of vector specificity (different responses of two vectors to the same host phenotype). In some cases, this tracks with vector efficiency. For instance, Chesnais et al. (2017) demonstrated that infection of *Camelina sativa* by *Turnip yellows virus* (a C-P-NProp virus in the Luteoviridae) enhanced palatability for *M. persicae*, an efficient vector, but did not enhance palatability for *Brevicoryne brassicae*, an inefficient vector. In other cases, there is no apparent relationship with vector efficiency. *M. persicae* and *B. brassicae* are both efficient vectors of the NC-NPer viruses, TuMV (Potyviridae), but TuMV infection in the same cultivar of turnips enhanced host quality for *M. persicae* and reduced host quality for *B. brassicae* (Hodgson, 1981). As discussed in Section 4, the robustness and consistency of vector responses to virus-induced changes in host phenotypes are expected to determine the frequency with which viruses having manipulative functions will be transmitted to new hosts. Viruses vectored by only one or two species may be more likely to evolve manipulative functions targeting multiple aspects of the transmission process (orientation, settling/feeding, and performance). But viruses vectored by many species (e.g., many NC-NPer viruses) may be less likely to evolve manipulative functions or may only evolve functions that target broad-spectrum cues (e.g., color or host odor enhancements that increase attraction of many vector species). Our quantitative synthesis shows that effects of NC-NPer viruses are more variable than those of C-P-Prop viruses or C-P-NProp viruses, which supports these expectations (Tables 2 and 3), but additional studies are needed to identify convergence or divergence in responses of multiple vectors to putative virus manipulations of host phenotypes.

6. CONCLUSIONS

Parasites shape the ecology of all life on earth and pose major threats to food security and human health. Understanding the selection pressures driving parasite evolution and emergence—including the evolution of manipulative traits—is essential from both basic and applied perspectives. There are now hundreds of reports of insect-vectored plant viruses inducing transmission-enhancing phenotypes in their host plants, and theoretical work elucidates the importance of these putative manipulations for disease spread in agricultural systems. Our quantitative synthesis of these reports, along with the limitations of the present body of work in the context of constraints on virus evolution, highlights numerous areas for future research to understand the relevance of manipulative plant viruses in real-world scenarios. Reverse genetics approaches coupled with phenotype profiling (metabolomics, transcriptomics) are beginning to reveal how virus proteins interact with host components to elicit changes in chemical cues or host quality for vectors. Artificial selection experiments with model pathogens will test the hypothesis that the evolution of secondary functions in virus proteins can be directed when selection is performed based on detectable differences in host phenotypes. Incorporation of multiple hosts, including wild-type and pathway mutants, will reveal trade-offs associated with manipulative functions and host factors involved in phenotype induction. And state-of-the-art phenotypic profiling techniques will provide a wealth of information about host cues and defenses mediating vector-feeding activities (most of which are presently unknown), and how virus infection augments these factors. Results of these studies will enrich our understanding of the evolution of parasite manipulation and facilitate development of new epidemiological models for predicting and mitigating virus threats to agriculture.

ACKNOWLEDGMENTS

We acknowledge the University of California Riverside for providing startup funding to K. Mauck.

REFERENCES

* Denotes references used in the quantitative synthesis (Tables 2 and 3).

Abisgold, J.D., Simpson, S.J., Douglas, A.E., 1994. Nutrient regulation in the pea aphid Acyrthosiphon pisum: application of a novel geometric framework to sugar and amino acid consumption. Physiol. Entomol. 19 (2), 95–102. https://doi.org/10.1111/j.1365-3032.1994.tb01081.x.

Agudelo-Romero, P., de la Iglesia, F., Elena, S.F., 2008. The pleiotropic cost of host-specialization in Tobacco etch potyvirus. Infect. Genet. Evol. 8 (6), 806–814. https://doi.org/10.1016/j.meegid.2008.07.010.

Aguilar, E., Cutrona, C., Del Toro, F.J., Vallarino, J.G., Osorio, S., Pérez-Bueno, M.L., Barón, M., Chung, B.-N., Canto, T., Tenllado, F., 2017. Virulence determines beneficial trade-offs in the response of virus-infected plants to drought via induction of salicylic acid. Plant Cell Environ. 40 (12), 2909–2930. https://doi.org/10.1111/pce.13028.

*Ajayi, O., Dewar, A.M., 1983. The effect of barley yellow dwarf virus on field populations of the cereal aphids Sitobion avenae and Metopolophium dirhodum. Ann. Appl. Biol. 103 (1), 1–11. https://doi.org/10.1111/j.1744-7348.1982.tb01932.x.

Alexander, H.M., Mauck, K.E., Whitfield, A.E., Garrett, K.A., Malmstrom, C.M., 2013. Plant-virus interactions and the agro-ecological interface. Eur. J. Plant Pathol. 138 (3), 529–547. https://doi.org/10.1007/s10658-013-0317-1.

Ali, A., Li, H., Schneider, W.L., Sherman, D.J., Gray, S., Smith, D., Roossinck, M.J., 2006. Analysis of genetic bottlenecks during horizontal transmission of Cucumber mosaic virus. J. Virol. 80 (17), 8345–8350. https://doi.org/10.1128/JVI.00568-06.

Angelella, G.M., Egel, D.S., Holland, J.D., Nemacheck, J.A., Williams, C.E., Kaplan, I., 2015. Differential life history trait associations of aphids with nonpersistent viruses in cucurbits. Environ. Entomol. 44 (3), 562–573. https://doi.org/10.1093/ee/nvv058.

Ángeles-López, Y.I., Rivera-Bustamante, R.F., Heil, M., 2016. Colonization by phloem-feeding herbivore overrides effects of plant virus on amino acid composition in phloem of chili plants. J. Chem. Ecol. 42 (10), 985–988. https://doi.org/10.1007/s10886-016-0747-2.

Ángeles-López, Y.I., Rivera-Bustamante, R., Heil, M., 2017. Fatal attraction of non-vector impairs fitness of manipulating plant virus. J. Ecol. 38 (1), 42–49. https://doi.org/10.1111/1365-2745.12838.

Anhalt, M.D., Almeida, R.P.P., 2008. Effect of temperature, vector life stage, and plant access period on transmission of banana bunchy top virus to banana. Phytopathology 98 (6), 743–748. https://doi.org/10.1094/PHYTO-98-6-0743.

Araya, J.E., Foster, J.E., 1987. Laboratory study on the effects of barley yellow dwarf virus on the life cycle of Rhopalosiphum padi (L.). Zeitsch. Pflanzenkrankheiten Pflanzenschutz 94 (6), 578–583.

*Bak, A., Cheung, A.L., Yang, C., Whitham, S.A., Casteel, C.L., 2017. A viral protease relocalizes in the presence of the vector to promote vector performance. Nat. Commun. 8, 14493. https://doi.org/10.1038/ncomms14493.

Barandoc-Alviar, K., Badillo-Vargas, I.E., Whitfield, A.E., 2016. Interactions between insect vectors and propagative plant viruses. In: Czosnek, H., Ghanim, M. (Eds.), Management of Insect Pests to Agriculture. Springer International Publishing, pp. 133–180. https://doi.org/10.1007/978-3-319-24049-7_6.

Bedhomme, S., Lafforgue, G., Elena, S.F., 2012. Multihost experimental evolution of a plant RNA virus reveals local adaptation and host-specific mutations. Mol. Biol. Evol. 29 (5), 1481–1492. https://doi.org/10.1093/molbev/msr314.

Bedhomme, S., Lafforgue, G., Elena, S.F., 2013. Genotypic but not phenotypic historical contingency revealed by viral experimental evolution. BMC Evol. Biol. 13, 46. https://doi.org/10.1186/1471-2148-13-46.

Bedhomme, S., Hillung, J., Elena, S.F., 2015. Emerging viruses: why they are not jacks of all trades? Curr. Opin. Virol. 10, 1–6. https://doi.org/10.1016/j.coviro.2014.10.006.

Belliure, B., Sabelis, M.W., Janssen, A., 2010. Vector and virus induce plant responses that benefit a non-vector herbivore. Basic Appl. Ecol. 11 (2), 162–169. https://doi.org/10.1016/j.baae.2009.09.004.

Belshaw, R., Pybus, O.G., Rambaut, A., 2007. The evolution of genome compression and genomic novelty in RNA viruses. Genome Res. 17 (10), 1496–1504. https://doi.org/10.1101/gr.6305707.

Belshaw, R., Gardner, A., Rambaut, A., Pybus, O.G., 2008. Pacing a small cage: mutation and RNA viruses. Trends Ecol. Evol. 23 (4), 188–193. https://doi.org/10.1016/j.tree.2007.11.010.

Betancourt, M., Fereres, A., Fraile, A., García-Arenal, F., 2008. Estimation of the effective number of founders that initiate an infection after aphid transmission of a multipartite plant virus. J. Virol. 82 (24), 12416–12421. https://doi.org/10.1128/JVI.01542-08.

*Blua, M.J., Perring, T.M., 1992a. Alatae production and population increase of aphid vectors on virus-infected host plants. Oecologia 92 (1), 65–70. https://doi.org/10.1007/BF00317263.

*Blua, M.J., Perring, T.M., 1992b. Effects of zucchini yellow mosaic virus on colonization and feeding behavior of Aphis gossypii (Homoptera: Aphididae) alatae. Environ. Entomol. 21 (1), 578–585. https://doi.org/10.1093/ee/21.3.578.

*Blua, M.J., Perring, T.M., Madore, M.A., 1994. Plant virus-induced changes in aphid population development and temporal fluctuations in plant nutrients. J. Chem. Ecol. 20 (3), 691–707. https://doi.org/10.1007/BF02059607.

Borer, E.T., Seabloom, E.W., Mitchell, C.E., Power, A.G., 2010. Local context drives infection of grasses by vector-borne generalist viruses. Ecol. Lett. 13 (7), 810–818. https://doi.org/10.1111/j.1461-0248.2010.01475.x.

Bosquee, E., Yin, R.L., Bragard, C., Yong, L., Chen, J.L., Francis, F., 2016. Transmission efficiency of cucumber mosaic virus by Myzus persicae according to virus strain and aphid clone from China. Asian J. Plant Pathol. 10 (4), 61–66. https://doi.org/10.3923/ajppaj.2016.61.66.

Bosque-Pérez, N.A., Eigenbrode, S.D., 2011. The influence of virus-induced changes in plants on aphid vectors: insights from luteovirus pathosystems. Virus Res. 159, 201–205. https://doi.org/10.1016/j.virusres.2011.04.020.

*Carmo-Sousa, M., Moreno, A., Garzo, E., Fereres, A., 2014. A non-persistently transmitted-virus induces a pull–push strategy in its aphid vector to optimize transmission and spread. Virus Res. 186, 38–46. https://doi.org/10.1016/j.virusres.2013.12.012.

*Carmo-Sousa, M., Moreno, A., Plaza, M., Garzo, E., Fereres, A., 2016. Cucurbit aphid-borne yellows virus (CABYV) modifies the alighting, settling and probing behaviour of its vector Aphis gossypii favouring its own spread. Ann. Appl. Biol. 169 (2), 284–297. https://doi.org/10.1111/aab.12300.

Carrasco, D., Larsson, M.C., Anderson, P., 2015. Insect host plant selection in complex environments. Curr. Opin. Insect Sci. 8, 1–7. https://doi.org/10.1016/j.cois.2015.01.014.

Carrière, Y., Degain, B., Liesner, L., Dutilleul, P., Palumbo, J.C., 2017. Validation of a landscape-based model for whitefly spread of the cucurbit yellow stunting disorder virus to fall melons. J. Econ. Entomol. 110 (5), 2002–2009. https://doi.org/10.1093/jee/tox213.

*Cassone, B.J., Michel, A.P., Stewart, L.R., Bansal, R., Mian, M.A.R., Redinbaugh, M.G., 2014. Reduction in fecundity and shifts in cellular processes by a native virus on an invasive insect. Genome Biol. Evol. 6 (4), 873–885. https://doi.org/10.1093/gbe/evu057.

Casteel, C.L., Falk, B.W., 2016. Plant virus-vector interactions: more than just for virus transmission. In: Wang, A., Zhou, X. (Eds.), Current Research Topics in Plant Virology. Springer International Publishing, Cham, pp. 217–240. https://doi.org/10.1007/978-3-319-32919-2_9.

*Casteel, C.L., Yang, C., Nanduri, A.C., De Jong, H.N., Whitham, S.A., Jander, G., 2014. The NIa-Pro protein of Turnip mosaic virus improves growth and reproduction of the aphid vector, Myzus persicae (green peach aphid). Plant J. 77 (4), 653–663. https://doi.org/10.1111/tpj.12417.

Casteel, C., De Alwis, M., Bak, A., Dong, H., Steven, A.W., Jander, G., 2015. Disruption of ethylene responses by Turnip mosaic virus mediates suppression of plant defense against the aphid vector, Myzus persicae. Plant Physiol. 169 (1), 209–218. https://doi.org/10.1104/pp.15.00332.

Cebrian, J., Shurin, J.B., Borer, E.T., Cardinale, B.J., Ngai, J.T., Smith, M.D., Fagan, W.F., 2009. Producer nutritional quality controls ecosystem trophic structure. PLoS One 4 (3), 1–5. https://doi.org/10.1371/journal.pone.0004929.

Chatzivassiliou, E.K., Peters, D., Katis, N.I., 2002. The efficiency by which Thrips tabaci populations transmit tomato spotted wilt virus depends on their host preference and reproductive strategy. Phytopathology 92 (6), 603–609. https://doi.org/10.1094/PHYTO.2002.92.6.603.

Chay, C.A., Gunasinge, U.B., Dinesh-Kumar, S.P., Miller, W.A., Gray, S.M., 1996. Aphid transmission and systemic plant infection determinants of barley yellow dwarf luteovirus-PAV are contained in the coat protein readthrough domain and 17-kDa protein, respectively. Virology 219 (1), 57–65. https://doi.org/10.1006/viro.1996.0222.

Chen, Y.H., Gols, R., Benrey, B., 2015a. Crop domestication and its impact on naturally selected trophic interactions. Annu. Rev. Entomol. 60, 35–58. https://doi.org/10.1146/annurev-ento-010814-020601.

Chen, Y., Lu, C., Li, M., Wu, W., Zhou, G., Wei, T., 2015b. Adverse effects of rice gall dwarf virus upon its insect vector Recilia dorsalis (Hemiptera: Cicadellidae). Plant Dis. 100 (4), 784–790. https://doi.org/10.1094/PDIS-06-15-0713-RE.

*Chesnais, Q., Couty, A., Uzest, M., Brault, V., Ameline, A., 2017. Plant infection by two different viruses induce contrasting changes of vectors fitness and behavior. Insect Sci (in press)https://doi.org/10.1111/1744-7917.12508.

Chung, B.Y.-W., Miller, W.A., Atkins, J.F., Firth, A.E., 2008. An overlapping essential gene in the Potyviridae. Proc. Natl. Acad. Sci. U.S.A. 105 (15), 5897–5902. https://doi.org/10.1073/pnas.0800468105.

*Costa, H.S., Brown, J.K., Byrne, D.N., H. S., 1991. Life history traits of the whitefly, Bemisia tabaci (Homoptera: Aleyrodidae) on six virus infected or healthy plant species. Environ. Entomol. 20, 1102–1107.

Cronin, J.P., Welsh, M.E., Dekkers, M.G., Abercrombie, S.T., Mitchell, C.E., 2010. Host physiological phenotype explains pathogen reservoir potential. Ecol. Lett. 13 (10), 1221–1232. https://doi.org/10.1111/j.1461-0248.2010.01513.x.

Cronin, J.P., Rúa, M.A., Mitchell, C.E., 2014. Why is living fast dangerous? Disentangling the roles of resistance and tolerance of disease. Am. Nat. 184 (2), 172–187. https://doi.org/10.1086/676854.

*Cui, H., Sun, Y., Chen, F., Zhang, Y., Ge, F., 2016. Elevated O3 and TYLCV infection reduce the suitability of tomato as a host for the whitefly Bemisia tabaci. Int. J. Mol. Sci. 17 (12), 1964. https://doi.org/10.3390/ijms17121964.

Dawkins, R., Krebs, J.R., 1979. Arms races between and within species. Proc. R. Soc. Lond. B Biol. Sci. 205 (1161), 489–511. Retrieved from:https://www.ncbi.nlm.nih.gov/pubmed/42057.

Douglas, A.E., 2003. The nutritional physiology of aphids. In: Advances in Insect Physiology. vol. 31. Academic Press, pp. 73–140. https://doi.org/10.1016/S0065-2806(03)31002-1.

Drake, J.W., Charlesworth, B., Charlesworth, D., Crow, J.F., 1998. Rates of spontaneous mutation. Genetics 148 (4), 1667–1686.

Dudareva, N., Klempien, A., Muhlemann, J.K., Kaplan, I., 2013. Biosynthesis, function and metabolic engineering of plant volatile organic compounds. New Phytol. 198 (1), 16–32. https://doi.org/10.1111/nph.12145.

Duffy, S., Holmes, E.C., 2008. Phylogenetic evidence for rapid rates of molecular evolution in the single-stranded DNA begomovirus tomato yellow leaf curl virus. J. Virol. 82 (2), 957–965. https://doi.org/10.1128/JVI.01929-07.

Duffy, S., Holmes, E.C., 2009. Validation of high rates of nucleotide substitution in geminiviruses: phylogenetic evidence from East African cassava mosaic viruses. J. Gen. Virol. 90, 1539–1547. https://doi.org/10.1099/vir.0.009266-0.

Duffy, S., Shackelton, L.A., Holmes, E.C., 2008. Rates of evolutionary change in viruses: patterns and determinants. Nat. Rev. Genet. 9 (4), 267–276. https://doi.org/10.1038/nrg2323.

Eigenbrode, S.D., Bosque-Perez, N.A., 2016. Chemical ecology of aphid-transmitted plant viruses. In: Brown, J.K. (Ed.), Vector-Mediated Transmission of Plant Pathogens. The American Phytopathological Society, pp. 3–19. https://doi.org/10.1094/9780890545355.001.

*Eigenbrode, S.D., Ding, H., Shiel, P., Berger, P.H., 2002. Volatiles from potato plants infected with potato leafroll virus attract and arrest the virus vector, Myzus persicae (Homoptera: Aphididae). Proc. R. Soc. B Biol. Sci. 269 (1490), 455–460. https://doi.org/10.1098/rspb.2001.1909.

Elena, S.F., 2016. Local adaptation of plant viruses: lessons from experimental evolution. Mol. Ecol. 26 (7), 1711–1719. https://doi.org/10.1111/mec.13836.

Elena, S.F., Lalić, J., 2013. Plant RNA virus fitness predictability: contribution of genetic and environmental factors. Plant Pathol. 62, 10–18. https://doi.org/10.1111/ppa.12102.

Elena, S.F., Fraile, A., García-Arenal, F., 2014. Evolution and emergence of plant viruses. Adv. Virus Res. 88, 161–191. https://doi.org/10.1016/B978-0-12-800098-4.00003-9.

Elser, J.J., Fagan, W.F., Kerkhoff, A.J., Swenson, N.G., Enquist, B.J., 2010. Biological stoichiometry of plant production: metabolism, scaling and ecological response to global change. New Phytol. 186 (3), 593–608. https://doi.org/10.1111/j.1469-8137.2010.03214.x.

Elzinga, D.A., Jander, G., 2013. The role of protein effectors in plant–aphid interactions. Curr. Opin. Plant Biol. 16 (4), 451–456. https://doi.org/10.1016/j.pbi.2013.06.018.

Escriu, F., Perry, K.L., García-Arenal, F., 2000. Transmissibility of cucumber mosaic virus by Aphis gossypii correlates with viral accumulation and is affected by the presence of its satellite RNA. Phytopathology 90 (10), 1068–1072. https://doi.org/10.1094/PHYTO.2000.90.10.1068.

Escriu, F., Fraile, A., García-Arenal, F., 2003. The evolution of virulence in a plant virus. Evolution 57 (4), 755–765. Retrieved from:https://www.ncbi.nlm.nih.gov/pubmed/12778545.

*Fang, Y., Jiao, X., Xie, W., Wang, S., Wu, Q., Shi, X., Chen, G., Su, Q., Yang, X., Pan, H., Zhang, Y., 2013. Tomato yellow leaf curl virus alters the host preferences of its vector Bemisia tabaci. Sci. Rep. 3, 2876. https://doi.org/10.1038/srep02876.

Fereres, A., 2016. Aphid behavior and the transmission of non-circulative viruses. In: Brown, J.K. (Ed.), Vector-Mediated Transmission of Plant Pathogens. The American Phytopathological Society, pp. 31–45. https://doi.org/10.1094/9780890545355.003.

Fereres, A., Collar, J.L., 2001. Analysis of non-circulative transmission by electrical penetration graphs. In: Harris, K.F., Smith, O.P., Duffus, J.E. (Eds.), Virus-Insect-Plant Interactions. Academic Press, pp. 87–109.

Fereres, A., Moreno, A., 2009. Behavioural aspects influencing plant virus transmission by homopteran insects. Virus Res. 141, 158–168https://doi.org/10.1016/j.virusres.2008.10.020.

Fereres, A., Raccah, B., 2009. Plant virus transmission by insects. In: Encyclopedia of Life Sciences. John Wiley & Sons, Ltd., Chichester, pp. 1–9

*Fereres, A., Lister, R.M., Araya, J.E., Foster, J.E., 1989. Development and reproduction of the English grain aphid (Homoptera: Aphididae) on wheat cultivars infected with barley yellow dwarf virus. Environ. Entomol. 18, 388–393. https://doi.org/10.1093/ee/18.3.388.

*Fereres, A., Shukle, R.H., Araya, J.E., Foster, J.E., 1990. Probing and feeding behavior of Sitobion avenae (F.) (Horn., Aphididae) on three wheat cultivars infected with barley yellow dwarf virus. J. Appl. Entomol. 109, 29–36. https://doi.org/10.1111/j.1439-0418.1990.tb00015.x.

*Fereres, A., Peñaflor, M.F.G.V., Favaro, C.F., Azevedo, K.E.X., Landi, C.H., Maluta, N.K.P., Bento, J.M., Lopes, J.R.S., 2016. Tomato infection by whitefly-transmitted circulative and non-circulative viruses induce contrasting changes in plant volatiles and vector behaviour. Viruses 8, 225. https://doi.org/10.3390/v8080225.

*Fiebig, M., Poehling, H.-M., Borgemeister, C., 2004. Barley yellow dwarf virus, wheat, and Sitobion avenae: a case of trilateral interactions. Entomol. Exp. Appl. 110 (1), 11–21. https://doi.org/10.1111/j.0013-8703.2004.00115.x.

Gabrys, B., Tjallingii, W.F., 2002. The role of sinigrin in host plant recognition by aphids during initial plant penetration. Entomol. Exp. Appl. 104 (1), 89–93. https://doi.org/10.1046/j.1570-7458.2002.00994.x.

Gaillard, M.D.P., Glauser, G., Robert, C.A.M., Turlings, T.C.J., 2018. Fine-tuning the "plant domestication-reduced defense" hypothesis: specialist vs generalist herbivores. New Phytol. 217 (1), 355–366. https://doi.org/10.1111/nph.14757.

García-Arenal, F., Fraile, A., 2013. Trade-offs in host range evolution of plant viruses. Plant Pathol. 62 (S1), 2–9. https://doi.org/10.1111/ppa.12104.

Geoghegan, J.L., Senior, A.M., Holmes, E.C., 2016. Pathogen population bottlenecks and adaptive landscapes: overcoming the barriers to disease emergence. Proc. Biol. Sci. 283, 20160727. https://doi.org/10.1098/rspb.2016.0727.

Gilbertson, R.L., Sudarshana, M., Jiang, H., Rojas, M.R., Lucas, W.J., 2003. Limitations on geminivirus genome size imposed by plasmodesmata and virus-encoded movement protein: insights into DNA trafficking. Plant Cell 15 (11), 2578–2591. https://doi.org/10.1105/tpc.015057.

Gray, S., Cilia, M., Ghanim, M., 2014. Circulative, "nonpropagative" virus transmission: an orchestra of virus-, insect-, and plant-derived instruments. Adv. Virus Res. 89, 141–199. https://doi.org/10.1016/B978-0-12-800172-1.00004-5.

Gutiérrez, S., Yvon, M., Thébaud, G., Monsion, B., Michalakis, Y., Blanc, S., 2010. Dynamics of the multiplicity of cellular infection in a plant virus. PLoS Pathog. 6 (9), e1001113. https://doi.org/10.1371/journal.ppat.1001113.

Gutiérrez, S., Yvon, M., Pirolles, E., Garzo, E., Fereres, A., Michalakis, Y., Blanc, S., 2012. Circulating virus load determines the size of bottlenecks in viral populations progressing within a host. PLoS Pathog. 8 (11), e1003009. https://doi.org/10.1371/journal.ppat.1003009.

Gutiérrez, S., Michalakis, Y., Van Munster, M., Blanc, S., 2013. Plant feeding by insect vectors can affect life cycle, population genetics and evolution of plant viruses. Funct. Ecol. 27 (3), 610–622. https://doi.org/10.1111/1365-2435.12070.

Hanley-Bowdoin, L., Settlage, S.B., Orozco, B.M., Nagar, S., Robertson, D., 2000. Geminiviruses: models for plant DNA replication, transcription, and cell cycle regulation. Crit. Rev. Biochem. Mol. Biol. 35 (2), 105–140.

Harrington, R., Verrier, P., Denholm, C., Hullé, M., Maurice, D., Bell, N., Knight, J., et al., 2004. EXAMINE (EXploitation of Aphid Monitoring In Europe): an European thematic network for the study of global change impacts on aphids. In: Aphids in a New Millennium. Proceedings of the Sixth International Symposium on Aphids, September, 2001, Institut National de la Recherche Agronomique, Rennes, France, pp. 45–49.

Hasegawa, D.K., Chen, W., Zheng, Y., Kaur, N., Wintermantel, W.M., Simmons, A.M., Fei, Z., Ling, K.-S., 2018. Comparative transcriptome analysis reveals networks of genes activated in the whitefly, Bemisia tabaci when fed on tomato plants infected with Tomato yellow leaf curl virus. Virology 513, 52–64. https://doi.org/10.1016/j.virol.2017.10.008.

Heil, M., 2016. Host manipulation by parasites: cases, patterns, and remaining doubts. Front. Ecol. Evol. 4, 80. https://doi.org/10.3389/fevo.2016.00080.

Higashi, C.H.V., Bressan, A., 2013. Infection rates and comparative population dynamics of Peregrinus maidis (Hemiptera: Delphacidae) on corn plants with and without symptoms of maize mosaic virus (Rhabdoviridae: Nucleorhabdovirus) infection. Environ. Entomol. 42 (5), 949–956. https://doi.org/10.1603/EN12321.

*Hily, J.M., Garcıa, A., Moreno, A., Plaza, M., Wilkinson, M.D., Fereres, A., Fraile, A., García-Arenal, F., 2014. The relationship between host lifespan and pathogen reservoir potential: an analysis in the system Arabidopsis thaliana-cucumber mosaic virus. PLoS Pathog. 10 (11), e1004492. https://doi.org/10.1371/journal.ppat.1004492.
*Hodgson, C.J., 1981. Effects of infection with the cabbage black ringspot strain of turnip mosaic virus on turnip as a host to Myzus persicae and Brevicoryne brassicae. Ann. Appl. Biol. 98, 1–14.
Hogenhout, S.A., Ammar, E.-D., Whitfield, A.E., Redinbaugh, M.G., 2008. Insect vector interactions with persistently transmitted viruses. Annu. Rev. Phytopathol. 46, 327–359. https://doi.org/10.1146/annurev.phyto.022508.092135.
Ingwell, L.L., Eigenbrode, S.D., Bosque-Pérez, N.A., 2012. Plant viruses alter insect behavior to enhance their spread. Sci. Rep. 2, 578. https://doi.org/10.1038/srep00578.
Jeger, M.J., Holt, J., Van Den Bosch, F., Madden, L.V., 2004. Epidemiology of insect-transmitted plant viruses: modelling disease dynamics and control interventions. Physiol. Entomol. 29 (3), 291–304. https://doi.org/10.1111/j.0307-6962.2004.00394.x.
Jeger, M.J., Chen, Z., Powell, G., Hodge, S., Bosch, F.V.D., 2011. Interactions in a host plant-virus-vector-parasitoid system: modelling the consequences for virus transmission and disease dynamics. Virus Res. 159 (2), 183–193. https://doi.org/10.1016/j.virusres.2011.04.027.
Ji, R., Yu, H., Fu, Q., Chen, H., Ye, W., Li, S., Lou, Y., 2013. Comparative transcriptome analysis of salivary glands of two populations of rice brown planthopper, Nilaparvata lugens, that differ in virulence. PLoS One 8 (11), e79612. https://doi.org/10.1371/journal.pone.0079612.
*Jiménez-Martínez, E.S., Bosque-Pérez, N.A., Berger, P.H., Zemetra, R.S., 2004a. Life history of the bird cherry-oat aphid, Rhopalosiphum padi (Homoptera: Aphididae), on transgenic and untransformed wheat challenged with Barley yellow dwarf virus. J. Econ. Entomol. 97 (2), 203–212. https://doi.org/10.1603/0022-0493-97.2.203.
*Jiménez-Martínez, E.S., Bosque-Pérez, N.A., Berger, P.H., Zemetra, R.S., Ding, H., Eigenbrode, S.D., 2004b. Volatile cues influence the response of Rhopalosiphum padi (Homoptera: Aphididae) to Barley yellow dwarf virus-infected transgenic and untransformed wheat. Environ. Entomol. 33, 1207–1216.
*Kersch-Becker, M.F., Thaler, J.S., 2013. Virus strains differentially induce plant susceptibility to aphid vectors and chewing herbivores. Oecologia 174 (3), 883–892. https://doi.org/10.1007/s00442-013-2812-7.
Lafferty, K.D., Shaw, J.C., 2013. Comparing mechanisms of host manipulation across host and parasite taxa. J. Exp. Biol. 216, 56–66. https://doi.org/10.1242/jeb.073668.
Lalić, J., Elena, S.F., 2013. Epistasis between mutations is host-dependent for an RNA virus. Biol. Lett. 9 (1), 20120396. https://doi.org/10.1098/rsbl.2012.0396.
Lalić, J., Cuevas, J.M., Elena, S.F., 2011. Effect of host species on the distribution of mutational fitness effects for an RNA virus. PLoS Genet. 7 (11), e1002378. https://doi.org/10.1371/journal.pgen.1002378, (ed D.S. Guttman).
Lefèvre, T., Lebarbenchon, C., Gauthier-Clerc, M., Missé, D., Poulin, R., Thomas, F., 2009. The ecological significance of manipulative parasites. Trends Ecol. Evol. 24 (1), 41–48. https://doi.org/10.1016/j.tree.2008.08.007.
*Legarrea, S., Barman, A., Marchant, W., Diffie, S., Srinivasan, R., 2015. Temporal effects of a begomovirus infection and host plant resistance on the preference and development of an insect vector, Bemisia tabaci, and implications for epidemics. PLoS One 10 (11), e0142114. https://doi.org/10.1371/journal.pone.0142114.
*Lei, W., Li, P., Han, Y., Gong, S., Yang, L., Hou, M., 2016. EPG recordings reveal differential feeding behaviors in Sogatella furcifera in response to plant virus infection and transmission success. Sci. Rep. 6, 30240. https://doi.org/10.1038/srep30240.
Lewsey, M., Surette, M., Robertson, F.C., Ziebell, H., Choi, S.H., Ryu, K.H., Canto, T., Palukaitis, P., Payne, T., Walsh, J.A., Carr, J.P., 2009. The role of the cucumber mosaic

virus 2b protein in viral movement and symptom induction. Mol. Plant Microbe Interact. 22 (6), 642–654. https://doi.org/10.1094/MPMI-22-6-0642.

Li, H., Roossinck, M.J., 2004. Genetic bottlenecks reduce population variation in an experimental RNA virus population. J. Virol. 78 (19), 10582–10587. https://doi.org/10.1128/JVI.78.19.10582-10587.2004.

Lind, E.M., Borer, E., Seabloom, E., Adler, P., Bakker, J.D., Blumenthal, D.M., Crawley, M., Davies, K., Firn, J., Gruner, D.S., Harpole, W.S., Hautier, Y., Hillebrand, H., Knops, J., Melourne, B., Mortenson, B., Risch, A.C., Schuetz, M., Stevens, C., Wragg, P.D., 2013. Life-history constraints in grassland plant species: a growth-defence trade-off is the norm. Ecol. Lett. 16 (4), 513–521. https://doi.org/10.1111/ele.12078.

*Liu, B., Preisser, E.L., Chu, D., Pan, H., Xie, W., Wang, S., Wu, Q., Zhou, X., Zhang, Y., 2013. Multiple forms of vector manipulation by a plant-infecting virus: Bemisia tabaci and tomato yellow leaf curl virus. J. Virol. 87 (9), 4929–4937. https://doi.org/10.1128/JVI.03571-12.

*Liu, X.-F., Hu, X.-S., Keller, M.A., Zhao, H.-Y., Wu, Y.-F., Liu, T.-X., 2014. Tripartite interactions of Barley yellow dwarf virus, Sitobion avenae and wheat varieties. PLoS One 9 (9), e106639. https://doi.org/10.1371/journal.pone.0106639.

*Lu, G., Zhang, T., He, Y., Zhou, G., 2016. Virus altered rice attractiveness to planthoppers is mediated by volatiles and related to virus titre and expression of defence and volatile-biosynthesis genes. Sci. Rep. 6, 38581. https://doi.org/10.1038/srep38581.

*Lu, S., Li, J., Wang, X., Song, D., Bai, R., Shi, Y., Gu, Q., Kuo, Y.-W., Falk, B.W., Yan, F., 2017. A semi-persistent plant virus differentially manipulates feeding behaviors of different sexes and biotypes of its whitefly vector. Viruses 9 (1), 4. https://doi.org/10.3390/v9010004.

*Luan, J., Yao, D., 2013. Suppression of terpenoid synthesis in plants by a virus promotes its mutualism with vectors. Ecol. Lett. 16, 390–398. https://doi.org/10.1111/ele.12055.

Luan, J.-B., Li, J.-M., Varela, N., Wang, Y.-L., Li, F.-F., Bao, Y.-Y., Zhang, C.-X., Liu, S.-S., Wang, X.-W., 2011. Global analysis of the transcriptional response of whitefly to tomato yellow leaf curl China virus reveals the relationship of coevolved adaptations. J. Virol. 85 (7), 3330–3340. https://doi.org/10.1128/JVI.02507-10.

Lucas, W.J., 2006. Plant viral movement proteins: agents for cell-to-cell trafficking of viral genomes. Virology 344 (1), 169–184. https://doi.org/10.1016/j.virol.2005.09.026.

Madden, L.V., Jeger, M.J., Van Den Bosch, F., 2000. A theoretical assessment of the effects of vector-virus transmission mechanism on plant virus disease epidemics. Phytopathology 90, 576–594. https://doi.org/10.1094/PHYTO.2000.90.6.576.

Mai, V.C., Bednarski, W., Borowiak-Sobkowiak, B., Wilkaniec, B., Samardakiewicz, S., Morkunas, I., 2013. Oxidative stress in pea seedling leaves in response to Acyrthosiphon pisum infestation. Phytochemistry 93, 49–62. https://doi.org/10.1016/j.phytochem.2013.02.011.

Malpica, J.M., Sacristán, S., Fraile, A., García-Arenal, F., 2006. Association and host selectivity in multi-host pathogens. PLoS One 1, e41. https://doi.org/10.1371/journal.pone.0000041.

*Mann, R.S., Sidhu, J.S., Butter, N.S., Sohi, A.S., Sekhon, P.S., 2008. Performance of Bemisia tabaci (Hemiptera: Aleyrodidae) on healthy and cotton leaf curl Virus infected cotton. Fla. Entomol. 91 (2), 249–255.

Mann, R.S., Ali, J.G., Hermann, S.L., Tiwari, S., Pelz-Stelinski, K.S., Alborn, H.T., Stelinski, L.L., 2012. Induced release of a plant-defense volatile "deceptively" attracts insect vectors to plants infected with a bacterial pathogen. PLoS Pathog. 8 (3), e1002610. https://doi.org/10.1371/journal.ppat.1002610.

Mar, T., Liu, W., Wang, X., 2014. Proteomic analysis of interaction between P7-1 of Southern rice black-streaked dwarf virus and the insect vector reveals diverse insect proteins

involved in successful transmission. J. Proteomics 102, 83–97. https://doi.org/10.1016/j.jprot.2014.03.004.
Margaritopoulos, J.T., Tsourapas, C., Tzortzi, M., Kanavaki, O.M., Tsitsipis, J.A., 2005. Host selection by winged colonisers within the Myzus persicae group: a contribution towards understanding host specialisation. Ecol. Entomol. 30 (4), 406–418. https://doi.org/10.1111/j.0307-6946.2005.00700.x.
Martin, B., Collar, J.L., Tjallingii, W.F., Fereres, A., Others, 1997. Intracellular ingestion and salivation by aphids may cause the acquisition and inoculation of non-persistently transmitted plant viruses. J. Gen. Virol. 78 (10), 2701–2705.
Martini, X., Hoffmann, M., Coy, M.R., Stelinski, L.L., Pelz-Stelinski, K.S., 2015. Infection of an insect vector with a bacterial plant pathogen increases Its propensity for dispersal. PLoS One 10 (6), e0129373. https://doi.org/10.1371/journal.pone.0129373.
Mas, F., Vereijssen, J., Suckling, D.M., 2014. Influence of the pathogen Candidatus Liberibacter solanacearum on tomato host plant volatiles and psyllid vector settlement. J. Chem. Ecol. 40 (11–12), 1197–1202. https://doi.org/10.1007/s10886-014-0518-x.
Mauck, K.E., 2016. Variation in virus effects on host plant phenotypes and insect vector behavior: what can it teach us about virus evolution? Curr. Opin. Virol. 21, 114–123. https://doi.org/10.1016/j.coviro.2016.09.002.
*Mauck, K.E., De Moraes, C.M., Mescher, M.C., 2010. Deceptive chemical signals induced by a plant virus attract insect vectors to inferior hosts. Proc. Natl. Acad. Sci. U.S.A. 107 (8), 3600–3605. https://doi.org/10.1073/pnas.0907191107.
Mauck, K.E., Bosque-Pérez, N.A., Eigenbrode, S.D., De Moraes, C.M., Mescher, M.C., 2012. Transmission mechanisms shape pathogen effects on host-vector interactions: evidence from plant viruses. Funct. Ecol. 26 (5), 1162–1175. https://doi.org/10.1111/j.1365-2435.2012.02026.x.
*Mauck, K.E., De Moraes, C.M., Mescher, M.C., 2014a. Evidence of local adaptation in plant virus effects on host–vector interactions. Integr. Comp. Biol. 54 (2), 193–209, https://doi.org/10.1093/icb/icu012.
Mauck, K.E., De Moraes, C.M., Mescher, M.C., 2014b. Biochemical and physiological mechanisms underlying effects of Cucumber mosaic virus on host-plant traits that mediate transmission by aphid vectors. Plant Cell Environ. 37, 1427–1439. https://doi.org/10.1111/pce.12249.
Mauck, K.E., Smyers, E., De Moraes, C.M., Mescher, M.C., 2015. Virus infection influences host plant interactions with non-vector herbivores and predators. Funct. Ecol. 29 (5), 662–673. https://doi.org/10.1111/1365-2435.12371.
Mauck, K.E., De Moraes, C.M., Mescher, M.C., 2016. Effects of pathogens on sensory-mediated interactions between plants and insect vectors. Curr. Opin. Plant Biol. 32, 53–61. https://doi.org/10.1016/j.pbi.2016.06.012.
Maxwell, D.J., Partridge, J.C., Roberts, N.W., Boonham, N., Foster, G.D., 2017. The effects of surface structure mutations in Arabidopsis thaliana on the polarization of reflections from virus-infected leaves. PLoS One 12 (3), e0174014. https://doi.org/10.1371/journal.pone.0174014.
McElhany, P., Real, L.A., Power, A.G., 1995. Vector preference and disease dynamics: a study of barley yellow dwarf virus. Ecology 76 (2), 444–457. https://doi.org/10.2307/1941203.
*McMenemy, L.S., Hartley, S.E., MacFarlane, S.A., Karley, A.J., Shepherd, T., Johnson, S.N., 2012. Raspberry viruses manipulate the behaviour of their insect vectors. Entomol. Exp. Appl. 144 (1), 56–68. https://doi.org/10.1111/j.1570-7458.2012.01248.x.
*Medina-Ortega, K.J., Bosque-Pérez, N.A., Ngumbi, E., Jiménez-Martínez, E.S., Eigenbrode, S.D., 2009. Rhopalosiphum padi (Hemiptera: Aphididae) responses to volatile cues from Barley yellow dwarf virus-infected wheat. Environ. Entomol. 38 (3), 836–845. https://doi.org/10.1603/022.038.0337.

Miyashita, Y., Atsumi, G., Nakahara, K.S., 2016. Trade-offs for viruses in overcoming innate immunities in plants. Mol. Plant Microbe Interact. 2016 (1), 1–4. https://doi.org/10.1094/MPMI-05-16-0103-CR.2016.1.test.

Mondal, S., Wenninger, E.J., Hutchinson, P.J.S., Weibe, M.A., Eigenbrode, S.D., Bosque-Pérez, N.A., 2016. Contribution of noncolonizing aphids to potato virus Y prevalence in potato in Idaho. Environ. Entomol. 45 (6), 1445–1462. https://doi.org/10.1093/ee/nvw131.

Monsion, B., Froissart, R., Michalakis, Y., Blanc, S., 2008. Large bottleneck size in Cauliflower Mosaic Virus populations during host plant colonization. PLoS Pathog. 4 (10), e1000174. https://doi.org/10.1371/journal.ppat.1000174.

Moore, J., 2013. An overview of parasite-induced behavioral alterations—and some lessons from bats. J. Exp. Biol. 216 (Pt 1), 11–17. https://doi.org/10.1242/jeb.074088.

Moreno, A., Tjallingii, W.F., Fernandez-Mata, G., Fereres, A., 2012. Differences in the mechanism of inoculation between a semi-persistent and a non-persistent aphid-transmitted plant virus. J. Gen. Virol. 93 (Pt 3), 662–667. https://doi.org/10.1099/vir.0.037887-0.

Moreno-Delafuente, A., Garzo, E., Moreno, A., Fereres, A., 2013. A plant virus manipulates the behavior of its whitefly vector to enhance its transmission efficiency and spread. PLoS One 8 (4), e61543. https://doi.org/10.1371/journal.pone.0061543.

Mutti, N.S., Louis, J., Pappan, L.K., Pappan, K., Begum, K., Chen, M.-S., Park, Y., Dittmer, N., Marshall, J., Reese, J.C., Reeck, G.R., 2008. A protein from the salivary glands of the pea aphid, Acyrthosiphon pisum, is essential in feeding on a host plant. Proc. Natl. Acad. Sci. U.S.A. 105 (29), 9965–9969. https://doi.org/10.1073/pnas.0708958105.

Ng, J.C.K., Falk, B.W., 2006. Virus-vector interactions mediating nonpersistent and semipersistent transmission of plant viruses. Annu. Rev. Phytopathol. 44, 183–212. https://doi.org/10.1146/annurev.phyto.44.070505.143325.

Ng, J.C.K., Zhou, J.S., 2015. Insect vector–plant virus interactions associated with non-circulative, semi-persistent transmission: current perspectives and future challenges. Curr. Opin. Virol. 15, 48–55. https://doi.org/10.1016/j.coviro.2015.07.006.

Ng, J.C.K., Josefsson, C., Clark, A.J., Franz, A.W.E., Perry, K.L., 2005. Virion stability and aphid vector transmissibility of Cucumber mosaic virus mutants. Virology 332 (1), 397–405. https://doi.org/10.1016/j.virol.2004.11.021.

*Ogada, P.A., Kiirika, L.M., Lorenz, C., Senkler, J., Braun, H.-P., Poehling, H.-M., 2017. Differential proteomics analysis of Frankliniella occidentalis immune response after infection with Tomato spotted wilt virus (Tospovirus). Dev. Comp. Immunol. 67, 1–7. https://doi.org/10.1016/j.dci.2016.10.013.

Pagán, I., González-Jara, P., Moreno-Letelier, A., Rodelo-Urrego, M., Fraile, A., Piñero, D., García-Arenal, F., 2012. Effect of biodiversity changes in disease risk: exploring disease emergence in a plant-virus system. PLoS Pathog. 8 (7), e1002796. https://doi.org/10.1371/journal.ppat.1002796.

Panter, S.N., Jones, D.A., 2002. Age-related resistance to plant pathogens. Adv. Bot. Res. 38, 251–280. https://doi.org/10.1016/S0065-2296(02)38032-7.

*Peñaflor, M.F.G.V., Mauck, K.E., Alves, K.J., De Moraes, C.M., Mescher, M.C., 2016. Effects of single and mixed infections of bean pod mottle virus and soybean mosaic virus on host plant chemistry and host-vector interactions. Funct. Ecol. 30, 1648–1659. https://doi.org/10.1111/1365-2435.12649.

Perring, T.M., Gruenhagen, N.M., Farrar, C.A., 1999. Management of plant viral diseases through chemical control of insect vectors. Annu. Rev. Entomol. 44, 457–481. https://doi.org/10.1146/annurev.ento.44.1.457.

Pirone, T.P., Perry, K.L., 2002. Aphids: non-persistent transmission. Adv. Bot. Res. 36, 1–19. https://doi.org/10.1016/S0065-2296(02)36056-7.

Poulin, R., 2010. Parasite manipulation of host behavior: an update and frequently asked questions. Adv. Stud. Behav. 41, 151–186. https://doi.org/10.1016/S0065-3454(10)41005-0.
Powell, G., Tosh, C.R., Hardie, J., 2006. Host plant selection by aphids: behavioral, evolutionary, and applied perspectives. Annu. Rev. Entomol. 51, 309–330. https://doi.org/10.1146/annurev.ento.51.110104.151107.
Power, A.G., 2000. Insect transmission of plant viruses: a constraint on virus variability. Curr. Opin. Plant Biol. 3 (4), 336–340.
Radcliffe, E.B., Ragsdale, D.W., 2002. Aphid-transmitted potato viruses: the importance of understanding vector biology. Am. J. Potato Res. 79 (5), 353–386. https://doi.org/10.1007/BF02870173.
*Rajabaskar, D., Ding, H., Wu, Y., Eigenbrode, S.D., 2013a. Different reactions of potato varieties to infection by potato leafroll virus, and associated responses by its vector, Myzus persicae (Sulzer). J. Chem. Ecol. 39, 1027–1035. https://doi.org/10.1007/s10886-013-0311-2.
*Rajabaskar, D., Wu, Y., Bosque-p, N.A., Eigenbrode, S.D., 2013b. Dynamics of Myzus persicae arrestment by volatiles from Potato leafroll virus -infected potato plants during disease progression. Entomol. Exp. Appl. 148, 172–181. https://doi.org/10.1111/eea.12087.
*Rajabaskar, D., Bosque-Pérez, N.A., Eigenbrode, S.D., 2014. Preference by a virus vector for infected plants is reversed after virus acquisition. Virus Res. 186, 32–37. https://doi.org/10.1016/j.virusres.2013.11.005.
Reich, P.B., 2014. The world-wide "fast–slow" plant economics spectrum: a traits manifesto. J. Ecol. 102 (2), 275–301. https://doi.org/10.1111/1365-2745.12211.
Rodelo-Urrego, M., Pagán, I., González-Jara, P., Betancourt, M., Moreno-Letelier, A., Ayllón, M.A., et al., 2013. Landscape heterogeneity shapes host-parasite interactions and results in apparent plant-virus codivergence. Mol. Ecol. 22 (8), 2325–2340. https://doi.org/10.1111/mec.12232.
Roosien, B.K., Gomulkiewicz, R., Ingwell, L.L., Bosque-Pérez, N.A., Rajabaskar, D., Eigenbrode, S.D., 2013. Conditional vector preference aids the spread of plant pathogens: results from a model. Environ. Entomol. 42 (6), 1299–1308. https://doi.org/10.1603/EN13062.
Roossinck, M.J., García-Arenal, F., 2015. Ecosystem simplification, biodiversity loss and plant virus emergence. Curr. Opin. Virol. 10, 56–62. https://doi.org/10.1016/j.coviro.2015.01.005.
Saad, K.A., Mohamad Roff, M.N., Hallett, R.H., ABd-Ghani, I.B., 2017. Effects of Cucumber mosaic virus-infected chilli plants on non-vector Bemisia tabaci (Hemiptera: Aleyrodidae). Insect Sci. https://doi.org/10.1111/1744-7917.12488. (in press).
Sadeghi, S.E., Bjur, J., Ingwell, L., Unger, L., Bosque-Pérez, N.A., Eigenbrode, S.D., 2016. Interactions between Metopolophium festucae cerealium (Hemiptera: Aphididae) and Barley yellow dwarf virus (BYDV-PAV). J. Insect Sci. 16 (1), 1–6. https://doi.org/10.1093/jisesa/iev160.
*Salvaudon, L., De Moraes, C.M., Mescher, M.C., 2013. Outcomes of co-infection by two potyviruses: implications for the evolution of manipulative strategies. Proc. Biol. Sci. 280 (1756), 20122959. https://doi.org/10.1098/rspb.2012.2959.
Sanjuán, R., Elena, S.F., 2006. Epistasis correlates to genomic complexity. Proc. Natl. Acad. Sci. U.S.A. 103 (39), 14402–14405. https://doi.org/10.1073/pnas.0604543103.
Sanjuán, R., Moya, A., Elena, S.F., 2004. The contribution of epistasis to the architecture of fitness in an RNA virus. Proc. Natl. Acad. Sci. U.S.A. 101 (43), 15376–15379. https://doi.org/10.1073/pnas.0404125101.
Sanjuán, R., Cuevas, J.M., Moya, A., Elena, S.F., 2005. Epistasis and the adaptability of an RNA virus. Genetics 170 (3), 1001–1008. https://doi.org/10.1534/genetics.105.040741.

Seidl-Adams, I., Richter, A., Boomer, K.B., Yoshinaga, N., Degenhardt, J., Tumlinson, J.H., 2015. Emission of herbivore elicitor-induced sesquiterpenes is regulated by stomatal aperture in maize (Zea mays) seedlings. Plant Cell Environ. 38 (1), 23–34. https://doi.org/10.1111/pce.12347.

Shaw, A.K., Peace, A., Power, A.G., Bosque-Pérez, N.A., 2017. Vector population growth and condition-dependent movement drive the spread of plant pathogens. Ecology 98 (8), 2145–2157. https://doi.org/10.1002/ecy.1907.

*Shi, X., Gao, Y., Yan, S., Tang, X., Zhou, X., Zhang, D., Liu, Y., 2016. Aphid performance changes with plant defense mediated by cucumber mosaic virus titer. Virol. J. 13 (70) https://doi.org/10.1186/s12985-016-0524-4.

Sigvald, R., 1985. Mature-plant resistance of potato plants against potato virus YO (PVYO). Potato Res. 28 (2), 135–143. https://doi.org/10.1007/BF02357439.

Simon, A.E., Roossinck, M.J., Havelda, Z., 2004. Plant virus satellite and defective interfering RNAs: new paradigms for a new century. Annu. Rev. Phytopathol. 42, 415–437. https://doi.org/10.1146/annurev.phyto.42.040803.140402.

Sisterson, M.S., 2008. Effects of insect-vector preference for healthy or infected plants on pathogen spread: insights from a model. J. Econ. Entomol. 101 (1), 1–8. https://doi.org/10.1603/0022-0493(2008)101[1:EOIPFH]2.0.CO;2.

Stafford, C.A., Walker, G.P., Ullman, D.E., 2011. Infection with a plant virus modifies vector feeding behavior. Proc. Natl. Acad. Sci. U.S.A. 108 (23), 9350–9355. https://doi.org/10.1073/pnas.1100773108.

Su, Y.-L., Li, J.-M., Li, M., Luan, J.-B., Ye, X.-D., Wang, X.-W., Liu, S.-S., 2012. Transcriptomic analysis of the salivary glands of an invasive whitefly. PLoS One 7 (6), e39303. https://doi.org/10.1371/journal.pone.0039303.

*Su, Q., Preisser, E.L., Zhou, X.M., Xie, W., Liu, B.M., Wang, S.L., Wu, W., Zhang, Y.J., 2015. Manipulation of host quality and defense by a plant virus improves performance of whitefly vectors. J. Econ. Entomol. 108 (1), 11–19. https://doi.org/10.1093/jee/tou012.

Thomas, P.E., Hang, A.N., Reed, G., Gilliland, G.C., Reisenauer, G., et al., 1993. Potential role of winter rapeseed culture on the epidemiology of potato leaf roll disease. Plant Dis. 77 (4), 420–423.

Thomas, F., Adamo, S., Moore, J., 2005. Parasitic manipulation: where are we and where should we go? Behav. Process. 68 (3), 185–199. https://doi.org/10.1016/j.beproc.2004.06.010.

Tjallingii, W.F., 1988. Electrical recording of stylet penetration activities. In: Minks, A.K., Harewijn, P. (Eds.), Aphids: Their Biology, Natural Enemies and Control. Elsevier Science Publishers B.V, Amsterdam, The Netherlands, pp. 95–108.

Torres-Barceló, C., Daròs, J.-A., Elena, S.F., 2010. Compensatory molecular evolution of HC-Pro, an RNA-silencing suppressor from a plant RNA virus. Mol. Biol. Evol. 27 (3), 543–551. https://doi.org/10.1093/molbev/msp272.

Tosh, C.R., Powell, G., Holmes, N.D., Hardie, J., 2003. Reproductive response of generalist and specialist aphid morphs with the same genotype to plant secondary compounds and amino acids. J. Insect Physiol. 49 (12), 1173–1182.

*Tungadi, T., Groen, S.C., Murphy, A.M., Pate, A.E., Iqbal, J., Bruce, T.J.A., Cunniffe, N.J., Carr, J.P., 2017. Cucumber mosaic virus and its 2b protein alter emission of host volatile organic compounds but not aphid vector settling in tobacco. Virol. J. 14 (1), 91. https://doi.org/10.1186/s12985-017-0754-0.

Wang, R.Y., Ghabrial, S.A., 2002. Effect of aphid behavior on efficiency of transmission of soybean mosaic virus by the soybean-colonizing aphid, Aphis glycines. Plant Dis. 86 (11), 1260–1264. https://doi.org/10.1094/PDIS.2002.86.11.1260.

Wang, J., Bing, X., Li, M., Ye, G., Liu, S., 2012. Infection of tobacco plants by a begomovirus improves nutritional assimilation by a whitefly. Entomol. Exp. Appl. 144, 191–201. https://doi.org/10.1111/j.1570-7458.2012.01278.x.

Webster, B., 2012. The role of olfaction in aphid host location. Physiol. Entomol. 37 (1), 10–18. https://doi.org/10.1111/j.1365-3032.2011.00791.x.
*Werner, B.J., Mowry, T.M., Bosque-Pérez, N.A., Ding, H., Eigenbrode, S.D., 2009. Changes in green peach aphid responses to potato leafroll virus-induced volatiles emitted during disease progression. Environ. Entomol. 38 (5), 1429–1438. https://doi.org/10.1603/022.038.0511.
*Westwood, J.H., Groen, S.C., Du, Z., Murphy, A.M., Tri, D., Carr, P., 2013a. A trio of viral proteins tunes aphid-plant interactions in Arabidopsis thaliana. PLoS One 8 (12), e83066. https://doi.org/10.1371/journal.pone.0083066.
Westwood, J.H., McCann, L., Naish, M., Dixon, H., Murphy, A.M., Stancombe, M.A., Bennett, M.H., Powell, G., Webb, A.A.R., Carr, J.P., 2013b. A viral RNA-silencing suppressor interferes with abscisic acid-mediated signalling and induces drought tolerance in Arabidopsis thaliana. Mol. Plant Pathol. 14 (2), 158–170. https://doi.org/10.1111/j.1364-3703.2012.00840.x.
Whitfield, A.E., Falk, B.W., Rotenberg, D., 2015. Insect vector-mediated transmission of plant viruses. Virology 479–480, 278–289. https://doi.org/10.1016/j.virol.2015.03.026.
Williams, C.T., 1995. Effects of plant age, leaf age and virus yellows infection on the population dynamics of Myzus persicae (Homoptera: Aphididae) on sugarbeet in field plots. Bull. Entomol. Res. 85 (4), 557–567. https://doi.org/10.1017/S000748530003306X.
*Wu, Y., Davis, T.S., Eigenbrode, S.D., 2014. Aphid behavioral responses to virus-infected plants are similar despite divergent fitness effects. Entomol. Exp. Appl. 153 (3), 246–255. https://doi.org/10.1111/eea.12246.
*Wu, D., Qi, T., Li, W.-X., Tian, H., Gao, H., Wang, J., Ge, J., Yao, R., Ren, C., Wang, X.-B., Liu, Y., Kang, L., Ding, S.-W., Xie, D., 2017. Viral effector protein manipulates host hormone signaling to attract insect vectors. Cell Res. 27, 402–415. https://doi.org/10.1038/cr.2017.2.
Xu, P., Chen, F., Mannas, J.P., Feldman, T., Sumner, L.W., Roossinck, M.J., 2008. Virus infection improves drought tolerance. New Phytol. 180 (4), 911–921. https://doi.org/10.1111/j.1469-8137.2008.02627.x.
*Zhang, T., Luan, J.-B., Qi, J.-F., Huang, C.-J., Li, M., Zhou, X.-P., Liu, S.-S., 2012. Begomovirus–whitefly mutualism is achieved through repression of plant defences by a virus pathogenicity factor. Mol. Ecol. 21 (5), 1294–1304. https://doi.org/10.1111/j.1365-294X.2012.05457.x.
*Ziebell, H., Murphy, A.M., Groen, S.C., Tungadi, T., Westwood, J.H., Lewsey, M.G., Moulin, M., Kleczkowski, A., Smith, A.G., Stevens, M., Powell, G., Carr, J.P., 2011. Cucumber mosaic virus and its 2b RNA silencing suppressor modify plant-aphid interactions in tobacco. Sci. Rep. 1 (187) https://doi.org/10.1038/srep00187.

FURTHER READING

References in this section were used in the quantitative synthesis (Tables 2 and 3) but not cited elsewhere in the text.

Abe, H., Tomitaka, Y., Shimoda, T., Seo, S., Sakurai, T., Kugimiya, S., Tsuda, S., Kobayashi, M., 2012. Antagonistic plant defense system regulated by phytohormones assists interactions among vector insect, thrips and a tospovirus. Plant Cell Physiol. 53 (1), 204–212. https://doi.org/10.1093/pcp/pcr173.
Alvarez, A.E., Garzo, E., Verbeek, M., Vosman, B., Dicke, M., Tjallingii, W.F., 2007. Infection of potato plants with potato leafroll virus changes attraction and feeding behaviour of Myzus persicae. Entomol. Exp. Appl. 1252005, 135–144. https://doi.org/10.1111/j.1570-7458.2007.00607.x.
Baker, P.F., 1960. Aphid behaviour on healthy and on yellows-virus-infected sugar beet. Ann. Appl. Biol. 48 (2), 384–391.

Bautista, R., Mau, R.F.L., Cho, J.J., Custer, D.M., 1995. Potential of tomato spotted wilt Tospovirus plant hosts in Hawaii as virus reservoirs for transmission by Frankliniella occidentalis (Thysanoptera: Thripidae). Phytopathology 85, 953–958.

Belliure, B., Janssen, A., Maris, P.C., Peters, D., Sabelis, M.W., 2005. Herbivore arthropods benefit from vectoring plant viruses. Ecol. Lett. 8 (1), 70–79. https://doi.org/10.1111/j.1461-0248.2004.00699.x.

Belliure, B., Janssen, A., Sabelis, M.W., 2008. Herbivore benefits from vectoring plant virus through reduction of period of vulnerability to predation. Oecologia 156 (4), 797–806. https://doi.org/10.1007/s00442-008-1027-9.

Boquel, S., Giordanengo, P., Ameline, A., 2011. Divergent effects of PVY-infected potato plant on aphids. Eur. J. Plant Pathol. 129, 507–510. https://doi.org/10.1007/s10658-010-9732-8.

Boquel, S., Delayen, C., Couty, A., Giordanengo, P., Ameline, A., 2012. Modulation of aphid vector activity by potato virus Y on in vitro potato plants. Plant Dis. 96 (1), 82–86. https://doi.org/10.1094/PDIS-06-11-0499.

Castle, S.J., Berger, P.H., 1993. Rates of growth and increase of Myzus persicae on virus-infected potatoes according to type of virus-vector relationship. Entomol. Exp. Appl. 69, 51–60.

Castle, S.J., Mowry, T.M., Berger, P.H., 1998. Differential settling by Myzus persicae (Homoptera: Aphididae) on various virus infected host plants. Ann. Entomol. Soc. Am. 91 (5), 661–667. https://doi.org/10.1093/aesa/91.5.661.

Chaisuekul, C., Riley, D.G., 2005. Host plant, temperature, and photoperiod effects on ovipositional preference of Frankliniella occidentalis and Frankliniella fusca (Thysanoptera: Thripidae). J. Econ. Entomol. 98 (6), 2107–2113.

Chen, G., Pan, H., Xie, W., Wang, S., Wu, Q., Fang, Y., Shi, X., Zhang, Y., 2013. Virus infection of a weed increases vector attraction to and vector fitness on the weed. Sci. Rep. 3, 2253. https://doi.org/10.1038/srep02253.

Chen, W., Tseng, C., Tsai, C., 2014. Effect of watermelon silver mottle virus on the life history and feeding preference of Thrips palmi. PLoS One 9 (7), e102021. https://doi.org/10.1371/journal.pone.0102021.

Colvin, J., Omongo, C.A., Govindappa, M.R., Stevenson, P.C., Maruthi, M.N., Gibson, G., Seal, S.E., Muniyappa, V., 2006. Host-plant viral infection effects on arthropod-vector population growth, development and behaviour: management and epidemiological implications. Adv. Virus Res. 67, 419–452. https://doi.org/10.1016/S0065-3527(06)67011-5.

Congdon, B.S.S., Coutts, B.A.A., Renton, M., Flematti, G.R.R., Jones, R.A.C., 2017. Establishing alighting preferences and species transmission differences for Pea seed-borne mosaic virus aphid vectors. Virus Res. 241, 145–155. https://doi.org/10.1016/j.virusres.2017.04.005.

Daimei, G., Raina, H.S., Devi, P.P., Saurav, G.K., Renukadevi, Malathi, V.G., Senthilraja, C., Mandal, B., Rajagopal, R., 2017. Influence of groundnut bud necrosis virus on the life history traits. Phytopathology 107 (11), 1440–1445.

Davis, T.S., Bosque-Pérez, N.A., Foote, N.E., Magney, T., Eigenbrode, S.D., 2015. Environmentally dependent host-pathogen and vector-pathogen interactions in the Barley yellow dwarf virus pathosystem. J. Appl. Ecol. 52 (5), 1392–1401.

Davis, T.S., Wu, Y., Eigenbrode, S.D., 2017. The effects of bean leafroll virus on life history traits and host selection behavior of specialized pea aphid (Acyrthosiphon pisum, Hemiptera: Aphididae) genotypes. Environ. Entomol. 46 (1), 68–74.

DeAngelis, J.D., Sether, D.M., Rossignol, P.A., 1993. Survival, development, and reproduction in western flower thrips (Thysanoptera: Thripidae) exposed to impatiens necrotic spot virus. Environ. Entomol. 22, 1308–1312. https://doi.org/10.1093/ee/22.6.1308.

Donaldson, J.R., Gratton, C., 2007. Antagonistic effects of soybean viruses on soybean aphid performance. Environ. Entomol. 36 (4), 918–925.

dos Santos, R.C., Peñaflor, M.F.G.V., Sanches, P.A., Nardi, C., Bento, J.M.S., 2015. The effects of Gibberella zeae, Barley Yellow Dwarf Virus, and co-infection on Rhopalosiphum padi olfactory preference and performance. Phytoparasitica 44 (February 2016), 47–54. https://doi.org/10.1007/s12600-015-0493-y.

Eckel, R.V.W., Lampert, E.P., 1996. Relative attractiveness of Tobacco etch virus-infected and healthy flue-cured tobacco plants to aphids (Homoptera: Aphididae). J. Econ. Entomol. 89 (4), 1017–1027.

Ellsbury, M.M., Pratt, R.G., Knight, W.E., 1985. Effects of single and combined infection of arrowleaf clover with bean yellow mosaic virus and a Phytophthora sp. on reproduction and colonization by pea aphids (Homoptera: Aphididae). Environ. Entomol. 14 (3), 356–359. https://doi.org/10.1093/ee/14.3.356.

Fereres, A., Kampmeier, G.E., Irwin, M.E., 1999. Aphid attraction and preference for soybean and pepper plants infected with Potyviridae. Ann. Entomol. Soc. Am. 92 (4), 542–548.

Ghosh, A., Das, A., Vijayanandraj, S., Mandal, B., 2015. Cardamom bushy dwarf virus infection in large cardamom alters plant selection preference, life stages, and fecundity of aphid vector, Micromyzus kalimpongensis (Hemiptera: Aphididae). Environ. Entomol. 45 (1), 178–184.

Guo, J., Ye, G., Dong, S., Liu, S., 2010. An invasive Whitefly feeding on a virus-infected plant increased its egg production and realized fecundity. PLoS One 5 (7), 1–8. https://doi.org/10.1371/journal.pone.0011713.

Hodge, S., Powell, G., 2008. Do plant viruses facilitate their aphid vectors by inducing symptoms that alter behavior and performance? Environ. Entomol. 37 (6), 1573–1581. Retrieved from: https://www.ncbi.nlm.nih.gov/pubmed/19161702.

Hodge, S., Powell, G., 2010. Conditional facilitation of an aphid vector, Acyrthosiphon pisum, by the plant pathogen, pea enation mosaic virus. J. Insect Sci. 10, 155. https://doi.org/10.1673/031.010.14115.

Jiu, M., Zhou, X.-P., Tong, L., Xu, J., Yang, X., Wan, F.-H., Liu, S.-S., 2007. Vector-virus mutualism accelerates population increase of an invasive whitefly. PLoS One 2 (1), e182. https://doi.org/10.1371/journal.pone.0000182.

Lei, W., Liu, D., Li, P., Hou, M., 2014. Interactive effects of southern rice black-streaked dwarf virus infection of host plant and vector on performance of the vector, Sogatella furcifera (Homoptera: Delphacidae). J. Econ. Entomol. 107 (5), 1721–1727. https://doi.org/10.1603/EC13569.

Li, M., Liu, J., Liu, S., 2011. Tomato yellow leaf curl virus infection of tomato does not affect the performance of the Q and ZHJ2 biotypes of the viral vector Bemisia tabaci. Insect Sci. 18, 40–49. https://doi.org/10.1111/j.1744-7917.2010.01354.x.

Li, J., Ding, T.B., Chi, H., Chu, D., 2017. Effects of Tomato chlorosis virus on the performance of its key vector, Bemisia tabaci, in China. J. Appl. Entomol. 142 (3), 296–304. https://doi.org/10.1111/jen.12477.

Lightle, D., Lee, J., 2014. Raspberry viruses affect the behavior and performance of Amphorophora agathonica in single and mixed infections. Entomol. Exp. Appl. 151, 57–64. https://doi.org/10.1111/eea.12170.

Liu, J., Zhao, H., Jiang, K., Zhou, X.-P., Liu, S.-S., 2009. Differential indirect effects of two plant viruses on an invasive and an indigenous whitefly vector: implications for competitive displacement. Ann. Appl. Biol. 155 (3), 439–448. https://doi.org/10.1111/j.1744-7348.2009.00366.x.

Liu, J., Li, M., Li, J., Huang, C., Zhou, X., Xu, F., Liu, S., 2010. Viral infection of tobacco plants improves performance of Bemisia tabaci but more so for an invasive than for an indigenous biotype of the whitefly. J. Zhejiang Univ. Sci. B 11 (1), 30–40. https://doi.org/10.1631/jzus.B0900213.

Lowe, S., Strong, F.E., 1962. The unsuitability of some viruliferous plants as hosts for the green peach aphid, Myzus persicae 1. J. Econ. Entomol. 56 (3), 307–309.

Macias, W., Mink, G.I., 1969. Preference of green peach aphids for virus-infected sugarbeet leaves. J. Econ. Entomol. 62 (1), 8–9.

Maluta, N.K.P., Garzo, E., Moreno, A., Lopes, J.R.S., Fereres, A., 2014. Tomato yellow leaf curl virus benefits population growth of the Q biotype of Bemisia tabaci (Gennadius) (Hemiptera: Aleyrodidae). Neotrop. Entomol. 43, 385–392. https://doi.org/10.1007/s13744-014-0223-z.

Maluta, N.K.P., Fereres, A., Lopes, J.R.S., 2017. Settling preferences of the whitefly vector Bemisia tabaci on infected plants varies with virus family and transmission mode. Entomol. Exp. Appl. 165, 138–147. https://doi.org/10.1111/eea.12631.

Maris, P.C., Joosten, N.N., Goldbach, R.W., Peters, D., 2004. Tomato spotted wilt virus infection improves host suitability for its vector Frankliniella occidentalis. Phytopathology 94 (7), 706–711. https://doi.org/10.1094/PHYTO.2004.94.7.706.

Matsuura, S.M., Hoshino, S.H., 2009. Effect of tomato yellow leaf curl disease on reproduction of Bemisia tabaci Q biotype (Hemiptera: Aleyrodidae) on tomato plants. Appl. Entomol. Zool. 44 (1), 143–148. https://doi.org/10.1303/aez.2009.143.

Mayer, R.T., Inbar, M., McKenzie, C.L., Shatters, R., Borowicz, V., Albrecht, U., Powell, C.A., Doostdar, H., 2002. Multitrophic interactions of the silverleaf whitefly, host plants, competing herbivores, and phytopathogens. Arch. Insect Biochem. Physiol. 51 (4), 151–169. https://doi.org/10.1002/arch.10065.

McKenzie, C.L., 2002. Effect of Tomato mottle virus (ToMoV) on Bemisia tabaci Biotype B (Homoptera: Aleyrodidae) oviposition and adult survivorship on healthy tomato. Fla. Entomol. 85 (2), 367–368.

Mesfin, T., Bosque-Pérez, N.A., 1998. Feeding behaviour of Cicadulina storeyi China (Homoptera: Cicadellidae) on maize varieties susceptible or resistant to maize streak virus. Afr. Entomol. 6 (2), 185–191.

Moiroux, J., Chesnais, Q., Spicher, F., Verrier, E., Ameline, A., Couty, A., 2018. Plant virus infection influences bottom-up regulation of a plant-aphid-parasitoid system. J. Pestic. Sci. 91, 361–372. https://doi.org/10.1007/s10340-017-0911-7.

Montllor, C.B., Gildow, F.E., 1986. Feeding responses of two grain aphids to barley yellow dwarf virus-infected oats. Entomol. Exp. Appl. 42 (1), 63–69.

Musser, R.O., Hum-Musser, S.M., Felton, G.W., Gergerich, R.C., 2003. Increased larval growth and preference for virus-infected leaves by the Mexican bean beetle, Epilachna varivestis Mulsant, a plant virus vector. J. Insect Behav. 16 (2), 247–256. https://doi.org/10.1023/A:1023919902976.

Ngumbi, E., Eigenbrode, S.D., Bosque-Pérez, N.A., Ding, H., Rodriguez, A., 2007. Myzus persicae is arrested more by blends than by individual compounds elevated in headspace of plrv-infected potato. J. Chem. Ecol. 33, 1733–1747. https://doi.org/10.1007/s10886-007-9340-z.

Ogada, P.A., Maiss, E., Poehling, H.-M., 2013. Influence of tomato spotted wilt virus on performance and behaviour of western flower thrips (Frankliniella occidentalis). J. Appl. Entomol. 137 (7), 488–498. https://doi.org/10.1111/jen.12023.

Pan, H., Chen, G., Li, F., Wu, Q., Wang, S., Xie, W., Liu, B., Xu, B., Zhang, Y., 2013. Tomato spotted wilt virus infection reduces the fitness of a nonvector herbivore on pepper. J. Econ. Entomol. 106 (2), 924–928.

Ponsen, M.B., 1969. The effect of potato leafroll virus on the biology of Myzus persicae. Neth. J. Pl. Path. 75, 360–368.

Power, A.G., 1996. Competition between viruses in a complex plant-pathogen system. Ecology 77 (4), 1004–1010.

Ren, G.-W., Wang, X.-F., Chen, D., Wang, X.-W., Fan, X.-J., Liu, X.-D., 2015. Potato virus Y-infected tobacco affects the growth, reproduction, and feeding behavior of a vector aphid, Myzus persicae (Hemiptera: Aphididae). Appl. Entomol. Zool. 50, 239–243. https://doi.org/10.1007/s13355-015-0328-9.

Rubinstein, G., Czosnek, H., 1997. Long-term association of tomato yellow leaf curl virus with its whitefly vector Bemisia tabaci: effect on the insect transmission capacity, longevity and fecundity. J. Gen. Virol. 78, 2683–2689.

Shalileh, S., Ogada, P.A., Moualeu, D.P., Poehling, H.-M., 2016. Behavior manipulation of Frankliniella occidentalis (Thysanoptera: Thripidae) by tomato spotted wilt virus (Tospovirus) via the host plant nutrients to enhance its transmission and spread. Environ. Entomol. 45, 1235–1242. https://doi.org/10.1093/ee/nvw102.

Shrestha, D., Mcauslane, H.J., Adkins, S.T., Smith, H.A., Dufault, N., Webb, S.E., 2016. Transmission of Squash vein yellowing virus to and from cucurbit weeds and effects on sweetpotato whitefly (Hemiptera: Aleyrodidae) behavior. 45, 967–973. https://doi.org/10.1093/ee/nvw086.

Shrestha, D., McAuslane, H.J., Adkins, S.T., Smith, H.A., Dufault, N., Colee, J., Webb, S.E., 2017. Host-mediated effects of semipersistently transmitted squash vein yellowing virus on sweetpotato Whitefly (Hemiptera: Aleyrodidae) behavior and fitness. J. Econ. Entomol. https://doi.org/10.1093/jee/tox161

Smith, C.M., Gedling, C.R., Wiebe, K.F., Cassone, B.J., 2017. A sweet story: bean pod mottle virus transmission dynamics by Mexican bean beetles (Epilachna varivestis). Genome Biol. Evol. 9, 714–725. https://doi.org/10.1093/gbe/evx033.

Srinivasan, R., Alvarez, J.M., 2007. Effect of mixed viral infections (potato virus Y–potato leafroll virus) on biology and preference of vectors Myzus persicae and Macrosiphum euphorbiae (Hemiptera: Aphididae). J. Econ. Entomol. 100 (3), 646–655. https://doi.org/10.1603/0022-04932007100[646:EOMVIP]2.0.CO;2.

Srinivasan, R., Alvarez, J.M., Eigenbrode, S.D., Bosque-Pérez, N.A., 2006. Influence of hairy nightshade Solanum sarrachoides (Sendtner) and potato leafroll virus (Luteoviridae: Polerovirus) on the host preference of Myzus persicae (Sulzer) (Homoptera: Aphididae). Environ. Entomol. 35 (2), 546–553. https://doi.org/10.1603/0046-225X-35.2.546.

Stumpf, C.F., Kennedy, G.G., 2007. Effects of tomato spotted wilt virus isolates, host plants, and temperature on survival, size, and development time of Frankliniella occidentalis. Entomol. Exp. Appl. 123 (2), 139–147. https://doi.org/10.1111/j.1570-7458.2007.00541.x.

Su, Q., Mescher, M.C., Wang, S., Chen, G., Xie, W., Wu, Q., Wang, W., Zhang, Y., 2016. Tomato yellow leaf curl virus differentially influences plant defence responses to a vector and a non-vector herbivore. Plant Cell Environ. 39 (3), 597–607. https://doi.org/10.1111/pce.12650.

Sun, Y.-C., Pan, L.-L., Ying, F.-Z., Li, P., Wang, X.-W., Liu, S.-S., 2017. Jasmonic acid-related resistance in tomato mediates interactions between whitefly and whitefly-transmitted virus. Sci. Rep. 7 (1), 566. https://doi.org/10.1038/s41598-017-00692-w.

Sylvester, E.S., 1973. Reduction of excretion, reproduction, and survival in Hyperomyzus lactucae fed on plants infected with isolates of sowthistle yellow vein virus. Virology 56 (2), 632–635.

Tomitaka, Y., Abe, H., Sakurai, T., Tsuda, S., 2015. Preference of the vector thrips Frankliniella occidentalis for plants infected with thrips-non-transmissible Tomato spotted wilt virus. J. Appl. Entomol. 139 (4), 250–259. https://doi.org/10.1111/jen.12165.

Tu, Z., Ling, B., Xu, D., Zhang, M., Zhou, G., 2013. Effects of southern rice black-streaked dwarf virus on the development and fecundity of its vector, Sogatella furcifera. Virol. J. 10, 145.

Wang, H., Xu, D., Pu, L., Zhou, G., 2014. Southern rice black-streaked dwarf virus alters insect vectors' host orientation preferences to enhance spread and increase rice ragged

stunt virus co-infection. Phytopathology 104 (2), 196–201. https://doi.org/10.1094/PHYTO-08-13-0227-R.
Wijkamp, I., Goldbach, R., Peters, D., 1996. Propagation of tomato spotted wilt virus in Frankliniella occidentalis does neither result in pathological effects nor in transovarial passage of the virus. Entomol. Exp. Appl. 81 (3), 285–292. https://doi.org/10.1046/j.1570-7458.1996.00098.x.
Wosula, E.N., Davis, J.A., 2014. Stylet penetration behaviors of Myzus persicae (Hemiptera: Aphididae) on four Ipomoea spp. infected or noninfected with sweet potato potyviruses. J. Econ. Entomol. 107 (2), 538–545.
Wosula, E.N., Davis, J.A., Clark, C.A., 2013. Population dynamics of three aphid species (Hemiptera: Aphididae) on four Ipomoea spp. infected or noninfected with sweetpotato potyviruses. J. Econ. Entomol. 106 (4), 1566–1573.
Zhang, J., Nie, X., Nanayakkara, U., Boquel, S., Giguère, M.A., Pelletier, Y., 2013. Detection of Potato virus Y from the stylets of a single aphid by one-step reverse transcription polymerase chain reaction. Entomol. Exp. Appl. 147 (1), 93–97. https://doi.org/10.1111/eea.12044.
Zhang, J., Zheng, X., Chen, Y., Hu, J., Dong, J., Su, X., Zhang, J.I.E., 2014. Southern rice black-streaked dwarf virus infection improves host suitability for its insect vector, Sogatella furcifera (Hemiptera: Delphacidae). J. Econ. Entomol. 107 (1), 92–97.

CHAPTER EIGHT

Emerging Viruses in Bees: From Molecules to Ecology

Dino P. McMahon[*,†,1], **Lena Wilfert**[‡], **Robert J. Paxton**[§,¶], **Mark J.F. Brown**[||]

[*]Institute of Biology, Freie Universität Berlin, Berlin, Germany
[†]Department for Materials and Environment, BAM Federal Institute for Materials Research and Testing, Berlin, Germany
[‡]Centre for Ecology and Conservation, University of Exeter, Penryn, United Kingdom
[§]Institute for Biology, Martin-Luther-Universität Halle-Wittenberg, Halle (Saale), Germany
[¶]German Centre for integrative Biodiversity Research (iDiv), Halle-Jena-Leipzig, Leipzig, Germany
[||]School of Biological Sciences, Royal Holloway University of London, Egham, United Kingdom
[1]Corresponding author e-mail address: dino.mcmahon@gmail.com

Contents

Advances in Virus Research, Volume 101
ISSN 0065-3527
https://doi.org/10.1016/bs.aivir.2018.02.008

Abstract

Emerging infectious diseases arise as a result of novel interactions between populations of hosts and pathogens, and can threaten the health and wellbeing of the entire spectrum of biodiversity. Bees and their viruses are a case in point. However, detailed knowledge of the ecological factors and evolutionary forces that drive disease emergence in bees and other host–pathogen communities is surprisingly lacking. In this review, we build on the fundamental insight that viruses evolve and adapt over timescales that overlap with host ecology. At the same time, we integrate the role of host community ecology, including community structure and composition, biodiversity loss, and human-driven disturbance, all of which represent significant factors in bee virus ecology. Both of these evolutionary and ecological perspectives represent major advances but, in most cases, it remains unclear how evolutionary forces actually operate across different biological scales (e.g., from cell to ecosystem). We present a molecule-to-ecology framework to help address these issues, emphasizing the role of molecular mechanisms as key bottom-up drivers of change at higher ecological scales. We consider the bee–virus system to be an ideal one in which to apply this framework. Unlike many other animal models, bees constitute a well characterized and accessible multispecies assemblage, whose populations and interspecific interactions can be experimentally manipulated and monitored in high resolution across space and time to provide robust tests of prevailing theory.

1. INTRODUCTION

The interrelationship between bees, their emerging pathogens, and ecology is a research area that is due to significant growth in years to come. Bee populations have declined significantly in recent decades, particularly in temperate zones, and this trend is thought to be at least in part attributable to the (re)emergence of pathogens including RNA viruses. Combined with their experimental tractability and ecological relevance, these circumstances make bees excellent models for investigating virus ecology. First, managed honeybees are often infested with the invasive mite, *Varroa destructor*, which vectors a wide range of viruses (de Miranda and Genersch, 2010; Martin et al., 2012; McMenamin and Genersch, 2015; Mondet et al., 2014), one of which, Deformed wing virus (DWV, family *Iflaviridae*), is thought to play a central role in colony mortality (Budge et al., 2015; Dainat et al., 2012; Di Prisco et al., 2016; Francis et al., 2013; Genersch et al., 2010; Highfield et al., 2009; Natsopoulou et al., 2017; Nazzi et al., 2012; Nguyen et al., 2011). Second, honeybee viruses are shared with sympatric wild bees; viral sequencing suggests on-going transmission between host bee species of the same viral variants (Fürst et al., 2014; McMahon et al., 2015; Radzevičiūtė et al., 2017), likely at flowers (McArt et al., 2014), opening up the possibility to experimentally manipulate transmission in an ecologically realistic context. Third, many bee viruses have been sequenced (Chen and Siede, 2007; Genersch and Aubert, 2010) and ultrastructural features are becoming available for DWV (Škubník et al., 2017) and other bee viruses (Kalynych et al., 2016, 2017), facilitating their study. Fourth, several bee species can be reared, permitting experimental investigation of host–parasite relations in the laboratory (e.g., Graystock et al., 2016; Manley et al., 2017; McMahon et al., 2016; Meeus et al., 2014; Natsopoulou et al., 2015). For example, honeybee larvae and pupae can be reared in the lab under controlled conditions and in high numbers, and virus evolution experiments consisting of tens of passages can be completed at the whole animal scale in a matter of weeks. Furthermore, while traditional transgenic technologies such as those available to *Drosophila* have not been available in bees, the emergence of gene manipulation methodologies such as RNA interference (RNAi)-mediated gene suppression which can be used to target both bees and ectoparasitic *Varroa* mites (Garbian et al., 2012) and novel gene-editing technologies in the form of CRISPR/cas are beginning to open the door to both loss- and gain-of-function studies in nonmodel eusocial organisms, such as ants (Yan et al., 2017). The advent of second- and third-generation sequencing technologies has further facilitated genome and transcriptome-based investigations in bees (Bigot et al., 2017; Doublet et al., 2017). That bee viruses are not known to be pathogenic to humans is an added advantage in laboratory studies. Fifth, most bee species can be readily collected from flowers in the field, allowing sampling of the commoner species and determination of viral presence (reviewed in Tehel et al., 2016) and prevalence. Finally, the economic and ecological relevance of bees, as described in Section 2.1, gives significant applied value to the investigation of bee–virus interactions.

In this review, we introduce the biology of bees and their viruses, and outline how they represent an ideal system with which to explore viral emergence in a realistic ecological setting. We put forward a framework that integrates molecular- and ecosystem-level approaches to virus ecology, arguing that the greatest insights will come from tackling questions at two or more levels of biological organization. We then discuss general evolutionary and ecological principles of infectious disease, drawing from human and other well-known animal systems to identify gaps in bee virus ecology, but also to highlight areas where research into bee viruses may bring distinct benefits. We finish by considering some critical challenges and open questions in bee virus research. Our intention in this review is to provide a basic introduction to bee virus research and a useful guide and source of ideas for forthcoming studies in this rapidly expanding field.

2. A BRIEF OVERVIEW OF BEES AND THEIR VIRUSES

2.1 Bees

The value of bees to agriculture, food security, and the wider environment through their role in pollination is indisputable. Approximately 70% of the world's most important crops are dependent to some extent on animal-mediated pollination (Klein et al., 2007), with an estimated value to the economy of US\$ $235–557 \times 10^9$ at today's prices (Gallai et al., 2009; Potts et al., 2016). They comprise a range of vegetable, fruit, nut, and seed crops that are not only of high-economic value but also rich in important micronutrients otherwise in short supply in many low-income countries (Eilers et al., 2011). Perhaps not surprisingly, world agriculture has witnessed significant growth in pollinator-dependent crops over the past five decades (Aizen et al., 2009; Potts et al., 2016). With over 87% of the world's angiosperm species also dependent on animal vectors to move their pollen from anther to stigma of the same or another conspecific plant (Ollerton et al., 2011; Willmer, 2011), pollination is placed high on the list of major ecosystem services (Costanza et al., 1997).

Though many insects, such as Coleoptera, Diptera, and Lepidoptera, and a diversity of vertebrates, including bats, birds, and lizards, can effect pollination, bees are by far the most important animal pollen vectors (Willmer et al., 2017). They comprise a monophyletic group of ca. 20,000 species divided into 7 taxonomic families (Danforth et al., 2013; see Table 1) and exhibit a range of social behaviors, from solitary (ca. 90%) to eusocial, as in the honeybees (*Apis* spp.) and neotropical stingless bees (tribe Meliponini; Kocher and Paxton, 2014). All collect nectar and pollen from flowers as larval provisions for their offspring (or act as cleptoparasites of other bee species) and therefore are avid flower visitors, accounting for their important role in pollination (Willmer et al., 2017). These facts and statistics underscore the importance of bees to agricultural and natural ecosystems.

Pollination is considered as an ecosystem service under threat because of the perceived decline of animal pollinators across the world, particularly bees (Brown and Paxton, 2009; Potts et al., 2016; Vanbergen and IPI Initiative, 2013). Though the number of colonies of the most important commercial pollinator, the Western honeybee (*Apis mellifera*), has increased exponentially

Table 1 The Bees, an Ecologically and Economically Important Monophyletic Group of Approximately 20,000 Insect Species Within the Order Hymenoptera (Danforth et al., 2013; Hedtke et al., 2013; Michener, 2007)

Family	Melittidae	Andrenidae	Halictidae	Colletidae	Stenotritidae	Apidae	Megachilidae
No. of described species	200	>2900	>4300	>2500	21	>5700	>4000
Distribution	Absent from Australia and S. America, well represented in southern Africa	Absent from Australia; otherwise worldwide	Worldwide	Worldwide, well represented in Southern Hemisphere	Australia only	Worldwide, well represented in the neotropical and oriental regions	Worldwide
Social organization	Solitary	Solitary	Solitary to eusocial, including socially polymorphic spp.; two independent origins of sociality	Solitary	Solitary (?)	Solitary to eusocial, including advanced eusocial spp., e.g., honey bees; two independent origins of sociality	Solitary
Ecological traits	Fossorial nesting; often annual	Fossorial nesting; often annual	Fossorial nesting; often annual; several cleptoparasitic spp.	Fossorial nesting; often annual	Fossorial nesting; annual (?)	Nesting below and above-ground and in self-made nests; advanced eusocial spp. are perennial; many cleptoparasitic spp.	Nesting above-ground in cavities using collected plant material; often annual; some cleptoparasitic spp.

Continued

Table 1 The Bees, an Ecologically and Economically Important Monophyletic Group of Approximately 20,000 Insect Species Within the Order Hymenoptera (Danforth et al., 2013; Hedtke et al., 2013; Michener, 2007)—cont'd

Family	Melittidae	Andrenidae	Halictidae	Colletidae	Stenotritidae	Apidae	Megachilidae

A maximum-likelihood estimate of phylogenetic relationships among bee species based on multiple DNA sequences (after Hedtke et al., 2013); bees arose from within the sphecid wasps. (?) Poorly studied, social and ecological traits unknown.

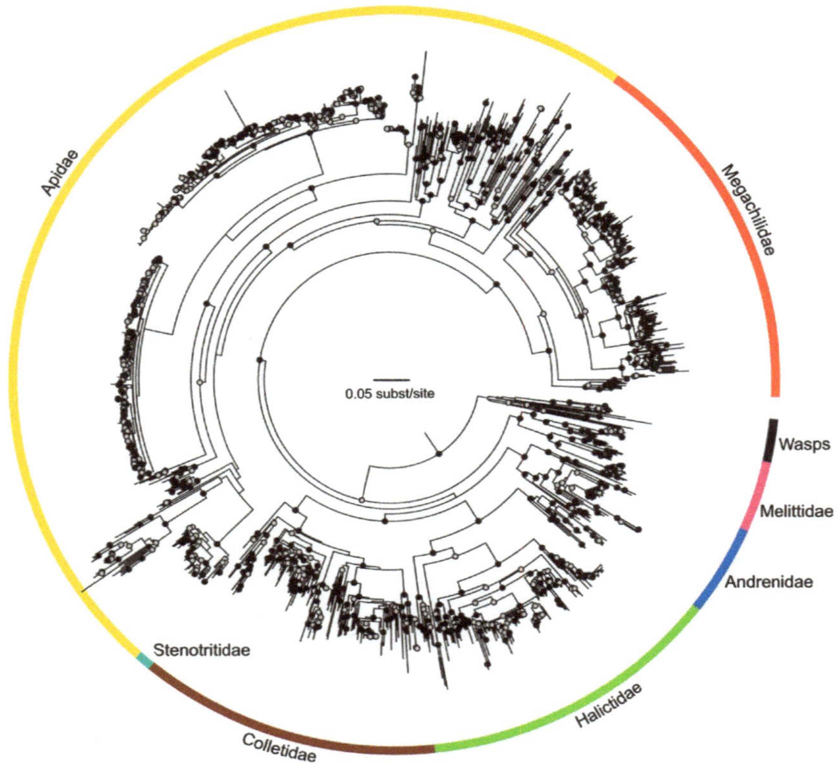

Figure reproduced in original form from Hedtke, S.M., Patiny, S., Danforth, B.N., 2013. The bee tree of life: a supermatrix approach to apoid phylogeny and biogeography. BMC Evol. Biol. 13, 138, (Publisher: Biomed Central) under a CC-BY-2.0 licence.

over the past 50 years (Potts et al., 2016)—reflecting the profitability of beekeeping and the wider socioeconomic climate (Moritz and Erler, 2016; Moritz et al., 2010)—this belies the high mortality of colonies at c.30% per year, particularly over winter, in Northern temperate regions (Neumann and Carreck, 2010). These losses are generally compensated for by beekeepers multiplying colonies in the subsequent season. The exotic ectoparasitic mite *Varroa destructor* and the viruses it transmits, particularly DWV, are considered the major cause of elevated honeybee mortality (Budge et al., 2015; Dainat et al., 2012; Di Prisco et al., 2016; Francis et al., 2013; Genersch et al., 2010; Highfield et al., 2009; Nazzi et al., 2012; Natsopoulou et al., 2017; Nguyen et al., 2011). DWV was present in honeybee populations prior to the arrival of *V. destructor*, but its epidemiology has altered dramatically since the global spread of the mite from southeast Asia since the 1950s (Martin et al., 2012; Oldroyd, 1999; Wilfert et al., 2016). Prior to *V. destructor*, DWV was associated with largely asymptomatic infections with occasional outbreaks of acute disease (Forsgren et al., 2012). Few other bee species are managed, though one, the largely European bumble bee *Bombus terrestris*, has been increasingly employed for glasshouse pollination since 1990 (Velthuis and van Doorn, 2006), and various other bumble bee species are now domesticated and commercialized as pollinators, along with some solitary bee species (Aizen et al., 2016). There is conclusive evidence for a decline of other, unmanaged bee species (often referred to as "wild bees") in Europe (Biesmeijer et al., 2006; Fitzpatrick et al., 2007; Nieto et al., 2014) and North America (Bartomeus et al., 2013; Cameron et al., 2011), which are increasingly acknowledged to play an important role in crop pollination and flowering plant biodiversity (Garibaldi et al., 2013). Degradation and loss of habitat is thought to be the major cause for the decline of wild bees in these regions (Winfree et al., 2009), though increasingly the role of pesticides (especially neonicotinoids; see Rundlöf et al., 2015; Woodcock et al., 2016, 2017) and pathogens, including viruses that have spilled over from managed honeybees (Fürst et al., 2014; McMahon et al., 2015; Radzevičiūtė et al., 2017), have been highlighted as potentially important drivers of decline. These different stressors may act synergistically in contributing to pollinator losses (Potts et al., 2016).

2.2 Bee Viruses

Because of their potential impact on apiculture, honeybee viruses have been intensely studied for many decades. At least 18 different RNA and DNA viruses had been identified using classical virological methods (Chen and

Siede, 2007). These viruses, reviewed in detail by Chen and Siede (2007) as well as by de Miranda et al. (2010a,b) and de Miranda and Genersch (2010) for individual viral groups, are predominantly positive single-stranded (+ss) RNA viruses, many of them belonging to the Picornavirales (picorna-like viruses). With the availability of next-generation sequencing, several additional RNA viruses, such as the Lake Sinai viruses (LSV) (assumed family *Noraviridae*), have been described (McMenamin and Genersch, 2015; Mordecai et al., 2016; Remnant et al., 2017; Runckel et al., 2011). We can expect many more to be added to this list as next-generation and third-generation sequencing becomes more widespread in the field of epidemiology (Bigot et al., 2017). The challenge with this powerful approach to virus discovery will in future lie first in determining whether these viruses in fact infect the bees they were isolated from, and second whether these viruses are virulent, i.e., whether they negatively affect the purported host's fitness. The first challenge can, at least partially, be addressed by sequencing host small RNAs; viral short RNA fragments show a peak at a characteristic length (21 bp in the case of bees), indicating that the host's antiviral RNAi system has attacked an actively replicating RNA virus (Obbard et al., 2009). The second issue is more difficult to address and will require a combination of detailed lab experiments, field/semifield studies and modeling to assess the cost of such infections to honeybees, and potentially other bee species. Importantly, it is now clear that many viruses previously considered "honeybee viruses" are in fact multihost pathogens that infect other bees and potentially even phylogenetically more distant groups (Manley et al., 2015; McMahon et al., 2015) (Table 2). Whether the arrival of the *V. destructor* mite in *A. mellifera* has been a driver of viral emergence in non-*Apis* bees represents a major outstanding question in bee virus ecology, although the positive correlation in virus prevalence and sequence identity between *Apis* and *Bombus* populations (Fürst et al., 2014; McMahon et al., 2015; Radzevičiūtė et al., 2017) indicates frequent and ongoing transmission between bee species. Regardless of virus origins, this means that for common bee viruses, including, e.g., DWV, black queen cell virus (BQCV), or acute bee paralysis virus (ABPV), the pathogen's evolutionary ecology has to be studied within a multihost pathogen framework.

2.3 Deformed Wing Virus

DWV is one of the most prevalent viruses in the western honeybee *A. mellifera*, which when combined with the vector *V. destructor* is a leading

Table 2 A List of Bee-Associated Viruses Sequenced to Date (All Positive Single-Stranded RNA Viruses), Their Families, and References

Family	Virus	Abbr.	References
Dicistroviridae	Acute bee paralysis virus[a]	ABPV	Bailey et al. (1963) and de Miranda et al. (2010a)
	Aphid lethal paralysis virus	ALPV	Runckel et al. (2011)
	Big Sioux River virus	BSRV	Runckel et al. (2011)
	Black queen cell virus	BQCV	Bailey and Woods (1977) and Leat et al. (2000)
	Kashmir bee virus[a]	KBV	Bailey and Woods (1977) and de Miranda et al. (2004)
	Israeli acute paralysis virus[a]	IAPV	Maori et al. (2007)
Iflaviridae	Deformed wing virus[b]	DWV	de Miranda and Genersch (2010)
	Kakugo virus[b]	KV	Fujiyuki et al. (2004)
	Sacbrood virus	SBV	Bailey and Fernando (1972) and Ghosh et al. (1999)
	Slow bee paralysis virus	SBPV	Bailey and Woods (1974) and de Miranda et al. (2010b)
	Varroa destructor virus-1[b]	VDV-1	Ongus et al. (2004)
Secoviridae	Tobacco ringspot virus	TRSV	Li et al. (2014)
Unassigned, related to *Nodaviridae* and *Tombusviridae*	Chronic bee paralysis virus	CBPV	Bailey et al. (1963), Olivier et al. (2008), and Ribière et al. (2010)
	Halictus scabiosae Adlikon virus	HsAV	Bigot et al. (2017)
	Lake Sinai virus 1 and 2[c]	LSV	Runckel et al. (2011)

[a]ABPV/IAPV/KBV.
[b]DWV/KV/VDV-1.
[c]LSV-1/LSV-2.
Note: Several viruses are so closely related that it is yet to be determined if they should be classified as different species.
Adapted from Manley, R., Boots, M., Wilfert, L., 2015. Emerging viral disease risk to pollinating insects: ecological, evolutionary and anthropogenic factors. J. Appl. Ecol. 52, 331–340.

cause of overt symptomatic infections typified by adult bees displaying crumpled wings and malformed bodies (Ball and Allen, 1988; Gisder et al., 2009; Möckel et al., 2011; Shen et al., 2005; Yue and Genersch, 2005). DWV is now known to comprise of at least three distinct genotypes (Mordecai et al., 2015) that are differentially virulent in honeybees (McMahon et al., 2016; Ryabov et al., 2014) and able to recombine in nature (Moore et al., 2011; Wang et al., 2013; Zioni et al., 2011). Interestingly, DWV is also known to naturally infect the managed species *Apis cerana* and has been detected in the nonmanaged species *Apis dorsata* and *Apis florea* (Zhang et al., 2012). Outside of honeybees, DWV has been found widely in bumblebees, including solitary bees and wasps (reviewed in Gisder and Genersch, 2017) and there is evidence that it can actively replicate in several *Bombus* and solitary bee species (Fürst et al., 2014; Genersch et al., 2006; Graystock et al., 2016; Levitt et al., 2013; Radzevičiūtė et al., 2017).

2.4 Black Queen Cell Virus

BQCV (family *Dicistroviridae*) was originally described from *A. mellifera* as the disease agent of decomposed black-patched queen pupae (Bailey and Woods, 1977) but has since been found to frequently infect workers. Alongside DWV, it is one of the most prevalent viruses of *A. mellifera*. As with DWV, BQCV is found frequently in non-*Apis* bees (McMahon et al., 2015; Radzevičiūtė et al., 2017; Singh et al., 2010), and there is evidence of active infection in at least one *Bombus* species (Peng et al., 2011) and several other solitary bees (Radzevičiūtė et al., 2017), although greater knowledge concerning its pathological effects on *Apis* and non-*Apis* bees is needed.

2.5 Acute Bee Paralysis Virus

ABPV (family *Dicistroviridae*) and its close relatives Kashmir bee virus (KBV) and Israeli acute bee paralysis virus (IAPV) are virulent and widespread viruses in *A. mellifera*. The arrival of the *V. destructor* mite may have increased the prevalence of ABPV in *A. mellifera* (Genersch and Aubert, 2010), but its high virulence (leading to reduced survival of infected pupae) could explain why ABPV is detected less frequently than DWV in *V. destructor*-infested colonies (Schroeder and Martin, 2012; Sumpter and Martin, 2004). Only a little is known about the ABPV-complex in non-*Apis* bees. For example, when orally infected with IAPV and KBV, *B. terrestris* microcolonies were seen to suffer slower colony establishment or lower offspring production (Meeus et al., 2014), while in the field, IAPV and KBV have been detected in non-*Apis* hymenopterans (Singh et al., 2010). Meanwhile, ABPV has been detected in *Bombus* species in the field as viable infections (Bailey and Gibbs, 1964). A UK survey showed that ABPV is significantly more prevalent in *Bombus lapidarius* than in other common bumblebee species and even than in *A. mellifera*, and can occur in bumblebees at putatively high viral loads (McMahon et al., 2015). Such species prevalence patterns might indicate significant variation in host susceptibility to viral infection (Ruiz-González et al., 2012).

2.6 Slow Bee Paralysis Virus

The situation for Slow bee paralysis virus (SBPV) (family *Iflaviridae*) is less complete, although both field (Carreck et al., 2010) and laboratory experiments (Santillán-Galicia et al., 2014) suggest that it can be transmitted between honeybees via *V. destructor* mites and that it may also be more virulent than DWV. With respect to wild bees, a recent laboratory study in *B. terrestris* found the virulence of SBPV infection to be condition (starvation) dependent (Manley et al., 2017), but virtually nothing is known of SBPV epidemiology. As with ABPV, SPBV is frequently detected in bumblebees (McMahon et al., 2015; Parmentier et al., 2016), and recent studies have confirmed its presence in several non-*Bombus* bees including *Melipona* (Ueira-Vieira et al., 2015) and *Anthophora* (Radzevičiūtė et al., 2017).

2.7 Sacbrood Virus

Sacbrood virus (SBV) (family *Iflaviridae*) is another common and globally prevalent virus of honeybees, which leads to queen larvae failing to pupate and acquiring a sac-like appearance. Although there is less evidence that *V. destructor* plays a role in SBV transmission, correlative studies have found SBV to be positively associated with *V. destructor*-infested honeybee colonies (Mondet et al., 2014). Although some studies have detected SBV in non-*Apis* bees, including bumblebees (Gamboa et al., 2015; McMahon et al., 2015; Singh et al., 2010) and the solitary bee *Andrena vaga* (Ravoet et al., 2014), it would appear that SBV may be more strictly associated with honeybees (Gisder and Genersch, 2017).

2.8 Chronic Bee Paralysis Virus

Chronic bee paralysis virus (CBPV) is an intriguing RNA virus because its genome consists of two separate positive-stranded RNA molecules. Its classification is unclear, although it shares characteristics with the plant virus family *Tombusviridae* and the *Nodaviridae* (Ribière et al., 2010). Much like SBV, CBPV can exert strong pathogenic effects, including trembling and clustering of crawling bees, although its frequency of detection is generally lower than several of the other common viruses discussed earlier. It is, though, apparently increasing in prevalence in US colonies (Traynor et al., 2016). Many of the symptoms associated with CBPV infection match the descriptions of "Isle of Wight Disease," which caused mass honeybee die-offs in Britain at the turn of the 20th century (Ribière et al., 2010). CBPV is thought to be able to infect non-*Apis* bees and other hymenopterans such as ants (Celle et al., 2008).

2.9 Other Viruses

In addition, a number of other viruses have recently been discovered in honeybees and some wild bees, including LSV, Big Sioux River virus (BSRV), aphid lethal paralysis virus (ALPV) (Runckel et al., 2011), tobacco ringspot virus (TRSV) (Li et al., 2014), and a novel "Halictivirus," Halictus scabiosae Adlikon virus (HsAV) (Bigot et al., 2017). Their pathogenicity across different bee species, including at the colony level in *A. mellifera* and other eusocial species, remains targets of ongoing research.

2.10 Bee Virus Evolutionary Ecology

For multihost pathogens including many bee-associated viruses, virus ecology in any one host will depend on the transmission networks between hosts, which in turn are determined by the frequency of transmission events within and between species. A species' transmission potential is determined by its abundance, the prevalence of disease within the species, and by its infectivity, i.e., how many infective particles it sheds (Streicker et al., 2013). In bees, between-species infectivity will also depend on realized transmission opportunities, as between-species transmission will depend on direct or indirect contacts, e.g., via contaminated flowers or through activities such as drifting of workers between colonies or robbing of food reserves from one colony by members of another colony and even another species (Manley et al., 2015). The evolutionary trajectory of a virus will in turn be determined by the different adaptive landscapes each host represents, as well as by the frequency and length of time each host is encountered by a viral isolate. This can result both in the evolution of viral generalists and in the diversification of viral specialists, depending on the underlying epidemiology and genetic differences in the hosts (see Rigaud et al., 2010 for a review of these ideas).

Viral ecology and evolution are in fact inseparable. RNA viruses are among the fastest evolving entities, with evolutionary rates of 10^{-3}–10^{-5} substitutions per site per year (Holmes, 2008); DWV, for example, has a mean evolutionary rate of 1.35×10^{-3} per site per year, with 10–15 substitutions expected per genome per year (Mordecai et al., 2015). For these viruses—in contrast to their eukaryotic hosts—the ecological and evolutionary timescales thus overlap, making virus evolution integral to any study of virus ecology. This rapid evolution allows RNA viruses to adapt quickly to novel host environments, having the highest risk factor for disease emergence among all classes of pathogen (Dobson and Foufopoulos, 2001; Taylor et al., 2001). Another key aspect to viral evolution is recombination or reassortment. These processes, exemplified by influenza viruses (Tumpey et al., 2005), but also common in nonsegmented viruses (Eden et al., 2013), can enable viruses to dramatically change their epidemiology and host range, being frequently associated with disease emergence such as in the recent case of H1N1 Swine flu (Dawood et al., 2009). In bee viruses, recombination between genotypes of DWV has been shown to be a common occurrence (McMahon et al., 2016; Moore et al., 2011; Wang et al., 2013; Zioni et al., 2011), with recombinants potentially under selection by transmission via the ectoparasitic viral vector *V. destructor* (Ryabov et al., 2014). Studying genome evolution is thus integral to understanding viral ecology in bees and the potential of these diseases to emerge in novel host species.

3. A UNIFIED MOLECULE-TO-ECOLOGY APPROACH TO BEE VIRUS RESEARCH

As we introduced above, evolution plays a fundamental role in virus ecology. Evolution is governed by conditions in the proximate environment of any organism, which in the case of a pathogen is the host. Both host and pathogen evolution are driven by environmental factors that influence population-scale processes, but the temporal and spatial scales at which hosts and pathogens operate can differ by several orders of magnitude. For example, RNA virus populations are defined by step-wise infection processes mediated by immune molecules and cell-level interactions inside individual hosts. In contrast, host population dynamics play out over much greater temporal and spatial scales. In the past these multiple spheres of host–pathogen activity have often been studied separately. Mechanistic interactions between pathogens and hosts have fallen largely in the domains of molecular and cell biology, while evolutionary focus has been more concerned with host-level outcomes and epidemiology. But as has recently been argued, a unification of these approaches can be beneficial for explaining variation in infectious disease traits between hosts (Hall et al., 2017). We argue, as others have recently (Hedrick, 2017), that such a unified framework is necessary to understand and ultimately predict the emergence, severity, and spread of infectious diseases.

For example, ecological considerations in the host can influence pathogen molecular adaptation because host immune response has been shaped by evolutionary events of the past, such as historical population exposure to disease. Such events can vary considerably between closely related host species, or even between populations of the same species (Black, 1992). Such variation can influence the ways in which hosts can respond to pathogens, in turn leading to significantly altered infection outcomes (Hedrick, 2011; Laayouni et al., 2014). At the most basic level, a host's immune response determines the likelihood and type of infection, which as discussed earlier can be species- or population-specific (Taylor et al., 2001; Woolhouse et al., 2005). Whether a host displays resistance or tolerance toward a novel pathogen can influence pathogen load and duration (acute vs chronic), infection success, and therefore onward transmission. Other direct consequences of infection and immune activation include factors such as host mortality, morbidity, and behavior (e.g., avoidance/fever), which can shape the probability and type of contact between infected and susceptible individuals. Furthermore, ecological considerations beyond the level of individual immunity can influence the dynamics of transmission, including host sociality, population demography (Faria et al., 2014), and density (Anderson and May, 1979); metapopulation structure; multispecies community composition (Johnson et al., 2015); biodiversity (Keesing et al., 2010); vector ecology (Randolph, 2001; Randolph and Rogers, 2000); and (human-driven) disturbance (Cable et al., 2017; Lafferty, 2009; Rogers and Randolph, 2000).

All of these factors combine to form multiple scales of biological organization within which pathogens, including bee viruses, must operate and evolve (Fig. 1). There are a number of areas in bee virus research that can benefit from a unified approach, with the added potential for such a framework in bees to serve as a general model for research in virus ecology. Key questions from these areas include: (Fig. 1(1)) What are the molecular features of the virion that drive viral pathogenesis? Molecular features of virion tertiary structure such as the VP3 domain of the DWV virion (indicated with an arrow) can play a role in directing cell entry and pathogenesis; (Fig. 1(2)) How do structural components mediate cell entry, and how do viral

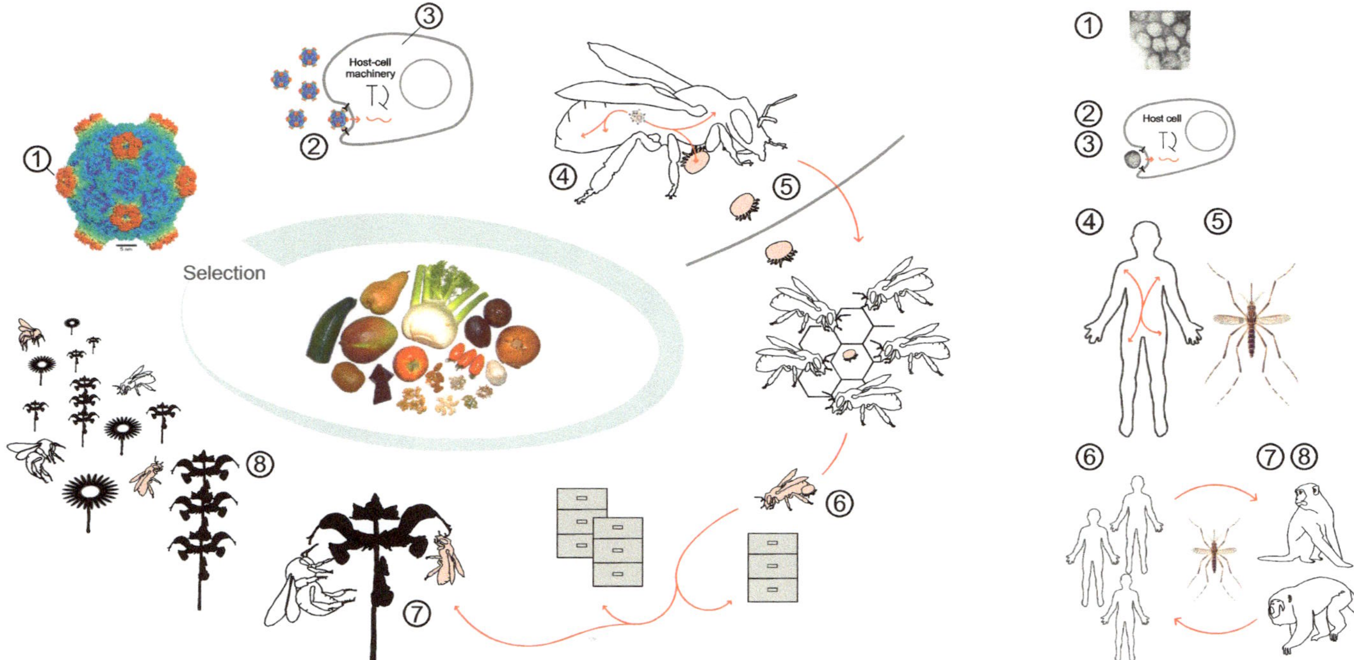

Fig. 1 *Left*: A molecule-to-ecology framework for bee–virus research. The framework integrates multiple levels of biological organization, demonstrated here by a series of levels (1–8). The scheme is intended to stimulate cross-disciplinary approaches to the study of bee viruses. The *gray ellipse* represents the strength of direct selection acting on viruses, which is reduced at higher levels of biological organization. Readers are referred to the main text (Section 3) for a detailed description of each level. Popular foods that depend on pollinators for successful fruit set are pictured in the center of the figure. *Right*: Equivalent steps in the molecule-to-ecology framework for human RNA viral diseases, such as the arboviruses Dengue virus (DENV) and Yellow fever virus (YFV) (family *Flaviviridae*). The viruses are transmitted by the mosquito *Aedes aegypti* both within and between human and monkey populations such as New World howler monkeys (YFV) and Old World rhesus macaques (DENV). *DWV image adapted from Škubník, K., Nováček, J., Füzik, T., Přidal, A., Paxton, R.J., Plevka, P., 2017. Structure of deformed wing virus, a major honey bee pathogen. Proc. Natl. Acad. Sci. U.S.A. 114, 3210–3215 under a CC BY-NC-ND-4.0 licence, (Publisher: National Academy of Sciences).*

receptors in host cells vary with host taxonomy and evolutionary history to dictate host susceptibility and virulence? (Table 1); (Fig. 1(3)) How does a virus suppress host antiviral defense and subvert host-cell machinery to its own benefit, and how important is population- and species-level variation in host immunity in mediating infections and virulence?; (Fig. 1(4)) How do viruses spread inside a host and are infections associated with chronic asymptomatic infections or severe acute infections? How are these infection outcomes related to mechanisms at steps 1–3? Furthermore, when two or more viral genotypes coinfect a host cell, does recombination lead to increased evolutionary scope?; (Fig. 1(5)) Does the *Varroa* mite not only vector virus but also act as an intermediary host, imposing its own selection on a virus? Furthermore, does the vector impact virus adaptation by allowing hemolymph-to-hemolymph indirect transmission. What impact does such adaptation in one host have on spillover and virulence in other hosts?; (Fig. 1(6)) How readily is virus transmitted between honeybee colonies and (Fig. 1(7)) between host bee species at flowers or via other routes (e.g., robbing)?; (Fig. 1(8)) How do pollinator-flower networks influence virus transmission, and what impact do emerging viruses have on stability of host populations, and the ecosystem service of pollination provision? Integrating across these questions will allow a far deeper understanding of the ecological and evolutionary forces driving viral emergence, with the prospect of developing models that can help to predict viral emergence under future socioeconomic and societal scenarios. In the next sections, we describe interactions between bees and their viruses at different levels within the molecule-to-ecology framework.

3.1 Host–Virus Molecular Mechanisms

At the most basic level, we do not know the mechanistic basis of cell entry and infection for any bee–virus system. Yet this knowledge is crucial for understanding the type and extent of molecular compatibility between a virus and a novel immune system or transmission route. This is relevant for understanding possible levels of maladaptation between novel host–virus combinations, and the ensuing impact on virus adaptation to a host's immune system, virulence evolution and transmission.

Only recently have the fine-detailed structures of some bee viruses been elucidated, with SBPV (Kalynych et al., 2016) and more recently DWV (Škubník et al., 2017) now characterized to less than 4 Å. Both of these iflaviruses contain similar surface topologies, with capsids containing protrusions of the so-called P-domain, which is derived from the VP3 protein subunit. While the P-domain differs in sequence identity between SBPV and DWV, both viruses contain eight highly conserved residues. They are shared with other iflaviruses and potentially involved in receptor-binding or catalytic disruption of the host-cell membrane, enabling the RNA genome to be transported into the cell cytoplasm (Škubník et al., 2017). Further evidence of involvement of the VP3 protein is suggested by interactions between the RNA genome and residues from the VP3 subunit. Unfortunately, the mechanistic basis of cell entry for iflaviruses—including DWV—is currently unknown. This represents a key area for future research, particularly for improving understanding of the potential similarities in host-mediated viral entry mechanisms between bee species. If mammalian picornaviruses are anything to go by (McKnight and Lemon, 2017), it is likely that at least some aspects of host cell entry mechanisms are conserved across different iflaviruses and bee-host species. One obvious first point of difference to explore between mammalian–picornavirus and invertebrate–iflavirus interactions is the key role for the RNAi machinery in bee antiviral defense, as has been extensively demonstrated in flies (Ding and Voinnet, 2007). However, a role for canonical immune pathways in antiviral defense has also been highlighted in insects (reviewed in Merkling and van Rij, 2013), with NF-κB signaling in the Toll pathway being implicated in honeybee antiviral defense against DWV (Di Prisco et al., 2013).

Although these represent relevant targets for forthcoming studies in bee–virus interactions, a robust approach requires exploration of the repertoire of mechanisms found at sequential steps in the infection process of the host (Ebert et al., 2016): from external membrane interactions at the gut or cuticle (which critically, are irrelevant in the case of *Varroa*-vectored viruses delivered directly into hemolymph) to host-cell transcriptional manipulation mechanisms, and eventually packaging and transport of virions to sites of transmission. A complete overview also requires characterization of potential points of difference between species, and identification of forms of selection pressure that may vary across stages of infection and transmission routes. Such processes can shape pathogen adaptation in various ways (Schmid-Hempel and Ebert, 2003). Depending on the mechanism involved, they could have potential repercussions for relevant disease traits, such as specificity and pathogenesis in novel hosts. Although it is hard to make general predictions, a possibility is that virus adaptation to vector-mediated transmission in honeybees may result in highly specific modifications to a small subset of host–virus mechanisms relating to virus replication.

From a mechanistic point of view, it is difficult to hypothesize what the consequences of vector-adaptation might be for wild bees exposed to viruses via the gut. Host specialization could lead to increased virulence in nonadapted hosts (Ebert, 1998), but it is also conceivable that relaxed selection on gut membrane and milieu associations leads to reduced infectiousness of a virus.

3.2 Between-Host Transmission and Impact of Novel Vectors

Transmission opportunities and routes play a key role in epidemiology. Disease emergence is generally facilitated by a change in host–parasite ecology (Woolhouse, 2002) that leads to novel transmission opportunities between species. Changes in host species range or biological niche, for example, through invasions or through global change, can bring potential host species into contact. In the case of bee diseases, globalization and the transport of nonnative bee species have led to the spread and emergence of disease. For example, the use of nonnative commercially produced bumble bees has led to the spread of nonnative infectious diseases or variants in South America (Maharramov et al., 2013; Schmid-Hempel et al., 2014) and Japan (Goka et al., 2006). The global transportation of honeybees has not only led to the spread of infectious diseases such as *Nosema ceranae* (Klee et al., 2007) but also to the emergence of the ectoparasitic mite *V. destructor* (Oldroyd, 1999) and the emergence of DWV as a global pandemic (Wilfert et al., 2016).

From human medicine, agriculture, and wildlife biology, it is clear that the emergence of novel transmission routes can have drastic effects on epidemiology, even though our understanding of the underlying theory in this area is not yet clearly developed. Most extant cases of novel transmission routes can be traced to human activities and have led to disease emergence or reemergence, beginning with diseases acquired during livestock domestication such as measles (Pearce-Duvet, 2006). For oral-fecal transmitted diseases, the introduction of novel food sources, or the contamination of existing ones can lead to the establishment of a novel transmission route as exemplified by the variant form of Creutzfeldt–Jakob disease in humans which arose as a result of consuming prion-contaminated beef (Collinge et al., 1996). In bees a parallel can be drawn with flowers, which may act as important points of interspecific transmission. Indirect vectors are also key transmission routes for emerging pathogens. For example, the use of unsterilized needles in health clinics is thought to have been a major contributor to the early spread of HIV (Faria et al., 2014). But biotic vectors have played additional major roles in disease emergence. Avian malaria in

Hawaii increased significantly following the introduction of malaria-carrying *Culex quinquefasciatus* mosquitoes (Woodworth et al., 2005), while the global expansion of rats and their *Yersinia pestis* carrying fleas (*Xenopsylla cheopis*)—or even human lice and fleas—could have led to the Black Death in Europe in the 14th century (Dean et al., 2018; Stenseth et al., 2008).

In the case of the honeybee and *V. destructor*, the situation is slightly different because viruses such as DWV were already present in bees prior to the arrival of the mite vector (Forsgren et al., 2012). As discussed in the introduction, this hemolymph-feeder has led to the emergence of a novel transmission route for bee viruses, leading to a dramatic increase in titer and prevalence of DWV-linked disease (de Miranda and Genersch, 2010; Martin, 2001; Martin et al., 2012; McMahon et al., 2016; Mondet et al., 2014; Wilfert et al., 2016). *V. destructor* was originally an ectoparasite of the honeybee *A. cerana*, native to south and east Asia, where it causes little damage to its host (Nazzi and Le Conte, 2016; Rosenkranz et al., 2010). Within the last century, *A. mellifera* colonies were brought from Europe to E. Asia, sympatric with *A. cerana*, which allowed *V. destructor* to successfully switch host to *A. mellifera* (Nazzi and Le Conte, 2016; Rosenkranz et al., 2010). Mite-infested colonies of *A. mellifera* were unwittingly returned to Europe, whereupon *V. destructor* dispersed into neighboring colonies and the mite was subsequently transported with *A. mellifera* colonies more or less worldwide. As a consequence, mites are nowadays found on all major land masses harboring *A. mellifera* except Australia, which maintains strict quarantine regulations. Mites feed on tissue of *A. mellifera* adults and pupae, from which they can acquire numerous host viruses that they can transmit with high efficiency to a subsequent host (Bowen-Walker et al., 1999; Gisder et al., 2009, Di Prisco et al., 2011), though they are only able to successfully reproduce on *A. mellifera* pupae.

An obvious question that arises and can be experimentally addressed in bees is: what impact did the introduction of indirect transmission have on virus evolution? Due to the nature of the *V. destructor* life cycle in the honeybee host, predictions can be made about the potential impact of the mite on virus virulence evolution (Fig. 2). Specifically, the developing honeybee host pupa should remain alive until close to the completion of metamorphosis to provide sufficient time for successful mite reproduction, including offspring mating. For optimal transmission, any virus found in a mature and mated daughter mite will hold a significant selective advantage over a virus found in an immature or unmated daughter mite—placing a cost on virus virulence that impacts honeybee pupae before mites can mate (Martin, 2001).

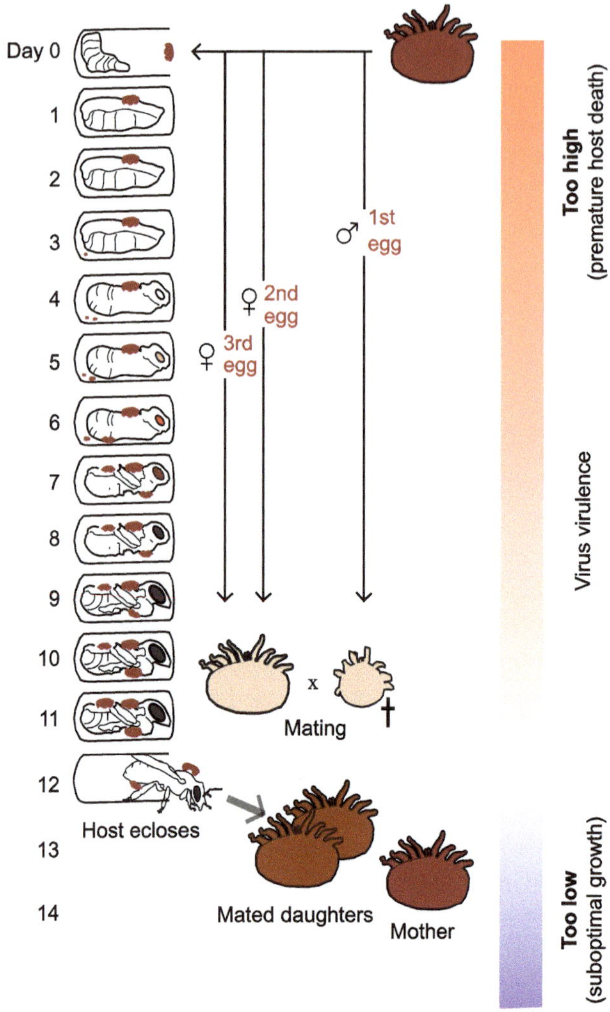

Fig. 2 Hypothetical impact of indirect vector transmission on virus virulence evolution in *A. mellifera*. The *V. destructor* reproductive life cycle is shown, with mites depicted at various developmental stages. Corresponding stages of honey bee pupal developmental (in days post-capping) are shown to the left of the figure. A mother mite lays maximally three eggs that can mature to adulthood before hosts emerge from the brood cell. Virulence should evolve to an intermediate level that maximizes transmission (*white region*). Too virulent (*red region*) and the virus will have suboptimal transmission due to high-host mortality preventing mite mating and host emergence (eclosion). Not virulent enough (*blue region*) and the virus will have suboptimal transmission due to low growth (fewer transmission units).

On the other side, viruses replicating too slowly and with delayed virulence effects will hold a selective handicap because fewer transmission units will be found in mated mites. We hypothesize that the evolution of virus virulence shifted following the arrival of *V. destructor*, with viruses such as DWV evolving a level of virulence in pupae (and likely also in adults) that optimizes the number of transmission units passed to uninfected susceptible vectors, and ultimately hosts. Addressing this issue in bees could provide key insight into how anthropogenic disturbance can drive pathogen adaptation in unpredictable ways, leading to unintended cascade effects across ecological communities.

3.3 Multihost–Multipathogen Community Ecology

A second interesting question that arises from the observation that viruses were already present in bees prior to the arrival of the mite vector, is: what impact did this new transmission route in one species have on other bee species? Addressing this question requires basic knowledge about host–virus dynamics in bee assemblages. Unfortunately, even knowledge of the species that are primary or natural hosts for the many so-called bee viruses is lacking, and this is partly attributable to a general lack of research in non-*Apis* species (Brown and Paxton, 2009; McMahon et al., 2015; Tehel et al., 2016). What can be stated with some degree of confidence is that the increased abundance and prevalence of vectored viruses in honeybees has resulted in increased virus transmission opportunities between bee species.

Which bee species are more or less susceptible to infection is not known, but in other host–virus systems (e.g., bats), there is some correlation between phylogeny and viral host range, with closely related host species more likely to share viruses (Faria et al., 2013; Streicker et al., 2010). Such relationships cannot be drawn from the current bee–virus literature due to lack of data, and anyhow do not necessarily reveal the reservoir host. However, we do know that some bee viruses are widely distributed across host bee species (e.g., DWV and BQCV) and that others have a narrower host range (e.g., SBV) (Gisder and Genersch, 2017; Tehel et al., 2016). Although *Bombus* species—which are phylogenetically close to *Apis* (Table 1)—have often been found to harbor viruses that are also in honeybees, these patterns most likely reflect investigator effort rather than real biological pattern. Clearly, more information on the primary or natural hosts is needed for understanding the historical exposure and future risks posed by RNA viruses to different populations or species of bee.

Nonetheless, some information and putative predictions can be made. Bee assemblages are integrated into complex multihost flower-visitor networks. Due to their inherent complexity, the relationship between pollinator networks and virus transmission both within and among bee species is unknown. Laboratory and semifield studies using bee pathogens, some including viruses, have demonstrated that transmission via flowers can occur (Dürrer and Schmid-Hempel, 1994), that flower visiting can lead to the movement of transmission stages from flower-to-flower by vectoring flower visitors (Graystock et al., 2015), and that different floral structures may alter the likelihood of transmission (Dürrer and Schmid-Hempel, 1994; McArt et al., 2014). Real field-scale studies have also shown that flower-visitor networks have some explanatory power when mapped onto patterns of parasite abundance across host species (Ruiz-González et al., 2012). However, flower visitation is not the only route of transmission for bee viruses. Nectar robbing of honeybee hives by other honeybees, bumble bees, or wasps—all of which are common events—and social parasitism and nectar robbing in bumble bees (Goulson et al., 2017), among others, are likely to play some, as yet unquantified role in viral transmission. Finally, in the social bees, transmission within nests is likely to be a major, again relatively unquantified mode of transmission (but see Otterstatter and Thomson, 2007; Rutrecht and Brown, 2008). Thus, while quantifying transmission via flower-visitor networks is important, it is also a key to determine what proportion of the patterns in viral prevalence across host species it can actually predict.

Quantifying these networks, however, is itself hugely challenging. Networks have a significant temporal component, which has been largely ignored by current snapshot studies (e.g., Ruiz-González et al., 2012). This temporal component is a combination of seasonal population dynamics and demographics, with different potential host and vector species being present at different times of year, and at different abundances across the year, including changes in the species presence and abundance of flowers themselves. While, in temperate regions, we know a lot about these changes individually, as yet nobody has mapped them onto each other at an annual scale. This is a major challenge for bee–virus biologists. Similar patterns are doubtless present in the tropics, but as yet no studies of transmission have occurred in this biome. In addition to simple changes in population abundance, the presence of different sexes and castes of bee, and other flower visitors, also change across the year. If sexes or castes vary in their resistance to viruses, which may be expected in social species given that for most of the year worker bees are the major component of the population, this could also have significant impacts on transmission networks. Finally, bees and other flower visitors, like wasps, exist in social and nonsocial forms (Table 1). As noted earlier, due to transmission within colonies, this could have dramatic impacts on transmission patterns via flowers through mass action effects. If pathogen prevalence builds up in colonies across the year (e.g., Rutrecht and Brown, 2008), then the contribution of social insects to transmission networks will also increase, and most likely disproportionately to the contribution of other nonsocial flower visitors.

The quantification of directionality in floral transmission networks may enable the identification of reservoir hosts, which would play a central role in transmission. It is at this point that the realms of ecology and evolution overlap again for these RNA viruses. If reservoir hosts exist in bee–virus systems, then we should expect adaptation of the virus to this reservoir host, with concomitant effects on infectivity and virulence within and across species (reviewed by Rigaud et al., 2010). Such reservoirs may be particularly important in temperate systems, where host population sizes decrease dramatically over winter for the majority of bee species.

A final but important consideration for natural populations of bees is that other stressors such as energetic stress and biocides can act synergistically with pathogens to negatively affect both individual and colony-level fitness (Bryden et al., 2013; Di Prisco et al., 2013; Doublet et al., 2015; Manley et al., 2017; Potts et al., 2016). These should also be taken into consideration as potential ecosystem-level factors that can further shape virus transmission and evolutionary trajectories across pollinator networks.

4. EVOLUTIONARY AND ECOLOGICAL PRINCIPLES OF DISEASE EMERGENCE: RELEVANCE FOR BEE RESEARCH

In the final section of this review, we shift emphasis to a discussion of general principles in disease ecology and their relevance to emerging bee viruses. We focus on insights derived from well-known study systems as well as theoretical developments. We pay particular attention to three different but overlapping topics in the molecule-to-ecology framework: (i) molecular mechanisms as determinants of pathogen life history, (ii) virulence as a balance between within- and between-host selection, and (iii) ecosystem disturbance as a driver of emergence.

4.1 Host–Pathogen Molecular Interactions Determine Virus Life History Attributes

Obtaining molecular-level mechanistic details of how pathogens interact with their hosts is essential for understanding key life history parameters of infection such as pathogen survival, which is achieved by evading the host's immune system (Schmid-Hempel, 2009), and pathogen growth, which is achieved by extracting resources from the host for replication. Properties such as host immune evasion and replication are often associated with damage to the host (pathogenesis). For example, HCV cell entry exploits a single phospholipid-modifying enzyme from the host, PLA2G16, which is recruited to the site of viral capsid attachment. This host-exploiting mechanism is used by diverse picornaviruses to enable genome delivery into mammalian host cell (Staring et al., 2017). Remarkably, the 2A protein of some picornaviruses is even derived from homologues of *PLA2G16* (Hughes and Stanway, 2000), suggesting these viruses have coopted host proteins into their genomes to accelerate host cell entry. Immune evasion mechanisms may also be passive, as in the case of HIV, where an extremely high mutation rate is sufficient to keep ahead of the human adaptive immune system (Rambaut et al., 2004). Or viruses may passively evade host immunity by entering a phase of latency, which could evolve as a successful evolutionary strategy under stable host population conditions (Sorrell et al., 2009). Virus latency is common in alphaherpesviruses such as Marek's disease virus (MDV) of poultry, which typically cause mild, chronic infections. Interestingly, human-driven disturbance in the form of widespread vaccination and agricultural intensification is thought to have driven the emergence of highly virulent viruses and a reduced role for latency in the infection cycle of the virus (Osterrieder et al., 2006; Trimpert et al., 2017).

Unlike vertebrates, insects do not have an adaptive immune system, outwardly making host immune interactions a comparatively simpler process to understand. Despite this, the precise mechanisms involved in host immune evasion remain poorly characterized in insect viruses, although some progress has been made in *Drosophila*. Here, evidence points to the exploitation of clathrin-mediated phagocytosis during cell entry in *Drosophila C virus* (DCV) (Bonning and Miller, 2010). DCV (family *Dicistroviridae*) is a relative of the widespread bee viruses BQCV and ABPV. In such viruses, translation is mediated by two IRESs (internal ribosomal entry sites) as opposed to one in other Picornavirales (such as DWV). Dicistrovirid IRESs fool the host ribosome by binding the 40S and 60S ribosomal subunits where the tRNAs and mRNA would normally be found (Pestova et al., 2004).

To counter this, insects including bees (Brutscher and Flenniken, 2015) employ RNAi as a key antiviral defense mechanism, which targets viral RNAs for intracellular destruction (in contrast to mammals; Bogerd et al., 2014; Seo et al., 2013; although see Li et al., 2016). Unsurprisingly, insect viruses have in turn evolved silencers of host RNAi. In DCV, a suppressor protein binds long dsRNAs, which prevents their destruction by inhibiting the enzyme dicer from processing dsRNAs into siRNAs. Interestingly, micro-RNAs (miRNAs) are not manipulated by DCV. This is interesting from a pathogen life-history perspective because miRNAs are required for normal insect development. The restricted targeting of dsRNA by DCV suggests that insect viruses have evolved sophisticated mechanisms to facilitate long-lasting persistent infections (van Rij et al., 2006). How the arrival of the *V. destructor* mite may have affected these carefully orchestrated host-manipulation mechanisms in bees (if present) represents an interesting open question, because a persistent infection strategy that is advantageous under stable host population conditions may suddenly become disadvantageous in the presence of a biotic vector. A further point here is that *V. destructor* mites vector virus directly into the host hemolymph during host feeding, thus side-stepping the ventricular and lower gut barrier, where environmental (e.g., gut microbiota composition), physical (e.g., cell lumen) as well as innate immune defenses can combine to impose major obstacles to infection (Engel et al., 2016; Kwong and Moran, 2016).

4.2 Virulence as a Balance Between Within- and Between-Host Selection

In animal systems, virulence is typically defined as the negative impact of a pathogen on host fitness. In natural coadapted host–pathogen systems, a well-established body of theory and empirical research places transmission between hosts as a key consideration in virulence evolution (Schmid-Hempel, 2011). Virulence is hypothesized to evolve to a level that maximizes R_0, the basic reproductive number of a pathogen, which must be above 1 for an infection to spread in a host population. However, selection at the within-host cellular level can shape and even be in conflict with selection for between-host transmission (Alizon et al., 2011; Mideo et al., 2008). Clearly, both levels are essential: while a pathogen must be able to infect a novel host, its fitness is zero unless it can also successfully invade other hosts. While the validity of the trade-off framework is generally supported by theory and experimental research (reviewed in Cressler et al., 2016), it is questionable whether a strict two-dimensional trade-off between transmission and virulence is realistic in most cases (Ebert and Bull, 2003), for example, in pathogens entering novel host populations (Weiss, 2002). As discussed in the previous section, a focus on the mechanism of pathogenesis, which is inherently system-specific, can provide significant benefits to the study of virulence evolution (Frank and Schmid-Hempel, 2008). In particular, the link between pathogenesis, the underlying process by which virulence is generated, and its role in host immune evasion needs to be known so as to predict the evolutionary trajectory of virulence—up, down, or stationary (Schmid-Hempel, 2009). And because a coevolved link between virulence and transmission is absent in emerging diseases, understanding the mechanisms that enable pathogens to survive and grow in potentially novel hosts (e.g., non-*Apis* bees) or host conditions (e.g., *V. destructor*-vectored conditions) is critical to explaining virus virulence. We have hypothesized that significant shifts in selection accompanied the arrival of the *V. destructor* mite (Fig. 2), potentially resulting in conditions that favored the emergence of differentially virulent viruses (McMahon et al., 2016). But the effect on other bee species of honeybee/*V. destructor*-driven changes in virus adaptation and prevalence remains unclear. More certain is the effect of increased prevalence of *V. destructor*-vectored viruses, which may enhance the frequency of encounters between viruses and a higher diversity of novel insect hosts, making epidemics in novel species potentially more likely.

In such cases, virulence is an incidental property resulting from (i) mechanistic and infection cycle similarities between the original and novel host environment and (ii) direct selection on pathogens to adapt to the immune system of the novel host. The strength of indirect selection at the population-level will initially be minor. This arises because host–pathogen interactions are sequential: selection to avoid host clearance precedes selection for optimal transmission to an uninfected host. Indeed, in the original formulation of the trade-off model, it is only indirect selection that results from between-host competition between pathogens that determines optimal population-level virulence (Anderson and May, 1982; Ewald, 1983).

As a possible example, in the recent West African Ebola epidemic, a single mutation originating in an ancestral node of the Ebola lineage (A82V in the glycoprotein) could have been the major driver of the vast majority of human infections during the 3-year (2013–16) epidemic. The mutation is associated with increased virus load (and virulence) in humans but not rodents or canines (Diehl et al., 2016). Furthermore, the A82V mutation—combined with subsequent mutations conferring primate-specific infectivity—carries the cost of reduced infectivity in cells from bats (Urbanowicz et al., 2016), which are the presumed original hosts of the 2013–16 human outbreak. Although it is not clear whether between- or within-host selection was the dominant driver of Ebola emergence, the fact that the predominant mutation emerged very early on in the epidemic shortly after the putative host-switch event implicates adaptation toward the novel human immune system as the primary cause of enhanced Ebola virulence (Bedford and Malik, 2016). Similar patterns of molecular adaptation have also been observed in the early epidemic dynamics of influenza (Chen et al., 2006).

A related component of this is whether host phylogenetic relatedness, and the assumed similarity between immune systems, could be a useful predictor of pathogen spread or virulence (Faria et al., 2013; Streicker et al., 2010). A comparative analysis of pathogens that have jumped species and even kingdom barriers shows that pathogenicity toward novel hosts may be based on traits that evolved to ensure survival in the original host, and/or are associated with the exploitation of highly conserved host defense molecules such as antimicrobial peptides (van Baarlen et al., 2007). Although studies in this area are lacking in bees, there is some evidence in *Drosophila* that while original-to-new host relatedness per se may not be correlated with virulence, host phylogeny does play a role in explaining variation in virulence (Longdon et al., 2015).

It is worth noting that novel host–pathogen interactions can be highly maladaptive, both for the host and pathogen, with hosts mounting immune responses that inflict major self-damage (Graham et al., 2005), or viruses undergoing short-sighted within host adaptation that inadvertently kills the host and prevents transmission (Levin and Bull, 1994). Clearly, outcomes are likely to be highly host–pathogen specific, reflecting complex interactions between molecular and evolutionary ecology (Schmid-Hempel and Ebert, 2003). Resolving molecular mechanisms for a wide range of host–virus systems and alternative transmission routes will help to generate predictions about how novel hosts and viruses play out in nature.

4.3 Human-Driven Ecological Disturbances as Drivers of Disease Emergence

Changes in the dynamics and ecology of host–parasite systems play important roles in the emergence of parasites, pathogens, and disease. While host–parasite systems are always changing in response to natural variation in weather, population dynamics, distribution, and other such factors, human impacts have led to much more rapid state changes in the epidemiology of disease agents. Major anthropogenic drivers of emergence include the global movement of crops and domesticated animals, invasive species, changes in landscape structure, the global use of agrochemicals (Mitchell et al., 2017), and climate change (Lafferty, 2009), which will lead to dramatic shifts in the distributional ranges of hosts and their parasites (Estrada-Peña et al., 2014). One of the earliest examples of such shifts was the inadvertent introduction of rinderpest virus to Africa from Asia, caused by the transfer of infected Asian cattle to Ethiopia. Rinderpest rapidly spread through direct transmission across sub-Saharan Africa, devastating African cattle populations as well as wild populations of buffalo, giraffe, and wildebeest (Normile, 2008). In bees, the transport of nonnative bee, including non-*Apis* bees, has led to the spread and emergence of disease. Examples include the use of nonnative commercially produced bumble bees in South America (Maharramov et al., 2013; Schmid-Hempel et al., 2014), Japan (Goka et al., 2006), and New Zealand (Dafni et al., 2010).

Such anthropogenically driven changes in the dynamics of directly transmitted viruses are only part of the picture. In fact, the dynamics of vector-transmitted viruses are both more complex and less easy to manage and predict. For example, environmentally driven shifts in the ranges of vectors such as the Asian tiger mosquito may enable the emergence of viruses in naïve host populations (Taber et al., 2017), while the rapid spread of the invasive mite *V. destructor* through its novel host *A. mellifera* has had dramatic impacts on the viral landscape within this host (Martin et al., 2012; Mondet et al., 2014; Wilfert et al., 2016). Vectors that share multiple host species can also be dramatically affected by anthropogenically driven changes, leading to host-switching opportunities in the viruses they transmit. Changes in host assemblages driven by invasive host species and habitat fragmentation can enhance encounter rates between vectors and primary hosts, leading to higher prevalence and transmission of the vectored pathogen (Allan et al., 2003; Hoyer et al., 2017). All of these changes in dynamics may have direct impacts on viral evolution because patterns of transmission and host encounter act as agents of natural selection on viral functional traits.

While our understanding of these selective pressures and their effects is in its infancy, we can make some broad predictions based on fundamental evolutionary principles. Shifts in host or vector range that disrupt current host–viral systems should select for more generalism in viruses with respect to host infection and exploitation strategies. In contrast, where landscape changes strengthen current vector–host interactions, higher viral specificity may result. The higher viral prevalence that may arise from both of these drivers is likely to produce higher genetic variability in the viral population, providing the substance on which further selection can act. However, how such trends might affect traits such as viral virulence is unclear and may depend upon the idiosyncratic details of particular host–vector–virus systems. It is for this reason that we believe major research effort should be focused on exploring the direct and indirect impacts of *V. destructor,* as discussed throughout this review. In the near term, we look forward to rapid advances in this area, which will enhance not only our understanding of bee–virus systems but also promise to shed light on viral evolutionary and ecological dynamics in general.

5. OPEN CHALLENGES AND CONCLUDING REMARKS

Two features currently hamper the study of the bee–virus system. First, the field lacks a bee cell culture in which to replicate bee viruses. Though infective clones (e.g., of DWV; see Lamp et al., 2016) may go some way to resolve the shortfall, the field would benefit from a system of in vitro viral cultivation. Second, though bees can be readily collected from flowers in the field and their individual movements even tracked in real time (Wolf et al., 2016), we have little idea of their long-distance natal dispersal (either before or after hibernation), making it difficult to evaluate the potential movement of them and the viruses they harbor over two or more host generations (Wilfert et al., 2016), although the genetic tools to measure bumblebee family lineages across landscapes over multiple seasons are now well established (Carvell et al., 2017). Combined with the advances in landscape genomics (Lozier and Zayed, 2017) and phylodynamics (Grenfell et al., 2004), these developments may help by allowing finer inference of virus dispersal and host-switching events. Given the increase in attention now being paid to bees and their pathogens, we are confident that many of these issues can be resolved. One prediction we can already make: research into the bee–virus system will surge in the coming years, and will bring with it a much richer and deeper understand of virus ecology and evolution, with relevance not only for the health and wellbeing of bees but of other host–virus systems, including *Homo sapiens.*

ACKNOWLEDGMENTS

We thank C.M. Malmstrom for inviting us to contribute a bee virus chapter to this *Advances in Virus Research* volume on Environmental Virology and Virus Ecology. R.J.P. acknowledges support from the DFG (PA632/10). M.J.F.B. and L.W. were supported by the BBSRC [BB/N000625/1].

REFERENCES

Aizen, M.A., Garibaldi, L.A., Cunningham, S.A., Klein, A.M., 2009. How much does agriculture depend upon pollinators? Lessons from long-term trends in crop production. Ann. Bot. 103, 1579–1588.

Aizen, M.A., Biesmeijer, J., Martins, D.J., Goka, K., Inouye, D.W., Jung, C., Medel, R., Pauw, A., Paxton, R.J., Seymour, C.L., 2016. The status and trends in pollinators, their functions and services. In: The Assessment Report of the Intergovernmental Science-Policy Platform on Biodiversity and Ecosystem Services (IPBES) on Pollinators, Pollination and Food Production. IPBES, Bonn, Germany.

Alizon, S., Luciani, F., Regoes, R.R., 2011. Epidemiological and clinical consequences of within-host evolution. Trends Microbiol. 19 (1), 24–32.

Allan, B.F., Keesing, F., Ostfeld, R.S., 2003. Effect of forest fragmentation on lyme disease risk. Conserv. Biol. 17, 267–272.

Anderson, R.M., May, R.M., 1979. Population biology of infectious diseases: part I. Nature 280, 361–367.

Anderson, R.M., May, R.M., 1982. Coevolution of hosts and parasites. Parasitology 85, 411–426.

Bailey, L., Fernando, E.F., 1972. Effects of sacbrood virus on adult honey bees. Ann. Appl. Biol. 72, 27–35.

Bailey, L., Gibbs, A.J., 1964. Acute infection of bees with paralysis virus. J. Insect Pathol. 6, 395–407.

Bailey, L., Woods, R.D., 1974. Three previously undescribed viruses from the honey bee. J. Gen. Virol. 25, 175–186.

Bailey, L., Woods, R.D., 1977. Two more small RNA viruses from honey bees and further observations on sacbrood and acute bee paralysis viruses. J. Gen. Virol. 37, 175–182.

Bailey, L., Gibbs, A.J., Woods, R.D., 1963. Two viruses from adult honey bees (Apis mellifera Linnaeus). Virology 21, 390–395.

Ball, B.V., Allen, M.F., 1988. The prevalence of pathogens in honey bee (*Apis mellifera*) colonies infested with the parasitic mite *Varroa jacobsoni*. Ann. Appl. Biol. 113, 237–244.

Bartomeus, I., Ascher, J.S., Gibbs, J., Danforth, B.N., Wagner, D.L., Hedtke, S.M., Winfree, R., 2013. Historical changes in northeastern US bee pollinators related to shared ecological traits. Proc. Natl. Acad. Sci. U.S.A. 110, 4656–4660.

Bedford, T., Malik, H.S., 2016. Did a single amino acid change make Ebola virus more virulent? Cell 167, 892–894.

Biesmeijer, J.C., Roberts, S.P.M., Reemer, M., Ohlemüller, R., Edwards, M., Peeters, T., Schaffers, A.P., Potts, S.G., Kleukers, R., Thomas, C.D., Settele, J., Kunin, W.E., 2006. Parallel declines in pollinators and insect-pollinated plants in Britain and the Netherlands. Science 313, 351–354.

Bigot, D., Dalmon, A., Roy, B., Hou, C., Germain, M., Romary, M., Deng, S., Diao, Q., Weinert, L.A., Cook, J.M., Herniou, E.A., Gayral, P., 2017. The discovery of halictivirus resolves the sinaivirus phylogeny. J. Gen. Virol. 98, 2864–2875.

Black, F.L., 1992. Why did they die? Science 258, 1739–1740.

Bogerd, H.P., Skalsky, R.L., Kennedy, E.M., Furuse, Y., Whisnant, A.W., Flores, O., et al., 2014. Replication of many human viruses is refractory to inhibition by endogenous cellular microRNAs. J. Virol. 88, 8065–8076.

Bonning, B.C., Miller, W.A., 2010. Dicistroviruses. Annu. Rev. Entomol. 55, 129–150.

Bowen-Walker, P.L., Martin, S.J., Gunn, A., 1999. The transmission of deformed wing virus between honeybees (Apis mellifera L.) by the ectoparasitic mite *Varroa jacobsoni* Oud. J. Invertebr. Pathol. 73, 101–106.

Brown, M.J.F., Paxton, R.J., 2009. The conservation of bees: a global perspective. Apidologie 40, 410–416.

Brutscher, L.M., Flenniken, M.L., 2015. RNAi and antiviral defense in the honey bee. J Immunol Res 2015, 941897.
Bryden, J., Gill, R.J., Mitton, R.A., Raine, N.E., Jansen, V.A., 2013. Chronic sublethal stress causes bee colony failure. Ecol. Lett. 16, 1463–1469.
Budge, G.E., Pietravalle, S., Brown, M., Laurenson, L., Jones, B., Tomkies, V., Delaplane, K.S., 2015. Pathogens as predictors of honey bee colony strength in England and Wales. PLoS One 10, e0133228.
Cable, J., Barber, I., Boag, B., Ellison, A.R., Morgan, E.R., Murray, K., et al., 2017. Global change, parasite transmission and disease control: lessons from ecology. Philos. Trans. R. Soc. Lond. Ser. B Biol. Sci. 372, 20160088.
Cameron, S.A., Lozier, J.D., Strange, J.P., Koch, J.B., Cordes, N., Solter, L.F., Griswold, T.L., 2011. Patterns of widespread decline in North American bumble bees. Proc. Natl. Acad. Sci. U.S.A. 108, 662–667.
Carreck, N.L., Ball, B.V., Martin, S.J., 2010. Honey bee colony collapse and changes in viral prevalence associated with *Varroa destructor*. J. Apic. Res. 49, 93–94.
Carvell, C., Bourke, A.F., Dreier, S., Freeman, S.N., Hulmes, S., Jordan, W.C., et al., 2017. Bumblebee family lineage survival is enhanced in high-quality landscapes. Nature 543, 547–549.
Celle, O., Blanchard, P., Olivier, V., Schurr, F., Cougoule, N., Faucon, J.P., Ribière, M., 2008. Detection of chronic bee paralysis virus (CBPV) genome and its replicative RNA form in various hosts and possible ways of spread. Virus Res. 133, 280–284.
Chen, Y., Siede, R., 2007. Honey bee viruses. Adv. Virus Res. 70, 33–80.
Chen, G.W., Chang, S.C., Mok, C.K., Lo, Y.L., Kung, Y.N., Huang, J.H., et al., 2006. Genomic signatures of human versus avian influenza A viruses. Emerg. Infect. Dis. 12, 1353.
Collinge, J., Sidle, K.C., Meads, J., Ironside, J., Hill, A.F., 1996. Molecular analysis of prion strain variation and the aetiology of 'new variant' CJD. Nature 383, 685–690.
Costanza, R., et al., 1997. The value of the world's ecosystem services and natural capital. Nature 387, 253–260.
Cressler, C.E., McLeod, D.V., Rozins, C., Van Den Hoogen, J., Day, T., 2016. The adaptive evolution of virulence: a review of theoretical predictions and empirical tests. Parasitology 143, 915–930.
Dafni, A., Kevan, P.G., Gross, C.L., Goka, K., 2010. Bombus terrestris, pollinator, invasive and pest: an assessment of problems associated with its widespread introductions for commercial purposes. Appl. Entomol. Zool. 45, 101–113.
Dainat, B., Evans, J.D., Chen, Y.P., Gauthier, L., Neumann, P., 2012. Dead or alive: deformed wing virus and *Varroa destructor* reduce the life span of winter honeybees. Appl. Environ. Microbiol. 78, 981–987.
Danforth, B.N., Cardinal, S., Praz, C., Almeida, E.A.B., Michez, D., 2013. The impact of molecular data on our understanding of bee phylogeny and evolution. Annu. Rev. Entomol. 58, 57–78.
Dawood, F.S., Jain, S., Finelli, L., Shaw, M.W., Lindstrom, S., Garten, R.J., Gubareva, L.V., Xu, X., Bridges, C.B., Uyeki, T.M., 2009. Emergence of a novel swine-origin influenza A (H1N1) virus in humans. N. Engl. J. Med. 360, 2605–2615.
de Miranda, J.R., Genersch, E., 2010. Deformed wing virus. J. Invertebr. Pathol. 103, S48–S61.
de Miranda, J.R., Drebot, M., Tyler, S., Shen, M., Cameron, C.E., Stoltz, D.B., Camazine, S.M., 2004. Complete nucleotide sequence of Kashmir bee virus and comparison with acute beeparalysis virus. J. Gen. Virol. 85, 2263–2270.
de Miranda, J.R., Cordoni, G., Budge, G., 2010a. The acute bee paralysis virus-Kashmir bee virus-Israeli acute paralysis virus complex. J. Invertebr. Pathol. 103, S30–S47.

de Miranda, J.R., Dainat, B., Locke, B., et al., 2010b. Genetic characterization of slow bee paralysis virus of the honeybee (*Apis mellifera* L.). J. Gen. Virol. 91, 2524–2530.

Dean, K.R., Krauer, F., Walløe, L., Lingjærde, O.C., Bramanti, B., Stenseth, N.C., Schmid, B.V., 2018. Human ectoparasites and the spread of plague in Europe during the second pandemic. Proc. Natl. Acad. Sci. U.S.A. https://doi.org/10.1073/pnas.1715640115.

Di Prisco, G., Pennacchio, F., Caprio, E., Boncristiani, H.F., Evans, J.D., Chen, Y., 2011. Varroa destructor is an effective vector of Israeli acute paralysis virus in the honeybee, *Apis mellifera*. J. Gen. Virol. 92, 151–155.

Di Prisco, G., Cavaliere, V., Annoscia, D., Varricchio, P., Caprio, E., Nazzi, F., et al., 2013. Neonicotinoid clothianidin adversely affects insect immunity and promotes replication of a viral pathogen in honey bees. Proc. Natl. Acad. Sci. U.S.A. 110 (46), 18466–18471.

Di Prisco, G., Annoscia, D., Margiotta, M., Ferrara, R., Varricchio, P., Zanni, V., Caprio, E., Nazzi, F., Pennacchio, F., 2016. A mutualistic symbiosis between a parasitic mite and a pathogenic virus undermines honey bee immunity and health. Proc. Natl. Acad. Sci. U.S.A. 113, 3203–3208.

Diehl, W.E., Lin, A.E., Grubaugh, N.D., Carvalho, L.M., Kim, K., Kyawe, P.P., et al., 2016. Ebola virus glycoprotein with increased infectivity dominated the 2013–2016 epidemic. Cell 167 (4), 1088–1098.

Ding, S.W., Voinnet, O., 2007. Antiviral immunity directed by small RNAs. Cell 130 (3), 413–426.

Dobson, A., Foufopoulos, J., 2001. Emerging infectious pathogens of wildlife. Philos. Trans. R. Soc. Lond. Ser. B Biol. Sci. 356, 1001–1012.

Doublet, V., Labarussias, M., de Miranda, J.R., Moritz, R.F.A., Paxton, R.J., 2015. Bees under stress: sublethal doses of a neonicotinoid pesticide and pathogens interact to elevate honey bee mortality across the life cycle. Environ. Microbiol. 17, 969–983.

Doublet, V., Poeschl, Y., Gogol-Döring, A., Alaux, C., Annoscia, D., Aurori, C., et al., 2017. Unity in defence: honeybee workers exhibit conserved molecular responses to diverse pathogens. BMC Genomics 18 (1), 207.

Dürrer, S., Schmid-Hempel, P., 1994. Shared use of flowers leads to horizontal pathogen transmission. Proc. R. Soc. B Biol. Sci. 258, 299–302.

Ebert, D., 1998. Experimental evolution of parasites. Science 282 (5393), 1432–1436.

Ebert, D., Bull, J.J., 2003. Challenging the trade-off model for the evolution of virulence: is virulence management feasible? Trends Microbiol. 11 (1), 15–20.

Ebert, D., Duneau, D., Hall, M.D., Luijckx, P., Andras, J.P., Du Pasquier, L., Ben-Ami, F., 2016. A population biology perspective on the stepwise infection process of the bacterial pathogen pasteuria ramosa in *Daphnia*. Adv. Parasitol. 91, 265–310.

Eden, J.-S., Tanaka, M.M., Boni, M.F., et al., 2013. Recombination within the pandemic norovirus GII.4 lineage. J. Virol. 87, 6270–6282.

Eilers, E.J., Kremen, C., Smith Greenleaf, S., Garber, A.K., Klein, A.-M., 2011. Contribution of pollinator-mediated crops to nutrients in the human food supply. PLoS One 6, e21363.

Engel, P., Kwong, W.K., McFrederick, Q., Anderson, K.E., Barribeau, S.M., Chandler, J.A., Cornman, R.S., Dainat, J., de Miranda, J.R., Doublet, V., Emery, O., Evans, J.D., Farinelli, L., Flenniken, M.L., Granberg, F., Grasis, J.A., Gauthier, L., Hayer, J., Koch, H., Kocher, S., Martinson, V.G., Moran, N., Munoz-Torres, M., Newton, I., Paxton, R.J., Powell, E., Sadd, B.M., Schmid-Hempel, P., Schmid-Hempel, R., Song, S.J., Schwarz, R.S., vanEngelsdorp, D., Dainat, B., 2016. The bee microbiome: impact on bee health and model for evolution and ecology of host-microbe interactions. MBio 7, e02164-02115.

Estrada-Peña, A., Ostfeld, R.S., Peterson, A.T., Poulin, R., de la Fuente, J., 2014. Effects of environmental change on zoonotic disease risk: an ecological primer. Trends Parasitol. 30, 205–214. https://doi.org/10.1016/j.pt.2014.02.003.

Ewald, P.W., 1983. Host-parasite relations, vectors, and the evolution of disease severity. Annu. Rev. Ecol. Syst. 14 (1), 465–485.

Faria, N.R., Suchard, M.A., Rambaut, A., Streicker, D.G., Lemey, P., 2013. Simultaneously reconstructing viral cross-species transmission history and identifying the underlying constraints. Philos. Trans. R. Soc. Lond. Ser. B Biol. Sci. 368, 20120196. https://doi.org/10.1098/rstb.2012.0196.

Faria, N.R., Rambaut, A., Suchard, M.A., Baele, G., Bedford, T., Ward, M.J., et al., 2014. The early spread and epidemic ignition of HIV-1 in human populations. Science 346, 56–61.

Fitzpatrick, Ú., Murray, T.E., Paxton, R.J., Breen, J., Cotton, D., Santorum, V., Brown, M.J.F., 2007. Rarity and decline in bumblebees—a test of causes and correlates in the Irish fauna. Biol. Conserv. 136, 185–194.

Forsgren, E., Fries, I., de Miranda, J.R., 2012. Adult honey bees (Apis mellifera) with deformed wings discovered in confirmed varroa-free colonies. J. Apic. Res. 51 (1), 136–138.

Francis, R.M., Nielsen, S.L., Kryger, P., 2013. Varroa-virus interaction in collapsing honey bee colonies. PLoS One 8, e57540.

Frank, S.A., Schmid-Hempel, P., 2008. Mechanisms of pathogenesis and the evolution of parasite virulence. J. Evol. Biol. 21 (2), 396–404.

Fujiyuki, T., Takeuchi, H., Ono, M., Ohka, S., Sasaki, T., Nomoto, A., Kubo, T., 2004. Novel insect picorna-like virus identified in the brains of aggressive worker honeybees. J. Virol. 78, 1093–1100.

Fürst, M.A., McMahon, D.P., Osborne, J.L., Paxton, R.J., Brown, M.J.F., 2014. Disease associations between honeybees and bumblebees as a threat to wild pollinators. Nature 506, 364–366.

Gallai, N., Salles, J.-M., Settele, J., Vaissière, B.E., 2009. Economic valuation of the vulnerability of world agriculture confronted to pollinator decline. Ecol. Econ. 68, 810–821.

Gamboa, V., Ravoet, J., Brunain, M., Smagghe, G., Meeus, I., Figueroa, J., et al., 2015. Bee pathogens found in bombus atratus from Colombia: a case study. J. Invertebr. Pathol. 129, 36–39.

Garbian, Y., Maori, E., Kalev, H., Shafir, S., Sela, I., 2012. Bidirectional transfer of RNAi between honey bee and varroa destructor: varroa gene silencing reduces varroa population. PLoS Pathog. 8 (12), e1003035.

Garibaldi, L.A., Steffan-Dewenter, I., Winfree, R., Aizen, M.A., Bommarco, R., Cunningham, S.A., Kremen, C., Carvalheiro, L.G., Harder, L.D., Afik, O., Bartomeus, I., Benjamin, F., Boreux, V., Cariveau, D., Chacoff, N.P., Dudenhöffer, J.H., Freitas, B.M., Ghazoul, J., Greenleaf, S., Hipólito, J., Holzschuh, A., Howlett, B., Isaacs, R., Javorek, S.K., Kennedy, C.M., Krewenka, K., Krishnan, S., Mandelik, Y., Mayfield, M.M., Motzke, I., Munyuli, T., Nault, B.A., Otieno, M., Petersen, J., Pisanty, G., Potts, S.G., Rader, R., Ricketts, T.H., Rundlöf, M., Seymour, C.L., Schüepp, C., Szentgyörgyi, H., Taki, H., Tscharntke, T., Vergara, C.H., Viana, B.F., Wanger, T.C., Westphal, C., Williams, N., Klein, A.M., 2013. Wild pollinators enhance fruit set of crops regardless of honey bee abundance. Science 339, 1608–1611.

Genersch, E., Aubert, M., 2010. Emerging and re-emerging viruses of the honey bee (*Apis mellifera* L.). Vet. Res. 41 (6), 54.

Genersch, E., Yue, C., Fries, I., de Miranda, J.R., 2006. Detection of deformed wing virus, a honey bee viral pathogen, in bumble bees (*Bombus terrestris* and *Bombus pascuorum*) with wing deformities. J. Invertebr. Pathol. 91 (1), 61–63.

Genersch, E., von der Ohe, W., Kaatz, H., Schroeder, A., Otten, C., Büchler, R., Berg, S., Ritter, W., Mühlen, W., Gisder, S., Meixner, M., Liebig, G., Rosenkranz, P., 2010. The German bee monitoring project: a long term study to understand periodically high winter losses of honey bee colonies. Apidologie 41, 332–352.

Ghosh, R.C., Ball, B.V., Willcocks, M.M., Carter, M.J., 1999. The nucleotide sequence of Sacbrood Virus of the honey bee: an insect picorna-like virus. J. Gen. Virol. 80, 1541–1549.

Gisder, S., Genersch, E., 2017. Viruses of commercialized insect pollinators. J. Invertebr. Pathol. 147, 51–59.

Gisder, S., Aumeier, P., Genersch, E., 2009. Deformed wing virus: replication and viral load in mites (varroa destructor). J. Gen. Virol. 90, 463–467.

Goka, K., Okabe, K., Yoneda, M., 2006. Worldwide migration of parasitic mites as a result of bumblebee commercialization. Popul. Ecol. 48 (4), 285–291.

Goulson, D., O'Connor, S., Park, K.J., 2017. The impacts of predators and parasites on wild bumblebee colonies. Ecol. Entomol. 43, 168–181. https://doi.org/10.1111/een.12482.

Graham, A.L., Allen, J.E., Read, A.F., 2005. Evolutionary causes and consequences of immunopathology. Annu. Rev. Ecol. Evol. Syst. 36, 373–397.

Graystock, P., Goulson, D., Hughes, W.H.O., 2015. Parasites in bloom: flowers aid dispersal and transmission of pollinator parasites within and between bee species. Proc. R. Soc. B Biol. Sci. 282, 20151371. https://doi.org/10.1098/rspb.2015.1371.

Graystock, P., Meeus, I., Smagghe, G., Goulson, D., Hughes, W.O., 2016. The effects of single and mixed infections of *Apicystis bombi* and deformed wing virus in *Bombus terrestris*. Parasitology 143 (3), 358–365.

Grenfell, B.T., Pybus, O.G., Gog, J.R., Wood, J.L., Daly, J.M., Mumford, J.A., Holmes, E.C., 2004. Unifying the epidemiological and evolutionary dynamics of pathogens. Science 303, 327–332.

Hall, M.D., Bento, G., Ebert, D., 2017. The evolutionary consequences of stepwise infection processes. Trends Ecol. Evol. 32, 612–623.

Hedrick, P.W., 2011. Population genetics of malaria resistance in humans. Heredity 107, 283–304.

Hedrick, S.M., 2017. Understanding immunity through the lens of disease ecology. Trends Immunol. 38, 888–903.

Hedtke, S.M., Patiny, S., Danforth, B.N., 2013. The bee tree of life: a supermatrix approach to apoid phylogeny and biogeography. BMC Evol. Biol. 13, 138.

Highfield, A.C., El Nagar, A., Mackinder, L.C.M., Noël, L.M.-L.J., Hall, M.J., Martin, S.J., Schroeder, D.C., 2009. Deformed wing virus implicated in overwintering honeybee colony losses. Appl. Environ. Microbiol. 75, 7212–7220.

Holmes, E.C., 2008. Evolutionary history and phylogeography of human viruses. Annu. Rev. Microbiol. 62, 307–328.

Hoyer, I.J., Blosser, E.M., Acevedo, C., Thompson, A.C., Reeves, L.E., Burkett-Cadena, N.D., 2017. Mammal decline, linked to invasive burmese python, shifts host use of vector mosquito towards reservoir hosts of a zoonotic disease. Biol. Lett. 13, 20170353. https://doi.org/10.1098/rsbl.2017.0353.

Hughes, P.J., Stanway, G., 2000. The 2A proteins of three diverse picornaviruses are related to each other and to the H-rev107 family of proteins involved in the control of cell proliferation. J. Gen. Virol. 81, 201–207.

Johnson, P.T., De Roode, J.C., Fenton, A., 2015. Why infectious disease research needs community ecology. Science 349, 1259504.

Kalynych, S., Přidal, A., Pálková, L., Levdansky, Y., de Miranda, J.R., Plevka, P., 2016. Virion structure of iflavirus slow bee paralysis virus at 2.6-angstrom resolution. J. Virol. 90, 7444–7455.

Kalynych, S., Füzik, T., Přidal, A., de Miranda, J., Plevka, P., 2017. Cryo-EM study of slow bee paralysis virus at low pH reveals iflavirus genome release mechanism. Proc. Natl. Acad. Sci. U.S.A. 114, 598–603.
Keesing, F., Belden, L.K., Daszak, P., Dobson, A., Harvell, C.D., Holt, R.D., et al., 2010. Impacts of biodiversity on the emergence and transmission of infectious diseases. Nature 468, 647–652.
Klee, J., Besana, A.M., Genersch, E., Gisder, S., Nanetti, A., Tam, D.Q., et al., 2007. Widespread dispersal of the microsporidian *Nosema ceranae*, an emergent pathogen of the western honey bee, *Apis mellifera*. J. Invertebr. Pathol. 96 (1), 1–10.
Klein, A.-M., Vaissiére, B.E., Cane, J.H., Steffan-Dewenter, I., Cunningham, S.A., Kremen, C., Tscharntke, T., 2007. Importance of pollinators in changing landscapes for world crops. Proc. R. Soc. Lond. B 274, 303–314.
Kocher, S.D., Paxton, R.J., 2014. Comparative methods offer powerful insights into social evolution in bees. Apidologie 45, 289–305.
Kwong, W.K., Moran, N.A., 2016. Gut microbial communities of social bees. Nat. Rev. Microbiol. 14, 374–384.
Laayouni, H., et al., 2014. Convergent evolution in European and Roma populations reveals pressure exerted by plague on toll-like receptors. Proc. Natl. Acad. Sci. U.S.A. 111, 2668–2673.
Lafferty, K.D., 2009. The ecology of climate change and infectious diseases. Ecology 90 (4), 888–900.
Lamp, B., Url, A., Seitz, K., Eichhorn, J., Riedel, C., Sinn, L.J., Indik, S., Köglberger, H., Rümenapf, T., 2016. Construction and rescue of a molecular clone of deformed wing virus (DWV). PLoS One 11, e0164639.
Leat, N., Ball, B.V., Govan, V., Davison, S., 2000. Analysis of the complete genome sequence of black queen cell virus, a picorna-like virus of honey bees. J. Gen. Virol. 81, 2111–2119.
Levin, B.R., Bull, J.J., 1994. Short-sighted evolution and the virulence of pathogenic microorganisms. Trends Microbiol. 2, 76–81.
Levitt, A.L., Singh, R., Cox-Foster, D.L., Rajotte, E., Hoover, K., Ostiguy, N., Holmes, E.C., 2013. Cross-species transmission of honey bee viruses in associated arthropods. Virus Res. 176 (1–2), 232–240.
Li, J.L., Cornman, R.S., Evans, J.D., Pettis, J.S., Zhao, Y., Murphy, C., Peng, W.J., Wu, J., Hamilton, M., Boncristiani, H.F.J., Zhou, L., Hammond, J., Chen, Y.P., 2014. Systemic spread and propagation of a plant-pathogenic virus in European honeybees, *Apis mellifera*. MBio 5, e00898-00813.
Li, Y., Basavappa, M., Lu, J., Dong, S., Cronkite, D.A., Prior, J.T., et al., 2016. Induction and suppression of antiviral RNA interference by influenza A virus in mammalian cells. Nat. Microbiol. 2, 16250.
Longdon, B., Hadfield, J.D., Day, J.P., Smith, S.C., McGonigle, J.E., Cogni, R., et al., 2015. The causes and consequences of changes in virulence following pathogen host shifts. PLoS Pathog. 11 (3), e1004728.
Lozier, J.D., Zayed, A., 2017. Bee conservation in the age of genomics. Conserv. Genet. 18, 713–729.
Maharramov, J., Meeus, I., Maebe, K., Arbetman, M., Morales, C., Graystock, P., et al., 2013. Genetic variability of the neogregarine *Apicystis bombi*, an etiological agent of an emergent bumblebee disease. PLoS One 8 (12), e81475.
Manley, R., Boots, M., Wilfert, L., 2015. Emerging viral disease risk to pollinating insects: ecological, evolutionary and anthropogenic factors. J. Appl. Ecol. 52, 331–340.
Manley, R., Boots, M., Wilfert, L., 2017. Condition-dependent virulence of slow bee paralysis virus in *Bombus terrestris*: are the impacts of honeybee viruses in wild pollinators underestimated? Oecologia 184, 305–315.

Maori, E., Lavi, S., Mozes-Koch, R., Gantman, Y., Peretz, Y., Edelbaum, O., Tanne, E., Sela, I., 2007. Isolation and characterization of Israeli acute paralysis virus, a discistrovirus affecting honeybees in Israel: evidence for diversity due to intra- and inter-species recombination. J. Gen. Virol. 88, 3428–3438.

Martin, S.J., 2001. The role of *Varroa* and viral pathogens in the collapse of honeybee colonies: a modelling approach. J. Appl. Ecol. 38 (5), 1082–1093.

Martin, S.J., Highfield, A.C., Brettell, L., Villalobos, E.M., Budge, G.E., Powell, M., Nikaido, S., Schroeder, D.C., 2012. Global honey bee viral landscape altered by a parasitic mite. Science 336, 1304–1306.

McArt, S.H., Koch, H., Irwin, R.E., Adler, L.S., 2014. Arranging the bouquet of disease: floral traits and the transmission of plant and animal pathogens. Ecol. Lett. 17, 624–636.

McKnight, K.L., Lemon, S.M., 2017. Virology: ins and outs of picornaviruses. Nature 541 (7637), 299–300.

McMahon, D.P., Fürst, M.A., Caspar, J., Theodorou, P., Brown, M.J.F., Paxton, R.J., 2015. A sting in the spit: widespread cross-infection of multiple RNA viruses across wild and managed bees. J. Anim. Ecol. 84, 615–624.

McMahon, D.P., Natsopoulou, M.E., Doublet, V., Fürst, M., Weging, S., Brown, M.J.F., Gogol-Döring, A., Paxton, R.J., 2016. Elevated virulence of an emerging viral genotype as a driver of honeybee loss. Proc. R. Soc. Lond. B Biol. Sci. 283, 20160811.

McMenamin, A.J., Genersch, E., 2015. Honey bee colony losses and associated viruses. Curr. Opin. Insect Sci. 8, 121–129.

Meeus, I., de Miranda, J.R., de Graaf, D.C., Wäckers, F., Smagghe, G., 2014. Effect of oral infection with Kashmir bee virus and Israeli acute paralysis virus on bumblebee (*Bombus terrestris*) reproductive success. J. Invertebr. Pathol. 121, 64–69.

Merkling, S.H., van Rij, R.P., 2013. Beyond RNAi: antiviral defense strategies in drosophila and mosquito. J. Insect Physiol. 59 (2), 159–170.

Michener, C.D., 2007. The Bees of the World, second ed. John Hopkins University Press, Baltimore, MD.

Mideo, N., Alizon, S., Day, T., 2008. Linking within- and between-host dynamics in the evolutionary epidemiology of infectious diseases. Trends Ecol. Evol. 23, 511–517.

Mitchell, E.A.D., Mulhauser, B., Mulot, M., Mutazabi, A., Glauser, G., Aebi, A., 2017. A worldwide survey of neonicotinoids in honey. Science 358, 109–111. https://doi.org/10.1126/science.aan3684.

Möckel, N., Gisder, S., Genersch, E., 2011. Horizontal transmission of deformed wing virus: pathological consequences in adult bees (*Apis mellifera*) depend on the transmission route. J. Gen. Virol. 92 (2), 370–377.

Mondet, F., de Miranda, J.R., Kretzschmar, A., Le Conte, Y., Mercer, A.R., 2014. On the front line: quantitative virus dynamics in honeybee (*Apis mellifera* L.) colonies along a new expansion front of the parasite *varroa destructor*. PLoS Pathog. 10, e1004323.

Moore, J., Jironkin, A., Chandler, D., et al., 2011. Recombinants between deformed wing virus and Varroa destructor virus-1 may prevail in *varroa destructor*-infested honeybee colonies. J. Gen. Virol. 92, 156–161.

Mordecai, G.J., Wilfert, L., Martin, S.J., et al., 2015. Diversity in a honey bee pathogen: first report of a third master variant of the deformed wing virus quasispecies. ISME J. 10, 1264–1273.

Mordecai, G.J., Brettell, L.E., Pachori, P., et al., 2016. Moku virus; a new Iflavirus found in wasps, honey bees and varroa. Sci. Rep. 6, 34983.

Moritz, R.F.A., Erler, S., 2016. Lost colonies found in a data mine: global honey trade but not pests or pesticides as a major cause of regional honeybee colony declines. Agric. Ecosyst. Environ. 216, 44–50.

Moritz, R.F.A., de Miranda, J., Fries, I., Le Conte, Y., Neumann, P., Paxton, R.J., 2010. Research strategies to improve honeybee health in Europe. Apidologie 41, 227–242.

Natsopoulou, M.E., McMahon, D.P., Doublet, V., Bryden, J., Paxton, R.J., 2015. Interspecific competition in honeybee intracellular gut parasites is asymmetric and favours the spread of an emerging infectious disease. Proc. Biol. Sci. 282, 20141896.

Natsopoulou, M.E., McMahon, D.P., Doublet, V., Frey, E., Rosenkranz, P., Paxton, R.J., 2017. The virulent, emerging genotype B of deformed wing virus is closely linked to overwinter honeybee worker loss. Sci. Rep. 7, 5242.

Nazzi, F., Le Conte, Y., 2016. Ecology of *Varroa destructor*, the major ectoparasite of the western honey bee, Apis mellifera. Annu. Rev. Entomol. 61, 417–432.

Nazzi, F., Brown, S.P., Annoscia, D., Del Piccolo, F., Di Prisco, G., Varricchio, P., Della Vedova, G., Cattonaro, F., Caprio, E., Pennacchio, F., 2012. Synergistic parasite-pathogen interactions mediated by host immunity can drive the collapse of honeybee colonies. PLoS Pathog. 8, e1002735.

Neumann, P., Carreck, N.L., 2010. Honey bee colony losses. J. Apic. Res. 49, 1–6.

Nguyen, B.K., Ribière, M., vanEngelsdorp, D., Snoeck, C., Saegerman, C., Kalkstein, A.L., Schurr, F., Brostaux, Y., Faucon, J.-P., Haubruge, E., 2011. Effects of honey bee virus prevalence, *Varroa destructor* load and queen condition on honey bee colony survival over the winter in Belgium. J. Apic. Res. 50, 195–202.

Nieto, A., Roberts, S.P.M., Kemp, J., Rasmont, P., Kuhlmann, M., García Criado, M., Biesmeijer, J.C., Bogusch, P., Dathe, H.H., De la Rúa, P., De Meulemeester, T., Dehon, M., Dewulf, A., Ortiz-Sánchez, F.J., Lhomme, P., Pauly, A., Potts, S.G., Praz, C., Quaranta, M., Radchenko, V.G., Scheuchl, E., Smit, J., Straka, J., Terzo, M., Tomozii, B., Window, J., Michez, D., 2014. European Red List of Bees. Publication Office of the European Commission, Luxembourg.

Normile, D., 2008. Driven to extinction. Science 319, 1606–1609.

Obbard, D.J., Gordon, K.H., Buck, A.H., Jiggins, F.M., 2009. The evolution of RNAi as a defence against viruses and transposable elements. Philos. Trans. R. Soc. Lond. Ser. B Biol. Sci. 364, 99–115.

Oldroyd, B.P., 1999. Coevolution while you wait: *Varroa jacobsoni*, a new parasite of western honeybees. Trends Ecol. Evol. 14, 312–315.

Olivier, V., Blanchard, P., Chaouch, S., Lallemand, P., Schurr, F., Celle, O., Dubois, E., Tordo, N., Thiery, R., Houlgatte, R., Ribiere, M., 2008. Molecular characterisation and phylogenetic analysis of chronic bee paralysis virus, a honey bee virus. Virus Res. 132, 59–68.

Ollerton, J., Winfree, R., Tarrant, S., 2011. How many flowering plants are pollinated by animals? Oikos 120, 321–326.

Ongus, J.R., Peters, D., Bonmatin, J., Bengsch, E., Vlak, J.M., van Oers, M.M., 2004. Complete sequence of a picorna-like virus of the genus Iflavirus replicating in the mite varroa destructor. J. Gen. Virol. 85, 3747–3755.

Osterrieder, N., Kamil, J.P., Schumacher, D., Tischer, B.K., Trapp, S., 2006. Marek's disease virus: from miasma to model. Nat. Rev. Microbiol. 4, 283–294.

Otterstatter, M.C., Thomson, J.D., 2007. Contact networks and transmission of an intestinal pathogen in bumble bee (*Bombus impatiens*) colonies. Oecologia 154, 411–421.

Parmentier, L., Smagghe, G., de Graaf, D.C., Meeus, I., 2016. *Varroa destructor* macula-like virus, lake sinai virus and other new RNA viruses in wild bumblebee hosts (*Bombus pascuorum, Bombus lapidarius* and *Bombus pratorum*). J. Invertebr. Pathol. 134, 6–11.

Pearce-Duvet, J.M.C., 2006. The origin of human pathogens: evaluating the role of agriculture and domestic animals in the evolution of human disease. Biol. Rev. 81, 369–382.

Peng, W., Li, J., Boncristiani, H., Strange, J.P., Hamilton, M., Chen, Y., 2011. Host range expansion of honey bee black queen cell virus in the bumble bee, *Bombus huntii*. Apidologie 42, 650–658.

Pestova, T.V., Lomakin, I.B., Hellen, C.U., 2004. Position of the CrPV IRES on the 40S subunit and factor dependence of IRES/80S ribosome assembly. EMBO Rep. 5, 906–913.

Potts, S.G., Imperatriz-Fonseca, V., Ngo, H.T., Aizen, M.A., Biesmeijer, J.C., Breeze, T.D., Dicks, L.V., Garibaldi, L.A., Hill, R., Settele, J., Vanbergen, A.J., 2016. Safeguarding pollinators and their values to human well-being. Nature 540, 220–229.

Radzevičiūtė, R., Theodorou, P., Husemann, M., Japoshvili, G., Kirkitadze, G., Zhusupbaeva, A., Paxton, R.J., 2017. Replication of honey bee-associated RNA viruses across multiple bee species in apple orchards of Georgia, Germany and Kyrgyzstan. J. Invertebr. Pathol. 146, 14–23.

Rambaut, A., Posada, D., Crandall, K.A., Holmes, E.C., 2004. The causes and consequences of HIV evolution. Nat. Rev. Genet. 5, 52–61.

Randolph, S.E., 2001. The shifting landscape of tick-borne zoonoses: tick-borne encephalitis and lyme borreliosis in Europe. Philos. Trans. R. Soc. Lond. Ser. B Biol. Sci. 356, 1045–1056.

Randolph, S.E., Rogers, D.J., 2000. Fragile transmission cycles of tick-borne encephalitis virus may be disrupted by predicted climate change. Proc. R. Soc. Lond. B Biol. Sci. 267, 1741–1744.

Ravoet, J., De Smet, L., Meeus, I., Smagghe, G., Wenseleers, T., de Graaf, D.C., 2014. Widespread occurrence of honey bee pathogens in solitary bees. J. Invertebr. Pathol. 122, 55–58.

Remnant, E.J., Shi, M., Buchmann, G., et al., 2017. A diverse range of novel RNA viruses in geographically distinct honey bee populations. J. Virol. 91, e00158-17.

Ribière, M., Olivier, V., Blanchard, P., 2010. Chronic bee paralysis: a disease and a virus like no other? J. Invertebr. Pathol. 103, S120–S131.

Rigaud, T., Perrot Minot, M.-J., Brown, M.J.F., 2010. Parasite and host assemblages: embracing the reality will improve our knowledge of parasite transmission and virulence. Proc. R. Soc. B Biol. Sci. 277, 3693–3702.

Rogers, D.J., Randolph, S.E., 2000. The global spread of malaria in a future, warmer world. Science 289 (5485), 1763–1766.

Rosenkranz, P., Aumeier, P., Ziegelmann, B., 2010. Biology and control of *varroa destructor*. J. Invertebr. Pathol. 103, S96–S119.

Ruiz-González, M.X., Bryden, J., Moret, Y., et al., 2012. Dynamic transmission, host quality, and population structure in a multihost parasite of bumblebees. Evolution 66, 3053–3066.

Runckel, C., Flenniken, M.L., Engel, J.C., et al., 2011. Temporal analysis of the honey bee microbiome reveals four novel viruses and seasonal prevalence of known viruses, *nosema*, and *crithidia*. PLoS One 6, e20656.

Rundlöf, M., Andersson, G.K.S., Bommarco, R., Fries, I., Hederstrom, V., Herbertsson, L., Jonsson, O., Klatt, B.K., Pedersen, T.R., Yourstone, J., Smith, H.G., 2015. Seed coating with a neonicotinoid insecticide negatively affects wild bees. Nature 521, 77–80.

Rutrecht, S.T., Brown, M.J.F., 2008. Within colony dynamics of *Nosema bombi* infections: disease establishment, epidemiology and potential vertical transmission. Apidologie 39, 504–514.

Ryabov, E.V., Wood, G.R., Fannon, J.M., et al., 2014. A virulent strain of deformed wing virus (DWV) of honeybees (*Apis mellifera*) prevails after varroa destructor-mediated, or in vitro, transmission. PLoS Pathog. 10, e1004230.

Santillán-Galicia, M.T., Ball, B.V., Clark, S.J., Alderson, P.G., 2014. Slow bee paralysis virus and its transmission in honey bee pupae by *varroa destructor*. J. Apic. Res. 53 (1), 146–154.

Schmid-Hempel, P., 2009. Immune defence, parasite evasion strategies and their relevance for 'macroscopic phenomena'such as virulence. Philos. Trans. R. Soc. Lond. Ser. B Biol. Sci. 364 (1513), 85–98.

Schmid-Hempel, P., 2011. Evolutionary Parasitology: The Integrated Study of Infections, Immunology, Ecology, and Genetics. Oxford University Press, Oxford.

Schmid-Hempel, P., Ebert, D., 2003. On the evolutionary ecology of specific immune defence. Trends Ecol. Evol. 18, 27–32.
Schmid-Hempel, R., Eckhardt, M., Goulson, D., Heinzmann, D., Lange, C., Plischuk, S., et al., 2014. The invasion of southern South America by imported bumblebees and associated parasites. J. Anim. Ecol. 83 (4), 823–837.
Schroeder, D.C., Martin, S.J., 2012. Deformed wing virus: the main suspect in unexplained honeybee deaths worldwide. Virulence 3 (7), 589–591.
Seo, G.J., Kincaid, R.P., Phanaksri, T., Burke, J.M., Pare, J.M., Cox, J.E., et al., 2013. Reciprocal inhibition between intracellular antiviral signaling and the RNAi machinery in mammalian cells. Cell Host Microbe 14 (4), 435–445.
Shen, M., Cui, L., Ostiguy, N., Cox-Foster, D., 2005. Intricate transmission routes and interactions between picorna-like viruses (Kashmir bee virus and sacbrood virus) with the honeybee host and the parasitic varroa mite. J. Gen. Virol. 86, 2281–2289.
Singh, R., Levitt, A.L., Rajotte, E.G., Holmes, E.C., Ostiguy, N., Lipkin, W.I., et al., 2010. RNA viruses in hymenopteran pollinators: evidence of inter-taxa virus transmission via pollen and potential impact on non-apis hymenopteran species. PLoS One 5 (12), e14357.
Škubník, K., Nováček, J., Füzik, T., Přidal, A., Paxton, R.J., Plevka, P., 2017. Structure of deformed wing virus, a major honey bee pathogen. Proc. Natl. Acad. Sci. U.S.A. 114, 3210–3215.
Sorrell, I., White, A., Pedersen, A.B., Hails, R.S., Boots, M., 2009. The evolution of covert, silent infection as a parasite strategy. Proc. R. Soc. Lond. B Biol. Sci. 276 (1665), 2217–2226.
Staring, J., von Castelmur, E., Blomen, V.A., van den Hengel, L.G., Brockmann, M., Baggen, J., et al., 2017. PLA2G16 represents a switch between entry and clearance of picornaviridae. Nature 541 (7637), 412–416.
Stenseth, N.C., Atshabar, B.B., Begon, M., Belmain, S.R., Bertherat, E., Carniel, E., et al., 2008. Plague: past, present, and future. PLoS Med. 5 (1), e3.
Streicker, D.G., Turmelle, A.S., Vonhof, M.J., Kuzmin, I.V., McCracken, G.F., Rupprecht, C.E., 2010. Host phylogeny constrains cross-species emergence and establishment of rabies virus in bats. Science 329, 676–679.
Streicker, D.G., Fenton, A., Pedersen, A.B., 2013. Differential sources of host species heterogeneity influence the transmission and control of multihost parasites. Ecol. Lett. 16, 975–984.
Sumpter, D.J., Martin, S.J., 2004. The dynamics of virus epidemics in varroa-infested honey bee colonies. J. Anim. Ecol. 73, 51–63.
Taber, E.D., Hutchinson, M.L., Smithwick, E.A.H., Blanford, J.I., 2017. A decade of colonization: the spread of the Asian tiger mosquito in Pennsylvania and implications for disease risk. J. Vector Ecol. 42, 3–12.
Taylor, L.H., Latham, S.M., Woolhouse, M.E.J., 2001. Risk factors for human disease emergence. Philos. Trans. R. Soc. Lond. Ser. B Biol. Sci. 356, 983–989.
Tehel, A., Brown, M.J., Paxton, R.J., 2016. Impact of managed honey bee viruses on wild bees. Curr. Opin. Virol. 19, 16–22.
Traynor, K.S., Rennich, K., Forsgren, E., Rose, R., Pettis, J., Kunkel, G., Madella, S., Evans, J., Lopez, D., vanEngelsdorp, D., 2016. Multiyear survey targeting disease incidence in US honey bees. Apidologie 47, 325–347.
Trimpert, J., Groenke, N., Jenckel, M., He, S., Kunec, D., Szpara, M.L., et al., 2017. A phylogenomic analysis of marek's disease virus (MDV) reveals independent paths to virulence in Eurasia and North America. Evol. Appl. 10, 1091–1101. https://doi.org/10.1111/eva.12515.
Tumpey, T.M., Basler, C.F., Aguilar, P.V., et al., 2005. Characterization of the reconstructed 1918 Spanish influenza pandemic virus. Science 310, 77–80.

Ueira-Vieira, C., Almeida, L.O., De Almeida, F.C., Amaral, I.M.R., Brandeburgo, M.A.M., Bonetti, A.M., 2015. Scientific note on the first molecular detection of the acute bee paralysis virus in Brazilian stingless bees. Apidologie 46 (5), 628–630.

Urbanowicz, R.A., McClure, C.P., Sakuntabhai, A., Sall, A.A., Kobinger, G., Müller, M.A., et al., 2016. Human adaptation of Ebola virus during the West African outbreak. Cell 167 (4), 1079–1087.

van Rij, R.P., Saleh, M.C., Berry, B., Foo, C., Houk, A., et al., 2006. The RNA silencing endonuclease Argonaute 2 mediates specific antiviral immunity in *Drosophila melanogaster*. Genes Dev. 20, 2985–2995.

van Baarlen, P., van Belkum, A., Summerbell, R.C., Crous, P.W., Thomma, B., 2007. Molecular mechanisms of pathogenicity: how do pathogenic microorganisms develop cross-kingdom host jumps? FEMS Microbiol. Rev. 31, 239–277.

Vanbergen, A.J., IPI Initiative, 2013. Threats to an ecosystem service: pressures on pollinators. Front. Ecol. Environ. 11, 251–259.

Velthuis, H.H.W., van Doorn, A., 2006. A century of advances in bumblebee domestication and the economic and environmental aspects of its commercialization for pollination. Apidologie 37, 421–451.

Wang, H., Xie, J., Shreeve, T.G., et al., 2013. Sequence recombination and conservation of varroa destructor virus-1 and deformed wing virus in field collected honey bees (*Apis mellifera*). PLoS One 8, e74508.

Weiss, R.A., 2002. Virulence and pathogenesis. Trends Microbiol. 10, 314–317.

Wilfert, L., Long, G., Leggett, H.C., Schmid-Hempel, P., Butlin, R., Martin, S.J.M., Boots, M., 2016. Deformed wing virus is a recent global epidemic in honeybees driven by *Varroa* mites. Science 351, 594–597.

Willmer, P., 2011. Pollination and Floral Ecology. Princeton University Press, Princeton, NJ.

Willmer, P.G., Cunnold, H., Ballantyne, G., 2017. Insights from measuring pollen deposition: quantifying the pre-eminence of bees as flower visitors and effective pollinators. Arthropod Plant Interact. 11, 411–425.

Winfree, R., Aguilar, R., Vázquez, D.P., LeBuhn, G., Aizen, M.A., 2009. A meta-analysis of bees' responses to anthropogenic disturbance. Ecology 90, 2068–2076.

Wolf, S., Nicholls, E., Reynolds, A.M., Wells, P., Lim, K.S., Paxton, R.J., Osborne, J.L., 2016. Optimal search patterns in honeybee orientation flights are robust against emerging infectious diseases. Sci. Rep. 6, 32612.

Woodcock, B.A., Isaac, N.J.B., Bullock, J.M., Roy, D.B., Garthwaite, D.G., Crowe, A., Pywell, R.F., 2016. Impacts of neonicotinoid use on long-term population changes in wild bees in England. Nat. Commun. 7, 12459.

Woodcock, B.A., Bullock, J.M., Shore, R.F., Heard, M.S., Pereira, M.G., Redhead, J., Ridding, L., Dean, H., Sleep, D., Henrys, P., Peyton, J., Hulmes, S., Hulmes, L., Sárospataki, M., Saure, C., Edwards, M., Genersch, E., Knäbe, S., Pywell, R.F., 2017. Country-specific effects of neonicotinoid pesticides on honey bees and wild bees. Science 356, 1393–1395.

Woodworth, B.L., Atkinson, C.T., LaPointe, D.A., Hart, P.J., Spiegel, C.S., Tweed, E.J., et al., 2005. Host population persistence in the face of introduced vector-borne diseases: Hawaii amakihi and avian malaria. Proc. Natl. Acad. Sci. U.S.A. 102, 1531–1536.

Woolhouse, M.E., 2002. Population biology of emerging and re-emerging pathogens. Trends Microbiol. 10, S3–S7.

Woolhouse, M.E., Haydon, D.T., Antia, R., 2005. Emerging pathogens: the epidemiology and evolution of species jumps. Trends Ecol. Evol. 20, 238–244.

Yan, H., Opachaloemphan, C., Mancini, G., Yang, H., Gallitto, M., Mlejnek, J., et al., 2017. An engineered orco mutation produces aberrant social behavior and defective neural development in ants. Cell 170, 736–747.

Yue, C., Genersch, E., 2005. RT-PCR analysis of deformed wing virus in honeybees (*Apis mellifera*) and mites (*Varroa destructor*). J. Gen. Virol. 86, 3419–3424.
Zhang, X., He, S.Y., Evans, J.D., Pettis, J.S., Yin, G.F., Chen, Y.P., 2012. New evidence that deformed wing virus and black queen cell virus are multi-host pathogens. J. Invertebr. Pathol. 109, 156–159.
Zioni, N., Soroker, V., Chejanovsky, N., 2011. Replication of varroa destructor virus 1 (VDV-1) and a vakrroa destructor virus 1-deformed wing virus recombinant (VDV-1-DWV) in the head of the honey bee. Virology 417, 106–112.

CHAPTER NINE

Ecological Complexity in Plant Virus Host Range Evolution

Michael J. McLeish, Aurora Fraile, Fernando García-Arenal[1]
Centro de Biotecnología y Genómica de Plantas UPM-INIA, and E.T.S.I. Agrícola, Alimentaria y de Biosistemas, Campus de Montegancedo, Universidad Politécnica de Madrid, Madrid, Spain
[1]Corresponding author: e-mail address: fernando.garciaarenal@upm.es

Contents

Abstract

The host range of a plant virus is the number of species in which it can reproduce. Most studies of plant virus host range evolution have focused on the genetics of host–pathogen interactions. However, the distribution and abundance of plant viruses and their hosts do not always overlap, and these spatial and temporal discontinuities in plant virus–host interactions can result in various ecological processes that shape host range evolution. Recent work shows that the distributions of pathogenic and resistant genotypes, vectors, and other resources supporting transmission vary widely in the environment, producing both expected and unanticipated patterns. The distributions of all of these factors are influenced further by competitive effects, natural enemies, anthropogenic disturbance, the abiotic environment, and herbivory to mention some. We suggest the need for further development of approaches that (i) explicitly consider resource use and the abiotic and biotic factors that affect the strategies by which viruses exploit resources; and (ii) are sensitive across scales. Host range and habitat specificity will largely determine which phyla are most likely to be new hosts, but predicting which host and when it is likely to be infected is enormously challenging because it is unclear how environmental heterogeneity affects the interactions of viruses and hosts.

Advances in Virus Research, Volume 101
ISSN 0065-3527
https://doi.org/10.1016/bs.aivir.2018.02.009

1. INTRODUCTION

Host range, defined as the number of host species used by a pathogen, is a simple metric that is central to understanding pathogen epidemiology and pathogenicity. Host range conditions the transmission dynamics and survival of pathogens and is predicted to be a major factor in their evolution. The host range of a pathogen affects how diversification and speciation occur, the incorporation of new species into the number of hosts a pathogen can infect, shifting among hosts in response to biotic and abiotic environmental variation, or the ability to persist in the environment between epidemics (Duffy et al., 2007; García-Arenal and Fraile, 2013; Moury et al., 2017; Parker and Gilbert, 2004; Weaver and Barrett, 2004; Woolhouse and Gowtage-Sequeria, 2005). From an anthropocentric viewpoint, host range determines pathogen reservoirs and inoculum sources, disease emergence and reemergence, population thresholds for disease invasion, and/or critical community size for disease persistence (Elena et al., 2014b; Frank, 1996; Haydon et al., 2002; Lajeunesse and Forbes, 2002; Woolhouse et al., 2001, 2005). Pathogens have been typically classified by host range either as specialists (which in nature can only infect, multiply efficiently, and be transmitted, in hosts from one or a few taxonomically related species) or as generalists or multihost pathogens (which infect, multiply, and are transmitted efficiently in hosts from different species, often from unrelated taxa). However, a pathogen host range is generally unknown in absolute terms and should not be treated as a fixed value in natural systems because ecological factors (broadly the distribution, abundance, and interaction of species) shape the range of host species with which a pathogen comes into contact (Arnold and Lutzoni, 2007). Moreover, this simple metric is difficult to estimate, largely because identifying all "nonhosts" is practically impossible. Changes in host–pathogen associations can lead to a number of evolutionary outcomes (de Vienne et al., 2013). A pathogen can shift, or change between hosts, with the number of hosts, i.e., host range, being unchanged. Alternatively, a pathogen species can expand its host range (host breadth expansion), or contract it (host breadth contraction) in response to ecological or other changes (Janz et al., 2001; Mitchell and Power, 2003; Ng and Perry, 2004; Petanidou et al., 2008; Sexton et al., 2017).

The ecological and evolutionary causes of specialization or generalization have fascinated biologists for decades (Futuyma and Moreno, 1988). Host range is determined by different sets of factors, some extrinsic to the pathogen, related to its ecology and epidemiology, and others intrinsic to the pathogen, such as genetic traits that determine its fitness in different hosts. Explanations of host range evolution have focused mostly on this second set of intrinsic, genetic factors. Plant viruses are largely generalists (García-Arenal and McDonald, 2003; Power and Flecker, 2003), as is true for many pathogens of humans and other animals (McDonald and Linde, 2002; Woolhouse et al., 2001).

Despite this evidence and the fact that a large host range provides more opportunities for pathogen transmission and survival, it is widely considered that evolution should favor specialization. The argument is that the evolution of host range breadth will be constrained by selection for specialization in host traits (Futuyma and Moreno, 1988): since the host provides the major environmental component of the pathogen, differential host-associated selection would limit the fitness of generalists in any one host, so that in any one host generalists would be outcompeted by specialists (Elena et al., 2009, 2014b; García-Arenal and Fraile, 2013; Kirchner and Roy, 2002; Woolhouse et al., 2001). This hypothesis is based on the concept of trade-offs in adaptation to different hosts, so that an increase of pathogen fitness in one host would result in a fitness decrease in others. As across-host fitness trade-offs will constrain host range expansion (Asplen et al., 2012), which is relevant for disease emergence or for developing sustainable disease control strategies, the analysis of across-host trade-offs has received much attention (Elena et al., 2014b; García-Arenal and Fraile, 2013). However, both theory and empirical evidence of across-host trade-offs derive from analyses of single-host–single-pathogen interactions in homogeneous environments. There is only limited empirical or experimental evidence about how environment modulates host range in plant viruses, and a need for studies that consider, jointly with genetic factors, the ecology of host range evolution.

Here we review what is known about the genetics of plant virus host range evolution, and how it may be modulated by environmental interactions, and then examine the ecology of host range evolution when increasing levels of biological complexity are considered. We also identify analytical and conceptual developments that specifically address some of the challenges ecology and plant virology face.

2. GENETICS OF HOST RANGE EVOLUTION IN PLANT VIRUSES

The search for intrinsic determinants of virus host ranges has led to analyses aimed at identifying virus traits that correlate with host range.

For example, for viruses infecting mammals, feature selection techniques have been used to identify sequences in viral genomes that are associated with the capacity to infect particular host species (Aguas and Ferguson, 2013). Variability in these alleles was then used to predict adaptation to new hosts. Similarly, locus-by-locus analyses of molecular variance have allowed the identification of the genes in potato virus Y (PVY) involved in a host jump (Vassilakos et al., 2016). The relationship between host range breadth and several viral traits (including the type of nucleic acid of the genome, the nonsegmented or segmented structure of the genome, and the nature of horizontal and vertical transmission) was analyzed for a set of 480 plant viruses with host ranges reported in the VIDE data base (Brunt et al., 1996). It was found that viruses with single-stranded genomes (either RNA or DNA), three-segmented genomes, and nematode transmitted had broader host ranges (Moury et al., 2017). It was also found that seed transmission was associated with host range breadth, but the nature of the association varied with the genome. A third important finding was that the widest host ranges were largely constrained to the taxonomic level of family in plant viruses (Moury et al., 2017). Specificity of infection is a major trait in host–pathogen interactions, and genetic studies of host range evolution have mostly focused on the coevolution of hosts and viruses, as developed in the rest of this section.

2.1 Genetic Models of Host Range Evolution

A major trait of host–pathogen interactions determining host range is genetic specificity, i.e., only a subset of pathogens can infect and multiply in a certain host, and for a particular host–pathogen interaction often only a subset of pathogen genotypes can infect and multiply in each host genotype (Agrawal and Lively, 2002, 2003; Antonovics et al., 2013; Zhan et al., 2014). The outcome (infection vs noninfection) of the host–pathogen interaction can be integrated into coevolutionary models that consider infection/noninfection as the result of the interaction between resistance and infectivity loci in the host and pathogen, respectively. Coevolutionary models differ in the underlying infection matrices (Dybdahl et al., 2014) and derive from two general ones—the gene-for-gene (GFG) and the matching-alleles (MA) models—initially proposed to explain plant–pathogen and invertebrate–pathogen interactions, respectively (Agrawal and Lively, 2002, 2003).

In the GFG model, there is a hierarchy of alleles determining resistance in the host or infectivity in the pathogen: some resistance alleles are intrinsically more effective than others, conferring resistance to a larger set of pathogen genotypes, and similarly, some infectivity alleles in the pathogen are intrinsically better than others, allowing infection of a larger set of host genotypes. In contrast, no such hierarchy of resistance/infectivity alleles occurs in MA systems. Instead a particular resistance allele is better at resisting a subset of pathogen genotypes and less effective at resisting the rest, and a pathogen infectivity allele is better at infecting a subset of host genotypes and worse on the rest. GFG and MA models correspond to different mechanisms of host–pathogen interaction: in the GFG model, infection occurs when the host genotype does not recognize the pathogen genotype, while in the MA model successful infection requires molecular matches between host and pathogen (Dybdahl et al., 2014). The above assumptions condition the evolution of resistance and infectivity loci. Notably, the GFG model predicts that pathogens will evolve to infect all host genotypes (universal infectivity), while this is not possible under the MA model (Agrawal and Lively, 2002, 2003; Frank, 1996).

Originally these models were developed to explain host genotype × pathogen genotype interactions, and so have been applied to the study of plant–virus interactions and plant virus host range evolution (reviewed in Fraile and García-Arenal, 2010; García-Arenal and Fraile, 2013). However, they can also be related to the evolution of host range at the interspecies level, a GFG-like system allowing host range expansion, and thus the evolution of generalists, and an MA-like determining host shifts and thus the evolution of specialists. Thus, the GFG and MA models have been used to interpret host range evolution in bacteriophages. The study of the interaction among large sets of bacteria and bacteriophage species suggests that the evolution of generalism would be due to a GFG-like interaction and results in a nested structure of infection networks, while the evolution of specialism will be driven by MA-like interactions and results in a modular structure of infection networks (Flores et al., 2011; Weitz et al., 2013). A similar analysis of infection network structure has been recently published based on a 37 virus × 28 plant species infectivity matrix. The matrix was significantly nested, but also included significant modules that corresponded to viruses infecting three particular plant families, indicating two groups of viruses, one built of generalists and the other of specialists on different host families (Moury et al., 2017).

Another important issue in which the predictions of GFG and MA models differ is that fitness penalties (costs) associated with host resistance and viral infectivity are required to maintain polymorphisms at resistance and infectivity loci in the host or pathogen population, respectively, if the interaction is according to the GFG model, but not if it is according to the MA one (Agrawal and Lively, 2002, 2003; Frank, 1996). Costs of infectivity will hinder host range expansions in GFG systems, while adaptation to a new host in an MA system will result in a host shift. It should be noted that pure GFG or MA systems may not exist in nature, and that the assumptions of MA models may be relaxed to allow for partial resistance/infectivity incorporating partial costs of resistance/infectivity (Agrawal and Lively, 2003). A good example of such intermediate systems in plant–virus interactions is that of PVY and recessive resistance in solanaceous hosts (Moury et al., 2014). In any case, the mechanistic assumptions of the MA model, and the costs of extended infectivity in the GFG model both result in the pathogen fitness varying across all its potential hosts. These fitness differences lead to the concept of adaptive trade-offs among hosts: because fitness is environment dependent and the host is a major component of a pathogen's environment, a pathogen cannot simultaneously maximize its fitness in all its potential hosts. Rather, it will become adapted to one or a few related hosts in which fitness will be maximal, i.e., it will evolve to specialization. If adaptation to a particular host implies a fitness cost in other hosts, an adaptive trade-off among hosts will be generated. As a corollary, a generalist would evolve to maximize its fitness across its hosts, but in each individual host its fitness would be lower than the corresponding maxima. Evolution would thus favor specialization rather than generalization.

2.2 Evidence for Across-Host Fitness Trade-Offs

Translating these general concepts about host range evolution to viruses, the hypothesis that generalism incurs a cost is supported by the observation that many plant and animal virus emergences result when a virus jumps from original (reservoir) hosts to a taxonomically related host (Elena et al., 2014b; Longdon et al., 2014). It is further supported by evidence that generalist plant viruses do not infect all hosts with equal efficiency. An analysis of the prevalence of 5 generalist viruses in 21 wild plant species showed significant host–virus associations. The more prevalent viruses were the more host selective (lack of homogeneity of the prevalence of a pathogen among hosts), indicating host selectivity (i.e., host specialization) as a successful strategy in generalist viruses (Malpica et al., 2006). Host selectivity by barley yellow dwarf viruses (BYDV) across different grass species was likewise reported (Power et al., 2011). Together, these results suggest that the multiplication and transmission of generalist viruses vary across hosts with possible adaptive trade-offs, as experimentally demonstrated in some cases (Betancourt et al., 2011; Power et al., 2011; Sacristán et al., 2005).

The nature of these trade-offs and how they affect transmission across host species is an important research area. The plant–virus literature abounds with experimental evidence of fitness penalties associated with the capacity to infect a new host, expressed as a reduced fitness in the primary host. We refer the reader to recent reviews on this subject (Elena et al., 2009, 2014b; García-Arenal and Fraile, 2013; Miyashita et al., 2016). The mechanistic causes of across-host fitness trade-offs have also been extensively analyzed. The simplest cause is antagonistic pleiotropy in which the phenotypic effect of mutations depends on the environment, and mutations that have a positive fitness effect on one given host are deleterious in another one (Whitlock, 1996). Evidence of antagonistic pleiotropy has been reported for several RNA viruses belonging to different taxa, such as PVY, turnip mosaic virus, soybean mosaic virus, hibiscus chlorotic ringspot virus, pepper mild mottle virus (PMMoV), tomato mosaic virus, or pelargonium flower break virus (e.g., Ayme et al., 2006; Fraile et al., 2011; Ishibashi et al., 2012; Janzac et al., 2010; Jenner et al., 2002; Liang et al., 2002; Montarry et al., 2012; Moury and Simon, 2011; Poulicard et al., 2010, 2012; Rico et al., 2006; Wang and Hajimorad, 2016). Antagonistic pleiotropy may be particularly relevant in plant viruses as an unavoidable consequence of their small genomes, which are highly compact, and encode multifunctional proteins. This genomic architecture makes it more difficult to improve one function without jeopardizing another. Epistatic interactions among mutations involved in host adaptation are another cause of across-host trade-offs across hosts (e.g., Ashby et al., 2014a), best studied in plant viruses through experimental evolution approaches (Hillung et al., 2015; Lalic and Elena, 2012a,b). For the role of antagonistic pleiotropy, epistatic, and higher order interactions of adaptive mutations in host range evolution, we refer the reader to an excellent recent review (Bedhomme et al., 2015).

Under a trade-off scenario, the evolution of generalism requires high rates of transmission among heterologous hosts (Wilke et al., 2006): if among-host transmission is infrequent, the virus will adapt to the current host; however, if heterologous transmission is frequent, the viral population behaves as if the fitness landscape were constant and equivalent to the average of each host. This prediction has been confirmed by experiments in which RNA viruses such as vesicular stomatitis virus, eastern encephalitis virus, or Venezuelan encephalitis virus were passaged alternating between different cells, in cell culture, or between different hosts (e.g., Coffey et al., 2008; Cuevas et al., 2003). Also, in a phylogenetic study of 23 bat species (Streicker et al., 2010), it was shown that rabies was transmitted at higher intensities between closely related, rather than distantly related species, and to a lesser extent, between geographically adjacent species. The work suggests that cross-species trade-offs were steeper between distantly related hosts, and is consistent with higher rates of transmission being necessary for adaptation to novel hosts. Thus, to explain the evolution of generalism under assumptions of genetic models, it is necessary to call on ecological parameters, such as transmission rates.

2.3 Environmental Heterogeneity and Across-Host Fitness Trade-Offs

Most evidence of across-host fitness trade-offs derives from experiments in which few virus–host interactions are tested. For instance, a classical experiment is to assay the effects of both a virus genotype that overcomes host resistance [a resistance-breaking (RB) genotype] and its non-RB parent in one resistant and one susceptible genotype of a single-host species. When experiments involve more host species, results are more complex and predictions about host range evolution become more difficult. For example, the assay of 20 single-nucleotide substitution mutants of tobacco etch virus (TEV) on 8 different plant host species showed that the distribution of mutational fitness effects depended on the host, and was more similar to taxonomically related host of the family Solanaceae, which include the natural hosts of this virus (Lalic et al., 2011). Most mutations in TEV genotypes had deleterious effects on the virus fitness in the solanaceous hosts, to which the virus was adapted, but a larger fraction of TEV mutations were classified as beneficial to the virus in hosts from unrelated families, in which the fitness of the wild-type TEV genotype was low. In another study, seven genotypes of PMMoV with coat protein mutations determining RB of different alleles at the *L* resistance locus of pepper were assayed in five susceptible pepper genotypes (Moreno-Pérez et al., 2016). These coat protein RB mutations had pleiotropic effects on within-host virus multiplication, but the sense and magnitude of these fitness effects depended on the specific mutation and on the susceptible host genotype and followed no evident pattern. These two studies underscore the fact that mutations for host adaptation are not associated with trade-offs in all instances of cross-species infection. If trade-offs exist at all, they will depend on the specific mutation and on the specific host. A corollary of this conclusion is that in heterogeneous host populations, such as those a virus will encounter in nature, the evolution of host range will be difficult to predict on the basis of such trade-offs.

In nature, virus populations are also heterogeneous and host plants are often coinfected by different virus species and genotypes (Malpica et al., 2006)—a fact for which deep sequencing analyses of the virome of individual plants or plant populations are providing compelling evidence (Mascia and Gallitelli, 2016). Coinfection results in virus interactions that span the range between synergism and antagonism, and that may determine the evolution of viral traits such as within-host multiplication, virulence, and host range (Elena et al., 2014a; Tollenaere et al., 2016). Costs associated with host range expansion may be affected by coinfection (Fraile et al., 2011). When the effects of RB mutations in PMMoV CP were assayed in coinfections combining different virus genotypes, it was shown that the sense and magnitude of the pleiotropic effects of these mutations on virus fitness depended on the type (single or multiple) of infection, and on the partner genotype in coinfection (Moreno-Pérez et al., 2016). This quite unexpected result unveiled the modulation of cross-host trade-offs by an environmental factor (multiple or single infection) not considered before.

Most published work on across-host fitness trade-offs considers the effects of host adaptation mutations on virus multiplication within the infected host, i.e., on the reproductive component of viral fitness. However, in order to fully understand the evolution of host range, it is necessary to consider potential costs on other fitness components. This is because the evolution of organisms may be constrained by conflicting trade-offs between different fitness components (Alizon et al., 2008; Goldhill and Turner, 2014; Kochin et al., 2010), and one such trade-off predicted by life history evolution models is that between increased reproduction and extended survival. When these theoretical considerations are applied to viruses, the reproduction–survival trade-off would be established between the intracellular (within-host multiplication) and extracellular (survival outside the host) stages of the virus life cycle (Goldhill and Turner, 2014). Unlike in other organisms, reproduction/survival trade-offs may not actually occur in viruses, as it is unclear what mechanistic basis could exist for it; virus survival in the extrahost environment is related to particle stability, which need not be related mechanistically to within-host multiplication (Goldhill and Turner, 2014). However, a trade-off between survival and reproduction does seem to be a general feature in bacteriophages (de Paepe and Taddei, 2006; Dessau et al., 2012; Keen, 2014), although apparently not so in viruses infecting eukaryotes (Handel et al., 2013, 2014;

McGee et al., 2014; Roche et al., 2014). A mechanistic link between reproduction and survival would perhaps not be infrequent in plant viruses, as in many systems host range mutations occur in the coat protein gene (Fraile and García-Arenal, 2010). Indeed, in PMMoV it has been shown that selection for an increased host range selects for altered particle stability and survival, i.e., for a trait unrelated to the plant–virus interaction (Fraile et al., 2014). The analysis of nine genotypes of PMMoV with host adaptation mutations in the coat protein showed that mutations had pleiotropic effects on particle stability (analyzed as a proxy for survival), whose sense and magnitude depended on the specific mutations, and was not correlated with within-host multiplication or host range breadth (Bera et al., 2017). These results do not support a reproduction–survival trade-off, but indicate, as do other studies discussed in this subsection, that across-host fitness trade-offs may not be general and depend on environmental conditions.

We should stress that experimental evidence of across-host trade-offs that are modulated by the environment is in line with analyses that incorporate ecological factors into genetic models of host–pathogen coevolution. Factors such as asynchrony in host and pathogen life cycles, or spatial structure in host and parasite populations or in disease severity also mitigate fitness costs (Ashby et al., 2014b; Brown and Tellier, 2011; Débarre, 2015; Tellier and Brown, 2011). All of these models predict that environmental heterogeneity may result in the maintenance of polymorphisms for host–pathogen specificity in the absence of fitness costs, and highlight the need to consider ecological factors for understanding host range evolution.

3. ECOLOGICAL EFFECTS ON HOST RANGE EVOLUTION

Most models of disease spread require assumptions about the breadth of a pathogen's host range, which are often inaccurate estimates of a pathogen's potential to move among hosts. Inaccurate estimates of host range will negatively affect how we interpret intrinsic (genetic) interactions between the host and the pathogen (Elena, 2017; Laine and Barrès, 2013; Roossinck and García-Arenal, 2015), discussed in Section 2, and the ability to predict the spread of disease (Dallas et al., 2017; García-Arenal and Fraile, 2013; Moury et al., 2017; Schulze-Lefert and Panstruga, 2011). A major concern motivating plant virologists to study host range evolution is the need to understand the process of viral emergence (Elena et al., 2014b; Woolhouse and Gowtage-Sequeria, 2005), which by definition first requires an encounter between the virus and the new host, then adaptation to the new host to ensure between-host transmission (usually by insects or other vectors), and finally optimization of between-host transmission. All three steps require nonexclusive molecular and ecological processes. Transmission is arguably the central process determining disease dynamics and the host ranges of viruses will strongly influence how transmission proceeds. In the previous sections of this review, we focused on the more abundant literature on molecular mechanisms that invoke evolutionary processes that are not independent of ecological processes. Critical knowledge gaps in the relationship between a virus and its hosts, and the influence the ecosystem has on this association, have been acknowledged in the field of plant virology (Malmstrom et al., 2011). In this section, the focus is on ecological processes that are implicitly spatial in nature—the area organisms occupy and the interactions within that space. Thus, we use the conceptual frameworks of ecology to describe spatial relationships and then introduce approaches for quantifying the effects of biological complexity on host range evolution. Unfortunately, most empirical examples explicitly examining scale relationships are from outside the plant virology community. Pioneering work in the field of ecology largely related to spatial relationships, which influences the interpretation of pattern process, is revisited in the first subsection. Specific examples from virology in relation to the ecological concepts presented are discussed at the end of this subsection. The relevance of observations made at different scales is then examined in Sections 3.2 and 3.3 in respect to trait variation and the biological complexity inherent in natural host–pathogen systems. We conclude by summarizing ecological interactions in terms of resource specialization and show that the conceptual frameworks used in ecology have widespread applications in plant virology. Although some of the working definitions presented here will continue to be the subject of debate, the essence of much of the early work by ecologists on scaling is still highly pertinent to future plant virology research. These tools will be necessary for establishing accurate biotic and abiotic contexts for developing realistic predictive models of disease spread in natural systems.

3.1 Spatial Scale and Host Range

Each pathogen and host has a limited spatial distribution and abundance. Generally, abiotic environmental variation constrains spatial distributions and ecological interactions at wide spatial scales because of variation among organisms in physiological, phenological, and behavioral traits. Environmental variation, or heterogeneity, is therefore linked to differences in encounter rates between pathogen and hosts, transmission, and the spread of disease at smaller scales. Observations made at different scales and by different studies may not match because researchers mistakenly treat the processes that produce the patterns they analyze as scale insensitive (Peterson et al., 1998; Ricklefs, 1987; Wiens, 1989). As Levin (1992) put it in reference to localized random disturbance: "[a]s the scale of description is increased beyond the scale of individual disturbances, variability declines, and predictability correspondingly increases." Put another way, the patterns of processes of interest at local scales are averaged over at regional scales, due to a negative variability–area relationship (Peterson et al., 1998). To further compound these discrepancies, there is often limited information about the distribution of pathogens, including plant viruses, in natural environments (Malmstrom et al., 2011; Roossinck and García-Arenal, 2015). Specifically, estimations of the number of hosts a pathogen can infect are prone to sampling bias because the accuracy of this parameter is dependent on the distribution and abundance of hosts (Dallas et al., 2017). Although appearing rather absolute under controlled conditions, host range in natural populations and communities is expected to vary widely (Bascompte, 2010; Bäumler and Fang, 2013; Leventhal et al., 2015; Roossinck, 2015; Woolhouse et al., 2001). This might be especially so when novel encounters occur at high rates in an agricultural ecosystem, or under anthropogenic disturbance, which promote ecological changes such that evolutionary constraints in host range may not explain disease risk (Elena et al., 2014b; Hoberg and Brooks, 2015). Host range is therefore central to quantifying the extent of species interactions, provides information about the potential of a pathogen to spread, and is subject to spatial dependencies. The potential for a pathogen to expand its host range has been examined from either a "phylogenetic relatedness" perspective, where the magnitude of shifts between hosts is measured by the genetic distances between them, or an ecological and environmental point of view, which may be assessed by the degree of niche conservatism (e.g., host range conservatism). Both these approaches rely on considering parameters of the community and environment in which the organism persists.

There is a long tradition of ecologists describing biotic processes on the basis of spatial patterns (Turner, 1989). It is widely recognized that different ecological processes (e.g., predation, herbivory, disturbance) shape species interactions at different scales. It is also the multiplicity of these processes among species that make the prediction of disease spread from spatial patterns alone, unrealistic (Real and McElhany, 1996). Fig. 1 presents a

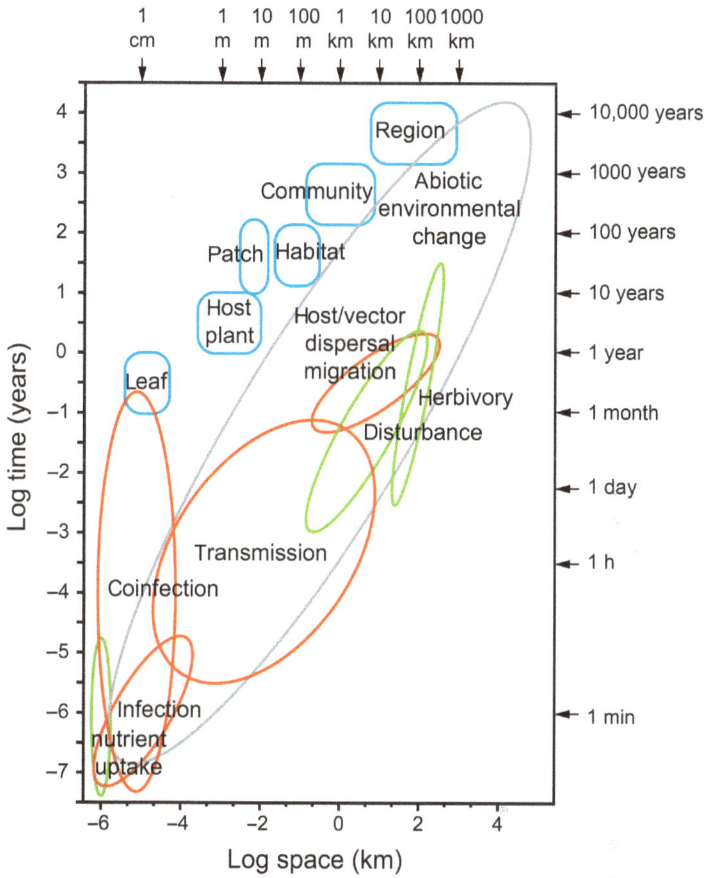

Fig. 1 A hypothetical scheme showing spatial and temporal scale relationships among physical properties and processes of a natural system. The following *color ellipses* represent: *orange*, species interactions; *green*, ecological processes; *blue*, properties of host plants; *gray*, climate and weather variation. *Modified from Peterson, G., Allen, C.R., Holling, C.S., 1998. Ecological resilience, biodiversity, and scale. Ecosystems 1, 6–18.*

hypothetical scheme of processes and the scales they may be applied to in terms of virus infection dynamics. For example, transmission and replication cycles of viruses occur at different spatial scales, and persistence strategies might be distinct between each scale (Handel et al., 2014). Generally, biophysical processes might dominate at small (and fast) scales, while competition for resources might occur at larger and slower scales. Disease pandemics might occur at still larger and even slower scales. At the largest scale, climatic and biogeographic factors are important to variation in and evolution of ecological processes across continents and over geological time. The working definitions used to define scales of ecological interaction (taken from Leibold et al., 2004) are helpful for consistency among studies. A *patch* is generally describing a discrete area of habitat and used interchangeably with *locality*. A locality comprises a *local* community. A region is a large area of habitat with multiple localities. A set of local communities linked by the movement ecology (how, why, when, and where they move; Nathan et al., 2008) of multiple interacting species are a metacommunity. Although the term "dispersal" is somewhat analogous to the movement ecology of an organism, the latter term is arguable more appropriate because plant virus transmission and seed dispersal often involve multiple species within and between communities. At the metacommunity scale, spatial variation in the distributions and abundances of host species and their viruses reveal whether they interact strongly with one another or not (Seabloom et al., 2015). Natural communities are filled with species that might share a given habitat at a given time and affect evolutionary processes among them (Laine, 2006; Thrall et al., 2002). It is entirely feasible that disease risk and host range evolution proceeds according to the community the population persists in.

Identifying key disease risk factors, or patterns, requires knowledge of the host range and distribution and abundance of species to determine the spatial scales at which processes such as pathogen transmission occurs. A pathogen can occupy multiple sites that are subject to changes in community composition, which determine species interactions, the distribution of host traits, and the possible extent of a pathogen's host range. For example, transmission at an ecosystem scale is subject to connectivity among crops and changes in host density (Margosian et al., 2009). Ecological processes are also dependent on the relationships between species traits and spatial scale. For instance, the size of the environment terrestrial mammals occupy has been explained by body mass (Fisher et al., 2011). So-called habitat selection was likewise species specific in mammalian hosts of rabies, but with the density of conspecifics and plant species composition of each community affecting the occupancy of a habitat by each host differently (Tardy et al., 2014). The latter example shows that spatial variation in species interactions can be determined by community composition. A different type of question is whether ecological processes are connected across scales, as is often assumed, or form discontinuities between them. Cross-scale analysis of habitat structure of reefs and the complexity of fish assemblages demonstrated scale-related discontinuities in fish body size distributions (Nash et al., 2013). To date it is unknown whether transmission is relevant to a particular scale or how it might connect other processes between them. Although these types of spatial dependencies have been recognized in the field of ecology since the 19th century (Connor and McCoy, 2001), they are only now being incorporated into a conceptual understanding of pathogen systems (Laine, 2006; Perkins et al., 2013; Thrall et al., 2016). By considering a high level of habitat complexity in multispecies systems, we have to assume that the co-occurrence of pathogens and hosts, and the potential for host range evolution, will be subject to scale patterns. The extent to which species richness changes in relation to area is called a species–area relationship and is a central concept in ecology (Connor and McCoy, 2001). Spatial dependencies influence ecological processes (and how they are interpreted) because "you will find more species if you sample a larger area" (Rosenzweig, 1995). When the abundance of a species is a positive function of sampling area, local extinction is more likely in small areas, and rates of colonization are higher in larger areas. None, some, or all of the host and pathogen species might be present in the area being studied. The composition of species in a given area is also subject to change over time. Species interactions at a local scale may not be representative of either their host range or generalizable ecological patterns. Incorporating explicit statements about scale is important to contextualize observations. As the dimensions of ecological interactions are often not self-evident, observations may or may not be consistent with general expectations.

Virus studies conducted at different scales illustrate the context-specific nature of transmission patterns. For instance, transmission patterns of West Nile virus by three avian host species at roughly a 24-km^2 scale varied among sites as a result of differences in how the mosquito vector fed on bird hosts (Hamer et al., 2011); thus, avian and vector traits associated with selective feeding by vectors on avian hosts had strong local effects on infection prevalence. Although this study did not directly address host range effects on variation in transmission, it demonstrates local effects that might alter perceptions of host range evolution. At still larger scales (ca., hundreds of square kilometers, 30-year period), transmission of canine distemper virus was seen first to be limited to exchanges between two host species (domestic dogs to lions), but later data indicated that other species were required to explain transmission patterns (Viana et al., 2015). Although the scale at which transmission occurs between hosts might be evident, the scope of the ecological interactions may not be. Within the plant virus field a most studied system is that of barley and cereal yellow dwarf viruses (B/CYDV) in annual and perennial grasses. In this system, Borer et al. (2010) addressed the effects of spatial scale on resource use by host species and on infection. The authors were careful to consider the spatial scale at which aphid-vector movement ecology was relevant to virus transmission. A key finding of the study was identifying the spatial scales (100–1000 m^2) at which local community context and infection differences were strongest. These examples highlight the difficulty in estimating the spatial and temporal scales at which species interactions become relevant to the analysis of host range and transmission dynamics, because the scale and nature of the interactions are specific to each study system. Knowing the host range of a virus is difficult to estimate without prior knowledge of the system and the movement ecology of the interacting species. The measurements encompassed at any scale, including those related to host range and transmission, must also be sensitive to an organism's responses to spatial heterogeneity.

Local ecological effects that produce an evolutionary response in a pathogen or host are considered to be features of ecological opportunity (Yoder et al., 2010). Much in the same way ecological opportunity might lead to diversification in herbivorous insects by shifting hosts (Winkler et al., 2009), viruses may also confront varying opportunities to shift hosts due to ecological factors (Parker and Gilbert, 2004; Pedersen and Davies, 2009). Ecological opportunity describes the concept that lineages exposed to a variety of available resources will diversify to utilize a different portion of those resources (Losos, 2010). For example, ecological interactions of humans with wild and domestic animal reservoirs possibly provide pathogens an opportunity to emerge (e.g., Woolhouse and Gaunt, 2007). The resources made available to viruses are a product of the ecological community they belong to because extinction, evolution, and colonization by hosts, vectors, and other viruses are subject to environmental heterogeneity. Local effects (e.g., predators, competitors, temperature, soil nutrients) will cause variation in host–pathogen interactions between local communities. For instance, Malmstrom et al. (2006) showed that clipping (to simulate herbivory) influenced water use of exotic *Bromus* species competing with native grasses. Virus infection reduced survivorship of the native grasses and this effect was intensified under competition. However, clipping ameliorated the resource-depleting effects of competition and survivorship of the native grasses increased. Ecological disturbance that upsets community equilibria (Fraile et al., 2017) also provides opportunity for novel encounters. Generally, when species coexist in the same environments or are closely related, the opportunity to interact is more likely than between spatially and taxonomically distant species (Moury et al., 2017).

Recent studies also show that local adaptation of host species at relatively small spatial scales might be common (Fournier-Level et al., 2011; Hendrick et al., 2016; Kawecki and Ebert, 2004; Müller et al., 2017). Local adaptation represents a source of heterogeneity in fitness effects on plant viruses and traits affecting vector behavior. For instance, the aggregation of fungal pathogens that cause powdery mildew in an herbaceous perennial has been related to the genetic structure of host populations, providing evidence of selection for resistance at local spatial scales (Laine, 2006). However, in pathogen systems, local adaptation can theoretically arise by several mechanisms (Richardson et al., 2014). These mechanisms ultimately rely on host susceptibility and competence, but will be shaped by habitat preferences of pathogens, hosts, and vectors. A hosts' competence to transmit a virus might depend on within-host mechanisms related to virus and vector multiplication (Parker et al., 2015; Seabloom et al., 2009) and/or on feeding preferences by vectors (Borer et al., 2010). An analysis of the effect of compositional changes among communities (relative surface of crops and unmanaged land) on the prevalence of PVY and its aphid vectors showed that percentage of crop land negatively impacted aphid species richness and positively affected virus prevalence. Effects occurred at a small spatial scale ($\leq$1500m) (Claflin et al., 2017a), which may be related to aphid dispersal ranges. It is clear that differences in infection patterns exist at the local community scale because species interactions are relatively variable at this scale compared to larger scales. The potential for host range variation may either be favored or inhibited by these ecological factors that have follow-on effects on emergence given specific characteristics of the community to which they belong.

Continental-scale studies demonstrate that heterogeneity in resource availability to hosts differs among communities and might be an important influence in the evolution of a diversity of virulence and resistance mechanisms (Young et al., 2017). For example, variation in ecological disruption to communities that support viruses infecting wild pepper (chiltepin) over its distribution in Mexico corresponded to differences in infection risk (Fraile et al., 2017). Rodelo-Urrego et al. (2013) showed that landscape variation over Mexico affected the genetic structure of begomoviruses infecting chiltepin populations, and suggested that the spatial patterns reflected vector movement ecology. Another large-scale study considered the relationship of pathogen richness with host species occupancy of habitat types, their geographic range sizes, and native-introduced range status in Europe and North America (Mitchell et al., 2010). The study established that accumulation of pathogen species on exotic hosts was slow compared to other ecological processes such as host shifting. Plant species adapted to higher nitrogen availability in their home ranges tend to experience elevated pathogen attack there and release from pathogens in exotic ranges (Blumenthal et al., 2009). These examples relate how variation in ecological and evolutionary processes that arise due to spatial separation might be most dramatic at larger scales. Differences in how host range evolution may proceed depending on the community are associated with processes propagating from distinct scales.

Process incompatibilities between sites also occur between scales. As the number of biotic interactions increases with the size of a study area, discontinuities among ecological processes that occur at different scales should become stronger. Wiens (1989) referred to these discontinuities as "domains" of scale relevant to particular ecological phenomena. For instance, a study conducted at the global scale on the distributions of three emerging pathogens from different kingdoms (*Batrachochytrium dendrobatidis*, *Borrelia burgdorferi*, and West Nile virus) showed that biotic (the reduction of host species diversity) processes affected ecological interactions at local scales (10^2–10^3 km^2), while abiotic (climate) processes affected ecological interactions at regional scales ($>10^4$ km^2), but the scale differential between processes also depended on the host–pathogen system (Cohen et al., 2016). The ecological processes of niche differentiation (in which competing species utilize the environment differently, e.g., anthropogenic disturbance) and ecological sorting (in which species abundances relate directly to environmental conditions, e.g., climate) can produce different patterns of spatial and temporal species richness similar to the previous example. However, it is unclear to what degree these processes are independent of one another. For example, the processes that result in variation among vertical soil profiles of mycorrhizal fungi communities might involve either deterministic factors affecting population dynamics at the microhabitat scale or stochastic factors generated by climate (Bahram et al., 2015). Together, these studies show that specific processes can be relevant to distinct scale domains, and may also interfere with one another to influence on how species occupy a space. Critically, to make predictions about the state (e.g., infection risk status) of a community, ecosystem, or region, it is important to understand how pattern–process relationships are distributed across scales (de Bello et al., 2013; Holling, 2001; Nash et al., 2014; Stoffels et al., 2005).

In summary, spatial scales are critical to realizing the extent of interactions that pathogens potentially encounter, and provide essential clues about the ways in which processes relate to one another. Dependencies between ecological processes at different scales remain a future challenge to research in ecology in general and are also pertinent to understanding host range evolution (Nash et al., 2013, 2014; O'Dwyer and Cornell, 2017). To achieve accurate predictions of disease spread, it is necessary to identify processes appropriate to particular domains of scale, and those that connect them. We must also be aware that processes are expected to vary among study systems and sites of comparable scale. Host range evolution by a particular plant virus is therefore likely to depend on molecular exchanges, local-scale species encounters, be influenced by the biotic community and regional effects of biodiversity, and likely to vary widely at regional and continental scales. At the very least, any steps taken to developing spatially explicit hypotheses (spatial pattern) will improve our ability to compare process relating to different study systems, and how processes at different scales relate to one another.

3.2 Effects of Ecological Community on Host Range Evolution

In biological communities it is necessary to identify complex species interactions to understand transmission, infection, and host range evolution. The paucity of studies on biologically complex host–pathogen systems reflects the challenge of accurately quantifying transmission in any multispecies community (Caron et al., 2015; Fenton et al., 2015; Johnson et al., 2015b). Studies of single-host–single-parasite interactions (Agrawal and Lively, 2002; Thrall et al., 2016) often assume homogeneous, randomly mixed populations of hosts and pathogens (Perkins et al., 2013), thus ignoring the effect biological complexity has on host range evolution. The likelihood and intensity of emerging disease outbreaks (Bedhomme et al., 2012; Rottstock et al., 2014; Strauss et al., 2015), the evolution of virulence in plant generalists (Betancourt et al., 2013), causes of widespread and persistent diseases (Johnson et al., 2015a), and pathogen species invasions (Mitchell et al., 2010) all require an understanding of multispecies interactions. For instance, species that dominate transmission (Streicker et al., 2013) and changes to communities that affect reservoir competence (Swei et al., 2011) influence a pathogen's movement among hosts. Community ecology offers plant virologists a way forward in understanding how these processes occur in biologically complex systems and how a virus's relationship with a community influences its host range.

An ecological community has been described as any "association of interacting populations, usually defined by the nature of their interaction, or by the place in which they live" (Ricklefs and Miller, 1999). Due to practicalities, plant virology studies have tended to define relatively few species associations as communities, and assume negligible host range effects. It is important to move beyond these "reduced" types of communities to those that clearly represent a higher biological complexity. Most experimental studies typically use model species, or those that are relatively easy to maintain either under controlled conditions in the greenhouse or under natural conditions. The most recent syntheses on host–pathogen biology advocate community-level approaches that address much wider biological organization and a higher order of ecological complexity (Alexander, 2015; Johnson et al., 2015a,b; Parker et al., 2015; Seabloom et al., 2015; Suzán et al., 2015; Thrall et al., 2016; Tollenaere et al., 2016; Viana et al., 2014; Young et al., 2017). In particular, knowledge about the ecology of viruses in wild plant and animal communities is critical to understanding the influence host range has on viral diseases in crops and humans (Alexander et al., 2014; Elena et al., 2014a; McLeish et al., 2017; Murray and Daszak, 2013; Roossinck and García-Arenal, 2015; Simpson et al., 2012). Biologically complex wild communities present many unique practical and empirical challenges for which the field of community ecology has developed robust methods to accommodate. Recent developments emphasize the distribution of traits, as opposed to species, in biological systems (Cavender-Bares et al., 2009; McGill et al., 2006). Groupings of species that share a trait or traits may be considered together as one functional group in the analysis of species interactions (Hooper and Vitousek, 1997; Leibold and McPeek, 2006). Species interactions within communities can therefore be quantified at a "higher" level of biological organization. Interactions are described in terms of traits such as a pathogen species specificity for a host-related trait (Gómez et al., 2010; Thrall et al., 2007), host disease resistance and tolerance (Barrett et al., 2009), and pathogen virulence (e.g., Alizon et al., 2013; Kirchner and Roy, 2002). Considering interactions between traits and phylogenetic relatedness of hosts and parasites/pathogens has also proven useful in generalizing about patterns of host range variation and pathogen diversification in communities (Schulze-Lefert and Panstruga, 2011). Parasitic and trophic interactions have also been investigated to understand their impact on transmission (Dáder et al., 2012) and herbivory (Mauck et al., 2015).

Johnson et al. (2015a) provide three motivations for the adoption by disease researchers of community ecology-based approaches: (i) to manage ecological complexity; (ii) to identify drivers of heterogeneities in hosts and pathogens; and (iii) to understand cross-scale processes. These motivations clearly point to translating analogous processes in community ecology to disease systems. Community ecology emphasizes four general processes and interactions among them: ecological drift, selection, dispersal, and speciation, all of which contribute to community dynamics (Vellend, 2010).

Stochastic processes, those with no deterministic effects on species abundances, produce *ecological drift* in community composition and are especially important in smaller communities. Ecological drift changes how interactions within a community occur and the frequency of species moving among habitats within those communities (Orrock and Watling, 2010). Evolutionary change in plant viruses can be linked to selection and speciation processes. Similarly, dispersal can be linked to transmission, as ecological drift can be used as a proxy for community composition changes. Species *coexistence* in communities (indefinite persistence of a defined set of species within a given space) largely depends on *selection*, due to fitness "trade-offs" under different conditions. Trade-offs may occur due to differences in host, pathogen, or vector densities, or in abiotic factors (Handel et al., 2014). *Dispersal* connects habitats and communities, and *speciation* adds species to communities. Each of these four processes connects species traits in the context of an ecological community and can be used to investigate a number of ecological and evolutionary mechanisms. Advocacy of community ecology is appropriate because local-scale processes, such as possible adaptive trade-offs of viruses across hosts, can be linked to those at larger scales such as transmission and emergence (Borer et al., 2013), as stressed in Section 2.2. In any multispecies community identifying connectivity between source and target populations of transmission within and between natural communities is challenging (Alexander et al., 2014). Pathogen reservoirs are expected to range from single-host populations to complex communities of mixed maintenance and nonmaintenance populations (Caron et al., 2015; Haydon et al., 2002; Murray and Daszak, 2013; Viana et al., 2014). Host range may affect the potential of pathogens to persist in species-rich reservoir communities. The movement ecology of vectors and pathogens, and direct and indirect factors affecting transmission, is therefore required to determine how disease spreads within and between reservoir communities and new hosts.

Individual pairs of species may interact directly through processes such as predation, competition, and symbioses. With the introduction of a third species, the fitness dependencies between the first pair of species change. Indirect interactions between them arise because they depend on the relative density of the third one. As such, these ecological processes influence viral transmission within communities via complex pathways. For example, the presence of parasitoids caused a reduction of aphid vectors that coincided with an increase in bean yellow mosaic virus infection due to a decrease in aphid resting times, thereby increasing their movement and thus virus transmission among plants (Hodge et al., 2011). Differences in the timing of transmission between cucumber mosaic virus (CMV) and cucumber aphid-borne yellows virus depended on the introduction of a parasitoid of their aphid vector (Dáder et al., 2012). The endosymbiotic bacterium *Hamiltonella* that is hosted by the whitefly *Bemisia tabaci*, which vectors tomato yellow leaf curl virus (TYLCV), has been implicated in the recent spread of TYLCV in China (Su et al., 2013). Regardless of the density of whiteflies, transmission frequency of TYLCV was significantly different when the bacterial symbiont was present in the vector. In all the above cases, virus transmission was dependent on the diversity of species and the distribution of their traits in the community.

In a community ecology context, a number of theoretical models have been invoked to explain the coexistence of species (Begon et al., 2002; Seabloom et al., 2015). Two models commonly used to explain disease transmission (Keesing et al., 2006) rely either on host contact rate as a function of population density (a density-dependent model) or on frequency of contact (in frequency-dependent models). The difference is essential that the contact rate is held constant in the frequency-dependent model regardless of the density of the population (Begon et al., 2002). In a two-host community of voles and mice, for example, frequency-dependent models best explained cowpox virus transmission within each species (Begon et al., 1999). However, the study also revealed negligible transmission between the two-host species, and an unexplained increase of infection in one of the host species, which suggested interactions with a third host occurred during the experiment. Gilbert et al. (2001) used a model parameterized with empirical data from wild populations of hosts (grouse, hare, and deer) of louping ill virus and its tick vectors, to predict virus persistence under changing host combinations and densities. The model was conditioned on density-dependent assumptions and showed that the addition of host species was related to increased virus persistence. However, the authors acknowledge that their model assumed that species densities were in equilibrium over time when this was not held up by the empirical data, and that model parameters (i.e., density-dependent constraints) were the same across sites when they were not. The study shows that density-dependent models of transmission are applicable only on a case-by-case basis. The probability of contact between an infectious and susceptible host in both models is assumed to be spatially homogeneous and constant. Other studies on transmission of zoonotic diseases in multihost systems have dispensed with the assumption of these models and instead focus on traits to average over species interactions when host heterogeneity becomes considerable. For example, asymmetrical infections between host species and differential vector and virus amplification suggest apparent competition between hosts (e.g., Alexander and Holt, 1998; Malmstrom, 1998). Apparent competition occurs when competition between two coexisting species occurs indirectly, where the interaction of a natural enemy mediates their abundances (Holt, 1977). Power and Mitchell (2004) relate this process to spillover and pathogens moving between reservoir and nonreservoir species. Likewise, Malmstrom et al. (2005a,b, 2006) have shown that displacement of native perennial grasses by exotic annual grasses in California grasslands is at least in part mediated by apparent competition due to the effects of infection by B/CYDV: the exotic grasses were more competent hosts and amplified aphid-vector populations, so that virus prevalence in natives was increased by their presence. In turn, infection substantially reduced plant growth, survivorship, and fecundity in natives, whose low-density population structure was more susceptible to disease-induced losses than were the exotics. Apparent competition between native and exotic grasses due to the effect of fungal seed pathogens has also been reported (Mordecai, 2013). These studies explain dynamics in terms of traits and their effects on species abundances, densities, and incidence and speculate about how changes to host diversity might have influenced these observations.

Disease dynamics in species diverse systems have generally related community parameters, such as richness and diversity, to disease prevalence (Johnson et al., 2013). Species diversity is typically described as a function of species richness (the number of species) and often also the species relative abundance (evenness) (Hurlbert, 1971; Mendes et al., 2008). Diversity is therefore a metric of community structure. Biodiversity ("biological variation," Gaston, 2000) has been identified as a key factor that shapes disease risk. Hypotheses propose that biodiversity may either increase or decrease the risk of infection (Keesing et al., 2006; Lacroix et al., 2014a; Ostfeld and Keesing, 2000). However, given the evidence that species distributions are shaped by multiple sources of variation at different scales, it is difficult to predict disease risk from metrics of biodiversity alone (Randolph and Dobson, 2012; Strauss et al., 2015). It seems essential that species identity and/or specific traits be considered as well. For instance, infection prevalence in three species of rodents (gerbils) depended on both the effect of their richness and the specificity traits of the microbial symbionts (Kedem et al., 2014). The work of Borer et al. (2009) illustrates how community parameters can be used in multispecies communities to assess their influence in plant virus infection patterns. Indirect effects of herbivores on the prevalence of B/CYDV in annual invasive grasses were quantified by evaluating virus prevalence in areas with and without a history of grazing by large herbivores.

Herbivory was associated with greater relative abundance of exotic annual grasses, which have greater competence for virus transmission than other resident species. The elevated relative abundance of competent hosts (an altered community composition) in grazed areas increased B/CYDV infection risk. This work underscores the value of assessing community interactions as a function of species traits and host diversity on infection patterns. Virus diversity within individual hosts in natural systems can also affect transmission in communities.

Coinfecting pathogens that share hosts have been defined as within-species communities that also represent another change in the scale of interactions and processes. Within-host processes have consequences for ecological community processes. Host range, variation in vector transmission efficiency, and within-host accumulation of pathogens in coinfected hosts are hypothesized to affect plant virus distributions (Elena et al., 2014a; Mascia and Gallitelli, 2016; Wintermantel et al., 2008). Coinfection communities of the aphid-transmitted viruses *Alfalfa mosaic virus* (AMV), *Beet western yellows virus* (BWYV), *CMV*, and *Watermelon mosaic virus* (WMV), and the thrips-transmitted *Tomato spotted wilt virus* (TSWV) in wild plant species were sampled from within an agroecosystem in Central Spain (Malpica et al., 2006). This study of the prevalence of infection by the 5 virus species in 21 wild plant species found that 18% of infected plants were infected by more than one virus. In another study, coinfections by BYDV and CYDV in a perennial grass species were common; among the 17% of plants found to have infection, 70% were infected by more than one virus (Seabloom et al., 2009). Community ecology approaches offer means of elucidating within-plant–pathogen interactions and the generation of novel genetic combinations and diversity, and host-mediated mechanisms (Tollenaere et al., 2016). Within-host interactions might have a role in viral host range evolution because coinfection has positive and negative influences on transmission (Pedersen and Fenton, 2007). For example, Seabloom et al. (2009) demonstrated spatial and temporal heterogeneity within communities of luteoviruses and poleroviruses. The study suggested that the community composition of vector species with different host ranges, and their relative degree of specialization, influences disease prevalence and affects cross-protection mechanisms. Thus, coinfections represent model systems for understanding small community dynamics (Seabloom et al., 2015), but also for investigating how processes at different scales interact. Vector ecology connects within-host processes with transmission and the structuring of viral reservoirs of vector-borne diseases in biologically complex communities.

Vector and host heterogeneity determine the intensity of interspecific transmission, which underlies mechanisms involved in host range evolution. Recapture experiments with whiteflies indicated bimodal ranges of movement with primarily dispersal flights of less than 500 m, with occasional migration flights of 2.5–10 km (Byrne, 1999). The study provided evidence for the spatial scale at which transmission can occur and how vector behavior might influence pathogen distributions in communities. For instance, the spatial scale of whitefly movement in another study was similar to that of genetic structure of their vectored begomoviruses (Rodelo-Urrego et al., 2013) and suggests the scales of virus and host evolution might be relatively similar. However, the diversity of vector species present in a community will affect transmission because of variation in their feeding behavior and movement ecology. For example, aphid movement increased with the initial density of aphid populations, but their community composition, not richness per se, increased the rate of PVY transmission (Claflin et al., 2017b). Analyses of models parameterized with experimental data on the transmission of BYDV and PVY suggest a positive correlation between population growth rate of aphid vectors and the rate of disease spread (Shaw et al., 2017). Together these studies show that vector movement ecology and diversity might have a strong effect on prevalence and rate of infection at scales between 500 m and 10 km. These examples are evidence that transmission dynamics results in infection heterogeneity at larger scales. This patchwork of virus distributions is embedded in a heterogeneous community of hosts with differences in their susceptibility and competence to transmit infection.

The trait-based perspective of community ecology is insightful because the distribution of traits relating to resource use (e.g., generalism) and host range evolution might be distinct from host species distributions (Barrett et al., 2015; Pilosof et al., 2014; Poisot et al., 2011). Intraspecific variation in virulence or resistance, or phenotypic plasticity in species more generally, is arguable better treated in terms of trait distributions instead of individual species. This is important to realize because host range evolution might be largely modulated by intra- and interspecific variation in traits associated with host resource use, not purely the number of hosts (abundance) or species (richness) that a community comprises. Therefore, virus host jumps might be both a function of phylogenetic relatedness among hosts (Alexander, 2015; Parker et al., 2015) and convergence on host traits that have evolved independently in multiple lineages (e.g., Tatineni et al., 2011; Vassilakos et al., 2016). Cronin et al. (2010) tested the hypothesis that reservoir potential was related to host physiological phenotype by assaying six grass species for susceptibility to BYDV infection, competence for transmission, and capacity to sustain vector populations. Short-lived hosts, with nutrient-rich and high metabolism tissues, ranged higher for these epidemiological traits, while phylogenetic relatedness or geographic provenance of the hosts had no explanatory effects. Subsequent model-based analyses determined that differences in resistance were important in explaining the higher reservoir potential of hosts with faster developmental tempo (Cronin et al., 2014). The relationship between the developmental traits of hosts and reservoir potential has also been addressed in an experimental study of the interaction of CMV with short- and long-lived genotypes of *Arabidopsis thaliana* (Hily et al., 2014). Results showed that host life span was not related to host competence in transmitting the virus, rather the short-lived host was more susceptible to infection than the long-lived host, and better able to sustain vector populations. The lower defenses of the short-lived host resulted in higher reservoir potential for infection, but this was balanced by the longer infectious period of the long-lived host. The balance depended on the density and genetic composition of the host population. The relationship between a host's life span and its reservoir potential illustrates how host traits might better predict transmission potential than might pathogen host ranges. Knowledge of trait distributions is thus a useful "shortcut" for summarizing complex biological interactions.

3.3 Habitat-Specific Host Range: Specialization of Generalist Viruses

Host range is a virus trait that only partly defines its specificity in host use, in addition to the qualities of the host that have been discussed earlier. In host–pathogen systems, the definitions of specialist and generalist are often given in terms of the number of hosts a pathogen uses to fulfill the needs of reproduction. There has been much interest in these forms of resource use across the whole of evolution and ecology (Agrawal and Lively, 2002; Ehrlich and Raven, 1964; Futuyma and Moreno, 1988; Hubbell, 2001; Hutchinson, 1957; Sexton et al., 2017). Hence, the conceptual framework goes much deeper than describing the host range of a pathogen or vector and rather emphasizes factors that determine resource use. Typically, the host range of a pathogen is enumerated as the number of hosts, rather than that of host-related resources. As such, distinctions between the host itself, the host's traits, and the host's environment tend to be ignored. For example, B/CYDV distributions in six grass species across a 700-km latitudinal gradient varied with host species identity, soil nutrient availability, and vector host species preferences (Seabloom et al., 2013). The differences in environmental conditions, hosts, or vectors associated with resource availability at each site explained variation in the virus community compositions. Distinguishing such traits that are associated with the breadth of resource use is tantamount to understanding the process of adaptation to a subset of possible environments: i.e., ecological specialization (Poisot et al., 2011). Similarly, niche theory (Hutchinson, 1957) explains the relationship between the resource breadth of a species (an n-dimensional hypervolume with axes representing environmental variables) and its resource exploitation strategy. A narrow range of resources used by a consumer reflects a small niche breadth (Ackermann and Doebeli, 2004). In other words, specialists are expected to exhibit highest fitness variation across habitats due to relatively specific requirements, whereas generalists do not. Host range has often been used analogously with niche breadth (e.g., Fellous et al., 2014) when examining variation in performance among available hosts. A major step in the development of niche theory has been the introduction of the "realized niche" concept, which refers to the subset of the available resources that a species actually uses, or the fundamental niche. The fundamental niche emphasizes genetic potentials in the face of environmental variation (Futuyma and Moreno, 1988; Soberón and Peterson, 2005). For a pathogen, the fundamental niche may include resources such as susceptible host tissue and the host genotype (Al-Naimi et al., 2005). The niche may also include ecological interactions and result in changes to the expression of those genes (MacArthur, 1970). Competition, predation, mutualism, herbivory, and parasitism affecting any phase of host range evolution (Coyte et al., 2015; Gilbert et al., 2001; Hily et al., 2016; Malmstrom et al., 2006; Orrock and Watling, 2010; Roossinck, 2015; Thrall et al., 2007) restrict the number of dimensions available in the hypervolume of an individual's niche. Thus, as the number of resources available for specialization (dimensionality of the niche) increases, the types of traits expected to be important to host range evolution at either local- or wider-scale processes (competition and local extinction vs colonization and patch connectivity) are expected to be different (Kneitel and Chase, 2004).

Pathogen host range restrictions have generally been assessed either at the species or population level, as a consequence of trade-offs in the performance and adaptation of pathogens to certain host genotypes (see Section 2.2). As pointed out earlier, fitness trade-offs are questionable and might not be a strong determinant in plant virus host range evolution (Egas et al., 2004; Elena, 2017; Lambrechts et al., 2006) because host genotypes that can be infected by a specific pathogen genotype vary between environments

(Wolinska and King, 2009). Preferences for particular hosts that result from differential costs between them (i.e., gene by gene) or their environments (i.e., gene by environment) might give rise to trade-offs that are hypothesized to result in specialization on a host (Agrawal, 2000) or habitat (Kassen, 2002). Broad host ranges are possibly maintained by selection across heterogeneous environments, while narrow host ranges are hypothesized to evolve in environmentally homogeneous and stable habitats (Kassen, 2002). Theoretically, species interactions are shaped by differentiation of the niches they are associated with, in response to spatial patchiness of suitable habitat. For instance, Seabloom et al. (2010), studying four different B/CYDV species, related within-host virus diversity to coinfection rates across a latitudinal gradient, with differences in the strength of niche differentiation between communities. The study showed that the ability of the four viruses to coexist was largely determined by their traits associated with niche assembly—where a species trait determines whether it is able to establish in a given locality (see Chase, 2007). The strength of such trait-based mechanisms was conditional on environmental factors including vector ecology. In other studies, it was shown that the prevalence and interactions of BYDV and CYDV in one host species were dependent on availability of limiting resources such as nitrogen and phosphorus (Lacroix et al., 2014b) that are determinants of their realized niche. Together, the distributions and abundances of traits, along with other processes such as intra- and interspecific competitive interactions, determine habitat quality for a pathogen. To explain the evolution of variation, maintenance, and stability of species niche breadths (e.g., variation in host range), genetic correlations with fitness in different environments must also be considered in terms of the amount of gene flow between them. Variation in habitat quality, due to environmental heterogeneity, makes host range evolution subject to spatial and temporal variation in gene flow, which in turn depends on the movement ecology of pathogens and hosts. For example, patterns of host range breadth among populations of the pathogenic fungus *Colletotrichum cereale* that infects exotic grasses in North America suggested that low levels of gene flow and environmental isolation affected the degree of specialization (Crouch et al., 2008). These studies demonstrate that not all host species of a pathogen are genetically or ecologically equivalent. Host specificity is a trait that is not only a measure of host range, but how closely hosts are genetically related to one another (Poulin and Mouillot, 2003), and so provides information about the lability of host traits influencing the degree of specialization. Therefore, the distinction between the niche breadth and host range of either "generalist" or "specialist" pathogens is not trivial. Differences in host specificity of plant viruses might be explained by variation in habitat quality and gene flow over their geographic range, which determines the availability of resources at a given space in time.

Differences in the breadth of resource use need not be explained as a function of species traits in a given environment, or by differential restrictions to resources depicted in high-dimensional hypervolumes. If fitness trade-offs are not strong among available host species (Bedhomme et al., 2012; Elena, 2017), the primary forces acting on pathogen–host range evolution might be driven by stochastic fluctuations in community composition, local extinctions and speciation, and the movement ecology of pathogens (e.g., dispersal characteristics of vectors). These processes have been examined under the neutral theory of ecology (Hubbell, 2001), which assumes functional equivalence between species at the community level (Hubbell, 2005). For example, neutral models of community assembly have been used to examine pathogen community composition in the absence of fitness trade-offs (Seabloom et al., 2015) and applied to cases of specialized and generalized resource use (Pandit et al., 2009). These studies suggest that, even in the absence of trade-offs, environmental heterogeneity and habitat connectivity have a part in the evolution of a species resource breadth. For instance, "generalist" and "specialist" plant viruses show particular resource use characteristics (Malpica et al., 2006) that might restrict or promote host range evolution (Woolhouse et al., 2005). In a multihost system, habitat heterogeneity and host range phenotypes of a suite of 11 generalist plant virus species strongly affected the prediction of infection prevalence (McLeish et al., 2017). Infection prevalence only became conditional on host diversity when habitat heterogeneity was considered. Predictive models were conditioned on spatial scale with a negative or positive prevalence–diversity relationship depending on host diversity of a given habitat. In this study, the realized host range estimates for each plant virus further altered the prevalence–diversity relationships. The changes to the realized host ranges of each plant virus across habitats were consistent with definitions of facultative generalism (Shipley et al., 2009). Facultative generalism has been reported in mammalian consumers (e.g., Sokos et al., 2015), but is also consistent with observations of facultative alterations of virulence in *Plasmodium* parasites (Taylor et al., 1998). Theoretically, facultative generalists have the broadest fundamental niches. This resource use strategy differs from obligate generalists essentially because preferences among resources change with time and become deterministic rather than being driven purely by resource encounter frequencies, as occurs in obligate generalists (Glasser, 1982). For example, preferences among hosts by vector-borne viruses develop in conjunction with their movement ecology. Viral gene products associated with host and vector manipulation alter transmission by changing host traits that affect the attractiveness, settlement, or feeding behavior (Fereres, 2015; Fereres and Moreno, 2009; Groen et al., 2017; Mauck et al., 2016). This type of manipulation has been observed in the homopteran and thrips vectors of viruses such as CMV (Mauck et al., 2010; Ziebell et al., 2011), BYDV and potato leafroll virus (Bosque-Pérez and Eigenbrode, 2011; Fereres et al., 1989), TYLCV (Fang et al., 2013), PVY (Boquel et al., 2012), and TSWV (Stafford et al., 2011). Overall, neutral model predictions highlight the contribution of stochastic factors to pathogen distributions in the environment, which are necessary to understand the evolution of host range. Resource exploitation strategies, like facultative generalism, imply conditional responses to environmental heterogeneity and the existence of niche differences among hosts and pathogens. Exploitation strategies often involve "trade-offs" for limited resources such as host species, and the ability of a pathogen species to exploit more than one resource or evolve to exploit another (Barrett and Heil, 2012) is often necessary to enable their coexistence.

In summary, selection for wider host ranges might involve both direct and indirect mechanisms. To understand the ecology and evolution of host range, it is critical to identify the traits responsible for specialist and generalist interactions involving molecular (Genin, 2010), species (Philibert et al., 2011), or environmental (Hily et al., 2016) factors. If many plant viruses infect more than one host species, then it is valuable to relate host range to the distribution of resources available to the viruses (e.g., as shaped by the distribution of resistance and tolerance traits), the vectors (e.g., diversity of the vector's host plants), and hosts (e.g., nutrients, soil moisture, pollinators) in the environment. Generally, a pathogen's host range or its habitat specificity will determine which phyla are more likely to be new hosts (Clark and Clegg, 2017). Given the ecological complexity of natural communities, establishing the niche dimensions of plant viruses is a useful approach to establishing factors important to host range evolution under specific conditions. The factors that restrict ecological interactions include biotic processes such as competition (e.g., coinfection), predation (e.g., on vectors), and herbivory (e.g., on hosts), and abiotic variables influencing environmental heterogeneity. Ecological interactions among species that mediate host range evolution can therefore be described between genes, between genes and the environment, and via deterministic and stochastic processes that govern changes to community composition.

4. FUTURE DIRECTIONS

Biological complexity in plant–virus interactions in natural environments inevitably produces analytical challenges such as multivariate data sets with dissimilar probability distributions (Warton et al., 2015), taxonomic anomalies (Blüthgen et al., 2008; Novotny et al., 2002), and spatial or temporal scale dependencies (Nash et al., 2014; O'Dwyer and Cornell, 2017; Peterson et al., 1998). The ability to generate larger data sets using data-mining or high-throughput approaches has spawned sophisticated methods for combining geographical, phenotypic, and phylogenetic variables to understand the role the environment has in structuring species interactions and communities. For example, the "metacommunity" approach is used to model interactions among local communities (Leibold et al., 2004) linked by dispersal at larger spatial scales (Declerck et al., 2013; Suzán et al., 2015). The conceptual overlap between ecology and epidemiology is complemented by niche modeling approaches that seek to determine the distribution of viruses using their niche requirements (Escobar and Craft, 2016). The availability of comprehensive climate and environmental data, and improved techniques for ecological niche modeling has been increasingly utilized in understanding environmental limitations on virus and host distributions at global (Samy and Peterson, 2016), interregional (Fraile et al., 2017), and regional (Sallam et al., 2016) scales. It therefore will be possible to separate the environmental conditions suited to a host and the viruses that infect it, and these conditions need not overlap because the physiological requirements of the pathogen need not correspond to those of hosts.

Network-based approaches quantify ecological interactions within communities by connecting species based on interdependencies among them and examine the architecture of this connectivity to understand ecological and evolutionary processes (Bascompte, 2010). Connecting hundreds of species in this way has been useful for interpreting the distribution of traits that are embedded in biologically complex interactions (Andreazzi et al., 2017; Lewinsohn et al., 2006). For instance, networks have been used to model microbial competition involved in stabilizing gut communities (Coyte et al., 2015), to infer the probability of transmission (May and Lloyd, 2001; Newman, 2002), to explore disease resistance among hosts (Jousimo et al., 2014), to compare antagonistic and mutualistic interactions (Olesen et al., 2007; Thébault and Fontaine, 2010; Welti and Joern, 2015), and to identify the causes of the spread of infectious disease in heterogeneous environments (Craft et al., 2011; Leventhal et al., 2015). Network approaches have also been used to identify keystone species that might have important structural or functional roles in a community (Jordano et al., 2003; Macfadyen et al., 2011). For instance, the presence of generalist species with high levels of connectivity in ecological networks implied that generalist species act as "hubs" that potentially stabilize communities (Ings et al., 2009; Solé and Montoya, 2001; Toju et al., 2017). Network analysis has also been used to compare the niche and "neutral" concepts explaining host–parasite interactions (Canard et al., 2014). Niche and neutral theories (Hubbell, 2001) have evolved as competing hypotheses used to explain the structure of natural communities and provide scope for comparing "null" or stochastic expectations against deterministic analyses (Leibold and McPeek, 2006). The most recent developments in the prediction of species distributions integrate quantitative parameters, derived from community ecology approaches, with variables related to regional processes such as climate data (Staniczenko et al., 2017; Vayssier-Taussat et al., 2014). These analytical frameworks have previously been treated separately due to scale incompatibilities. The need for scale-sensitive methods in the analysis of nonlinear dynamical systems is arguably driving the forefront of development in elucidating on distribution and abundance patterns and species interactions (Bjørnstad, 2015; Bokler, 1993; O'Dwyer and Cornell, 2017; Sugihara and May, 1990). Neutral theory (Hubbell, 2001) also represents a suitable candidate for cross-scale spatial modeling (Rosindell and Cornell, 2013). A number of techniques to resolve relationships between discontinuities in processes at different scales have been available for some time (outlined in Purse et al., 2015; Stow et al., 2007).

A substantial step in understanding large- and small-scale ecological processes in disease systems has come with the advent of high-throughput sequencing technologies. Information on the host range of a virus is essential to predicting avenues for transmission, the influence of maintenance species on virus persistence, and the effective implementation of disease control strategies. Bulk sampling approaches for the detection of the virus (or viruses) are becoming increasingly available (Ho and Tzanetakis, 2014; Massart et al., 2017) and can be used to expand the scope of measurements of ecological interactions evaluated in a study (Vayssier-Taussat et al., 2014). Metagenomics approaches are commonly used for the purposes of identification (Massart et al., 2017), detection (Massart et al., 2014), and discovery (Massart et al., 2017) of plant viruses and these can be improved in combination with traditional techniques (Blouin et al., 2016). Metagenomic approaches supersede the prerequisite of amplifying cDNA/DNA of a target virus (Blawid et al., 2017; Picard et al., 2017; Roossinck et al., 2015). The taxonomic scope, spatial, and temporal scales of investigative studies can be considerably enlarged using an approach that pools samples from identified host plants gathered from larger areas known as "ecogenomics" (Roossinck et al., 2010). Metagenomics has also resulted in considerable development and revision of how virus taxonomy is conducted (Simmonds et al., 2017). (For additional discussion of metagenomics applications with plant viruses, see Claverie et al., 2018.)

In conclusion, the application of spatially and temporally sensitive as well as high-throughput approaches that integrate biological complexity in plant virus systems is still in its infancy. Readily available approaches can be used to understand how the distribution and abundance of hosts, vectors, and plant viruses result in particular ecological interactions at different scales. The ability to rapidly gather data over many host and vector taxa will go a long way to unlocking the influence species interactions, and the environment, have on the evolution of host range in the enormous numbers of known and unknown plant viruses.

We trust that this review of the role of ecological complexity in plant virus host range evolution will convince the reader this recent field of research addresses a topic with an exciting future. Understanding the role of ecological complexity in plant–virus interactions is complementary to earlier studies of the molecular genetics, population genetics, and evolutionary factors that shape virus host ranges. Unlocking scale relationships among these factors will provide scope for future lines of research. The time is ripe for collaboration with plant community ecologists at well-inventoried sites to examine viral dynamics in community assembly at the global scale.

ACKNOWLEDGMENT

This work was funded by Plan Estatal de I+D+i, MINECO, Spain (BFU2015-64018-R).

REFERENCES

Ackermann, M., Doebeli, M., 2004. Evolution of niche width and adaptive diversification. Evolution 58, 2599–2612.

Agrawal, A.A., 2000. Host-range evolution: adaptation and trade-offs in fitness of mites on alternative hosts. Ecology 81, 500–508.

Agrawal, A., Lively, C.M., 2002. Infection genetics: gene-for-gene versus matching-alleles models and all points in between. Evol. Ecol. Res. 4, 91–107.

Agrawal, A.F., Lively, C.M., 2003. Modelling infection as a two-step process combining gene-for-gene and matching-allele genetics. Proc. Biol. Sci. 270, 323–334.

Aguas, R., Ferguson, N.M., 2013. Feature selection methods for identifying genetic determinants of host species in RNA viruses. PLoS Comput. Biol. 9 (10), e1003254.

Alexander, H.M., 2015. Ecology: shared ancestry predicts disease levels. Nature 520, 446–447.

Alexander, H.M., Holt, R.D., 1998. The interaction between plant competition and disease. Perspect. Plant Ecol. Evol. Syst. 1, 206–220.

Alexander, H.M., Mauck, K.E., Whitfield, A.E., Garrett, K.A., Malmstrom, C.M., 2014. Plant-virus interactions and the agro-ecological interface. Eur. J. Plant Pathol. 138, 529–547.

Alizon, S., Hurford, A., Mideo, N., van Baalen, M., 2008. Virulence evolution and the trade-off hypothesis: history, current state of affairs and the future. J. Evol. Biol. 22, 245–259.

Alizon, S., de Roode, J.C., Michalakis, Y., 2013. Multiple infections and the evolution of virulence. Ecol. Lett. 16, 556–567.

Al-Naimi, F.A., Garrett, K.A., Bockus, W.W., 2005. Competition, facilitation, and niche differentiation in two foliar pathogens. Oecologia 143, 449–457.

Andreazzi, C.S., Thompson, J.N., Guimarães Jr., P.R., 2017. Network structure and selection asymmetry drive coevolution in species-rich antagonistic interactions. Am. Nat. 190, 99–115.

Antonovics, J., Boots, M., Ebert, D., Koskella, B., Poss, M., Sadd, B.M., 2013. The origin of specificity by means of natural selection: evolved and nonhost resistance in host–pathogen interactions. Evolution 67, 1–9.

Arnold, A.E., Lutzoni, F., 2007. Diversity and host range of foliar fungal endophytes: are tropical leaves biodiversity hotspots? Ecology 88, 541–549.

Ashby, B., Gupta, S., Buckling, A., 2014a. Effects of epistasis on infectivity range during host-parasite coevolution. Evolution 68, 2972–2982.

Ashby, B., Gupta, S., Buckling, A., 2014b. Spatial structure mitigates fitness costs in host-parasite coevolution. Am. Nat. 183, E64–E74.

Asplen, M.K., Bruns, E., David, A.S., Denison, R.F., Epstein, B., Kaiser, M.C., et al., 2012. Do trade-offs have explanatory power for the evolution of organismal interactions? Evolution 66, 1297–1307.

Ayme, V., Souche, S., Caranta, C., Jacquemond, M., Chadoeuf, J., Palloix, A., et al., 2006. Different mutations in the genome-linked protein VPg of *Potato virus Y* confer virulence on *the* $pvr2^3$ resistance in pepper. Mol. Plant-Microbe Interact. 19, 557–563.

Bahram, M., Peay, K.G., Tedersoo, L., 2015. Local-scale biogeography and spatiotemporal variability in communities of mycorrhizal fungi. New Phytol. 205, 1454–1463.

Barrett, L.G., Heil, M., 2012. Unifying concepts and mechanisms in the specificity of plant–enemy interactions. Trends Plant Sci. 17, 282–292.

Barrett, L.G., Kniskern, J.M., Bodenhausen, N., Zhang, W., Bergelson, J., 2009. Continua of specificity and virulence in plant host–pathogen interactions: causes and consequences. New Phytol. 183, 513–529.

Barrett, L., Encinas-Viso, F., Thrall, P.H., Burdon, J., 2015. Specialization for resistance in wild host-pathogen interaction networks. Front. Plant Sci. 6, 761.

Bascompte, J., 2010. Structure and dynamics of ecological networks. Science 329, 765–766.

Bäumler, A., Fang, F.C., 2013. Host specificity of bacterial pathogens. Cold Spring Harb. Perspect. Med. 3, a010041.

Bedhomme, S., Lafforgue, G., Elena, S.F., 2012. Multihost experimental evolution of a plant RNA virus reveals local adaptation and host-specific mutations. Mol. Biol. Evol. 29, 1481–1492.

Bedhomme, S., Hillung, J., Elena, S.E., 2015. Emerging viruses: why they are not jacks of all trades? Curr. Opin. Virol. 10, 1–6.

Begon, M., Hazel, S.M., Baxby, D., Bown, K., Cavanagh, R., Chantrey, J., et al., 1999. Transmission dynamics of a zoonotic pathogen within and between wildlife host species. Proc. Biol. Sci. 266, 1939–1945.

Begon, M., Bennett, M., Bowers, R.G., French, N.P., Hazel, S.M., Turner, J., 2002. A clarification of transmission terms in host-microparasite models: numbers, densities and areas. Epidemiol. Infect. 129, 147–153.

Bera, S., Moreno-Pérez, M.G., García-Figuera, S., Pagán, I., Fraile, A., Pacios, L.F., et al., 2017. Pleiotropic effects of resistance-breaking mutations on particle stability provide insight into life history evolution of a plant RNA virus. J. Virol. 91, e00435-17.

Betancourt, M., Fraile, A., García-Arenal, F., 2011. *Cucumber mosaic virus* satellite RNAs that induce similar symptoms in melon plants show large differences in fitness. J. Gen. Virol. 92, 1930–1938.

Betancourt, M., Escriu, F., Fraile, A., García-Arenal, F., 2013. Virulence evolution of a generalist plant virus in a heterogeneous host system. Evol. Appl. 6, 875–890.

Bjørnstad, O.N., 2015. Nonlinearity and chaos in ecological dynamics revisited. Proc. Natl. Acad. Sci. U.S.A. 112, 6252–6253.

Blawid, R., Silva, J.M.F., Nagata, T., 2017. Discovering and sequencing new plant viral genomes by next-generation sequencing: description of a practical pipeline. Ann. Appl. Biol. 170, 301–314.

Blouin, A.G., Ross, H.A., Hobson-Peters, J., O'Brien, C.A., Warren, B., MacDiarmid, R., 2016. A new virus discovered by immunocapture of double-stranded RNA, a rapid method for virus enrichment in metagenomic studies. Mol. Ecol. Resour. 16, 1255–1263.

Blumenthal, D., Mitchell, C.E., Pyšek, P., Jarošík, V., 2009. Synergy between pathogen release and resource availability in plant invasion. Proc. Natl. Acad. Sci. U.S.A. 106, 7899–7904.

Blüthgen, N., Fründ, J., Vázquez, D.P., Menzel, F., 2008. What do interaction network metrics tell us about specialization and biological traits. Ecology 89, 3387–3399.

Bokler, B., 1993. Chaos and complexity in measles models: a comparative numerical study. Math. Med. Biol. 10, 83–95.

Boquel, S., Delayen, C., Couty, A., Giordanengo, P., Ameline, A., 2012. Modulation of aphid vector activity by *Potato virus Y* on in vitro potato plants. Plant Dis. 96 (1), 82–86.

Borer, E.T., Mitchell, C.E., Power, A.G., Seabloom, E.W., 2009. Consumers indirectly increase infection risk in grassland food webs. Proc. Natl. Acad. Sci. U.S.A. 106, 503–506.

Borer, E.T., Seabloom, E.W., Mitchell, C.E., Power, A.G., 2010. Local context drives infection of grasses by vector-borne generalist viruses. Ecol. Lett. 13, 810–818.

Borer, E.T., Kinkel, L.L., May, G., Seabloom, E.W., 2013. The world within: quantifying the determinants and outcomes of a host's microbiome. Basic Appl. Ecol. 14, 533–539.

Bosque-Pérez, N.A., Eigenbrode, S.D., 2011. The influence of virus-induced changes in plants on aphid vectors: insights from luteovirus pathosystems. Virus Res. 159, 201–205.

Brown, J.K.M., Tellier, A., 2011. Plant-parasite coevolution: bridging the gap between genetics and ecology. Annu. Rev. Phytopathol. 49, 345–367.

Brunt, A., Crabtree, K., Dallwitz, M., Gibbs, A., Watson, L., 1996. Viruses of Plants: Descriptions and Lists From the VIDE Database. CAB International, UK. 1484 pp.

Byrne, D.N., 1999. Migration and dispersal by the sweet potato whitefly, *Bemisia tabaci*. Agric. For. Meteorol. 97, 309–316.

Canard, E.F., Mouquet, N., Mouillot, D., Stanko, M., Miklisova, D., Gravel, D., 2014. Empirical evaluation of neutral interactions in host-parasite networks. Am. Nat. 183, 468–479.

Caron, A., Cappelle, J., Cumming, G.S., de Garine-Wichatitsky, M., Gaidet, N., 2015. Bridge hosts, a missing link for disease ecology in multi-host systems. Vet. Res. 46, 83.

Cavender-Bares, J., Kozak, K.H., Fine, P.V., Kembel, S.W., 2009. The merging of community ecology and phylogenetic biology. Ecol. Lett. 12, 693–715.

Chase, J.M., 2007. Drought mediates the importance of stochastic community assembly. Proc. Natl. Acad. Sci. U.S.A. 104, 17430–17434.

Claflin, S.B., Jones, L.E., Thaler, J.S., Power, A.G., 2017a. Crop-dominated landscapes have higher vector-borne plant virus prevalence. J. Appl. Ecol. 54, 1190–1198.

Claflin, S.B., Power, A.G., Thaler, J.S., 2017b. Aphid density and community composition differentially affect apterous aphid movement and plant virus transmission. Ecol. Entomol. 42, 245–254.
Clark, N.J., Clegg, S.M., 2017. Integrating phylogenetic and ecological distances reveals new insights into parasite host specificity. Mol. Ecol. 26, 3074–3086.
Claverie, S., Bernardo, P., Kraberger, S., Hartnady, P., Lefeuvre, P., Lett, J.-M., et al., 2018. From spatial metagenomics to molecular characterization of plant viruses: a geminivirus case study. Adv. Virus Res. 101, 55–83.
Coffey, L.L., Vasilakis, N., Brault, A.C., Powers, A.M., Tripet, F., Weaver, S.C., 2008. Arbovirus evolution in vivo is constrained by host alternation. Proc. Natl. Acad. Sci. U.S.A. 105, 6970–6975.
Cohen, J.M., Civitello, D.J., Brace, A.J., Feichtinger, E.M., Ortega, C.N., Richardson, J.C., et al., 2016. Spatial scale modulates the strength of ecological processes driving disease distributions. Proc. Natl. Acad. Sci. U.S.A. 113, E3359–E3364.
Connor, E.F., McCoy, E.D., 2001. Species-area relationships. In: Encyclopedia of Biodiversity. vol. 5. Academic Press (Elsevier Science), San Diego, CA, pp. 397–411.
Coyte, K.Z., Schluter, J., Foster, K.R., 2015. The ecology of the microbiome: networks, competition, and stability. Science 350, 663–666.
Craft, M.E., Volz, E., Packer, C., Meyers, L.A., 2011. Disease transmission in territorial populations: the small-world network of Serengeti lions. J. R. Soc. Interface 8, 776–786.
Cronin, J.P., Welsh, M.E., Dekkers, M.G., Abercrombie, S.T., Mitchell, C.E., 2010. Host physiological phenotype explains pathogen reservoir potential. Ecol. Lett. 13, 1221–1232.
Cronin, J.P., Rúa, M.A., Mitchell, C.E., 2014. Why is living fast dangerous? Disentangling the roles of resistance and tolerance of disease. Am. Nat. 184, 172–187.
Crouch, J.A., Tredway, L.P., Clarke, B.B., Hillman, B.I., 2008. Phylogenetic and population genetic divergence correspond with habitat for the pathogen *Colletotrichum cereale* and allied taxa across diverse grass communities. Mol. Ecol. 18, 123–135.
Cuevas, J.M., Moya, A., Elena, S.F., 2003. Evolution of RNA virus in spatially structured heterogeneous environments. J. Evol. Biol. 16, 456–466.
Dáder, B., Moreno, A., Viñuela, E., Fereres, A., 2012. Spatio-temporal dynamics of viruses are differentially affected by parasitoids depending on the mode of transmission. Viruses 4, 3069–3089.
Dallas, T., Huang, S., Nunn, C., Park, A.W., Drake, J.M., 2017. Estimating parasite host range. Proc. Biol. Sci. 284, 20171250.
de Bello, F.D., Lavorel, S., Lavergne, S., Albert, C.H., Boulangeat, I., Mazel, F., et al., 2013. Hierarchical effects of environmental filters on the functional structure of plant communities: a case study in the French Alps. Ecography 36, 393–402.
Débarre, F., 2015. Fitness costs in spatially structured environments. Evolution 69, 1329–1335.
Declerck, S.A., Winter, C., Shurin, J.B., Suttle, C.A., Matthews, B., 2013. Effects of patch connectivity and heterogeneity on metacommunity structure of planktonic bacteria and viruses. ISME J. 7, 533–542.
de Paepe, M., Taddei, F., 2006. Viruses' life history: towards a mechanistic basis of a trade-off between survival and reproduction among phages. PLoS Biol. 4, e193.
Dessau, M., Goldhill, D., McBride, R.C., Turner, P.E., 2012. Selective pressure causes an RNA virus to trade reproductive fitness for increased structural and thermal stability of a viral enzyme. PLoS Genet. 8, e10003102.
de Vienne, D.M., Refrégier, G., López-Villavicencio, M., Tellier, A., Hood, M.E., Giraud, T., 2013. Cospeciation vs host-shift speciation: methods for testing, evidence from natural associations and relation to coevolution. New Phytol. 198, 347–385.
Duffy, S., Burch, C.L., Turner, P.E., 2007. Evolution of host specificity drives reproductive isolation among RNA viruses. Evolution 61, 2614–2622.

Dybdahl, M.F., Jenkins, C.E., Nuismer, S.L., 2014. Identifying the molecular basis of host-parasite coevolution: merging models and mechanisms. Am. Nat. 184, 1–13.
Egas, M., Dieckmann, U., Sabelis, M.W., 2004. Evolution restricts the coexistence of specialists and generalists: the role of trade-off structure. Am. Nat. 163, 518–531.
Ehrlich, P.R., Raven, P.H., 1964. Butterflies and plants: a study in coevolution. Evolution 18, 586–608.
Elena, S.F., 2017. Local adaptation of plant viruses: lessons from experimental evolution. Mol. Ecol. 26, 1711–1719.
Elena, S.F., Agudelo-Romero, P., Lalic, J., 2009. The evolution of viruses in multi-host fitness landscapes. Open Virol. J. 3, 1–6.
Elena, S.F., Bernet, G.P., Carrasco, J.L., 2014a. The games plant viruses play. Curr. Opin. Virol. 8, 62–67.
Elena, S.F., Fraile, A., García-Arenal, F., 2014b. Evolution and emergence of plant viruses. Adv. Virus Res. 88, 161–191.
Escobar, L.E., Craft, M.E., 2016. Advances and limitations of disease biogeography using ecological niche modeling. Front. Microbiol. 7, 1174.
Fang, Y., Jiao, X., Xie, W., Wang, S., Wu, Q., Shi, X., … Zhang, Y., 2013. Tomato yellow leaf curl virus alters the host preferences of its vector *Bemisia tabaci*. Sci. Rep. 3, 2876.
Fellous, S., Angot, G., Orsucci, M., Migeon, A., Auger, P., Olivieri, I., et al., 2014. Combining experimental evolution and field population assays to study the evolution of host range breadth. J. Evol. Biol. 27, 911–919.
Fenton, A., Streicker, D.G., Petchey, O.L., Pedersen, A.B., 2015. Are all hosts created equal? Partitioning host species contributions to parasite persistence in multihost communities. Am. Nat. 186, 610–622.
Fereres, A., 2015. Insect vectors as drivers of plant virus emergence. Curr. Opin. Virol. 10, 42–46.
Fereres, A., Moreno, A., 2009. Behavioural aspects influencing plant virus transmission by homopteran insects. Virus Res. 141, 158–168.
Fereres, A., Lister, R.M., Araya, J.E., Foster, J.E., 1989. Development and reproduction of the English grain aphid (Homoptera: Aphididae) on wheat cultivars infected with barley yellow dwarf virus. Environ. Entomol. 18, 388–393.
Fisher, J.T., Anholt, B., Volpe, J.P., 2011. Body mass explains characteristic scales of habitat selection in terrestrial mammals. Ecol. Evol. 1, 517–528.
Flores, C.O., Meyer, J.R., Valverde, S., Farr, L., Weitz, J.S., 2011. Statistical structure of host-phage interactions. Proc. Natl. Acad. Sci. U.S.A. 108, E288–E297.
Fournier-Level, A., Korte, A., Cooper, M.D., Nordborg, M., Schmitt, J., Wilczek, A.M., 2011. A map of local adaptation in Arabidopsis thaliana. Science 334, 86–89.
Fraile, A., García-Arenal, F., 2010. The coevolution of plants and viruses: resistance and pathogenicity. Adv. Virus Res. 76, 1–32.
Fraile, A., Pagán, I., Anastasio, G., Sáez, E., García-Arenal, F., 2011. Rapid genetic diversification and high fitness penalties associated with pathogenicity evolution in a plant virus. Mol. Biol. Evol. 28, 1425–1437.
Fraile, A., Hily, J.-M., Pagán, I., Pacios, L.F., García-Arenal, F., 2014. Host resistance selects for traits unrelated to resistance-breaking that affect fitness in a plant virus. Mol. Biol. Evol. 31, 928–939.
Fraile, A., McLeish, M.J., Pagán, I., González-Jara, P., Piñero, D., García-Arenal, F., 2017. Environmental heterogeneity and the evolution of plant-virus interactions: viruses in wild pepper populations. Virus Res. 241, 68–76.
Frank, S.A., 1996. Models of parasite virulence. Q. Rev. Biol. 71, 37–78.
Futuyma, D.J., Moreno, G., 1988. The evolution of ecological specialization. Annu. Rev. Ecol. Syst. 19, 207–233.

García-Arenal, F., Fraile, A., 2013. Trade-offs in host range evolution of plant viruses. Plant Pathol. 62, 2–9.

García-Arenal, F., McDonald, B.A., 2003. An analysis of the durability of resistance to plant viruses. Phytopathology 93, 941–952.

Gaston, K.J., 2000. Global patterns in biodiversity. Nature 405, 220–227.

Genin, S., 2010. Molecular traits controlling host range and adaptation to plants in *Ralstonia solanacearum*. New Phytol. 187, 920–928.

Gilbert, L., Norman, R., Laurenson, K.M., Reid, H.W., Hudson, P.J., 2001. Disease persistence and apparent competition in a three-host community: an empirical and analytical study of large-scale, wild populations. J. Anim. Ecol. 70, 1053–1061.

Glasser, J.W., 1982. A theory of trophic strategies: the evolution of facultative specialists. Am. Nat. 119, 250–262.

Goldhill, D.H., Turner, P.E., 2014. The evolution of life history trade-offs in viruses. Curr. Opin. Virol. 8, 79–84.

Gómez, J.M., Verdú, M., Perfectti, F., 2010. Ecological interactions are evolutionarily conserved across the entire tree of life. Nature 465, 918–921.

Groen, S.C., Wamonje, F.O., Murphy, A.M., Carr, J.P., 2017. Engineering resistance to virus transmission. Curr. Opin. Virol. 26, 20–27.

Hamer, G.L., Chaves, L.F., Anderson, T.K., Kitron, U.D., Brawn, J.D., Ruiz, M.O., et al., 2011. Fine-scale variation in vector host use and force of infection drive localized patterns of West Nile virus transmission. PLoS One 6, e23767.

Handel, A., Brown, J., Stallknecht, D., Rohani, P., 2013. A multi-scale analysis of influenza A virus fitness trade-offs due to temperature-dependent virus persistence. PLoS Comput. Biol. 9, e1002989.

Handel, A., Lebarbenchon, C., Stallknecht, D., Rohani, P., 2014. Trade-offs between and within scales: environmental persistence and within-host fitness of avian influenza viruses. Proc. Biol. Sci. 281, 20133051.

Haydon, D.T., Cleaveland, S., Taylor, L.H., Laurenson, M.K., 2002. Identifying reservoirs of infection: a conceptual and practical challenge. Emerg. Infect. Dis. 8, 1468–1473.

Hendrick, M.F., Finseth, F.R., Mathiasson, M.E., Palmer, K.A., Broder, E.M., Breigenzer, P., et al., 2016. The genetics of extreme microgeographic adaptation: an integrated approach identifies a major gene underlying leaf trichome divergence in Yellowstone *Mimulus guttatus*. Mol. Ecol. 25, 5647–5662.

Hillung, J., Cuevas, J.M., Elena, S.F., 2015. Evaluating the within-host fitness effects of mutations fixed during virus adaptation to different ecotypes of a new host. Philos. Trans. R. Soc. Lond. Ser. B Biol. Sci. 370, 9.

Hily, J.M., García, A., Moreno, A., Plaza, M., Wilkinson, M.D., Fereres, A., et al., 2014. The relationship between host lifespan and pathogen reservoir potential: an analysis in the system Arabidopsis thaliana-Cucumber mosaic virus. PLoS Pathog. 10, e1004492.

Hily, J.M., Poulicard, N., Mora, M.Á., Pagán, I., García-Arenal, F., 2016. Environment and host genotype determine the outcome of a plant–virus interaction: from antagonism to mutualism. New Phytol. 209, 812–822.

Ho, T., Tzanetakis, I.E., 2014. Development of a virus detection and discovery pipeline using next generation sequencing. Virology 471, 54–60.

Hoberg, E.P., Brooks, D.R., 2015. Evolution in action: climate change, biodiversity dynamics and emerging infectious disease. Philos. Trans. R. Soc. Lond. Ser. B Biol. Sci. 370, 20130553.

Hodge, S., Hardie, J., Powell, G., 2011. Parasitoids aid dispersal of a nonpersistently transmitted plant virus by disturbing the aphid vector. Agric. For. Entomol. 13, 83–88.

Holling, C.S., 2001. Understanding the complexity of economic, ecological, and social systems. Ecosystems 4, 390–405.

Holt, R.D., 1977. Predation, apparent competition, and the structure of prey communities. Theor. Popul. Biol. 12, 197–229.
Hooper, D.U., Vitousek, P.M., 1997. The effects of plant composition and diversity on ecosystem processes. Science 277, 1302–1305.
Hubbell, S.P., 2001. The Unified Neutral Theory of Biodiversity and Biogeography. Princeton University Press, Princeton, NJ.
Hubbell, S.P., 2005. Neutral theory in community ecology and the hypothesis of functional equivalence. Funct. Ecol. 19, 166–172.
Hurlbert, S.H., 1971. The nonconcept of species diversity: a critique and alternative parameters. Ecology 52, 577–586.
Hutchinson, G.E., 1957. Concluding remarks. Cold Spring Harb. Symp. Quant. Biol. 22, 415–427.
Ings, T.C., Montoya, J.M., Bascompte, J., Blüthgen, N., Brown, L., Dormann, C.F., et al., 2009. Ecological networks—beyond food webs. J. Anim. Ecol. 78, 253–269.
Ishibashi, K., Mawatari, N., Miyashita, S., Kishino, H., Meshi, T., Ishikawa, M., 2012. Coevolution and hierarchical interactions of *Tomato mosaic virus* and the resistance gene *Tm-1*. PLoS Pathog. 8, e10002975.
Janz, N., Nyblom, K., Nylin, S., 2001. Evolutionary dynamics of host-plant specialization: a case study of the tribe Nymphalini. Evolution 55, 783–796.
Janzac, B., Montarry, J., Palloix, A., Navaud, O., Moury, B., 2010. A point mutation in the polymerase of *Potato virus Y* confers virulence toward the *pvr4* resistance of pepper and a high competitiveness cost in susceptible cultivar. Mol. Plant-Microbe Interact. 23, 823–830.
Jenner, C.E., Wang, X., Ponz, F., Walsh, J.A., 2002. A fitness cost for *Turnip mosaic virus* to overcome host resistance. Virus Res. 86, 1–6.
Johnson, P.T., Preston, D.L., Hoverman, J.T., Richgels, K.L., 2013. Biodiversity decreases disease through predictable changes in host community competence. Nature 494, 230–233.
Johnson, P.T., De Roode, J.C., Fenton, A., 2015a. Why infectious disease research needs community ecology. Science 349, 1259504.
Johnson, P.T., Ostfeld, R.S., Keesing, F., 2015b. Frontiers in research on biodiversity and disease. Ecol. Lett. 18, 1119–1133.
Jordano, P., Bascompte, J., Olesen, J.M., 2003. Invariant properties in coevolutionary networks of plant–animal interactions. Ecol. Lett. 6, 69–81.
Jousimo, J., Tack, A.J., Ovaskainen, O., Mononen, T., Susi, H., Tollenaere, C., et al., 2014. Ecological and evolutionary effects of fragmentation on infectious disease dynamics. Science 344, 1289–1293.
Kassen, R., 2002. The experimental evolution of specialists, generalists, and the maintenance of diversity. J. Evol. Biol. 15, 173–190.
Kawecki, T.J., Ebert, D., 2004. Conceptual issues in local adaptation. Ecol. Lett. 7, 1225–1241.
Kedem, H., Cohen, C., Messika, I., Einav, M., Pilosof, S., Hawlena, H., 2014. Multiple effects of host-species diversity on coexisting host-specific and host-opportunistic microbes. Ecology 95, 1173–1183.
Keen, E.C., 2014. Trade-offs in bacteriophage life histories. Bacteriophage 4, e28365.
Keesing, F., Holt, R.D., Ostfeld, R.S., 2006. Effects of species diversity on disease risk. Ecol. Lett. 9, 485–498.
Kirchner, J.W., Roy, B.A., 2002. Evolutionary implications of host–pathogen specificity: fitness consequences of pathogen virulence traits. Evol. Ecol. Res. 4, 27–48.
Kneitel, J.M., Chase, J.M., 2004. Trade-offs in community ecology: linking spatial scales and species coexistence. Ecol. Lett. 7, 69–80.

Kochin, B.F., Bull, J.J., Antia, R., 2010. Parasite evolution and life history theory. PLoS Biol. 8, e1000524.
Lacroix, C., Jolles, A., Seabloom, E.W., Power, A.G., Mitchell, C.E., Borer, E.T., 2014a. Non-random biodiversity loss underlies predictable increases in viral disease prevalence. J. R. Soc. Interface 11, 20130947.
Lacroix, C., Seabloom, E.W., Borer, E.T., 2014b. Environmental nutrient supply alters prevalence and weakens competitive interactions among coinfecting viruses. New Phytol. 204, 424–433.
Laine, A.L., 2006. Evolution of host resistance: looking for coevolutionary hotspots at small spatial scales. Proc. Biol. Sci. 273, 267–273.
Laine, A.L., Barrès, B., 2013. Epidemiological and evolutionary consequences of life-history trade-offs in pathogens. Plant Pathol. 62, 96–105.
Lajeunesse, M.J., Forbes, M.R., 2002. Host range and local parasite adaptation. Proc. Biol. Sci. 269, 703–710.
Lalic, J., Elena, S.F., 2012a. Epistasis between mutations is host-dependent for an RNA virus. Biol. Lett. 9, 20120396.
Lalic, J., Elena, S.F., 2012b. Magnitude and sign epistasis among deleterious mutations in a positive-sense plant RNA virus. Heredity 109, 71–77.
Lalic, J., Cuevas, J.M., Elena, S.F., 2011. Effect of host species on the distribution of mutational fitness effects for an RNA virus. PLoS Genet. 7, e1002378.
Lambrechts, L., Fellous, S., Koella, J.C., 2006. Coevolutionary interactions between host and parasite genotypes. Trends Parasitol. 22, 12–16.
Leibold, M.A., McPeek, M.A., 2006. Coexistence of the niche and neutral perspectives in community ecology. Ecology 87, 1399–1410.
Leibold, M.A., Holyoak, M., Mouquet, N., Amarasekare, P., Chase, J.M., Hoopes, M.F., et al., 2004. The metacommunity concept: a framework for multi-scale community ecology. Ecol. Lett. 7, 601–613.
Leventhal, G.E., Hill, A.L., Nowak, M.A., Bonhoeffer, S., 2015. Evolution and emergence of infectious diseases in theoretical and real-world networks. Nat. Commun. 6, 6101.
Levin, S.A., 1992. The problem of pattern and scale in ecology: the Robert H. MacArthur award lecture. Ecology 73, 1943–1967.
Lewinsohn, T.M., Inácio Prado, P., Jordano, P., Bascompte, J., Olesen, J.M., 2006. Structure in plant–animal interaction assemblages. Oikos 113, 174–184.
Liang, X.-Z., Lee, B.T.K., Wong, S.-M., 2002. Covariation in the capsid protein of hibiscus chlorotic ringspot virus induced by serial passaging in a host that restricts movement leads to avirulence in its systemic host. J. Virol. 76, 12320–12324.
Longdon, B., Brockhurst, M.A., Russell, C.A., Welch, J.J., Jiggins, F.M., 2014. The evolution and genetics of virus host shifts. PLoS Pathog. 10, e1004395.
Losos, J.B., 2010. Adaptive radiation, ecological opportunity, and evolutionary determinism: American Society of Naturalists EO Wilson award address. Am. Nat. 175, 623–639.
MacArthur, R.H., 1970. Species packing and competitive equilibria for many species. Theor. Popul. Biol. 1, 1–11.
Macfadyen, S., Gibson, R.H., Symondson, W.O., Memmott, J., 2011. Landscape structure influences modularity patterns in farm food webs: consequences for pest control. Ecol. Appl. 21, 516–524.
Malmstrom, C.M., 1998. Barley yellow dwarf virus in native California grasses. Grasslands 8, 6–10.
Malmstrom, C.M., Hughes, C.C., Newton, A., Stoner, C.J., 2005a. Virus infection in remnant native bunchgrasses from invaded California grasslands. New Phytol. 168, 217–230.

Malmstrom, C.M., McCullough, A.J., Johnson, H.A., Newton, L.A., Borer, E.T., 2005b. Invasive annual grasses indirectly increase virus incidence in California native perennial bunchgrasses. Oecologia 145, 153–164.
Malmstrom, C.M., Stoner, C.J., Brandenburg, S., Newton, L.A., 2006. Virus infection and grazing exert counteracting influences on survivorship of native bunchgrass seedlings competing with invasive exotics. J. Ecol. 94, 264–275.
Malmstrom, C.M., Melcher, U., Bosque-Pérez, N.A., 2011. The expanding field of plant virus ecology: historical foundations, knowledge gaps, and research directions. Virus Res. 159, 84–94.
Malpica, J.M., Sacristán, S., Fraile, A., García-Arenal, F., 2006. Association and host selectivity in multi-host pathogens. PLoS One 1, e41.
Margosian, M.L., Garrett, K.A., Hutchinson, J.S., With, K.A., 2009. Connectivity of the American agricultural landscape: assessing the national risk of crop pest and disease spread. Bioscience 59, 141–151.
Mascia, T., Gallitelli, D., 2016. Synergies and antagonisms in virus interactions. Plant Sci. 252, 176–192.
Massart, S., Olmos, A., Jijakli, H., Candresse, T., 2014. Current impact and future directions of high throughput sequencing in plant virus diagnostics. Virus Res. 188, 90–96.
Massart, S., Candresse, T., Gil, J., Lacomme, C., Predajna, L., Ravnikar, M., et al., 2017. A framework for the evaluation of biosecurity, commercial, regulatory, and scientific impacts of plant viruses and viroids identified by NGS technologies. Front. Microbiol. 8, 45.
Mauck, K.E., De Moraes, C.M., Mescher, M.C., 2010. Deceptive chemical signals induced by a plant virus attract insect vectors to inferior hosts. Proc. Natl. Acad. Sci. U.S.A. 107 (8), 3600–3605.
Mauck, K.E., Smyers, E., De Moraes, C.M., Mescher, M.C., 2015. Virus infection influences host plant interactions with non-vector herbivores and predators. Funct. Ecol. 29, 662–673.
Mauck, K.E., De Moraes, C.M., Mescher, M.C., 2016. Effects of pathogens on sensory-mediated interactions between plants and insect vectors. Curr. Opin. Plant Biol. 32, 53–61.
May, R.M., Lloyd, A.L., 2001. Infection dynamics on scale-free networks. Phys. Rev. E Stat. Nonlinear Soft Matter Phys. 64, 066112.
McDonald, B.A., Linde, C., 2002. Pathogen population genetics, evolutionary potential, and durable resistance. Annu. Rev. Phytopathol. 40, 349–379.
McGee, L.W., Aitchison, E.W., Caudle, S.B., Morrison, A.J., Zheng, L., Yang, W., et al., 2014. Payoffs, not tradeoffs, in the adaptation of a virus to ostensibly conflicting selective pressures. PLoS Genet. 10, e1004611.
McGill, B.J., Enquist, B.J., Weiher, E., Westoby, M., 2006. Rebuilding community ecology from functional traits. Trends Ecol. Evol. 21, 178–185.
McLeish, M.J., Sacristán, S., Fraille, A., García-Arenal, F., 2017. Scale dependencies and generalism in host use shape virus prevalence. Proc. Biol. Sci. 284, 20172066.
Mendes, R.S., Evangelista, L.R., Thomaz, S.M., Agostinho, A.A., Gomes, L.C., 2008. A unified index to measure ecological diversity and species rarity. Ecography 31, 450–456.
Mitchell, C.E., Power, A.G., 2003. Release of invasive plants from fungal and viral pathogens. Nature 421, 625–627.
Mitchell, C.E., Blumenthal, D., Jarošík, V., Puckett, E.E., Pyšek, P., 2010. Controls on pathogen species richness in plants' introduced and native ranges: roles of residence time, range size and host traits. Ecol. Lett. 13, 1525–1535.
Miyashita, Y., Atsumi, G., Nakahara, K.S., 2016. Trade-offs for viruses in overcoming innate immunities in plants. Mol. Plant-Microbe Interact. 29, 595–598.

Montarry, J., Cartier, E., Jacquemond, M., Palloix, A., Moury, B., 2012. Virus adaptation to quantitative plant resistance: erosion or breakdown? J. Evol. Biol. 25, 2242–2252.
Mordecai, E.A., 2013. Despite spillover, a shared pathogen promotes native plant persistence in a cheatgrass-invaded grassland. Ecology 94, 2744–2753.
Moreno-Pérez, M.G., García-Luque, I., Fraile, A., García-Arenal, F., 2016. Mutations that determine resistance breaking in a plant RNA virus have pleiotropic effects on its fitness that depend on the host environment and on the type, single or mixed, of infection. J. Virol. 90, 9128–9137.
Moury, B., Simon, V., 2011. dN/dS-based methods detect positive selection linked to trade-offs between different fitness traits in the coat protein of *Potato virus Y*. Mol. Biol. Evol. 28, 2707–2717.
Moury, B., Janzac, B., Ruellan, Y., Simon, V., Ben Khalifa, M., Fakhfakh, H., et al., 2014. Interaction patterns between Potato virus Y and *eIF4E*-mediated recessive resistance in the Solanaceae. J. Virol. 88, 9799–9807.
Moury, B., Fabre, F., Hébrard, E., Froissart, R., 2017. Determinants of host species range in plant viruses. J. Gen. Virol. 98, 862–873.
Müller, C.M., Schulz, B., Lauterbach, D., Ristow, M., Wissemann, V., Gemeinholzer, B., 2017. *Geropogon hybridus* (L.) Sch. Bip. (Asteraceae) exhibits micro-geographic genetic divergence at ecological range limits along a steep precipitation gradient. Plant Syst. Evol. 303, 91–104.
Murray, K.A., Daszak, P., 2013. Human ecology in pathogenic landscapes: two hypotheses on how land use change drives viral emergence. Curr. Opin. Virol. 3, 79–83.
Nash, K.L., Graham, N.A., Wilson, S.K., Bellwood, D.R., 2013. Cross-scale habitat structure drives fish body size distributions on coral reefs. Ecosystems 16, 478–490.
Nash, K.L., Allen, C.R., Angeler, D.G., Barichievy, C., Eason, T., Garmestani, A.S., et al., 2014. Discontinuities, cross-scale patterns, and the organization of ecosystems. Ecology 95, 654–667.
Nathan, R., Getz, W.M., Revilla, E., Holyoak, M., Kadmon, R., Saltz, D., et al., 2008. A movement ecology paradigm for unifying organismal movement research. Proc. Natl. Acad. Sci. U.S.A. 105, 19052–19059.
Newman, M.E., 2002. Spread of epidemic disease on networks. Phys. Rev. E 66, 016128.
Ng, J.C., Perry, K.L., 2004. Transmission of plant viruses by aphid vectors. Mol. Plant Pathol. 5, 505–511.
Novotny, V., Basset, Y., Miller, S.E., Weiblen, G.D., Bremer, B., Cizek, L., et al., 2002. Low host specificity of herbivorous insects in a tropical forest. Nature 416, 841–844.
O'Dwyer, J.P., Cornell, S.J., 2017. Cross-scale neutral ecology and the maintenance of biodiversity. arXiv, preprint arXiv:1705.07856.
Olesen, J.M., Bascompte, J., Dupont, Y.L., Jordano, P., 2007. The modularity of pollination networks. Proc. Natl. Acad. Sci. U.S.A. 104, 19891–19896.
Orrock, J.L., Watling, J.I., 2010. Local community size mediates ecological drift and competition in metacommunities. Proc. Biol. Sci. 277, 2185–2191.
Ostfeld, R.S., Keesing, F., 2000. Biodiversity series: the function of biodiversity in the ecology of vector-borne zoonotic diseases. Can. J. Zool. 78, 2061–2078.
Pandit, S.N., Kolasa, J., Cottenie, K., 2009. Contrasts between habitat generalists and specialists: an empirical extension to the basic metacommunity framework. Ecology 90, 2253–2262.
Parker, I.M., Gilbert, G.S., 2004. The evolutionary ecology of novel plant-pathogen interactions. Annu. Rev. Ecol. Evol. Syst. 35, 675–700.
Parker, I.M., Saunders, M., Bontrager, M., Weitz, A.P., Hendricks, R., Magarey, R., et al., 2015. Phylogenetic structure and host abundance drive disease pressure in communities. Nature 520, 542–544.

Pedersen, A.B., Davies, T.J., 2009. Cross-species pathogen transmission and disease emergence in primates. EcoHealth 6, 496–508.
Pedersen, A.B., Fenton, A., 2007. Emphasizing the ecology in parasite community ecology. Trends Ecol. Evol. 22, 133–139.
Perkins, T.A., Scott, T.W., Le Menach, A., Smith, D.L., 2013. Heterogeneity, mixing, and the spatial scales of mosquito-borne pathogen transmission. PLoS Comput. Biol. 9, e1003327.
Petanidou, T., Kallimanis, A.S., Tzanopoulos, J., Sgardelis, S.P., Pantis, J.D., 2008. Long-term observation of a pollination network: fluctuation in species and interactions, relative invariance of network structure and implications for estimates of specialization. Ecol. Lett. 11, 564–575.
Peterson, G., Allen, C.R., Holling, C.S., 1998. Ecological resilience, biodiversity, and scale. Ecosystems 1, 6–18.
Philibert, A., Desprez-Loustau, M.L., Fabre, B., Frey, P., Halkett, F., Husson, C., et al., 2011. Predicting invasion success of forest pathogenic fungi from species traits. J. Appl. Ecol. 48, 1381–1390.
Picard, C., Dallot, S., Brunker, K., Berthier, K., Roumagnac, P., Soubeyrand, S., et al., 2017. Exploiting genetic information to trace plant virus dispersal in landscapes. Annu. Rev. Phytopathol. 55, 139–160.
Pilosof, S., Fortuna, M.A., Cosson, J.F., Galan, M., Kittipong, C., Ribas, A., et al., 2014. Host–parasite network structure is associated with community-level immunogenetic diversity. Nat. Commun. 5, 6172.
Poisot, T., Bever, J.D., Nemri, A., Thrall, P.H., Hochberg, M.E., 2011. A conceptual framework for the evolution of ecological specialisation. Ecol. Lett. 14, 841–851.
Poulicard, N., Pinel-Galzi, A., Hebrard, E., Fargette, D., 2010. Why *Rice yellow mottle virus*, a rapidly evolving RNA plant virus, is not efficient at breaking *rymv1-2* resistance. Mol. Plant Pathol. 11, 145–154.
Poulicard, N., Pinel-Galzi, A., Traore, O., Vignols, F., Ghesquiere, A., Konate, G., et al., 2012. Historical contingencies modulate the adaptability of *Rice yellow mottle virus*. PLoS Pathog. 8, e1002482.
Poulin, R., Mouillot, D., 2003. Parasite specialization from a phylogenetic perspective: a new index of host specificity. Parasitology 126, 473–480.
Power, A.G., Flecker, A.S., 2003. Virus specificity in disease systems: are species redundant? In: Kareiva, P., Levin, S.A. (Eds.), The Importance of Species: Perspectives on Expendability and Triage. Princeton University Press, Princeton, NJ, pp. 330–347.
Power, A.G., Mitchell, C.E., 2004. Pathogen spillover in disease epidemics. Am. Nat. 164, S79–S89.
Power, A.G., Borer, E.T., Hosseini, P., Mitchell, C.E., Seabloom, E.W., 2011. The community ecology of barley/cereal yellow dwarf viruses in Western US grasslands. Virus Res. 159, 95–100.
Purse, B.V., Carpenter, S., Venter, G.J., Bellis, G., Mullens, B.A., 2015. Bionomics of temperate and tropical Culicoides midges: knowledge gaps and consequences for transmission of Culicoides-borne viruses. Annu. Rev. Entomol. 60, 373–392.
Randolph, S.E., Dobson, A.D.M., 2012. Pangloss revisited: a critique of the dilution effect and the biodiversity-buffers-disease paradigm. Parasitology 139, 847–863.
Real, L.A., McElhany, P., 1996. Spatial pattern and process in plant-pathogen interactions. Ecology 77, 1011–1025.
Richardson, J.L., Urban, M.C., Bolnick, D.I., Skelly, D.K., 2014. Microgeographic adaptation and the spatial scale of evolution. Trends Ecol. Evol. 29, 165–176.
Ricklefs, R.E., 1987. Community diversity: relative roles of local and regional processes. Science 235, 167–171.

Ricklefs, R.E., Miller, G.L., 1999. Ecology, fourth ed. WH Freeman and Company, New York City, NY.
Rico, P., Ivars, P., Elena, S.F., Hernandez, C., 2006. Insights into the selective pressures restricting Pelargonium flower break virus genome variability: evidence for host adaptation. J. Virol. 80, 8124–8132.
Roche, B., Drake, J.M., Brown, J., Stallknecht, D.E., Bedford, T., Rohani, P., 2014. Adaptive evolution and environmental durability jointly structure phylodynamic patterns in Avian influenza viruses. PLoS Biol. 12, e1001931.
Rodelo-Urrego, M., Pagán, I., González-Jara, P., Betancourt, M., Moreno-Letelier, A., Ayllón, M.A., et al., 2013. Landscape heterogeneity shapes host-parasite interactions and results in apparent plant-virus codivergence. Mol. Ecol. 22, 2325–2340.
Roossinck, M.J., 2015. Plants, viruses and the environment: ecology and mutualism. Virology 479, 271–277.
Roossinck, M.J., García-Arenal, F., 2015. Ecosystem simplification, biodiversity loss and plant virus emergence. Curr. Opin. Virol. 10, 56–62.
Roossinck, M.J., Saha, P., Wiley, G.B., Quan, J., White, J.D., Lai, H., Chavarria, F., Shen, G., Roe, B.A., 2010. Ecogenomics: using massively parallel pyrosequencing to understand virus ecology. Mol. Ecol. 19, 81–88.
Roossinck, M.J., Martin, D.P., Roumagnac, P., 2015. Plant virus metagenomics: advances in virus discovery. Phytopathology 105, 716–727.
Rosenzweig, M.L., 1995. Species Diversity in Space and Time. Cambridge University Press, Cambridge, UK.
Rosindell, J., Cornell, S.J., 2013. Universal scaling of species-abundance distributions across multiple scales. Oikos 122, 1101–1111.
Rottstock, T., Joshi, J., Kummer, V., Fischer, M., 2014. Higher plant diversity promotes higher diversity of fungal pathogens, while it decreases pathogen infection per plant. Ecology 95, 1907–1917.
Sacristán, S., Fraile, A., Malpica, J.M., Garcia-Arenal, F., 2005. An analysis of host adaptation and its relationship with virulence in *Cucumber mosaic virus*. Phytopathology 95, 827–833.
Sallam, M.F., Xue, R.D., Pereira, R.M., Koehler, P.G., 2016. Ecological niche modelling of mosquito vectors of West Nile virus in St. John's County, Florida, USA. Parasit. Vectors 9, 371.
Samy, A.M., Peterson, A.T., 2016. Climate change influences on the global potential distribution of bluetongue virus. PLoS One 11, e0150489.
Schulze-Lefert, P., Panstruga, R., 2011. A molecular evolutionary concept connecting nonhost resistance, pathogen host range, and pathogen speciation. Trends Plant Sci. 16, 117–125.
Seabloom, E.W., Hosseini, P.R., Power, A.G., Borer, E.T., 2009. Diversity and composition of viral communities: coinfection of barley and cereal yellow dwarf viruses in California grasslands. Am. Nat. 173, E79–E98.
Seabloom, E.W., Borer, E.T., Mitchell, C.E., Power, A.G., 2010. Viral diversity and prevalence gradients in north American Pacific Coast grasslands. Ecology 91, 721–732.
Seabloom, E.W., Borer, E.T., Lacroix, C., Mitchell, C.E., Power, A.G., 2013. Richness and composition of niche-assembled viral pathogen communities. PLoS One 8, e55675.
Seabloom, E.W., Borer, E.T., Gross, K., Kendig, A.E., Lacroix, C., Mitchell, C.E., et al., 2015. The community ecology of pathogens: coinfection, coexistence and community composition. Ecol. Lett. 18, 401–415.
Sexton, J.P., Montiel, J., Shay, J.E., Stephens, M.R., Slatyer, R.A., 2017. Evolution of ecological niche breadth. Annu. Rev. Ecol. Evol. Syst. 48, 183–206.
Shaw, A.K., Peace, A., Power, A.G., Bosque-Pérez, N., 2017. Vector population growth and condition-dependent movement drive the spread of plant pathogens. Ecology 98, 2145–2157.

Shipley, L.A., Forbey, J.S., Moore, B.D., 2009. Revisiting the dietary niche: when is a mammalian herbivore a specialist? Integr. Comp. Biol. 49, 274–290.
Simmonds, P., Adams, M.J., Benkő, M., Breitbart, M., Brister, J.R., Carstens, E.B., et al., 2017. Consensus statement: virus taxonomy in the age of metagenomics. Nat. Rev. Microbiol. 15, 161–168.
Simpson, J.E., Hurtado, P.J., Medlock, J., Molaei, G., Andreadis, T.G., Galvani, A.P., et al., 2012. Vector host-feeding preferences drive transmission of multi-host pathogens: West Nile virus as a model system. Proc. Biol. Sci. 279, 925–933.
Soberón, J., Peterson, A.T., 2005. Interpretation of models of fundamental ecological niches and species' distributional areas. Biodivers. Inform. 2, 1–10.
Sokos, C., Andreadis, K., Papageorgiou, N., 2015. Diet adaptability by a generalist herbivore: the case of brown hare in a Mediterranean agroecosystem. Zool. Stud. 54, 27.
Solé, R.V., Montoya, M., 2001. Complexity and fragility in ecological networks. Proc. Biol. Sci. 268, 2039–2045.
Stafford, C.A., Walker, G.P., Ullman, D.E., 2011. Infection with a plant virus modifies vector feeding behavior. Proc. Natl. Acad. Sci. U.S.A. 108, 9350–9355.
Staniczenko, P., Sivasubramaniam, P., Suttle, K.B., Pearson, R.G., 2017. Linking macroecology and community ecology: refining predictions of species distributions using biotic interaction networks. Ecol. Lett. 20, 693–707.
Stoffels, R.J., Clarke, K.R., Closs, G.P., 2005. Spatial scale and benthic community organisation in the littoral zones of large oligotrophic lakes: potential for cross-scale interactions. Freshw. Biol. 50, 1131–1145.
Stow, C., Allen, C., Garmestani, A., 2007. Evaluating discontinuities in complex systems: toward quantitative measures of resilience. Ecol. Soc. 12, 26.
Strauss, A.T., Civitello, D.J., Cáceres, C.E., Hall, S.R., 2015. Success, failure and ambiguity of the dilution effect among competitors. Ecol. Lett. 18, 916–926.
Streicker, D.G., Turmelle, A.S., Vonhof, M.J., Kuzmin, I.V., McCracken, G.F., Rupprecht, C.E., 2010. Host phylogeny constrains cross-species emergence and establishment of rabies virus in bats. Science 329, 676–679.
Streicker, D.G., Fenton, A., Pedersen, A.B., 2013. Differential sources of host species heterogeneity influence the transmission and control of multihost parasites. Ecol. Lett. 16, 975–984.
Su, Q., Pan, H., Liu, B., Chu, D., Xie, W., Wu, Q., et al., 2013. Insect symbiont facilitates vector acquisition, retention, and transmission of plant virus. Sci. Rep. 3, 1367.
Sugihara, G., May, R.M., 1990. Applications of fractals in ecology. Trends Ecol. Evol. 5, 79–86.
Suzán, G., García-Peña, G.E., Castro-Arellano, I., Rico, O., Rubio, A.V., Tolsá, M.J., et al., 2015. Metacommunity and phylogenetic structure determine wildlife and zoonotic infectious disease patterns in time and space. Ecol. Evol. 5, 865–873.
Swei, A., Ostfeld, R.S., Lane, R.S., Briggs, C.J., 2011. Impact of the experimental removal of lizards on Lyme disease risk. Proc. Biol. Sci. 278, 2970–2978.
Tardy, O., Massé, A., Pelletier, F., Mainguy, J., Fortin, D., 2014. Density-dependent functional responses in habitat selection by two hosts of the raccoon rabies virus variant. Ecosphere 5, 1–16.
Tatineni, S., Robertson, C.J., Garnsey, S.M., Dawson, W.O., 2011. A plant virus evolved by acquiring multiple nonconserved genes to extend its host range. Proc. Natl. Acad. Sci. U.S.A. 108, 17366–17371.
Taylor, L.H., Mackinnon, M.J., Read, A.F., 1998. Virulence of mixed-clone and single-clone infections of the rodent malaria *Plasmodium chabaudi*. Evolution 52, 583–591.
Tellier, A., Brown, J.K.M., 2011. Spatial heterogeneity, frequency-dependent selection and polymorphism in host-parasite interactions. BMC Evol. Biol. 11, 319.

Thébault, E., Fontaine, C., 2010. Stability of ecological communities and the architecture of mutualistic and trophic networks. Science 329, 853–856.
Thrall, P.H., Burdon, J.J., Bever, J.D., 2002. Local adaptation in the Linum marginale–Melampsora lini host-pathogen interaction. Evolution 56, 1340–1351.
Thrall, P.H., Hochberg, M.E., Burdon, J.J., Bever, J.D., 2007. Coevolution of symbiotic mutualists and parasites in a community context. Trends Ecol. Evol. 22, 120–126.
Thrall, P.H., Barrett, L.G., Dodds, P.N., Burdon, J.J., 2016. Epidemiological and evolutionary outcomes in gene-for-gene and matching allele models. Front. Plant Sci. 6, 1084.
Toju, H., Yamamichi, M., Guimarães Jr., P.R., Olesen, J.M., Mougi, A., Yoshida, T., et al., 2017. Species-rich networks and eco-evolutionary synthesis at the metacommunity level. Nat. Ecol. Evol. 1, 0024.
Tollenaere, C., Susi, H., Laine, A.L., 2016. Evolutionary and epidemiological implications of multiple infection in plants. Trends Plant Sci. 21, 80–90.
Turner, M.G., 1989. Landscape ecology: the effect of pattern on process. Annu. Rev. Ecol. Syst. 20, 171–197.
Vassilakos, N., Simon, V., Tzima, A., Johansen, E., Moury, B., 2016. Genetic determinism and evolutionary reconstruction of a host jump in a plant virus. Mol. Biol. Evol. 33, 541–553.
Vayssier-Taussat, M., Albina, E., Citti, C., Cosson, J.F., Jacques, M.A., Lebrun, M.H., et al., 2014. Shifting the paradigm from pathogens to pathobiome: new concepts in the light of meta-omics. Front. Cell. Infect. Microbiol. 4, 29.
Vellend, M., 2010. Conceptual synthesis in community ecology. Q. Rev. Biol. 85, 183–206.
Viana, M., Mancy, R., Biek, R., Cleaveland, S., Cross, P.C., Lloyd-Smith, J.O., et al., 2014. Assembling evidence for identifying reservoirs of infection. Trends Ecol. Evol. 29, 270–279.
Viana, M., Cleaveland, S., Matthiopoulos, J., Halliday, J., Packer, C., Craft, M.E., et al., 2015. Dynamics of a morbillivirus at the domestic–wildlife interface: canine distemper virus in domestic dogs and lions. Proc. Natl. Acad. Sci. U.S.A. 112, 1464–1469.
Wang, Y., Hajimorad, M.R., 2016. Gain of virulence by *Soybean mosaic virus* on *Rsv4*-genotype soybeans is associated with a relative fitness loss in a susceptible host. Mol. Plant Pathol. 17, 1154–1159.
Warton, D.I., Blanchet, F.G., O'Hara, R.B., Ovaskainen, O., Taskinen, S., Walker, S.C., et al., 2015. So many variables: joint modeling in community ecology. Trends Ecol. Evol. 30, 766–779.
Weaver, S.C., Barrett, A.D., 2004. Transmission cycles, host range, evolution and emergence of arboviral disease. Nat. Rev. Microbiol. 2, 789.
Weitz, J.S., Poisot, T., Meyer, J.R., Flores, C.O., Valverde, S., Sullivan, M.B., et al., 2013. Phage-bacteria infection networks. Trends Microbiol. 21, 82–91.
Welti, E.A., Joern, A., 2015. Structure of trophic and mutualistic networks across broad environmental gradients. Ecol. Evol. 5, 326–334.
Whitlock, M.C., 1996. The red queen beats the jack-of-all-trades: the limitations on the evolution of phenotypic plasticity and niche breadth. Am. Nat. 148, S65–S77.
Wiens, J.A., 1989. Spatial scaling in ecology. Funct. Ecol. 3, 385–397.
Wilke, C.O., Forster, R., Novella, I.S., 2006. Quasispecies in time-dependent environments. Curr. Top. Microbiol. Immunol. 299, 33–50.
Winkler, I.S., Mitter, C., Scheffer, S.J., 2009. Repeated climate-linked host shifts have promoted diversification in a temperate clade of leaf-mining flies. Proc. Natl. Acad. Sci. U.S.A. 106, 18103–18108.
Wintermantel, W.M., Cortez, A.A., Anchieta, A.G., Gulati-Sakhuja, A., Hladky, L.L., 2008. Co-infection by two criniviruses alters accumulation of each virus in a host-specific manner and influences efficiency of virus transmission. Phytopathology 98, 1340–1345.

Wolinska, J., King, K.C., 2009. Environment can alter selection in host–parasite interactions. Trends Parasitol. 25, 236–244.
Woolhouse, M., Gaunt, E., 2007. Ecological origins of novel human pathogens. Crit. Rev. Microbiol. 33, 231–242.
Woolhouse, M.E., Gowtage-Sequeria, S., 2005. Host range and emerging and reemerging pathogens. Emerg. Infect. Dis. 11, 1842.
Woolhouse, M.E., Taylor, L.H., Haydon, D.T., 2001. Population biology of multihost pathogens. Science 292, 1109–1112.
Woolhouse, M.E., Haydon, D.T., Antia, R., 2005. Emerging pathogens: the epidemiology and evolution of species jumps. Trends Ecol. Evol. 20, 238–244.
Yoder, J.B., Clancey, E., Des Roches, S., Eastman, J.M., Gentry, L., Godsoe, W., et al., 2010. Ecological opportunity and the origin of adaptive radiations. J. Evol. Biol. 23, 1581–1596.
Young, H.S., Parker, I.M., Gilbert, G.S., Guerra, A.S., Nunn, C.L., 2017. Introduced species, disease ecology, and biodiversity–disease relationships. Trends Ecol. Evol. 32, 41–54.
Zhan, J.S., Thrall, P.H., Burdon, J.J., 2014. Achieving sustainable plant disease management through evolutionary principles. Trends Plant Sci. 19, 570–575.
Ziebell, H., Murphy, A.M., Groen, S.C., Tungadi, T., Westwood, J.H., Lewsey, M.G., Moulin, M., Kleczkowski, A., Smith, A.G., Stevens, M., Powell, G., 2011. Cucumber mosaic virus and its 2b RNA silencing suppressor modify plant-aphid interactions in tobacco. Sci. Rep. 1, 187.

FURTHER READING

Laine, A.L., 2008. Temperature-mediated patterns of local adaptation in a natural plant–pathogen metapopulation. Ecol. Lett. 11, 327–337.